Advanced Structured Materials

Volume 42

More information about this series at http://www.springer.com/series/8611

Holm Altenbach · Samuel Forest
Editors

Generalized Continua as Models for Classical and Advanced Materials

Springer

Editors
Holm Altenbach
Otto-von-Guericke Universität Magdeburg
Magdeburg
Germany

Samuel Forest
Mines ParisTech
Paris
France

ISSN 1869-8433 ISSN 1869-8441 (electronic)
Advanced Structured Materials
ISBN 978-3-319-31719-9 ISBN 978-3-319-31721-2 (eBook)
DOI 10.1007/978-3-319-31721-2

Library of Congress Control Number: 2016935580

Printed on acid-free paper

This Springer imprint is published by Springer Nature
The registered company is Springer International Publishing AG Switzerland

Preface

Generalized Continua have been in the focus of scientists from the end of the nineteenth century. A first summary was given in 1909 by the Cosserat brothers. After World War II, a true renaissance in this field occurred with the publication of Ericksen and Truesdell in 1958. Further developments were connected with the fundamental contributions of scientists from Germany, Russia, and France. During the past years the centennial of the Cosserat book was celebrated by two colloquia, both held in Paris in 2009. In addition, previous trilateral seminars *Mechanics of Generalized Continua—from Micromechanical Basics to Engineering Applications* (Wittenberg 2010, 2012) and the CISM Course *Generalized Continua—from the Theory to the Engineering Applications* (Udine 2011) discussed problems related to the theory and applications. During a new Advanced Seminar (Magdeburg, September 2015), attention was paid to the most recent research items, i.e., new generalized models, materials with a significant microstructure, multi-field loadings, or identification of constitutive equations. Last but not least, a comparison of discrete modeling approaches have been discussed.

This book contains 21 papers submitted to the Advanced Seminar *Generalized Continua as Models for Materials with Multi-Scale Effects or Under Multi-Field Actions* or discussed during the seminar. Finally, after reviewing and acceptance they were collected as a unique collection of papers. The authors are from France, Germany, and Russia, the traditional countries of the previous trilateral seminars, completed by authors from Egypt, Estonia, Finland, Great Britain, Italy, and the United States.

The editors thank the anonymous reviewers for their carefully performed job, Mrs. Johanna Eisenträger and Mrs. Barbara Renner for proofreading, Mr. Marcus Aßmus for the technical support in latex processing and compiling the whole book. Special thanks to Christoph Baumann from Springer Publisher for supporting this project.

Magdeburg, Germany
Paris, France
February 2016

Holm Altenbach
Samuel Forest

Contents

Contributors

Holm Altenbach Lehrstuhl für Technische Mechanik, Institut für Mechanik, Fakultät für Maschinenbau, Otto-von-Guericke-Universtät Magdeburg, Magdeburg, Germany

Hossein Aminpour Università Degli Studi Roma Tre, Rome, Italy

Viacheslav Balobanov Department of Civil and Structural Engineering, Aalto University, Aalto, Espoo, Finland

Petr Belov Institute of Research, Development and Technology Transfer, Moscow, Russia

Stéphane Berbenni Laboratoire d'Etude des Microstructures et de Mécanique des Matériaux, LEM3, UMR CNRS 7239, University of Lorraine, Metz, France

Albrecht Bertram Lehrstuhl für Festigkeitslehre, Institut für Mechanik, Fakultät für Maschinenbau, Otto-von-Guericke-Universität Magdeburg, Magdeburg, Germany

Laurent Capolungo George Woodruff School of Mechanical Engineering, Georgia Institute of Technology, Metz, France

Maria Laura De Bellis Department of Innovation Engineering, University of Salento, Lecce, Italy

Alessandro Della Corte Department of Mechanical and Aerospace Engineering, University La Sapienza of Rome, Rome, Italy

Francesco dell'Isola Department of Structural and Gentechnical Engineering, University La Sapienza of Rome, Rome, Italy; International Research Center on Mathematics and Mechanics of Complex Systems (M&MoCS), University of L'Aquila, L'Aquila, Italy

Francisco Dos Reis LEMTA, Université de Lorraine, Vandoeuvre-lès-Nancy Cedex, France

Martin A. Dunn Department of Mechanical and Aerospace Engineering, University of Strathclyde, Glasgow, UK

Victor A. Eremeyev Faculty of Mechanical Engineering, Rzeszów University of Technology, Rzeszów, Poland

Vladimir I. Erofeev Mechanical Engineering Research Institute of Russian Academy of Sciences, Nizhny Novgorod Lobachevsky State University, Nizhny Novgorod, Russia

Raffaele Esposito International Research Center on Mathematics and Mechanics of Complex Systems (M&MoCS), University of L'Aquila, L'Aquila, Italy

Jamie C. Frame Department of Biomedical Engineering, University of Strathclyde, Glasgow, UK

Claude Fressengeas Laboratoire d'Etude des Microstructures et de Mécanique des Matériaux, LEM3, UMR CNRS 7239, University of Lorraine, Metz, France

Larisa Yu. Frolenkova State University - Education-Science-Production Complex, Orel, Russia

Jean-François Ganghoffer LEMTA, Université de Lorraine, Vandoeuvre-lès-Nancy Cedex, France

Rainer Glüge Lehrstuhl für Festigkeitslehre, Institut für Mechanik, Fakultät für Maschinenbau, Otto-von-Guericke-Universität Magdeburg, Magdeburg, Germany

Ibrahim Goda LEMTA, Université de Lorraine, Vandoeuvre-lès-Nancy Cedex, France; Faculty of Engineering, Department of Industrial Engineering, Fayoum University, Fayoum, Egypt

Heiko Herrmann Centre for Nonlinear Studies, Institute of Cybernetics at Tallinn University of Technology, Tallinn, Estonia

Jan Kalisch Graduiertenkolleg 1554 "Micro-Macro-Interactions in Structured Media and Particle Systems", Otto-von-Guericke-Universität Magdeburg, Magdeburg, Germany

Vladimir V. Kazhaev Mechanical Engineering Research Institute of Russian Academy of Sciences, Nizhny Novgorod Lobachevsky State University, Nizhny Novgorod, Russia

Sergei Khakalo Department of Civil and Structural Engineering, Aalto University, Aalto, Espoo, Finland

Christian Liebold Institute of Mechanics, Berlin University of Technology, Berlin, Germany

Sergey Lurie Institute for Problem of Mechanics of RAS and Institute of Applied Mechanics of RAS, Moscow, Russia

Wolfgang H. Müller Institute of Mechanics, Berlin University of Technology, Berlin, Germany

Jarkko Niiranen Department of Civil and Structural Engineering, Aalto University, Aalto, Espoo, Finland

Mikhail U. Nikabadze Lomonosov Moscow State University, Moscow, Russia

Martin Ostoja-Starzewski Department of Mechanical Science and Engineering, University of Illinois at Urbana-Champaign, Urbana, IL, USA

Igor S. Pavlov Mechanical Engineering Research Institute of Russian Academy of Sciences, Nizhny Novgorod Lobachevsky State University, Nizhny Novgorod, Russia

Victoria Yu. Presnetsova State University - Education-Science-Production Complex, Orel, Russia

Philip E. Riches Department of Biomedical Engineering, University of Strathclyde, Glasgow, UK

Nicola Rizzi Università Degli Studi Roma Tre, Roma, Italy

Sergey N. Romashin State University - Education-Science-Production Complex, Orel, Russia

Luciano Rosati Dipartimento di Strutture per l'Ingegneria e l'Architettura (DIST), Universitá degli Studi di Napoli Federico II, Naples, Italy; Department of Mathematics, University Tor Vergata of Rome, Rome, Italy

Roberto Serpieri Dipartimento di Ingegneria, Università degli Studi del Sannio, Benevento, Italy

Denis N. Sheydakov South Scientific Center of Russian Academy of Sciences, Rostov-on-Don, Russia

Nikolay E. Sheydakov Rostov State University of Economics, Rostov-on-Don, Russia

Vladimir S. Shorkin State University - Education-Science-Production Complex, Orel, Russia

Anatoli Stulov Institute of Cybernetics at Tallinn University of Technology, Tallinn, Estonia

Vincent Taupin Laboratoire d'Etude des Microstructures et de Mécanique des Matériaux, LEM3, UMR CNRS 7239, University of Lorraine, Metz, France

Francesco Travascio Biomechanics Research Laboratory, Department of Industrial Engineering, University of Miami, Coral Gables, FL, USA

Patrizia Trovalusci Department of Structural Engineering and Geotechnics, Sapienza University of Rome, Rome, Italy

Armine R. Ulukhanyan Bauman State Technical University, Moscow, Russia

Marcus A. Wheel Department of Mechanical and Aerospace Engineering, University of Strathclyde, Glasgow, UK

On Strain Rate Tensors and Constitutive Equations of Inelastic Micropolar Materials

Holm Altenbach and Victor A. Eremeyev

Abstract Following Altenbach and Eremeyev (Int J Plast 63:3–17, 2014) we introduce a new family of strain rate tensors for micropolar materials. With the help of introduced strain rates we discuss the possible forms of constitutive equations of the nonlinear inelastic micropolar continuum, that is micropolar viscous and viscoelastic fluids and solids, hypo-elastic and viscoelastoplastic materials. Considering the fact that some of strain rates are not true tensors but pseudotensors we obtain some constitutive restrictions following from the material frame indifference principle. Using the theory of tensorial invariants we present the general form of constitutive equations of some types of inelastic isotropic micropolar materials including several new constitutive equations.

Keywords Micropolar continua · Strain rate · Constitutive equations · Finite deformations

1 Introduction

Nonlinear micropolar continuum model allows to describe complex micro-structured media, for example, polycrystals, foams, cellular solids, lattices, masonries, particle assemblies, magnetic rheological fluids, liquid crystals, etc., for which the rotational degrees of freedom of material particles are important. In the case of inelastic behavior the constitutive equations of the micropolar continuum have more complicated structure, the stress and couple stress tensors as well as other quantities depend

H. Altenbach
Lehrstuhl für Technische Mechanik, Institut für Mechanik,
Fakultät für Maschinenbau, Otto-von-Guericke-Universtät Magdeburg,
Universitätsplatz 2, 39106 Magdeburg, Germany
e-mail: holm.altenbach@ovgu.de

V.A. Eremeyev (✉)
Faculty of Mechanical Engineering, Rzeszów University of Technology,
al. Powstańców Warszawy 8, 35-959 Rzeszów, Poland
e-mail: veremeyev@prz.edu.pl

H. Altenbach and S. Forest (eds.), *Generalized Continua as Models for Classical and Advanced Materials*, Advanced Structured Materials 42,
DOI 10.1007/978-3-319-31721-2_1

on the history of strain measures. For the basics of micropolar mechanics we refer to Nowacki (1986), Eringen (1999), Eringen and Kafadar (1976), Eringen (2001), Eremeyev et al. (2013), Łukaszewicz (1999).

Below we discuss the constitutive equations of the nonlinear micropolar continuum considering strain rates. The discussion of strain measures for polar elastic materials presented by Pietraszkiewicz and Eremeyev (2009a, b), where the natural Lagrangian and Eulerian strain measures are introduced. Strain–stress pairs within the framework of micropolar mechanics are discussed in Ramezani and Naghdabadi (2007). Using strain rate tensors incremental equations of the micropolar hypo-elasticity were presented in Ramezani and Naghdabadi (2010), Ramezani et al. (2008). The interrelations between strain rates in discrete and continual models are discussed in Trovalusci and Masiani (1997), Pau and Trovalusci (2012). Here we discuss the Rivlin–Ericksen analogues of strain rate tensors for micropolar mechanics and several types of the constitutive equations of inelastic micropolar solids are summarized.

2 Basic Relations of the Micropolar Mechanics

Following Eringen (1999), Pietraszkiewicz and Eremeyev (2009a), Eremeyev et al. (2013) let us recall the basic equations of micropolar mechanics under finite deformations. In what follows we use the standard direct tensor notations (Lebedev et al. 2010; Truesdell 1966; Truesdell and Noll 2004). For example, the gradient and divergence operators in the actual and reference configurations are defined as follows

$$\operatorname{grad}(\bullet) = \frac{\partial(\bullet)}{\partial x_k} \otimes \mathbf{r}^k, \quad \operatorname{div}(\bullet) = \frac{\partial(\bullet)}{\partial x_k} \cdot \mathbf{r}^k, \quad \mathbf{r}_i = \frac{\partial \mathbf{r}}{\partial x_i}, \quad \mathbf{r}_i \cdot \mathbf{r}^k = \delta_i^k,$$
$$\operatorname{Grad}(\bullet) = \frac{\partial(\bullet)}{\partial X_k} \otimes \mathbf{R}^k, \operatorname{Div}(\bullet) = \frac{\partial(\bullet)}{\partial X_k} \cdot \mathbf{R}^k, \mathbf{R}_i = \frac{\partial \mathbf{R}}{\partial x_i}, \quad \mathbf{R}_i \cdot \mathbf{R}^k = \delta_i^k,$$

where x_i and X_i are the Eulerian and Lagrangian coordinates, respectively, and δ_i^j is the Kronecker symbol.

2.1 Kinematics

We describe the micropolar continuum deformation by the following relations:

$$\mathbf{r} = \mathbf{r}(\mathbf{R}, t), \quad \mathbf{H} \equiv \mathbf{d}_k \otimes \mathbf{D}_k = \mathbf{H}(\mathbf{R}, t). \tag{1}$$

Vector $\mathbf{r}(t)$ describes the position of the particle of the continuum at time t, whereas $\mathbf{H}(t)$ defines its rotation. The linear velocity is given by the relation

$$\mathbf{v} = \frac{\mathrm{d}\mathbf{r}}{\mathrm{d}t}, \tag{2}$$

the angular velocity vector $\boldsymbol{\omega}$ can be presented by

$$\frac{\mathrm{d}\mathbf{d}_k}{\mathrm{d}t} = \boldsymbol{\omega} \times \mathbf{d}_k, \quad k = 1, 2, 3,$$

where $\times$ is the vector (cross) product. $\boldsymbol{\omega}$ can be also expressed using the derivative of $\mathbf{H}$ as follows

$$\boldsymbol{\omega} = -\frac{1}{2}\left(\frac{\mathrm{d}\mathbf{H}}{\mathrm{d}t} \cdot \mathbf{H}^T\right)_\times, \tag{3}$$

where subindex $\times$ stands for the vectorial invariant of second-order tensor (Lebedev et al. 2010).

2.2 *Motion Equations*

The Eulerian equations of motion of micropolar media are

$$\rho\frac{\mathrm{d}\mathbf{v}}{\mathrm{d}t} = \operatorname{div}\mathbf{T} + \rho\mathbf{f}, \quad j\frac{\mathrm{d}\boldsymbol{\omega}}{\mathrm{d}t} = \operatorname{div}\mathbf{M} - \mathbf{T}_\times + \rho\mathbf{m}, \tag{4}$$

where $\mathbf{T}$ and $\mathbf{M}$ are the stress and couple stress tensors of Cauchy type which are non-symmetric, in general, ρ is the density in the actual configuration, j is the measure of rotatory inertia of particles of micropolar medium, $\mathbf{f}$ and $\mathbf{m}$ are the external forces and couples, respectively.

2.3 *Constitutive Equations*

In the pure mechanical theory of the micropolar continuum with memory the constitutive equations consist of dependence of the stress and the couple stress tensors of the history of deformations. As a result, $\mathbf{T}$ and $\mathbf{M}$ take the following form:

$$\mathbf{T}(t) = \mathscr{A}_1[\mathbf{F}^t(s), \mathbf{H}^t(s), \operatorname{Grad}\mathbf{H}^t(s)], \quad \mathbf{M}(t) = \mathscr{A}_2[\mathbf{F}^t(s), \mathbf{H}^t(s), \operatorname{Grad}\mathbf{H}^t(s)], \tag{5}$$

where we introduced the histories of the deformation gradient

$$\mathbf{F}^t(s) = \mathbf{F}(t-s), \quad \mathbf{F}(t) = \operatorname{Grad}\mathbf{r}(t), \quad s \geq 0,$$

and of the microrotation tensor

$$\mathbf{H}^t(s) = \mathbf{H}(t - s), \quad s \geq 0.$$

Here $\mathscr{A}_1$ and $\mathscr{A}_2$ are operators describing the micropolar material behavior.

The further reduction of (5) is possible using the principle of material frame-indifference. The stress measures $\mathbf{T}$ and $\mathbf{M}$ should be indifferent (objective) quantities. In classical mechanics, two motions $\mathbf{r}$ and $\mathbf{r}^*$ are called equivalent if they relate as follows

$$\mathbf{r}^* = \mathbf{a}(t) + \mathbf{O}(t) \cdot (\mathbf{r} - \mathbf{r}_0), \tag{6}$$

where $\mathbf{O}(t)$ is an arbitrary orthogonal tensor, $\mathbf{a}(t)$ is an arbitrary vector function and the constant vector $\mathbf{r}_0$ represents a fixed point position (a pole). We assume that in the equivalent motion the directors $\mathbf{d}_k$ transform similarly to $\mathbf{r}$:

$$\mathbf{d}_k^* = \mathbf{O}(t) \cdot \mathbf{d}_k \quad \text{or} \quad \mathbf{H}^* = \mathbf{O}(t) \cdot \mathbf{H}. \tag{7}$$

Denoting by superscript "*" the stress tensors in the equivalent motions we formulate the property of objectivity for $\mathbf{T}$ and $\mathbf{M}$ as follows

$$\mathbf{T}^* = \mathbf{O}(t) \cdot \mathbf{T} \cdot \mathbf{O}(t)^T, \quad \mathbf{M}^* = \det \mathbf{O}(t)\, \mathbf{O}(t) \cdot \mathbf{M} \cdot \mathbf{O}(t)^T \tag{8}$$

for any orthogonal tensor $\mathbf{O}(t)$. Let us note that $\mathbf{M}$ is an pseudotensor, that is a reason of difference in transformation rules (8).

Thus, operators $\mathscr{A}_1$ and $\mathscr{A}_2$ satisfy the relations

$$\begin{aligned} &\mathscr{A}_1 \left[\mathbf{O}^t(s) \cdot \mathbf{F}^t(s), \mathbf{O}^t(s) \cdot \mathbf{H}^t(s), \mathbf{O}^t(s) \cdot \operatorname{Grad} \mathbf{H}^t(s)\right] \\ &\quad = \mathbf{O}(t) \cdot \mathscr{A}_1[\mathbf{F}^t(s), \mathbf{H}^t(s), \operatorname{Grad} \mathbf{H}^t(s)] \cdot \mathbf{O}^T(t), \end{aligned} \tag{9}$$

$$\begin{aligned} &\mathscr{A}_2 \left[\mathbf{O}^t(s) \cdot \mathbf{F}^t(s), \mathbf{O}^t(s) \cdot \mathbf{H}^t(s), \mathbf{O}^t(s) \cdot \operatorname{Grad} \mathbf{H}^t(s)\right] \\ &\quad = \det \mathbf{O}(t)\, \mathbf{O}(t) \cdot \mathscr{A}_2[\mathbf{F}^t(s), \mathbf{H}^t(s), \operatorname{Grad} \mathbf{H}^t(s)] \cdot \mathbf{O}^T(t) \end{aligned} \tag{10}$$

Finally, we can prove that (5) can be represented as follows

$$\mathbf{T}(t) = \mathbf{H}(t) \cdot \mathscr{B}_1[\mathbf{E}^t(s), \mathbf{K}^t(s)] \cdot \mathbf{H}^T(t), \quad \mathbf{M}(t) = \mathbf{H}(t) \cdot \mathscr{B}_2[\mathbf{E}^t(s), \mathbf{K}^t(s)] \cdot \mathbf{H}^T(t), \tag{11}$$

where $\mathscr{B}_1$ and $\mathscr{B}_2$ are operators depending on histories of two Lagrangian strain measures $\mathbf{E}$ and $\mathbf{K}$ defined by formulas (Pietraszkiewicz and Eremeyev 2009a)

$$\mathbf{E} = \mathbf{H}^T \cdot \mathbf{F} - \mathbf{I}, \quad \mathbf{K} = -\frac{1}{2}\boldsymbol{\epsilon} : (\mathbf{H}^T \cdot \operatorname{Grad} \mathbf{H}). \tag{12}$$

2.4 Elastic Materials

In the case of elastic behaviour Eq. (11) reduce to

$$\mathbf{T}(t) = \mathbf{H}(t) \cdot f_1[\mathbf{E}(t), \mathbf{K}(t)] \cdot \mathbf{H}^T(t), \quad \mathbf{M}(t) = \mathbf{H}(t) \cdot f_2[\mathbf{E}(t), \mathbf{K}(t)] \cdot \mathbf{H}^T(t), \quad (13)$$

where vector functions f_1 and f_2 can be expressed with use of the strain energy function $W = W(\mathbf{E}, \mathbf{K})$. For isotropic micropolar elastic solids W is considered as a function of two strain measures which can be represented as a scalar function depending on 15 joint invariants of $\mathbf{E}$ and $\mathbf{K}$, see Eringen and Kafadar (1976), Eremeyev and Pietraszkiewicz (2012)

$$W = W(I_1, \ldots, I_{15}).$$

In particular, assuming W in the form $W = W_1(\mathbf{E}) + W_2(\mathbf{K})$ one obtains that functions W_1 and W_2 depend on six invariants of $\mathbf{E}$ and $\mathbf{K}$. An isotropic scalar-valued function of one non-symmetric tensor $\mathbf{E}$ can be constructed as a function of six invariants I_n, $n = 1, \ldots, 6$, where

$$I_1 = \operatorname{tr}\mathbf{E}, I_2 = \operatorname{tr}\mathbf{E}^2, I_3 = \operatorname{tr}\mathbf{E}^3, I_4 = \operatorname{tr}(\mathbf{E} \cdot \mathbf{E}^T), I_5 = \operatorname{tr}(\mathbf{E}^2 \cdot \mathbf{E}^T), I_6 = \operatorname{tr}(\mathbf{E}^2 \cdot \mathbf{E}^{T2}). \quad (14)$$

An isotropic scalar-valued function of two non-symmetric tensors $\mathbf{E}$ and $\mathbf{K}$ depends on the following 15 invariants:

$$\begin{array}{llll} I_1 = \operatorname{tr}\mathbf{E}, & I_2 = \operatorname{tr}\mathbf{E}^2, & I_3 = \operatorname{tr}\mathbf{E}^3, \\ I_4 = \operatorname{tr}(\mathbf{E} \cdot \mathbf{E}^T), & I_5 = \operatorname{tr}(\mathbf{E}^2 \cdot \mathbf{E}^T), & I_6 = \operatorname{tr}(\mathbf{E}^2 \cdot \mathbf{E}^{T2}), \\ I_7 = \operatorname{tr}(\mathbf{E} \cdot \mathbf{K}), & I_8 = \operatorname{tr}(\mathbf{E}^2 \cdot \mathbf{K}), & I_9 = \operatorname{tr}(\mathbf{E} \cdot \mathbf{K}^2), \\ I_{10} = \operatorname{tr}\mathbf{K}, & I_{11} = \operatorname{tr}\mathbf{K}^2, & I_{12} = \operatorname{tr}\mathbf{K}^3, \\ I_{13} = \operatorname{tr}(\mathbf{K} \cdot \mathbf{K}^T), & I_{14} = \operatorname{tr}(\mathbf{K}^2 \cdot \mathbf{K}^T), & I_{15} = \operatorname{tr}(\mathbf{K}^2 \cdot \mathbf{K}^{T2}) \end{array} \quad (15)$$

3 Relative Strain Measures

For using the fading memory concept let us introduce relative strain measures. Within the framework of so-called relative description of deformation of continuum we consider the actual configuration χ at instant t as the reference one while the actual configuration χ at instant τ is considered as actual one. The relative deformation gradient and the relative microrotation tensor are introduced by formulas

$$\mathbf{F}_t(\tau) = \mathbf{F}(\tau) \cdot \mathbf{F}^{-1}(t), \quad \mathbf{H}_t(\tau) = \mathbf{d}_k(\tau) \times \mathbf{d}_k(t) = \mathbf{H}(\tau) \cdot \mathbf{H}^{-1}(t). \quad (16)$$

Obviously, $\mathbf{F}_t(t) = \mathbf{I}$, $\mathbf{H}_t(t) = \mathbf{I}$. Using $\mathbf{F}_t(\tau)$ and $\mathbf{H}_t(\tau)$ we introduce the relative strain measures $\mathbf{E}_t(\tau)$, $\mathbf{K}_t(\tau)$ by the relations

$$\mathbf{E}_t(\tau) = \mathbf{H}_t(\tau)^T \cdot \mathbf{F}_t(\tau) - \mathbf{I}, \quad \mathbf{K}_t(\tau) = -\frac{1}{2}\boldsymbol{\epsilon} : \left[\mathbf{H}_t(\tau)^T \cdot \operatorname{Grad} \mathbf{H}_t(\tau)\right]. \tag{17}$$

In what follows we denote the histories of relative tensors $\mathbf{F}_t(\tau)$, $\mathbf{H}_t(\tau)$, etc., as follows

$$\mathbf{F}_t^t(s) = \mathbf{F}_t(t-s), \quad \mathbf{H}_t^t(s) = \mathbf{H}_t(t-s).$$

From (16) and (17) it follows the following relations:

$$\mathbf{U}^t(s) = \mathbf{H}^T(t) \cdot \mathbf{U}_t^t(s) \cdot \mathbf{H}(t) \cdot \mathbf{U}(t), \quad \mathbf{Y}^t(s) = \mathbf{H}^T(t) \cdot \mathbf{Y}_t^t(s) \cdot \mathbf{H}(t) \cdot \mathbf{U}(t), \tag{18}$$

$$\mathbf{E}_t^t(s) = \mathbf{U}_t^t(s) - \mathbf{I}, \quad \mathbf{K}_t^t(s) = \mathbf{Y}_t^t(s) - \mathbf{B}(t) \tag{19}$$

and

$$\mathbf{U}_t^t(0) = \mathbf{I}, \quad \mathbf{Y}_t^t(0) = \mathbf{B}(t), \quad \mathbf{E}_t^t(0) = \mathbf{0}, \quad \mathbf{K}_t^t(0) = \mathbf{0},$$

where we introduced the histories

$$\mathbf{U}_t^t(s) = \mathbf{U}_t(t-s), \mathbf{Y}_t^t(s) = \mathbf{Y}_t(t-s), \mathbf{E}_t^t(s) = \mathbf{E}_t(t-s), \mathbf{K}_t^t(s) = \mathbf{K}_t(t-s).$$

4 Relations of Isotropic Materials with Relative Strain Measures

Substituting Eqs. (18) and (19) into (11) we obtain

$$\begin{aligned}
\mathbf{T}(t) = \mathbf{H}(t) \cdot \mathscr{B}_1 \big[&\mathbf{H}^T(t) \cdot \big(\mathbf{U}_t^t(s) - \mathbf{I}\big) \cdot \mathbf{H}(t) \cdot \mathbf{U}(t), \\
&\mathbf{H}^T(t) \cdot \big(\mathbf{Y}_t^t(s) - \mathbf{B}(t)\big) \cdot \mathbf{H}(t) \cdot \mathbf{U}(t)\big] \cdot \mathbf{H}^T(t), \\
\mathbf{M}(t) = \mathbf{H}(t) \cdot \mathscr{B}_2 \big[&\mathbf{H}^T(t) \cdot \big(\mathbf{U}_t^t(s) - \mathbf{I}\big) \cdot \mathbf{H}(t) \cdot \mathbf{U}(t), \\
&\mathbf{H}^T(t) \cdot \big(\mathbf{Y}_t^t(s) - \mathbf{B}(t)\big) \cdot \mathbf{H}(t) \cdot \mathbf{U}(t)\big] \cdot \mathbf{H}^T(t).
\end{aligned}$$

The latter relations transform to

$$\begin{aligned}
\mathbf{T}(t) &= \mathbf{H}(t) \cdot \mathscr{C}_1 \left[\mathbf{U}(t), \mathbf{Y}(t), \mathbf{H}^T(t) \cdot \mathbf{U}_t^t(s) \cdot \mathbf{H}(t), \mathbf{H}^T(t) \cdot \mathbf{Y}_t^t(s) \cdot \mathbf{H}(t)\right] \cdot \mathbf{H}^T(t), \\
\mathbf{M}(t) &= \mathbf{H}(t) \cdot \mathscr{C}_2 \left[\mathbf{U}(t), \mathbf{Y}(t), \mathbf{H}^T(t) \cdot \mathbf{U}_t^t(s) \cdot \mathbf{H}(t), \mathbf{H}^T(t) \cdot \mathbf{Y}_t^t(s) \cdot \mathbf{H}(t)\right] \cdot \mathbf{H}^T(t)
\end{aligned}$$

with new operators $\mathscr{C}_1$ and $\mathscr{C}_2$.

Further reduction of constitutive equation is possible assuming some type of anisotropy as was done by Eremeyev and Pietraszkiewicz (2012). In what follows we restrict ourselves by isotropic behavior. In this case $\mathscr{C}_1$ and $\mathscr{C}_2$ should satisfy the restrictions

$$
\begin{aligned}
\mathbf{O}^T \cdot \mathscr{C}_1 & \left[\mathbf{U}(t), \mathbf{Y}(t), \mathbf{H}^T(t) \cdot \mathbf{U}_t^t(s) \cdot \mathbf{H}(t), \mathbf{H}^T(t) \cdot \mathbf{Y}_t^t(s) \cdot \mathbf{H}(t)\right] \cdot \mathbf{O} \\
& = \mathscr{C}_1 \left[\mathbf{O}^T \cdot \mathbf{U}(t) \cdot \mathbf{O}, (\det \mathbf{O})\, \mathbf{O}^T \cdot \mathbf{Y}(t) \cdot \mathbf{O}, \right. \\
& \left. \mathbf{O}^T \cdot \mathbf{H}^T(t) \cdot \mathbf{U}_t^t(s) \cdot \mathbf{H}(t) \cdot \mathbf{O}, (\det \mathbf{O})\, \mathbf{O}^T \cdot \mathbf{H}^T(t) \cdot \mathbf{Y}_t^t(s) \cdot \mathbf{H}(t) \cdot \mathbf{O}\right], \\
(\det \mathbf{O})\mathbf{O}^T \cdot \mathscr{C}_2 & \left[\mathbf{U}(t), \mathbf{Y}(t), \mathbf{H}^T(t) \cdot \mathbf{U}_t^t(s) \cdot \mathbf{H}(t), \mathbf{H}^T(t) \cdot \mathbf{Y}_t^t(s) \cdot \mathbf{H}(t)\right] \cdot \mathbf{O} \\
& = \mathscr{C}_2 \left[\mathbf{O}^T \cdot \mathbf{U}(t) \cdot \mathbf{O}, (\det \mathbf{O})\, \mathbf{O}^T \cdot \mathbf{Y}(t) \cdot \mathbf{O}, \mathbf{O}^T \cdot \mathbf{H}^T(t) \cdot \mathbf{U}_t^t(s) \cdot \mathbf{H}(t) \cdot \mathbf{O}, \right. \\
& \left. (\det \mathbf{O})\, \mathbf{O}^T \cdot \mathbf{H}^T(t) \cdot \mathbf{Y}_t^t(s) \cdot \mathbf{H}(t) \cdot \mathbf{O}\right]
\end{aligned}
\tag{20}
$$

for all orthogonal tensors $\mathbf{O}, \mathbf{O}^{-1} = \mathbf{O}^T$. As the result we obtain

$$
\begin{aligned}
\mathbf{T}(t) &= \mathscr{C}_1 \left[\mathbf{H} \cdot \mathbf{U}(t) \cdot \mathbf{H}^T(t), \mathbf{H} \cdot \mathbf{Y}(t) \cdot \mathbf{H}^T(t), \mathbf{U}_t^t(s), \mathbf{Y}_t^t(s)\right], \\
\mathbf{M}(t) &= \mathscr{C}_2 \left[\mathbf{H} \cdot \mathbf{U}(t) \cdot \mathbf{H}^T(t), \mathbf{H} \cdot \mathbf{Y}(t) \cdot \mathbf{H}^T(t), \mathbf{U}_t^t(s), \mathbf{Y}_t^t(s)\right].
\end{aligned}
$$

Thus, the constitutive equations of any isotropic micropolar medium with memory take the following form

$$
\mathbf{T}(t) = \mathscr{D}_1[\mathbf{e}(t), \mathbf{k}(t), \mathbf{U}_t^t(s), \mathbf{Y}_t^t(s)], \quad \mathbf{M}(t) = \mathscr{D}_2[\mathbf{e}(t), \mathbf{k}(t), \mathbf{U}_t^t(s), \mathbf{Y}_t^t(s)], \tag{21}
$$

where $\mathscr{D}_1$ and $\mathscr{D}_2$ are isotropic operators and the Eulerian strain measures defined by

$$
\mathbf{e} = \mathbf{I} - \mathbf{H} \cdot \mathbf{F}^{-1}, \quad \mathbf{k} = \mathbf{H} \cdot \mathbf{K} \cdot \mathbf{F}^{-1}, \quad \mathbf{u} = \mathbf{H} \cdot \mathbf{F}^{-1}. \tag{22}
$$

5 Rivlin–Ericksen Tensors

The history of $\mathbf{U}_t(\tau)$ and $\mathbf{Y}_t(\tau)$ can be represented as series with respect of two families of tensors $\mathbf{A}_k$ and $\mathbf{B}_k$ as follows

$$
\mathbf{U}_t^t(s) = \sum_{k=0}^{\infty} \frac{(-1)^k}{k} s^k \mathbf{A}_k(t), \quad \mathbf{Y}_t^t(s) = \sum_{k=0}^{\infty} \frac{(-1)^k}{k} s^k \mathbf{B}_k(t). \tag{23}
$$

In the micropolar continuum tensors $\mathbf{A}_k$ and $\mathbf{B}_k$ play a role of the Rivlin–Ericksen tensors used in the nonlinear viscoelasticity of simple materials. They are given by the recurrent relations

$$
\begin{aligned}
\mathbf{A}_{k+1} &= \frac{\mathrm{d}}{\mathrm{d}t}\mathbf{A}_k + \mathbf{A}_k \cdot \operatorname{grad}\mathbf{v} - \boldsymbol{\omega} \times \mathbf{A}_k, \quad \mathbf{A}_0 = \mathbf{I}, \\
\mathbf{B}_{k+1} &= \frac{\mathrm{d}}{\mathrm{d}t}\mathbf{B}_k + \mathbf{B}_k \cdot \operatorname{grad}\mathbf{v} - \boldsymbol{\omega} \times \mathbf{B}_k, \quad \mathbf{B}_0 = \mathbf{B}.
\end{aligned}
$$

Tensors $\mathbf{A}_k$ and $\mathbf{B}_k$ can be also represented using the derivative of $\mathbf{U}$ and $\mathbf{Y}$ as follows

$$\mathbf{A}_k = \mathbf{H} \cdot \frac{\mathrm{d}^k\mathbf{U}}{\mathrm{d}t^k} \cdot \mathbf{F}^{-1}, \quad \mathbf{B}_k = \mathbf{H} \cdot \frac{\mathrm{d}^k\mathbf{Y}}{\mathrm{d}t^k} \cdot \mathbf{F}^{-1}, \tag{24}$$

or by formulae

$$\mathbf{A}_{k+1} = \mathbf{A}_k^\circ, \quad \mathbf{B}_{k+1} = \mathbf{B}_k^\circ, \tag{25}$$

where the corotational derivative is defined by the relations

$$(\ldots)^\circ = \mathbf{H} \cdot \frac{\mathrm{d}}{\mathrm{d}t}\left[\mathbf{H}^T \cdot (\ldots) \cdot \mathbf{F}\right] \cdot \mathbf{F}^{-1} \equiv \frac{\mathrm{d}}{\mathrm{d}t}(\ldots) + (\ldots) \cdot \mathrm{grad}\mathbf{v} - \boldsymbol{\omega} \times (\ldots). \tag{26}$$

Let us note that $\mathbf{A}_1$ and $\mathbf{B}_1$ coincide with the strain rates used in the theory of micropolar continuum

$$\mathbf{A}_1 = \boldsymbol{\varepsilon} \equiv \mathrm{grad}\,\mathbf{v} - \mathbf{I} \times \boldsymbol{\omega}, \quad \mathbf{B}_1 = \boldsymbol{\kappa} \equiv \mathrm{grad}\,\boldsymbol{\omega}. \tag{27}$$

For example, stress power in the micropolar continuum is given by $w = \mathbf{T} : \boldsymbol{\varepsilon} + \mathbf{M} : \boldsymbol{\kappa}$.

6 Examples of Constitutive Equations

6.1 Linear Viscous Micropolar Fluid

The simplest example of an inelastic micropolar material is the micropolar viscous fluid with the constitutive equations (Aero et al. 1965; Eringen 1966)

$$\mathbf{T} = -p(\rho)\mathbf{I} + \alpha_1\boldsymbol{\varepsilon} + \alpha_2\boldsymbol{\varepsilon}^T + \alpha_3\mathbf{I}\,\mathrm{tr}\,\boldsymbol{\varepsilon}, \quad \mathbf{M} = \beta_1\boldsymbol{\kappa} + \beta_2\boldsymbol{\kappa}^T + \beta_3\mathbf{I}\,\mathrm{tr}\,\boldsymbol{\kappa}, \tag{28}$$

where p is the pressure, ρ is the density, α_1, α_2, α_3 and β_1, β_2, β_2 are viscosities.

6.2 Non-linear Viscous Micropolar Fluid

The further generalization of (28) is non-linear viscous micropolar fluid with the following constitutive equations:

$$\mathbf{T} = -p(\rho)\mathbf{I} + \mathbf{T}_v(\boldsymbol{\varepsilon}, \boldsymbol{\kappa}), \quad \mathbf{M} = \mathbf{M}_v(\boldsymbol{\varepsilon}, \boldsymbol{\kappa}), \tag{29}$$

where $\mathbf{T}_v(\boldsymbol{\varepsilon}, \boldsymbol{\kappa})$ and $\mathbf{M}_v(\boldsymbol{\varepsilon}, \boldsymbol{\kappa})$ are non-linear isotropic functions of two non-symmetric 2nd order tensors. Such model may be applied for highly viscous suspensions or ferrofluids. Assuming the existence of a dissipative potential that is a scalar isotropic function $\Phi(\boldsymbol{\varepsilon}, \boldsymbol{\kappa}) \geq 0$ such that

$$\mathbf{T}_v = \frac{\partial \Phi}{\partial \boldsymbol{\varepsilon}}, \quad \mathbf{M}_v = \frac{\partial \Phi}{\partial \boldsymbol{\kappa}},$$

we can apply the theory of invariants for representation of Φ. As a result Φ depends on 15 invariants of $I_j(\boldsymbol{\varepsilon}, \boldsymbol{\kappa}), j = 1, \ldots, 15$ and should satisfy the requirement

$$\begin{aligned} &\Phi(I_1, I_2, I_3, I_4, I_5, I_6, I_7, I_8, I_9, I_{10}, I_{11}, I_{12}, I_{13}, I_{14}, I_{15}) \\ &= \Phi(I_1, I_2, I_3, I_4, I_5, I_6, -I_7, -I_8, I_9, -I_{10}, I_{11}, -I_{12}, I_{13}, -I_{14}, I_{15}), \end{aligned} \tag{30}$$

since I_7, I_8, I_{10}, I_{12}, I_{14} are the relative invariants and change sign during the non-proper transformations. For linear viscous fluid Φ is a quadratic potential

$$\Phi = \alpha_1 I_1^2 + \alpha_2 I_2 + \alpha_3 I_4 + \beta_1 I_{10}^2 + \beta_2 I_{11} + \beta_3 I_{13}$$

and Eq. (29) reduce to the linear case (28).

6.3 *Viscoelastic Micropolar Fluids*

The model of viscous micropolar fluid can be generalized to the case of viscoelastic behaviour. The viscoelastic micropolar fluid has the following constitutive relations (Yeremeyev and Zubov 1999):

$$\mathbf{T} = \mathscr{H}_1[\rho(t), \mathbf{B}(t), \mathbf{E}_t^t(s), \mathbf{K}_t^t(s)], \quad \mathbf{M} = \mathscr{H}_2[\rho(t), \mathbf{B}(t), \mathbf{E}_t^t(s), \mathbf{K}_t^t(s)], \tag{31}$$

where $\mathscr{H}_1$ and $\mathscr{H}_2$ are isotropic operators. In particular, we define the viscoelastic micropolar fluid of differential type of order (m, n) as a micropolar fluid with following constitutive dependencies:

$$\mathbf{T} = \mathbf{h}_1(\rho, \mathbf{B}, \mathbf{A}_1 \ldots \mathbf{A}_m, \mathbf{B}_1 \ldots \mathbf{B}_n), \quad \mathbf{M} = \mathbf{h}_2(\rho, \mathbf{B}, \mathbf{A}_1 \ldots \mathbf{A}_m, \mathbf{B}_1 \ldots \mathbf{B}_n), \tag{32}$$

where $\mathbf{h}_1$ and $\mathbf{h}_2$ are tensor-valued isotropic functions of $m + n + 1$ tensorial arguments. Since $\mathbf{M}, \mathbf{B}_k, k = 0, 1, \ldots n$ are pseudo-tensors we prove that $\mathbf{h}_1$ and $\mathbf{h}_2$ satisfy the relations

$$\begin{aligned} \mathbf{h}_1(\rho, \mathbf{B}, \mathbf{A}_1 \ldots \mathbf{A}_m, \mathbf{B}_1 \ldots \mathbf{B}_n) &= \mathbf{h}_1(\rho, -\mathbf{B}, \mathbf{A}_1 \ldots \mathbf{A}_m, -\mathbf{B}_1 \ldots - \mathbf{B}_n), \\ -\mathbf{h}_2(\rho, \mathbf{B}, \mathbf{A}_1 \ldots \mathbf{A}_m, \mathbf{B}_1 \ldots \mathbf{B}_n) &= \mathbf{h}_2(\rho, -\mathbf{B}, \mathbf{A}_1 \ldots \mathbf{A}_m, -\mathbf{B}_1 \ldots - \mathbf{B}_n). \end{aligned}$$

6.4 Micropolar Hypo-elasticity

Original model of hypoelastic material was introduced in Truesdell (1963). For micropolar solids it was generalized in Tejchman and Bauer (2005), Ramezani et al. (2008), Ramezani and Naghdabadi (2010), Surana and Reddy (2015). Within the framework of the hypo-elastic micropolar solids the constitutive equations for **T** and **M** are formulated as follows

$$\mathbf{T}^{\circ} = \boldsymbol{\eta}_1(\mathbf{T}, \boldsymbol{\varepsilon}, \boldsymbol{\kappa}), \quad \mathbf{M}^{\circ} = \boldsymbol{\eta}_2(\mathbf{M}, \boldsymbol{\varepsilon}, \boldsymbol{\kappa}), \tag{33}$$

where $^{\circ}$ denotes an objective time derivative, and $\boldsymbol{\eta}_1$ and $\boldsymbol{\eta}_2$ are isotropic functions of their arguments and linear with respect to strain rates $\boldsymbol{\varepsilon}$ and $\boldsymbol{\kappa}$. As a result, Eq. (33) take the form

$$\mathbf{T}^{\circ} = \mathbf{C}_1(\mathbf{T}) : \boldsymbol{\varepsilon} + \mathbf{C}_2(\mathbf{T}) : \boldsymbol{\kappa}, \quad \mathbf{M}^{\circ} = \mathbf{C}_3(\mathbf{M}) : \boldsymbol{\varepsilon} + \mathbf{C}_4(\mathbf{M}) : \boldsymbol{\kappa}, \tag{34}$$

where $\mathbf{C}_1, \mathbf{C}_2, \mathbf{C}_3, \mathbf{C}_4$ are 4th-order tensors, which depend on stress and couple stress tensors, in general. The following restrictions for $\boldsymbol{\eta}_1$ and $\boldsymbol{\eta}_2$ and $\mathbf{C}_k$:

$$\boldsymbol{\eta}_1(\mathbf{T}, \boldsymbol{\varepsilon}, \boldsymbol{\kappa}) = \boldsymbol{\eta}_1(\mathbf{T}, \boldsymbol{\varepsilon}, -\boldsymbol{\kappa}), \quad -\boldsymbol{\eta}_2(\mathbf{M}, \boldsymbol{\varepsilon}, \boldsymbol{\kappa}) = \boldsymbol{\eta}_2(\mathbf{M}, \boldsymbol{\varepsilon}, -\boldsymbol{\kappa}), \tag{35}$$

which lead to constraints $\mathbf{C}_2 = \mathbf{0}$, $\mathbf{C}_3(\mathbf{M}) = -\mathbf{C}_3(-\mathbf{M})$, $\mathbf{C}_4(\mathbf{M}) = \mathbf{C}_4(-\mathbf{M})$.

Constitutive equations (33) or (34) can be extended as follows

$$\begin{aligned} &\mathbf{T}^{\circ} = \boldsymbol{\eta}_1(\mathbf{T}, \mathbf{M}, \boldsymbol{\varepsilon}, \boldsymbol{\kappa}), && \mathbf{M}^{\circ} = \boldsymbol{\eta}_2(\mathbf{T}, \mathbf{M}, \boldsymbol{\varepsilon}, \boldsymbol{\kappa}), \\ &\mathbf{T}^{\circ} = \mathbf{C}_1(\mathbf{T}, \mathbf{M}) : \boldsymbol{\varepsilon} + \mathbf{C}_2(\mathbf{T}, \mathbf{M}) : \boldsymbol{\kappa}, && \mathbf{M}^{\circ} = \mathbf{C}_3(\mathbf{T}, \mathbf{M}) : \boldsymbol{\varepsilon} + \mathbf{C}_4(\mathbf{T}, \mathbf{M}) : \boldsymbol{\kappa} \end{aligned} \tag{36}$$

with the following constraints

$$\begin{gathered} \boldsymbol{\eta}_1(\mathbf{T}, \mathbf{M}, \boldsymbol{\varepsilon}, \boldsymbol{\kappa}) = \boldsymbol{\eta}_1(\mathbf{T}, -\mathbf{M}, \boldsymbol{\varepsilon}, -\boldsymbol{\kappa}), \quad -\boldsymbol{\eta}_2(\mathbf{T}, \mathbf{M}, \boldsymbol{\varepsilon}, \boldsymbol{\kappa}) = \boldsymbol{\eta}_2(\mathbf{T}, -\mathbf{M}, \boldsymbol{\varepsilon}, -\boldsymbol{\kappa}), \\ \mathbf{C}_1(\mathbf{T}, \mathbf{M}) = \mathbf{C}_1(\mathbf{T}, -\mathbf{M}), \quad -\mathbf{C}_2(\mathbf{T}, \mathbf{M}) = \mathbf{C}_2(\mathbf{T}, -\mathbf{M}), \\ -\mathbf{C}_3(\mathbf{T}, \mathbf{M}) = \mathbf{C}_3(\mathbf{T}, -\mathbf{M}), \quad \mathbf{C}_4(\mathbf{T}, \mathbf{M}) = \mathbf{C}_4(\mathbf{T}, -\mathbf{M}). \end{gathered} \tag{37}$$

6.5 Viscoelastic Materials

Considered finite approximation of series (23) we obtain the model of micropolar material of order (m, n)

$$\begin{aligned} \mathbf{T}(t) &= \mathscr{F}_1[\mathbf{e}(t), \mathbf{k}(t), \mathbf{B}(t), \mathbf{A}_1(t) \ldots \mathbf{A}_m(t), \mathbf{B}_1(t) \ldots \mathbf{B}_n(t)], \\ \mathbf{M}(t) &= \mathscr{F}_2[\mathbf{e}(t), \mathbf{k}(t), \mathbf{B}(t), \mathbf{A}_1(t) \ldots \mathbf{A}_m(t), \mathbf{B}_1(t) \ldots \mathbf{B}_n(t)], \end{aligned} \tag{38}$$

where $\mathscr{F}_1$ and $\mathscr{F}_2$ are isotropic operators.

Finally, in order to consider various form of rate-type constitutive equations of micropolar materials we introduce implicit constitutive equations of differential type in the following form

$$
\begin{aligned}
\mathbf{g}_1[\mathbf{T}^{\circ\{M\}} \ldots \mathbf{T}^{\circ}, \mathbf{M}^{\circ\{N\}} \ldots \mathbf{M}^{\circ}, \mathbf{M}(t), \mathbf{e}(t), \mathbf{k}(t), \mathbf{B}, \mathbf{A}_1 \ldots \mathbf{A}_m, \mathbf{B}_1 \ldots \mathbf{B}_n] = \mathbf{0},\\
\mathbf{g}_2[\mathbf{T}^{\circ\{M\}} \ldots \mathbf{T}^{\circ}, \mathbf{M}^{\circ\{N\}} \ldots \mathbf{M}^{\circ}, \mathbf{M}(t), \mathbf{e}(t), \mathbf{k}(t), \mathbf{B}, \mathbf{A}_1 \ldots \mathbf{A}_m, \mathbf{B}_1 \ldots \mathbf{B}_n] = \mathbf{0},
\end{aligned}
\tag{39}
$$

where $\mathbf{g}_1$ and $\mathbf{g}_2$ are isotropic tensor-valued functions of $M + N + m + n + 3$ tensorial arguments, and $\circ\{k\}$ stands for kth objective derivative.

The constitutive equation (21) include various forms of micropolar viscoelastic behaviour under finite deformations. Here we present few examples of constitutive equations of differential type. The constitutive equations of the form

$$\mathbf{T} = \boldsymbol{\Phi}_1(\mathbf{e}, \mathbf{k}, \boldsymbol{\varepsilon}, \boldsymbol{\kappa}), \quad \mathbf{M} = \boldsymbol{\Phi}_2(\mathbf{e}, \mathbf{k}, \boldsymbol{\varepsilon}, \boldsymbol{\kappa}), \tag{40}$$

$$\tau_1 \mathbf{T}^{\circ} + \mathbf{T} = \boldsymbol{\Psi}_1(\boldsymbol{\varepsilon}, \boldsymbol{\kappa}), \quad \tau_2 \mathbf{M}^{\circ} + \mathbf{M} = \boldsymbol{\Psi}_2(\boldsymbol{\varepsilon}, \boldsymbol{\kappa}), \tag{41}$$

$$\tau_1 \mathbf{T}^{\circ} + \mathbf{T} = \boldsymbol{\Omega}_1(\mathbf{e}, \mathbf{k}, \boldsymbol{\varepsilon}, \boldsymbol{\kappa}), \quad \tau_2 \mathbf{M}^{\circ} + \mathbf{M} = \boldsymbol{\Omega}_2(\mathbf{e}, \mathbf{k}, \boldsymbol{\varepsilon}, \boldsymbol{\kappa}) \tag{42}$$

play a role of Kelvin–Voigt, Maxwell and standard models in micropolar viscoelasticity, respectively. Here $\circ$ denotes an objective time derivative, τ_1 and τ_2 are relaxation time parameters and $\boldsymbol{\Phi}_1$, $\boldsymbol{\Phi}_2$, $\boldsymbol{\Psi}_1$, $\boldsymbol{\Psi}_2$, $\boldsymbol{\Omega}_1$ and $\boldsymbol{\Omega}_2$ are constitutive tensor-valued functions. Using higher order objective time derivatives and tensors $\mathbf{A}_k$, $\mathbf{B}_k$ one can present constitutive equations of differential type of any order.

7 Conclusions and Discussion

Following Altenbach and Eremeyev (2014) we presented a family of non-symmetric strain rate tensors for micropolar materials and discussed constitutive equations of inelastic micropolar materials. Using basics principles of continuum mechanics that is the principle of equipresence and the material frame indifference we discussed the constraints for the constitutive equations. In particular, considering the fact that part of strain rates are not true tensors but pseudotensors we obtain some constitutive restrictions following from the material frame indifference principle.

Considering difference between models of classic continua and Cosserat continua let us note that some classic methods of constitutive modelling used in the Cauchy mechanics can be straightforward extended for the micropolar continuum. Among them are the theory of local material symmetry group, invariance properties applied for the micropolar elasticity (Eringen 1999; Ramezani et al. 2009; Pietraszkiewicz and Eremeyev 2009a; Eremeyev and Pietraszkiewicz 2012, 2016), micropolar hypoelasticity (Ramezani and Naghdabadi 2010), and mechanics of viscous micropolar fluids (Aero et al. 1965; Eringen 1966, 2001). Several generalizations of yield criterium for elasto-plastic materials and other models for micropolar elastoplasticity are given by (Lippmann 1969; de Borst 1993; Steinmann 1994; Grammenoudis and Tsakmakis 2007, 2009). But in some case such straightforward extensions are impossible, let us mention the logarithmic Hencky's strain measure and related logarithmic strain rate (Xiao et al. 1997a, b; Bruhns 2014). Introduction of similar strain tensors based on logarithmic objective derivative in micropolar mechanics is more difficult or impossible, in general.

Similar to introduced above non-symmetric strain measures and strain rates are also used for description of two-level deformations of inelastic materials considering independent spin (Trusov et al. 2015) for derivation of generalized models of elasticity (Lurie et al. 2005).

References

Aero EL, Bulygin AN, Kuvshinskii EV (1965) Asymmetric hydromechanics. J Appl Math Mech 29(2):333–346

Altenbach H, Eremeyev VA (2014) Strain rate tensors and constitutive equations of inelastic micropolar materials. Int J Plast 63:3–17

Bruhns OT (2014) The Prandtl-Reuss equations revisited. ZAMM 94(3):187–202

de Borst R (1993) A generalization of J_2-flow theory for polar continua. Comput Methods Appl Mech Eng 103(3):347–362

Eremeyev VA, Pietraszkiewicz W (2012) Material symmetry group of the non-linear polar-elastic continuum. J Solids Struct 49(14):1993–2005

Eremeyev VA, Pietraszkiewicz W (2016) Material symmetry group and constitutive equations of micropolar anisotropic elastic solids. Math Mech Solids 21(2):210–221

Eremeyev VA, Lebedev LP, Altenbach H (2013) Foundations of micropolar mechanics. Springerbriefs in applied sciences and technologies. Springer, Heidelberg

Eringen AC (1966) Theory of micropolar fluids. J Math Mech 16(1):1–18

Eringen AC (1999) Microcontinuum field theory. I. Foundations and solids. Springer, New York

Eringen AC (2001) Microcontinuum field theory. II. Fluent media. Springer, New York

Eringen AC, Kafadar CB (1976) Polar field theories. In: Eringen AC (ed) Continuum physics, vol IV. Academic Press, New York, pp 1–75

Grammenoudis P, Tsakmakis C (2007) Micropolar plasticity theories and their classical limits. Part I: resulting model. Acta Mech 189(3–4):151–175

Grammenoudis P, Tsakmakis C (2009) Isotropic hardening in micropolar plasticity. Arch Appl Mech 79(4):323–334

Lebedev LP, Cloud MJ, Eremeyev VA (2010) Tensor analysis with applications in mechanics. World Scientific, New Jersey

Lippmann H (1969) Eine Cosserat-Theorie des plastischen Fließens. Acta Mech 8(3–4):93–113

Łukaszewicz G (1999) Micropolar fluids: theory and applications. Birkhäuser, Boston

Lurie S, Belov P, Tuchkova N (2005) The application of the multiscale models for description of the dispersed composites. Compos Part A: Appl Sci Manuf 36(2):145–152

Nowacki W (1986) Theory of asymmetric elasticity. Pergamon-Press, Oxford

Pau A, Trovalusci P (2012) Block masonry as equivalent micropolar continua: the role of relative rotations. Acta Mech 223(7):1455–1471

Pietraszkiewicz W, Eremeyev VA (2009a) On natural strain measures of the non-linear micropolar continuum. Int J Solids Struct 46(3–4):774–787

Pietraszkiewicz W, Eremeyev VA (2009b) On vectorially parameterized natural strain measures of the non-linear Cosserat continuum. Int J Solids Struct 46(11–12):2477–2480

Ramezani S, Naghdabadi R (2007) Energy pairs in the micropolar continuum. Int J Solids Struct 44(14–15):4810–4818
Ramezani S, Naghdabadi R (2010) Micropolar hypo-elasticity. Arch Appl Mech 80(12):1449–1461
Ramezani S, Naghdabadi R, Sohrabpour S (2008) Non-linear finite element implementation of micropolar hypo-elastic materials. Comput Methods Appl Mech Eng 197(49–50):4149–4159
Ramezani S, Naghdabadi R, Sohrabpour S (2009) Constitutive equations for micropolar hyper-elastic materials. Int J Solids Struct 46(14–15):2765–2773
Steinmann P (1994) A micropolar theory of finite deformation and finite rotation multiplicative elastoplasticity. Int J Solids Struct 31(8):1063–1084
Surana KS, Reddy JN (2015) A more complete thermodynamic framework for solid continua. J Therm Eng 1(6):446–459
Tejchman J, Bauer E (2005) Modeling of a cyclic plane strain compression-extension test in granular bodies within a polar hypoplasticity. Granul Matter 7(4):227–242
Trovalusci P, Masiani R (1997) Strain rates of micropolar continua equivalent to discrete systems. Meccanica 32(6):581–583
Truesdell CA (1963) Remarks on hypo-elasticity. J Res Natl Bur Stand—B Math Math Phys 67(3):141–143
Truesdell CA (1966) The elements of continuum mechanics. Springer, Berlin
Truesdell CA, Noll W (2004) The non-linear field theories of mechanics, 3rd edn. Springer, Berlin
Trusov PV, Volegov PS, Yanz AY (2015) Two-level models of polycrystalline elastoviscoplasticity: complex loading under large deformations. ZAMM 95(10):1067–1080
Xiao H, Bruhns OT, Meyers A (1997a) Hypo-elasticity model based upon the logarithmic stress rate. J Elast 47(1):51–68
Xiao H, Bruhns OT, Meyers A (1997b) Logarithmic strain, logarithmic spin and logarithmic rate. Acta Mech 124(1–4):89–105
Yeremeyev VA, Zubov LM (1999) The theory of elastic and viscoelastic micropolar liquids. J Appl Math Mech 63(5):755–767

On the Modelling of Carbon Nano Tubes as Generalized Continua

Hossein Aminpour and Nicola Rizzi

Abstract A 1D continuum endowed with internal structure, previously introduced by the authors in order to describe some nonlinear behaviours of Carbon Nano Tubes (CNTs), is extended and generalised by giving a procedure for constructing the constitutive functions. Starting from the reference configuration of a Carbon Nano Sheet (CNS), a Representative Elementary Volume (REV) is chosen. The deformation measures of the REV are identified with the change of the length of the Carbon-Carbon (C-C) bonds and the angle variation between each pair of adjacent bonds. The strain energy density of the REV is given as a standard function of the microscopic strain measures. A relationship between the micro and the continuum strain measures is then put down, this leading to an expression of the strain energy density of the REV in terms of the latter strains. Making the derivative of this energy with respect to its argument the constitutive functions for the 1D continuum are obtained. The geometric and mechanical properties of a graphene nano sheet are used to construct its equivalent continuum and some numerical comparisons are discussed. Although the procedure is set up for a CNS, its extension to Carbon Nano Tubes (CNTs) involves only simple geometric computations.

Keywords Nonlinear elasticity · Bifurcation analysis · Generalized continua · Carbon nano sheets

1 Introduction

CNTs have been given a large attention due to the fact that they show very peculiar mechanical properties (Shima and Sato 2013). In addition, as they can undergo very large deformations without losing the elastic behaviour, nonlinear models must be

H. Aminpour · N. Rizzi (✉)
Università Degli Studi Roma Tre, Via Della Madonna Dei Monti 40, 00184 Roma, Italy
e-mail: nicolaluigi.rizzi@uniroma3.it

H. Aminpour
e-mail: hossein.aminpour@uniroma3.it

H. Altenbach and S. Forest (eds.), *Generalized Continua as Models for Classical and Advanced Materials*, Advanced Structured Materials 42,
DOI 10.1007/978-3-319-31721-2_2

constructed in order to have a fair description of a number of very relevant phenomena. Even though the molecular dynamic approach has been and is still largely used as a simulation tool, it has been recognized as cumbersome in many circumstances. For this reason, the attention of many researchers has been focused on the continuum modelling making recourse to both 3D and shell theories.

Following this line the authors (Aminpour et al. 2014; Aminpour and Rizzi 2015), in accordance with the approach in Antman (2005) and Podio-Guidugli (1982), proposed to use a 1D continuum endowed with a suitable microstructure for the modelling of the mechanical behaviour of CNTs. With respect to a Cosserat beam (Pietraszkiewicz et al. 2007; Bîrsan et al. 2012; Altenbach et al. 2012), that model has one more scalar field which is introduced with the aim of accounting for the deformation of the cross-sections in their own plane. It has been shown that by adopting very simple constitutive relationships—that were polynomial functions up to grade two in the components of the deformation measures—the model is able to capture the relevant phenomena of necking and kinking that appear when a CNT is loaded by two axial forces or two bending moments, at its ends (see Shima and Sato 2013).

The aim of this work is to extend the theory in Aminpour et al. (2014); Aminpour and Rizzi (2015) by proposing a procedure that leads to identify the constitutive relationships of an *equivalent* 1D continuum, starting from the analysis of the microscopic behaviour of a carbon nanomaterial (see e.g. dell'Isola et al. 2012). In view of the results obtained previously, these relationships are given an approximate polynomial expansion of grade two. The analysis is performed for a CNS and, even though it is not presented here, its generalization to CNTs is only matter of simple geometry.

Given a reference configuration of the sheet, a REV including two atoms connected with five C-C bonds connecting themselves and their first neighbours, is considered. By assuming a generic displacement of the six atoms, the change of the lengths of the C C bonds, together with the angle variation between the adjacent bonds, are assumed as deformation measures of the microscopic model. The strain energy density of the REV is then given in terms of those deformation measures. By assuming a suitable map between the kinematic of the microscopic model and that of the 1D model, the deformation measures of the former are given as functions of the deformation measures of the latter. This allows to write the strain energy density of the REV as a function of the deformation measures of the 1D model. This energy is then assumed as the strain energy density of the 1D continuum *equivalent* to the CNS. Successively, this energy is given a power expansion with respect to the deformation measures, up to the third order. The derivative of the strain energy density with respect to its arguments gives the constitutive relationships of the 1D stress measures.

As a numerical example, the analysis of the mechanical behaviour of a graphene sheet studied in Pei et al. (2010) is performed by means of the geometric model introduced here for two cases of atoms arrangements, namely the so called armchair and zigzag. A trivial nonlinear equilibrium solution is found and the Young's modulus tangent in the origin determined for both the atoms arrangement. The results are found to be in agreement with the values reported in the literature (Xiao et al. 2005;

Lee et al. 2008; WenXing et al. 2004). In addition, a bifurcation analysis has been performed and the bifurcation point on the trivial equilibrium paths for both the armchair and zigzag cases, determined. It is shown that the branched solutions describe a necking phenomenon. Finally we want to stress that only interactions between each atom and its first neighbours have been considered and that to account also for long range interactions would lead to a strain energy density for the equivalent continuum containing gradients of the deformation measures (Alibert et al. 2003).

2 One-Dimensional Beam Model

A beam is thought as a one dimensional continuous body, whose kinematics is described through the placement of the points of a line that we call the beam axis, and the placement of a couple of orthogonal vectors attached to each point of it. They are allowed to change their length in order to describe the cross section deformation. We will consider here a reference configuration φ_0 in which the axis is straight and the sections are orthogonal to it. The axis is described by the function

$$\boldsymbol{q}(s) \qquad s \in [0, 1], \tag{1}$$

where s is its arc length parameter. The unit vector field tangent to the axis of the beam is

$$\boldsymbol{q}'(s), \tag{2}$$

where the prime denotes differentiation with respect to s. Let φ_t be the configuration assumed by the beam during a motion at time t. Such a configuration is described by

- the function $\boldsymbol{p}(s, t)$, providing the present position of $\boldsymbol{q}(s)$;
- a proper orthogonal tensor field $\boldsymbol{R}(s, t)$, providing the rotation of the cross-sections when passing from φ_0 to φ_t ;
- a scalar field $\delta(s, t)$, providing a coarse description of the cross-sections deformation superimposed to the rotation $\boldsymbol{R}(s, t)$.

The tangent vector to the axis of the beam in the present configuration is given by

$$\boldsymbol{p}'(s, t), \tag{3}$$

and the velocity fields are

$$\boldsymbol{w} = \dot{\boldsymbol{p}}, \quad \boldsymbol{W} = \dot{\boldsymbol{R}}\boldsymbol{R}^{\mathrm{T}}, \quad \omega = \dot{\delta}, \tag{4}$$

where $\boldsymbol{W}$ is a skew tensor and the dot denotes differentiation with respect to t. A change of placement is rigid when

$$\boldsymbol{R}' = \boldsymbol{0}, \quad \boldsymbol{p}' = \boldsymbol{R}\boldsymbol{q}', \quad \delta = 0, \tag{5}$$

so that in a rigid motion

$$\dot{\boldsymbol{p}}' = \dot{\boldsymbol{R}}\boldsymbol{q}', \quad \dot{\delta} = 0, \quad \forall t \tag{6}$$

2.1 Strain Measures

Deformation is defined as the difference between the given transplacement and a rigid one. A suitable choice of strain measures with respect to φ_0 is

$$\boldsymbol{e} = \boldsymbol{R}^{\mathrm{T}}\boldsymbol{p}' - \boldsymbol{q}', \quad \boldsymbol{E} = \boldsymbol{R}^{\mathrm{T}}\boldsymbol{R}', \quad \delta, \quad \delta', \tag{7}$$

where $\boldsymbol{E}$ is a skew tensor.

We limit ourselves to the case of a beam whose motion can be described in a 2D subspace of a 3D Euclidean space. Given a fixed orthonormal basis

$$\boldsymbol{D}_1, \boldsymbol{D}_2, \boldsymbol{D}_3 = \boldsymbol{D}_1 \times \boldsymbol{D}_2$$

we assume that $\boldsymbol{q}'(s) = \boldsymbol{D}_1$ so that

$$\boldsymbol{e} = \varepsilon\boldsymbol{D}_1 + \gamma\boldsymbol{D}_2, \quad \boldsymbol{E} = \kappa\boldsymbol{D}_2 \wedge \boldsymbol{D}_1, \tag{8}$$

where ε is the axial strain and γ is the shearing strain, κ denotes the bending curvature.

It is useful to write the strain measures in terms of the displacement field $\boldsymbol{u}$ defined as

$$\boldsymbol{u} = \boldsymbol{p} - \boldsymbol{q} = u\boldsymbol{D}_1 + v\boldsymbol{D}_2 \tag{9}$$

Thus, in view of the definitions (8), (9), the components of the deformation measures (7), read

$$\begin{aligned} \varepsilon &= (1 + u')\cos\vartheta + v'\sin\vartheta - 1, \\ \gamma &= v'\cos\vartheta - (1 + u')\sin\vartheta, \\ \kappa &= \vartheta', \\ &\delta, \ \delta', \end{aligned} \tag{10}$$

where ϑ is the angle of rotation about $\boldsymbol{D}_3$.

2.2 Balance

The interaction of the beam with the surrounding environment is defined as a linear functional of the velocities (6) and of their first-order derivatives with respect to s, and the external power can be expressed as

$$P_e = \int_0^l (\boldsymbol{b} \cdot \boldsymbol{w} + \boldsymbol{B} \cdot \boldsymbol{W} + \beta \dot{\delta}) \mathrm{d}s + \left[\boldsymbol{t} \cdot \boldsymbol{w} + \boldsymbol{T} \cdot \boldsymbol{W} + \Omega \dot{\delta}\right]_0^l \tag{11}$$

$\boldsymbol{b}, \boldsymbol{B}, \beta$ and $\boldsymbol{t}, \boldsymbol{T}, \Omega$ being bulk and contact actions, respectively.

The power expended by the contact actions is a linear functional of the velocity fields and of their first derivatives with respect to X_1, that is

$$P_i = \int_0^l (\boldsymbol{c}_0 \cdot \boldsymbol{w} + \boldsymbol{C}_0 \cdot \boldsymbol{W} + \Delta \dot{\delta} + \boldsymbol{c}_1 \cdot \boldsymbol{w}' + \boldsymbol{C}_1 \cdot \boldsymbol{W}' + \vartheta \dot{\delta}') \mathrm{d}s \tag{12}$$

Following Germain (1973a, b); di Carlo (1996), we require that $P_i = 0$ for any rigid motion, that is true if and only if

$$P_i = \int_0^l (\boldsymbol{c}_1 \cdot \boldsymbol{w}' - (\boldsymbol{p}' \wedge \boldsymbol{c}_1) \cdot \boldsymbol{W} + \boldsymbol{C}_1 \cdot \boldsymbol{W}' + \Delta \dot{\delta} + \vartheta \dot{\delta}') \mathrm{d}s \tag{13}$$

Besides, by requiring that

$$P_e = P_i \qquad \forall t \tag{14}$$

the expressions (11), (13) and (14) give $\boldsymbol{c}_1 = \boldsymbol{t}$, $\boldsymbol{C}_1 = \boldsymbol{T}$ and $\vartheta = \Omega$ and

$$\begin{aligned} \boldsymbol{t}' + \boldsymbol{b} &= \boldsymbol{0}, \\ \boldsymbol{T}' + \boldsymbol{p}' \wedge \boldsymbol{t} + \boldsymbol{B} &= \boldsymbol{0}, \\ \Omega' - \Delta + \beta &= 0 \end{aligned} \tag{15}$$

Equation (15) can be conveniently rewritten in the form

$$\begin{aligned} \boldsymbol{s}' + \boldsymbol{E}\boldsymbol{s} + \boldsymbol{a} &= \boldsymbol{0}, \\ \boldsymbol{S}' + \boldsymbol{E}\boldsymbol{S} - \boldsymbol{S}\boldsymbol{E} + (\boldsymbol{q}' + \boldsymbol{e}) \wedge \boldsymbol{s} + \boldsymbol{A} &= \boldsymbol{0}, \\ \Omega' - \Delta + \beta &= 0, \end{aligned} \tag{16}$$

where $\boldsymbol{s} = \boldsymbol{R}^{\mathrm{T}}\boldsymbol{t}$, $\boldsymbol{S} = \boldsymbol{R}^{\mathrm{T}}\boldsymbol{T}\boldsymbol{R}$, $\boldsymbol{a} = \boldsymbol{R}^{\mathrm{T}}\boldsymbol{b}$, $\boldsymbol{A} = \boldsymbol{R}^{\mathrm{T}}\boldsymbol{B}\boldsymbol{R}$, whose components in the fixed basis are

$$\begin{aligned} \boldsymbol{s} &= N\boldsymbol{D}_1 + Q\boldsymbol{D}_2, \\ \boldsymbol{S} &= M\boldsymbol{D}_2 \wedge \boldsymbol{D}_1, \\ \boldsymbol{a} &= a_1\boldsymbol{D}_1 + a_2\boldsymbol{D}_2, \\ \boldsymbol{A} &= A\boldsymbol{D}_2 \wedge \boldsymbol{D}_1 \end{aligned} \tag{17}$$

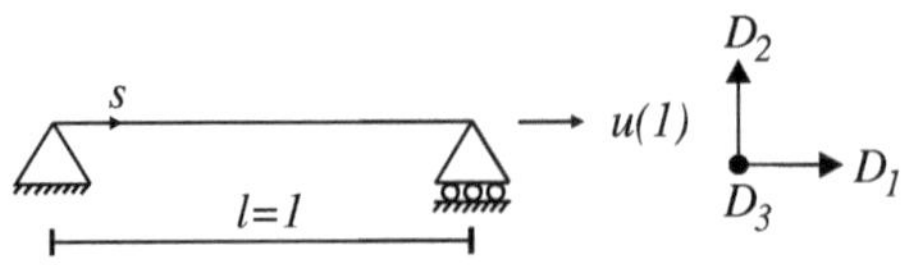

Fig. 1 Axial end displacement

2.3 *Beam Subjected to an Axial End Displacement*

In the following we will consider the case of the strut shown in Fig. 1 and look for solutions in which $\vartheta \equiv v \equiv 0$ so that expressions (10) reduce to

$$\begin{aligned} &\varepsilon = u', \\ &\delta, \quad \delta' \end{aligned} \tag{18}$$

In addition, we will assume that the continuum is hyperelastic and its strain energy density written as

$$\pi(\varepsilon, \delta, \delta'), \tag{19}$$

which means that the only stress measures different from zero will be

$$N(\varepsilon, \delta, \delta') = \frac{\partial \pi}{\partial \varepsilon}, \quad \Delta(\varepsilon, \delta, \delta') = \frac{\partial \pi}{\partial \delta}, \quad \Omega(\varepsilon, \delta, \delta') = \frac{\partial \pi}{\partial \delta'}, \tag{20}$$

while $Q \equiv M \equiv 0$ and the definitions (17) if the body actions vanish, becomes

$$\begin{aligned} N' &= 0, \\ \Omega' - \Delta &= 0 \end{aligned} \tag{21}$$

By assuming the boundary conditions

$$\begin{aligned} &\Omega(0) = \Omega(1) = 0, \\ &u(0) = 0, \quad u(1) = \text{assigned} \end{aligned} \tag{22}$$

the Eqs. (18), (20), (21) and (22) constitute a nonlinear boundary value problem (see Aminpour et al. 2014; Aminpour and Rizzi 2015; Antman 2005) that will be discussed in the following.

3 Graphene

Graphite is a layered 3D material which is made up of a successive series of parallel two-dimensional sheets, called graphene sheets. A graphene sheet is a single layer of graphite which has only one atom in the thickness. Each graphene sheet is composed of a regular hexagonal network of strongly bonded carbon atoms. Within each graphene sheet the distance between two adjacent carbon atoms is (Saito et al. 1998)

$$a_{\mathrm{c-c}} = 0.142 \times 10^{-9}\ \mathrm{m} = 1.42\ \text{Å}, \tag{23}$$

while its thickness is assumed to be

$$t = 3.4\ \text{Å} \tag{24}$$

A two-dimensional graphene sheet can be described as a lattice of regular hexagons, whose vertices show the position of the atoms and the edges describe the bonds (see Fig. 2). In the same figure, the parallelogram drawn in shadow is the unit cell (or REV) of the sheet and the vectors $\boldsymbol{a}_1$ and $\boldsymbol{a}_2$ are the basis vectors of the unit cell. The whole lattice can be generated by translations of the unit cell by the vectors

$$\boldsymbol{C}_{\mathrm{h}} = n\boldsymbol{a}_1 + m\boldsymbol{a}_2, \tag{25}$$

where n, m are integers. The vector $\boldsymbol{C}_{\mathrm{h}}$ is called the chiral vector of the sheet.

Two atoms whose positions (say $\boldsymbol{B}_1$ and $\boldsymbol{B}_2$) can be expressed by means of a chiral vector—which means that there are two integers n, m such that

$$\boldsymbol{B}_2 - \boldsymbol{B}_1 = \boldsymbol{C}_{\mathrm{h}}$$

are called equivalent. It is clear that the two atoms in the unit cell are not equivalent as

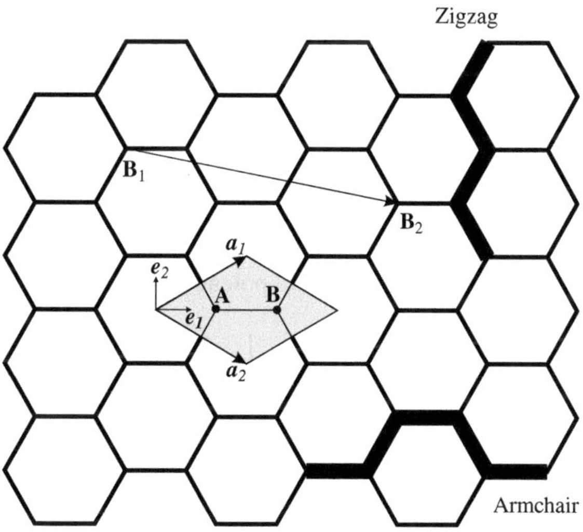

Fig. 2 Graphene sheet and its unit cell (or REV)

$$\boldsymbol{B} - \boldsymbol{A} = \frac{\boldsymbol{a}_1 + \boldsymbol{a}_2}{3} \tag{26}$$

In view of the following applications, it is useful to introduce an orthonormal basis $(\boldsymbol{e}_1, \boldsymbol{e}_2)$ whose two vectors are directed along the so called armchair and zigzag directions, respectively (see Fig. 2). Through the new basis, the basis vectors of the cell can be written as

$$\begin{aligned} \boldsymbol{a}_1 &= a\left(\frac{\sqrt{3}}{2}\boldsymbol{e}_1 + \frac{1}{2}\boldsymbol{e}_2\right), \\ \boldsymbol{a}_2 &= a\left(\frac{\sqrt{3}}{2}\boldsymbol{e}_1 - \frac{1}{2}\boldsymbol{e}_2\right), \end{aligned} \tag{27}$$

which give

$$\begin{aligned} \boldsymbol{a}_1 \cdot \boldsymbol{a}_1 &= \boldsymbol{a}_2 \cdot \boldsymbol{a}_2 = a^2, \\ \boldsymbol{a}_1 \cdot \boldsymbol{a}_2 &= a^2/2, \end{aligned} \tag{28}$$

where a is the lattice constant which is related to the carbon-carbon bond length a_{c-c} by the relationship

$$a = \sqrt{3}a_{c-c} = 0.246 \times 10^{-9}\ \text{m} \tag{29}$$

4 Energy

The total potential energy of the nanostructure may be given by the sum of energies duc to the interatomic interactions (Rappe et al. 1992)

$$U = U_r + U_\theta + U_\tau + U_\omega + U_{\text{vdW}} + U_{\text{es}}, \tag{30}$$

where U_r is the bond stretching energy, U_θ is the bond angle bending energy, U_τ is the bond torsion (or dihedral angle variation) energy, U_ω is the bond inversion (or out of plane angle variation) energy, U_{vdW} is van der Waals interaction energy, and U_{es} is the electrostatic interaction energy.

Following Leamy (2007) for the terms $U_r + U_\theta$ in Eq. (30) we will assume the Modified Morse interatomic potential, that is

$$\begin{aligned} U_r &= D_e\left\{\left(1 - \mathrm{e}^{-\beta_e(r-r_0)}\right)^2 - 1\right\}, \\ U_\theta &= \frac{1}{2}k_\theta\,(\theta - \theta_0)^2\left\{1 + k_{\text{sextic}}(\theta - \theta_0)^4\right\}, \end{aligned} \tag{31}$$

in which D_e, β_e, k_θ, k_{sextic} are constitutive constants, r is the length of a bond and θ is the angle between two adjacent bonds in the present configuration, while r_0 and θ_0 are the corresponding values in the reference configuration. The terms U_τ, U_ω,

U_{vdW} and U_{es}, on the contrary, will be neglected according to Xiao et al. (2005). This means that the expression (30) reduces to

$$U = U_r + U_\theta \tag{32}$$

5 Atomic Model

Figure 3 shows the reference configuration of the unit cells directed along the armchair and zigzag directions, respectively. The REV has been enlarged in order to consider the first neighbours of the two atoms in it. Let us assume that the sheet is stretched in the direction of the y axis in each one of the cases in Fig. 3 that are referred to as armchair and zigzag, respectively.

The displacement of an atom which occupies the reference position E, is described by the vector (see Fig. 4)

$$\boldsymbol{d}_E = v_E \boldsymbol{e}_1 + u_E \boldsymbol{e}_2 \tag{33}$$

The reference configuration of the enlarged REV is obtained by specifying the coordinates of each one of the six points in which are located the atoms. By assuming a local Cartesian coordinate system with origin at the middle point of the segment AB and basis $(\boldsymbol{e}_1, \boldsymbol{e}_2)$, the coordinates of a point E, will be denoted $E \equiv (x_E, y_E)$.

The reference configuration of the REV can also be completely specified when we fix the position of an atom and give the five bond lengths and the six angles between adjacent bonds. Note that the six angles are not independent because the sum of the three angles formed by the bonds that meet at the positions A and B, must be 2π. In the deformed configuration the new position occupied by the atom that in the reference configuration was in E, will be denoted by $e \equiv (x_e, y_e)$, where

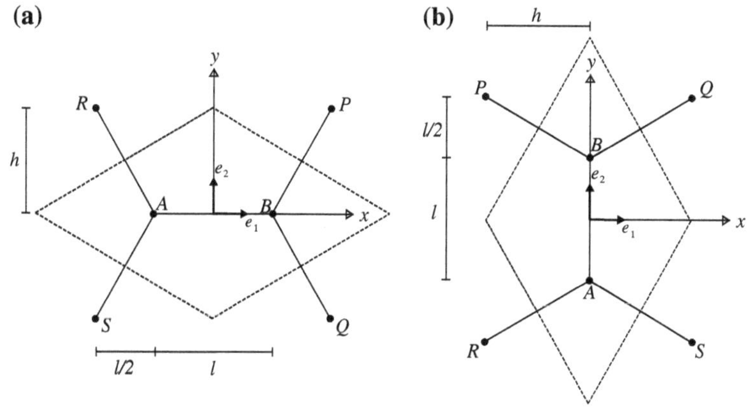

Fig. 3 The armchair (a) and zigzag (b) REVs

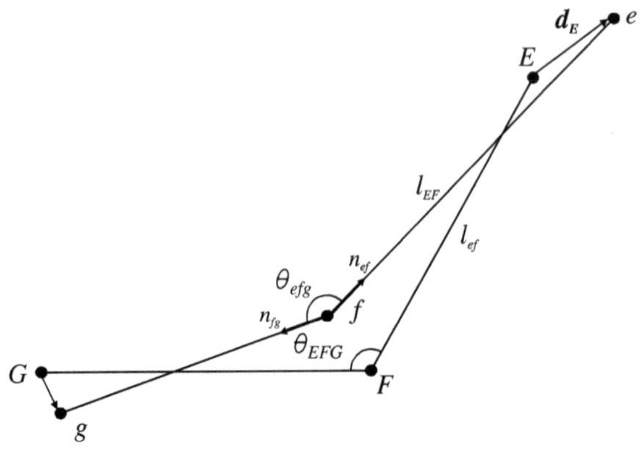

Fig. 4 Reference and present configuration of two adjacent bonds

$$x_e = x_E + v_E,$$
$$y_e = y_E + u_E \tag{34}$$

In this way, the new length of a bond, say EF, results to be

$$\begin{aligned}\ell_{ef} &= |f - e| = \sqrt{(x_f - x_e)^2 + (y_f - y_e)^2} \\ &= \sqrt{[(x_F - x_E) + (v_F - v_E)]^2 + [(y_F - y_E) + (u_F - u_E)]^2},\end{aligned} \tag{35}$$

while the angle between two adjacent bonds, say ef and fg is obtained from the relationship (see Fig. 4)

$$\cos \vartheta_{efg} = \boldsymbol{n}_{ef} \cdot \boldsymbol{n}_{fg}, \tag{36}$$

where

$$\begin{aligned}\boldsymbol{n}_{ef} &= \frac{e - f}{\ell_{ef}}, \\ \boldsymbol{n}_{fg} &= \frac{g - f}{\ell_{fg}}\end{aligned} \tag{37}$$

are the unit vectors along the bonds ef, and fg in the present configuration, respectively.

In view of the Eqs. (35)–(37), we define as deformation measures, the bonds' stretching and the angles' variation

$$\begin{aligned}\Delta r_{EF} &= \ell_{ef} - \ell_{EF}, \\ \Delta\vartheta_{EFG} &= \vartheta_{efg} - \vartheta_{EFG}\end{aligned} \tag{38}$$

Now, if we put

$$\begin{aligned} \Delta v_{EF} &= v_F - v_E, \\ \Delta u_{EF} &= u_F - u_E \end{aligned} \tag{39}$$

we can write

$$\begin{aligned} \ell_{ef} &= \sqrt{[(x_F - x_E) + \Delta v_{EF}]^2 + [(y_F - y_E) + \Delta u_{EF}]^2}, \\ \cos\vartheta_{efg} &= \frac{1}{\ell_{ef}\ell_{fg}}[(x_F - x_E + \Delta v_{EF})(x_G - x_F + \Delta v_{GF}) \\ &\quad + (y_F - y_E + \Delta u_{EF})(y_G - y_F + \Delta u_{GF})] \end{aligned} \tag{40}$$

and expressions (38) can be rewritten in terms of the components of the relative displacements.

6 1D Continuum Equivalent to the Atomic Model

The kinematic of the REV described in Sect. 5 and that of the continuum described in Sect. 2, are related by postulating the following relationships

$$\begin{aligned} u_E &= y_E\varepsilon(s), \\ v_E &= [\delta(s) + y_E\delta'(s)]x_E, \end{aligned} \tag{41}$$

where E stands for the position of a generic atom in the reference configuration of the REV and the abscissa of the 1D continuum has been chosen to coincide with the y axis on the REV, that is $s = y$.

Using the relationships (41), the components of the relative displacements of two atoms which form a C-C bond and in the reference configuration are in the positions E, F, will be

$$\begin{aligned} \Delta u_{EF} &= (y_F - y_E)\,\varepsilon, \\ \Delta v_{EF} &= (x_F - x_E)\delta + (x_F y_F - x_E y_E)\delta' \end{aligned} \tag{42}$$

Now, in view of the relationships (42), (40) and (38), the energy (31) is given in terms of the deformation measures of the continuum model

$$\begin{aligned} U_{EF}(\varepsilon, \delta, \delta') &= D_e\left\{\left(1 - \mathrm{e}^{-\beta_e(\ell_{ef} - \ell_{EF})}\right)^2 - 1\right\}, \\ U_{EFG}(\varepsilon, \delta, \delta') &= \frac{1}{2}k_\vartheta\left(\vartheta_{efg} - \vartheta_{EFG}\right)^2\left\{1 + k_{\text{sextic}}(\vartheta_{efg} - \vartheta_{EFG})^4\right\} \end{aligned} \tag{43}$$

Summing up the contribution of all the bonds of the enlarged REV, one obtains

$$\begin{aligned}\hat{\pi}(\varepsilon,\delta,\delta') = U_{AB} + \frac{1}{2}U_{BP} + \frac{1}{2}U_{BQ} + \frac{1}{2}U_{AR} + \frac{1}{2}U_{AS} \\ + U_{RAB} + U_{ABP} + U_{SAB} + U_{ABQ} + U_{PBQ} + U_{RAS}\end{aligned} \tag{44}$$

By giving a power expansion up to the third order of $\hat{\pi}(\varepsilon,\delta,\delta')$ with respect to its arguments, we denote by $\breve{\pi}(\varepsilon,\delta,\delta')$ the approximation obtained.

If we denote by V the volume of the REV, the strain energy density of the 1D continuum *equivalent* to the nanomaterial is assumed to be

$$\pi(\varepsilon,\delta,\delta') = \frac{1}{V}\breve{\pi}(\varepsilon,\delta,\delta') \tag{45}$$

Making the derivatives of the strain energy density with respect to the components of the deformation measures, one obtains the components of the stress measures.

They result to have the following expressions for both the armchair and zigzag cases

$$\begin{aligned} N &= \frac{1}{V}\left(A\delta + \frac{1}{2}B\delta^2 + \frac{1}{2}C\delta'^2 + F\varepsilon + E\varepsilon\delta + \frac{1}{2}G\varepsilon^2\right), \\ \Delta &= \frac{1}{V}\left(P\delta + \frac{1}{2}R\delta^2 + \frac{1}{2}S\delta'^2 + A\varepsilon + B\varepsilon\delta + \frac{1}{2}E\varepsilon^2\right), \\ \Omega &= \frac{1}{V}(H\delta' + S\delta\delta' + C\varepsilon\delta') \end{aligned} \tag{46}$$

The explicit expressions for the constitutive coefficients have been evaluated for the following four cases using the MATHEMATICA software (Wolfram 2015)

- case a_1 armchair REV considering the sole stretching energy;
- case a_2 armchair REV considering stretching and angle variation energy;
- case b_1 zigzag REV considering the sole stretching energy;
- case b_2 zigzag REV considering stretching and angle variation energy.

As the resulting expressions are very cumbersome, in the following we give only the constitutive functions for the case a_1, that are

$$\begin{aligned} N = \frac{\partial\pi}{\partial\varepsilon} = \frac{1}{V}\Bigg[& 2\frac{h^4\beta_e^2 D_e\varepsilon}{\ell^2} + \frac{1}{2}h^2\beta_e^2 D_e\delta \\ & + \left(\frac{3}{4}\frac{h^4\beta_e^2 D_e}{\ell^2} - 3\frac{h^6\beta_e^3 D_e}{\ell^3}\right)\varepsilon^2 \\ & + \left(-\frac{1}{8}h^2\beta_e^2 D_e + \frac{1}{4}\frac{h^4\beta_e^2 D_e}{\ell^2} - \frac{3}{16}h^2\beta_e^3 D_e\ell\right)\delta^2 \\ & + \left(\frac{1}{8}h^2\beta_e^2 D_e - \frac{h^4\beta_e^2 D_e}{\ell^2} - \frac{3}{2}\frac{h^4\beta_e^3 D_e}{\ell}\right)\varepsilon\delta \end{aligned} \tag{47}$$

$$
+\left(-\frac{1}{2}h^2\beta_e^2 D_e + \frac{h^4\beta_e^2 D_e}{\ell^2} - \frac{3}{4}h^2\beta_e^3 D_e\ell\right)\delta'^2\Bigg],
$$

$$
\begin{aligned}
\Delta = \frac{\partial\pi}{\partial\delta} = \frac{1}{V}\Bigg[&\frac{1}{2}h^2\beta_e^2 D_e\varepsilon \\
&+\left(\frac{1}{8}\beta_e^2 D_e\ell^2 + \frac{1}{4}\beta_e^2 D_e\ell^2 + h^2\beta_e^2 D_e\right)\delta \\
&+\left(\frac{1}{16}h^2\beta_e^2 D_e - \frac{1}{2}\frac{h^4\beta_e^2 D_e}{\ell^2} - \frac{3}{4}\frac{h^4\beta_e^3 D_e}{\ell}\right)\varepsilon^2 \\
&+\left(\frac{3}{16}h^2\beta_e^2 D_e - \frac{3}{64}\beta_e^3 D_e\ell^3 - \frac{3}{2}\beta_e^3 D_e\ell^3\right)\delta^2 \\
&-\left(\frac{1}{4}h^2\beta_e^2 D_e\ell^4 + \frac{1}{2}\frac{h^4\beta_e^2 D_e}{\ell^2} - \frac{3}{8}h^2\beta_e^3 D_e\ell\right)\varepsilon\delta \\
&+\left(\frac{3}{4}h^2\beta_e^2 D_e - \frac{3}{16}\beta_e^3 D_e\ell^3\right)\delta'^2\Bigg],
\end{aligned} \tag{48}
$$

$$
\begin{aligned}
\Omega = \frac{\partial\pi}{\partial\delta'} = \frac{1}{V}\Bigg[&\frac{1}{2}\beta_e^2 D_e\delta' + \left(\frac{3}{2}h^2\beta_e^2 D_e - \frac{3}{8}\beta_e^3 D_e\ell^3\right)\delta\delta' \\
&-\left(h^2\beta_e^2 D_e + 2\frac{h^4\beta_e^2 D_e}{\ell^2} - \frac{3}{2}h^2\beta_e^3 D_e\ell\right)\varepsilon\delta'\Bigg],
\end{aligned} \tag{49}
$$

where h and ℓ are shown in Fig. 3.

7 Beam with End Displacement: Trivial Solution

Let us consider now the case of a beam described in Sect. 2.3 and let $u_\ell = u(\ell)$ be the axial displacement assigned to the end section. The boundary value problem (18), (21), (22), (46), admits the following trivial solution

$$
\begin{aligned}
&\Omega_o \equiv 0, \quad \Delta_o \equiv 0, \quad \delta'_o \equiv 0, \quad \varepsilon'_o \equiv 0, \\
&\varepsilon_o \; = u_\ell/\ell = u_\ell \quad (\text{as } \ell = 1), \\
&\delta_o \; = \delta(u_\ell), \\
&N_o = N(\varepsilon(u_\ell), \delta(u_\ell))
\end{aligned} \tag{50}
$$

By using the chain rule it can be seen that

$$
\begin{aligned}
\frac{dN_o}{du_\ell} &= \frac{dN_o}{d\varepsilon_o} = (\frac{\partial N}{\partial\varepsilon})_o + (\frac{\partial N}{\partial\delta})_o(\frac{\partial\delta}{\partial\varepsilon})_o, \\
\frac{d\Delta_o}{du_\ell} &= \frac{d\Delta_o}{d\varepsilon_o} = (\frac{\partial\Delta}{\partial\varepsilon})_o + (\frac{\partial\Delta}{\partial\delta})_o(\frac{\partial\delta}{\partial\varepsilon})_o,
\end{aligned} \tag{51}
$$

where $(\,)_o$ means that the expression is evaluated along the solution (50).

As on the trivial path,

$$\Delta_o = 0 \quad \Rightarrow \quad \frac{\mathrm{d}\Delta_o}{\mathrm{d}\varepsilon_o} = 0 \tag{52}$$

Equation $(51)_2$ gives

$$(\frac{\partial \delta}{\partial \varepsilon})_o = -\frac{(\frac{\partial \Delta}{\partial \varepsilon})_o}{(\frac{\partial \Delta}{\partial \delta})_o} \tag{53}$$

and by substituting (53) in the equation $(51)_1$ one obtains

$$\frac{\mathrm{d}N_o}{\mathrm{d}\varepsilon_o} = \frac{(\frac{\partial N}{\partial \varepsilon})_o(\frac{\partial \Delta}{\partial \delta})_o - (\frac{\partial N}{\partial \delta})_o(\frac{\partial \Delta}{\partial \varepsilon})_o}{(\frac{\partial \Delta}{\partial \delta})_o} \tag{54}$$

8 Perturbation Method

We are interested in looking for another solution branching off from the given (fundamental) one. Besides, we assume that the branching solution can be expressed in terms of a suitable parameter η, in the form

$$\Xi^b(s, \eta) = \Xi_o(s, p(\eta)) + \Xi(s, \eta) \tag{55}$$

where Ξ stands for a generic field and p is the parameter chosen to describe the trivial solution that, in general, is different from η.

We admit that the functions can be represented by the following power expansions near the bifurcation point that corresponds to $\eta = 0$

$$\begin{aligned} \Xi^b(s, \eta) &= \Xi_c(s) + \Xi_1^b(s)\,\eta + \frac{1}{2}\Xi_2^b(s)\,\eta^2 + o(\eta^2), \\ \Xi_o(s, \eta) &= \Xi_c(s) + \Xi_{o1}(s)\,\eta + \frac{1}{2}\Xi_{o2}(s)\,\eta^2 + o(\eta^2), \end{aligned} \tag{56}$$

$$p(\eta) = p_c + p_1\,\eta + \frac{1}{2}p_2\,\eta^2 + o(\eta^2), \tag{57}$$

in which the subscript c denotes the value of a function at the bifurcation point ($\eta = 0$) while the other subscripts denote differentiation with respect to η evaluated at $\eta = 0$, as well.

Using the expansion (56) for both $\Xi^b(s,\eta)$, and $\Xi_o(s,p(\eta))$, Eq. (55) gives

$$\Xi(s,\eta)=\Xi_1(s)\,\eta+\frac{1}{2}\Xi_2(s)\,\eta^2+o(\eta^2) \tag{58}$$

The aim of the analysis is to obtain the coefficients of the series expansion in terms of η, of the sliding variables ε, δ, N, Δ, Ω and of the parameter p, up to a given order. To do that we use a perturbation technique so that, making use of the expansions (56) and (58), the *nonlinear* BVP stated in Sect. 2.2 is transformed in a sequence of *linear* BVPs, one for each power of η.

9 Beam with End Displacement: Bifurcation Analysis

The first order (linear) counterpart of the nonlinear BVP results to be

$$\varepsilon_1=u_1', \tag{59}$$

$$\begin{aligned} N_1'&=0,\\ \Omega_1'-\Delta_1&=0, \end{aligned} \tag{60}$$

$$\begin{aligned} N_1&=A\delta_1+B\delta_c\delta_1+F\varepsilon_1+E\delta_c\varepsilon_1+E\varepsilon_c\delta_1+G\varepsilon_c\varepsilon_1,\\ \Delta_1&=P\delta_1+R\delta_c\delta_1+A\varepsilon_1+B\delta_c\varepsilon_1+B\varepsilon_c\delta_1+E\varepsilon_c\varepsilon_1,\\ \Omega_1&=H\delta_1'+S\delta_c\delta_1'+C\varepsilon_c\delta_1', \end{aligned} \tag{61}$$

$$\begin{aligned} \Omega_1(0)&=\Omega_1(1)=0,\\ u_1(0)&=u_1(1)=0 \end{aligned} \tag{62}$$

Now by substituting (62) into Eq. (60) one obtains

$$\begin{aligned} A\delta_1'+B\delta_c\delta_1'+F\varepsilon_1'+E\delta_c\varepsilon_1'+E\varepsilon_c\delta_1'+G\varepsilon_c\varepsilon_1'&=0,\\ H\delta_1''+S\delta_c\delta_1''+C\varepsilon_c\delta_1''-(P\delta_1+R\delta_c\delta_1+A\varepsilon_1+B\delta_c\varepsilon_1+B\varepsilon_c\delta_1+E\varepsilon_c\varepsilon_1)&=0 \end{aligned} \tag{63}$$

Equation $(63)_1$, then gives

$$\varepsilon_1'=-\frac{A+B\delta_c+E\varepsilon_c}{F+E\delta_c+G\varepsilon_c}\,\delta_1' \tag{64}$$

and, by taking the first derivative of Eq. $(63)_2$ with respect to s, making use of Eqs. $(62)_3$ and (64), one obtains

$$\Omega_1''+q(u_\ell)\Omega_1=0, \tag{65}$$

where

$$q(u_\ell) = -\frac{(F + E\delta_c + G\varepsilon_c)(P + R\delta_c + B\varepsilon_c) - (A + B\delta_c + E\varepsilon_c)^2}{(H + S\delta_c + C\varepsilon_c)(F + E\delta_c + G\varepsilon_c)}, \tag{66}$$

which results to be

$$q(u_\ell) = -\frac{(\frac{\partial N}{\partial \varepsilon})_o(\frac{\partial \Delta}{\partial \delta})_o - (\frac{\partial N}{\partial \delta})_o(\frac{\partial \Delta}{\partial \varepsilon})_o}{(\frac{\partial N}{\partial \varepsilon})_o(\frac{\partial \Omega}{\partial \delta'})_o} \tag{67}$$

Note that Eq. (65) is the one reported by Antman (2005). It admits a non trivial solution for any positive integer n such that

$$-\frac{(\frac{\partial N}{\partial \varepsilon})_o(\frac{\partial \Delta}{\partial \delta})_o - (\frac{\partial N}{\partial \delta})_o(\frac{\partial \Delta}{\partial \varepsilon})_o}{(\frac{\partial N}{\partial \varepsilon})_o(\frac{\partial \Omega}{\partial \delta'})_o} = n^2\pi^2 \tag{68}$$

Equation (68) determines the values $u_{\ell c}$ corresponding to the bifurcation points. The eigenmode associated to the first eigenvalue, is

$$\Omega_1 = C_1 \sin \pi s \tag{69}$$

Then we can integrate the expressions $(62)_3$ to obtain

$$\delta_1 = -\frac{C_1}{\pi}\frac{\cos \pi s}{H + S\delta_c + C\varepsilon_c} + C_2 \tag{70}$$

In addition, from Eq. $(60)_2$

$$\Delta_1 = \Omega_1' = C_1\pi \cos \pi s \tag{71}$$

and, from the expression $(62)_2$

$$\varepsilon_1 = \frac{C_1\pi \cos \pi s}{A + B\delta_c + E\varepsilon_c} + \frac{P + R\delta_c + B\varepsilon_c}{A + B\delta_c + E\varepsilon_c}\left(\frac{C_1 \cos \pi s}{\pi(H + S\delta_c + C\varepsilon_c)} - C_2\right) \tag{72}$$

and Eq. (59) gives then the displacement in the form

$$u_1 = \frac{C_1 \sin \pi s}{A + B\delta_c + E\varepsilon_c} + \frac{P + R\delta_c + B\varepsilon_c}{A + B\delta_c + E\varepsilon_c}\left(\frac{C_1 \sin \pi s}{\pi^2(H + S\delta_c + C\varepsilon_c)} - C_2 s\right) + C_3 \tag{73}$$

Finally, the boundary condition $(62)_3$ gives

$$C_3 = 0 \tag{74}$$

and, the boundary condition $(62)_4$ gives C_2. The constant C_1 is the amplitude of the eigenmode and can be assigned a value by choosing a normalization condition.

10 Beam with End Displacement: Numerical Results

We consider the REV in Fig. 3a and assume that the atoms, in the reference configuration, have the following coordinates

$$\begin{array}{llll} x_A = -\ell/2 & y_A = 0 & x_B = \ell/2 & y_B = 0 \\ x_R = -\ell & y_R = h & x_P = \ell & y_P = h \\ x_S = -\ell & y_S = -h & x_Q = \ell & y_Q = -h \end{array} \tag{75}$$

Similarly, the coordinates of the atoms in the REV in Fig. 3b are

$$\begin{array}{llll} x_A = 0 & y_A = -\ell/2 & x_B = 0 & y_B = \ell/2 \\ x_R = -h & y_R = -\ell & x_P = -h & y_P = \ell \\ x_S = h & y_S = -\ell & x_Q = h & y_Q = \ell \end{array} \tag{76}$$

Following Belytschko et al. (2002) the geometric and constitutive constants in the energy expressions (43) are given the following values

$$\begin{array}{lll} \ell = 0.142 \text{ nm} & h = 0.123 \text{ nm} & \theta_0 = 2.094 \text{ rad} \\ D_e = 0.603105 \text{ nN}\times\text{nm} & k_\theta = 0.9 \text{ nN}\times\text{nm} \times \text{rad}^{-2} & \\ \beta_e = 26.25 \text{ nm}^{-1} & k_{\text{sextic}} = 0.754 \text{ rad}^{-4} & \end{array} \tag{77}$$

From the coordinates (76) its is evident that the REV has width $w = 3\ell$ and height $2h$. Denoting by t its thickness, the volume appearing in (49) is $V = 3\ell h t$ and, for $t = 0.34$ nm, $V = 0.0178$ nm^3. Using the preceding data the numerical values for the constitutive coefficients in (62) have been evaluated for the four cases considered and are given in Table 1. In addition, the normal force as a function of u_ℓ in the trivial solution has been determined and is plotted in Fig. 5 only for the cases a_2 and b_2.

Table 1 Constitutive coefficients

	A	B	C	E	F	P	G	R	S	H
case a1	6.28	−15.99	−0.96	−60.58	18.86	18.85	−144.1	−188.54	−0.06	0.12
case a2	4.26	−17	−0.94	−55.52	20.88	20.87	−153.2	−191.58	−0.17	0.16
case b1	6.04	−32.65	−0.65	−0.81	19.41	17.16	−181.92	−111.92	−2.25	0.34
case b2	4.88	−29.71	−0.59	−1.7	20.69	18.42	−183.74	−116.28	−2.34	0.37

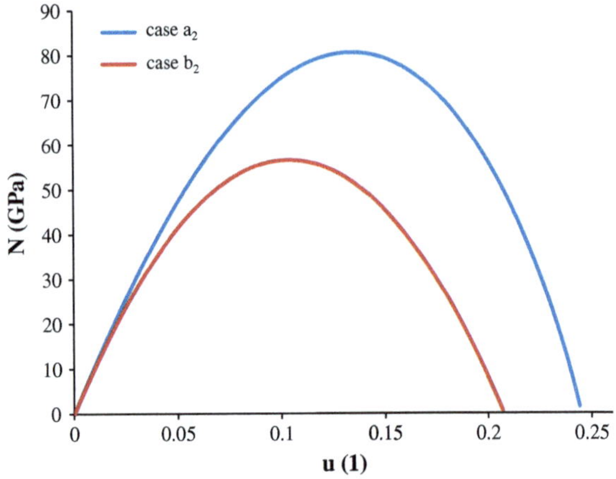

Fig. 5 Trivial path

Table 2 Values of the apparent Young's modulus

	Young's modulus (TPa)
case a1, case b1	0.94
case a2, case b2	1.12

Finally the values of the *apparent* Young's modulus—which is defined as the slope at the origin of the trivial path—that is

$$E = \left(\frac{\partial N_o}{\partial \varepsilon}\right)_{\varepsilon=0},$$

have been calculated. The values obtained for all the four cases are reported in Table 2. In Table 3 the results obtained by some other authors are reported for the sake of comparison. It can be seen that they are very close to those given in Table 2.

The bifurcation points along the trivial path can be obtained from Eq. (68) and the smaller values for u_ℓ are obtained by putting $n = 1$. In Fig. 6 the curve $q(u_\ell) - \pi^2$ for the case a_2, is plotted. Bifurcation occurs when the curve crosses the u_ℓ axis, that is when $u_\ell = 0.136$. Now, as the maximum value of the normal force is reached for $u_\ell = 0.133$ (see Fig. 5) the bifurcation occurs just a little bit after the limit point.

Table 3 Comparison of Young's modulus

Reference	Modeling method	Young's modulus (TPa)
Lee et al. (2008)	Experimental	1 ± 0.1
van Lier et al. (2000)	Density functional theory	1.11
WenXing et al. (2004)	Molecular dynamic	1.02
Xiao et al. (2005)	Nano scale- Continuum	1.13
Wu et al. (2006)	Nano scale- Continuum	1.06

We want to stress that the asymptotes in Fig. 6 correspond to the values of u_ℓ at which

$$(\frac{\partial N}{\partial \varepsilon})_o$$

and

$$\left(\frac{\partial \Omega}{\partial \delta'}\right)_o$$

do vanish, respectively. This means that the constitutive functions obtained for the equivalent beam are not strongly elliptic and this is due, in turn, to the fact that the Modified Morse potential is not convex. In Fig. 7 the curve $q(u_\ell) - \pi^2$ for the case b_2, is plotted. In this case bifurcation occurs when $u_\ell = 0.106$.

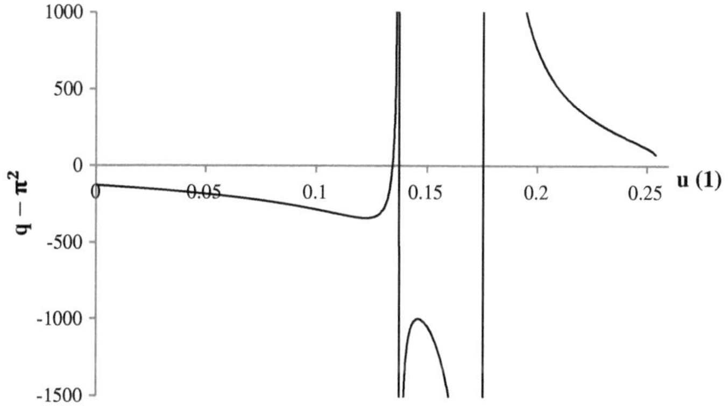

Fig. 6 Armchair arrangement: bifurcation plot

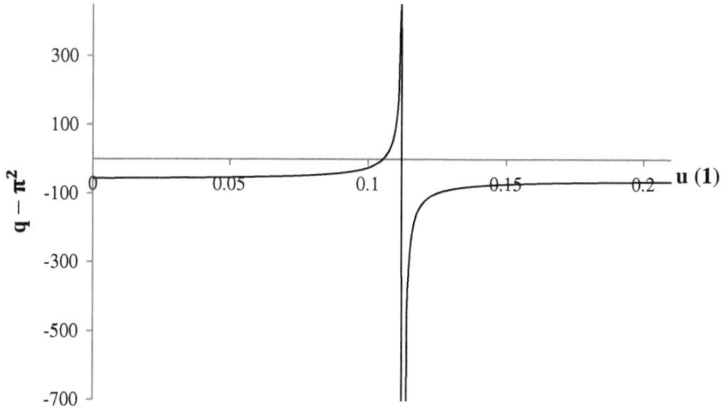

Fig. 7 Zigzag arrangement: bifurcation plot

The maximum value of the normal force, instead, is reached for $u_\ell = 0.101$ (see Fig. 5) so that, as before, bifurcation occurs after the limit point. In this case only one asymptote is present. It is located at the value of u_ℓ at which

$$(\frac{\partial N}{\partial \varepsilon})_o = 0$$

We want to stress that, looking at the values of the end displacement reported before, one can conclude that in both examined cases the bifurcation point practically coincides with the limit point of the curve $N(u_\ell)$. Looking at the shape of the component δ_1 of the buckling mode, given by the expressions (74), it is clear that such a solution describes a necking phenomenon.

11 Conclusions

A procedure that leads to the identification of constitutive functions for a hyperelastic 1D continuum endowed with a suitable internal structure able to describe the mechanics behaviour of carbon nano structures, has been introduced. A graphene sheet with armchair and zigzag structures have been considered and a representative elementary volume identified. An equilibrium configuration is assumed as a reference configuration for the REV. When the atoms are displaced in a generic configuration, the variation of the energy of the REV is evaluated by using the Modified Morse interatomic potential. A displacement field of the 1D model is put in relation with displacement of the atoms in the REV. By using this map, the energy variation of the REV is written on terms of the deformation measures of the 1D model. This expression, divided by the volume of the REV is assumed as the strain energy density of the continuum. The stress measures and their constitutive functions are then obtained by differentiating the strain energy density with respect to the 1D deformation measures. As an example, the Young's modulus of a carbon nano sheet has been evaluated and compared with the values reported it the literature. The two cases of a sheet stretched in the armchair or zigzag direction, respectively, have been studied and a bifurcation which describes a necking phenomenon, detected.

References

Alibert J, Seppecher P, dell'Isola F, (2003) Truss modular beams with deformation energy depending on higher displacement gradients. Math Mech Solids 8(1):51–73

Altenbach H, Bîrsan M, Eremeyev V (2012) On a thermodynamic theory of rods with two temperature fields. Acta Mech 223(8):1583–1596

Aminpour H, Rizzi N (2015) A 1D continuum with microstructure for single-wall CNTs bifurcation analysis. Math Mech Solids 114. doi:10.1177/1081286515577037

Aminpour H, Rizzi N, Salerno G (2014) A one-dimensional beam model for single-wall carbon nano tube column buckling. In: Topping B, Ivnyi P (eds) Proceedings of the twelfth international conference on computational structures technology. Civil-Comp Press, Stirlingshire
Antman S (2005) Nonlinear problems of elasticity, applied mathematical sciences, vol 107. Springer, New York
Belytschko T, Xiao S, Schatz G, Ruoff R (2002) Atomistic simulations of nanotube fracture. Phys Rev B 65(23):235–430
Bîrsan M, Altenbach H, Sadowski T, Eremeyev V, Pietras D (2012) Deformation analysis of functionally graded beams by the direct approach. Compos Part B: Eng 43(3):1315–1328
di Carlo A (1996) A non-standard format for continuum mechanics. In: Batra R, Beatty M (eds) Contemporary research in the mechanics and mathematics of materials. CIMNE, Barcelona
dell'Isola F, Seppecher P, Madeo A, (2012) How contact interactions may depend on the shape of Cauchy cuts in Nth gradient continua: approach 'à la D'Alembert'. ZAMP 63(6):1119–1141
Germain P (1973a) La méthode des puissance virtuelles en mécanique des milieux continus, 1ère partie: la théorie du second gradient. J de Mécanique 12(2):235–274
Germain P (1973b) The method of virtual power in continuum mechanics, II: microstructure. SIAM J Appl Math 25(3):555–575
Leamy MJ (2007) Bulk dynamic response modeling of carbon nanotubes using an intrinsic finite element formulation incorporating interatomic potentials. Int J Solids Struct 44(3–4):874–894
Lee C, Wei X, Kysar J, Hone J (2008) Measurement of the elastic properties and intrinsic strength of monolayer graphene. Science 321(5887):385–388
van Lier G, van Alsenoy C, van Doren V, Geerlings V (2000) Ab initio study of the elastic properties of single-walled carbon nanotubes and graphene. Chem Phys Lett 326(1–2):181–185
Pei Q, Zhang Y, Shenoy V (2010) A molecular dynamics study of the mechanical properties of hydrogen functionalized graphene. Carbon 48(3):898–904
Pietraszkiewicz W, Eremeyev V, Konopińska V (2007) Extended non-linear relations of elastic shells undergoing phase transitions. ZAMM 87(2):150–159
Podio-Guidugli P (1982) Flexural instabilities of elastic rods. J Elast 12(1):3–17
Rappe AK, Casewit CJ, Colwell KS, Goddard WA III, Skiff WM (1992) UFF, a rule-based full periodic table force field for molecular mechanics and molecular dynamics simulations. J Am Chem Soc 114(25):10,024–10,035
Saito R, Dresselhaus G, Dresselhaus MS (1998) Physical properties of carbon nanotubes. Imperial College Press, London
Shima H, Sato M (2013) Elastic and plastic deformation of carbon nanotubes. CRC Press, Boca Raton
WenXing B, ChangChun Z, WanZhao C (2004) Simulation of Young's modulus of single-walled carbon nanotubes by molecular dynamics. Physica B 352(1–4):156–163
Wolfram (2015) Mathematica, Version 10.3. Imperial College Press, Champaign
Wu Y, Zhang X, Leung A, Zhang W (2006) An energy-equivalent model on studying the mechanical properties of single-walled carbon nanotubes. Thin-Walled Struct 44(6):667–676
Xiao JR, Gamaa BA, Gillespie JW Jr, (2005) An analytical molecular structural mechanics model for the mechanical properties of carbon nanotubes. Int J Solids Struct 42(11–12):3075–3092

Isogeometric Analysis of Gradient-Elastic 1D and 2D Problems

Viacheslav Balobanov, Sergei Khakalo and Jarkko Niiranen

Abstract In the present contribution, isogeometric methods are used to analyze the statics and dynamics of rods as well as plane strain and plane stress problems based on a simplified version of the form II of Mindlin's strain gradient elasticity theory. The adopted strain gradient elasticity models, in particular, include only two length scale parameters enriching the classical energy expressions and resulting in fourth order partial differential equations instead of the corresponding second order ones based on the classical elasticity. The problems are discretized by an isogeometric non-uniform rational B-splines (NURBS) based C^{p-1} continuous Galerkin method. Computational results for benchmark problems demonstrate the applicability of the method and verify the implementation.

Keywords Gradient elasticity · Bar · Plane strain/stress · Isogeometric analysis

1 Introduction. Basic Formulae of the Mindlin's Gradient Elasticity Theory

Classical linear theory of elasticity is not capable to describe multi-scale phenomena as effects of meso-scale, micro-scale or nano-scale in primarily macro-scale problems because it leaves out of account that materials have microstructure. A lot of improvements of classical elasticity theory have been done in order to explain such effects. One of the first significant contributions was done by Mindlin (1964).

V. Balobanov (✉) · S. Khakalo · J. Niiranen
Department of Civil and Structural Engineering, Aalto University, P. O. Box 12100, 00076 Aalto, Espoo, Finland
e-mail: viacheslav.balobanov@aalto.fi

S. Khakalo
e-mail: sergei.khakalo@aalto.fi

J. Niiranen
e-mail: jarkko.niiranen@aalto.fi

H. Altenbach and S. Forest (eds.), *Generalized Continua as Models for Classical and Advanced Materials*, Advanced Structured Materials 42,
DOI 10.1007/978-3-319-31721-2_3

His first strain gradient theory of linear elasticity implies the existence of an additional term in the definition of the potential energy W:

$$W = \int_B \left[\frac{1}{2} \boldsymbol{\tau} : \boldsymbol{\varepsilon} + \frac{1}{2} \boldsymbol{\mu} \vdots \boldsymbol{\kappa} \right] \mathrm{d}V, \tag{1}$$

where $\boldsymbol{\varepsilon}$ stands for the classical strain tensor, $\boldsymbol{\tau}$ is the Cauchy stress tensor, $\boldsymbol{\mu}$ is the double stress tensor and $\boldsymbol{\kappa}$ is the micro-deformation gradient tensor. The nabla-operator is denoted by ∇ and can be defined in Cartesian coordinate system as $\nabla = \boldsymbol{e}_i \frac{\partial}{\partial x_i}$, $\int_B \mathrm{d}V$ designates integration over the volume of a body B.

Form II Mindlin's strain gradient elasticity theory proposes to define tensor $\boldsymbol{\kappa}$ as the gradient of the macroscopic strain:

$$\boldsymbol{\kappa} = \nabla \boldsymbol{\varepsilon}. \tag{2}$$

The simplest variant of this theory widely used in the literature, with roots in Aifantis (1992), Altan and Aifanis (1997), defines the double stress tensor as follows

$$\boldsymbol{\mu} = l_s^2 \nabla \boldsymbol{\tau}, \tag{3}$$

where l_s denotes gradient elasticity parameter which has dimension of length.

In the framework of considering gradient theory, the well known expressions for classical stress and strain tensors are valid:

$$\boldsymbol{\varepsilon} = \frac{1}{2} (\nabla \boldsymbol{u} + (\nabla \boldsymbol{u})^T), \tag{4}$$

$$\boldsymbol{\tau} = \boldsymbol{C} : \boldsymbol{\varepsilon}. \tag{5}$$

Here $\boldsymbol{u}$ stands for the displacement vector and $\boldsymbol{C}$ is the fourth-order tensor of elastic moduli.

By substitution (2) and (3) into (1), one can obtain the expression for the potential energy by using only the classical stress and strain tensors:

$$W = \int_B \left[\frac{1}{2} \boldsymbol{\tau} : \boldsymbol{\varepsilon} + \frac{1}{2} l_s^2 \nabla \boldsymbol{\tau} \vdots \nabla \boldsymbol{\varepsilon} \right] \mathrm{d}V. \tag{6}$$

The kinetic energy has an additional term as well:

$$T = \int_B \left[\frac{1}{2} \rho \dot{\boldsymbol{u}} \cdot \dot{\boldsymbol{u}} + \frac{1}{2} \rho l_d^2 \nabla \dot{\boldsymbol{u}} : \nabla \dot{\boldsymbol{u}} \right] \mathrm{d}V, \tag{7}$$

where ρ stands for the mass density, upper dots indicate the differentiation with respect to time and l_d is the second gradient coefficient called micro inertia coefficient with the dimension of length.

The work done by external forces alongside with two classical terms has two additional ones:

$$W_{\text{ext}} = \int\limits_{B} \boldsymbol{F} \cdot \boldsymbol{u} \mathrm{d}V + \int\limits_{\partial B_P} \boldsymbol{P} \cdot \boldsymbol{u} \mathrm{d}S + \int\limits_{\partial B_R} \boldsymbol{R} \cdot (\boldsymbol{n} \cdot \nabla \boldsymbol{u}) \mathrm{d}S + \oint\limits_{\partial\partial B_E} \boldsymbol{E} \cdot \boldsymbol{u} \mathrm{d}l, \tag{8}$$

where $\boldsymbol{F}$ stands for the body force per unit volume, $\boldsymbol{P}$ is the traction force, $\boldsymbol{R}$ is the double traction force, $\int_{\partial B} \mathrm{d}S$ denotes integration over the surface of the body, $\boldsymbol{n}$ is the unit vector normal to the surface, $\boldsymbol{E}$ is the force distributed on the wedges $\partial\partial B_E$ of the body surface, $\oint_{\partial\partial B_E} \mathrm{d}l$ denotes integration over these wedges. For simplicity, external loadings at possible corners of the wedges (see, for instance, Polizzotto 2012) have been excluded in the formulation above.

By substitution of (6)–(8) into Hamilton's principle for independent variation $\delta \boldsymbol{u}$ between fixed limits of $\boldsymbol{u}$ at times t_0 and t_1

$$\delta \int\limits_{t_0}^{t_1} (T - W) \mathrm{d}t + \int\limits_{t_0}^{t_1} \delta W_{\text{ext}} \mathrm{d}t = 0, \tag{9}$$

one can obtain the equation of motion of elastic continuum with micro-structure:

$$\nabla \cdot \boldsymbol{\tau} - l_s^2 \nabla \cdot \Delta \boldsymbol{\tau} + \boldsymbol{F} = \rho \ddot{\boldsymbol{u}} - \rho l_d^2 \Delta \ddot{\boldsymbol{u}} \;\; \text{in} \;\; B \tag{10}$$

and expressions for the external forces (Mindlin 1964; Polizzotto 2012):

$$\boldsymbol{P} = \boldsymbol{n} \cdot (\boldsymbol{\tau} - l_s^2 \Delta \boldsymbol{\tau}) - l_s^2 \nabla_s \cdot (\boldsymbol{n} \cdot \nabla \boldsymbol{\tau}) + l_s^2 (\nabla_s \cdot \boldsymbol{n}) \boldsymbol{n} \otimes \boldsymbol{n} : \nabla \boldsymbol{\tau} + \rho l_d^2 \boldsymbol{n} \cdot \nabla \ddot{\boldsymbol{u}} \, \text{on} \, \partial B_P, \tag{11a}$$

$$\boldsymbol{R} = l_s^2 \boldsymbol{n} \otimes \boldsymbol{n} : \nabla \boldsymbol{\tau} \; \text{on} \; \partial B_R, \tag{11b}$$

$$\boldsymbol{E} = l_s^2 \boldsymbol{s} \cdot \boldsymbol{L} : [\boldsymbol{n} \otimes \boldsymbol{n} \cdot \nabla \boldsymbol{\tau}] \; \text{on} \; \partial\partial B_E, \tag{11c}$$

where $\Delta = \nabla^2$ stands for the Laplacian, ∇_s being the surface part of the nabla-operator: $\nabla_s = \nabla - \boldsymbol{n} \otimes \boldsymbol{n} \cdot \nabla$, $\boldsymbol{s}$ being the unit vector tangent to C and $\boldsymbol{L}$ is the third order Levi-Civita tensor. The bold face brackets in (11c) indicate that the enclosed quantity is the difference of its values, at $\partial\partial B$ on the two portions of ∂B that intersect at $\partial\partial B$.

2 Weak Form

Vector equation of motion (10) within a framework of the gradient elasticity theory is partial differential equation with high order derivatives. It can be solved analytically only in the simplest cases. The most common way of solving continuum mechanics problems numerically is to use the family of Finite Element Methods. In order to get the numerical solution, it is necessary to begin with definition of the weak form of the equation:

Definition 3.1 For a given loading $\boldsymbol{F} \in [L^2(B)]^3$ find the deformation vector $\boldsymbol{u} \in \boldsymbol{U}$ such that

$$a(\boldsymbol{u}, \boldsymbol{w}) = l(\boldsymbol{w}) \;\forall\, \boldsymbol{w} \in \boldsymbol{W} \subset [H^2(B)]^3, \tag{12}$$

where the bilinear form $a : \boldsymbol{U} \times \boldsymbol{W} \to \mathbb{R}^3$ and the load functional $l : \boldsymbol{W} \to \mathbb{R}^3$, respectively, are defined as

$$a(\boldsymbol{u}, \boldsymbol{w}) = \int\limits_B [\boldsymbol{\tau}(\boldsymbol{\epsilon}(\boldsymbol{u})) : \boldsymbol{\epsilon}(\boldsymbol{w}) + l_s^2 \nabla \boldsymbol{\tau}(\boldsymbol{\epsilon}(\boldsymbol{u}))\,\vdots\, \nabla \boldsymbol{\epsilon}(\boldsymbol{w}) + \rho \ddot{\boldsymbol{u}} \cdot \boldsymbol{w} + \rho l_d^2 \nabla \ddot{\boldsymbol{u}} : \nabla \boldsymbol{w}] dV, \tag{13}$$

$$l(\boldsymbol{w}) = \int\limits_B \boldsymbol{F} \cdot \mathbf{w} \mathrm{d}V + \int\limits_{\partial B_P} \boldsymbol{P} \cdot \mathbf{w} \mathrm{d}S + \int\limits_{\partial B_R} \boldsymbol{R} \cdot (\boldsymbol{n} \cdot \nabla \boldsymbol{w}) \mathrm{d}S + \oint\limits_{\partial\partial B_E} \boldsymbol{E} \cdot \mathbf{w} \mathrm{d}l. \tag{14}$$

Here $\boldsymbol{U}$ is a subspace of Sobolev space $[H^2(B)]^3$ and each function from this subspace satisfies the Dirichlet boundary conditions:

$$\boldsymbol{U} = \{\boldsymbol{u} \in [H^2(B)]^3 \mid \boldsymbol{u}_{|\partial B_u} = \boldsymbol{u}_0,\ \boldsymbol{u}_{|C_u} = \boldsymbol{u}_c,\ \boldsymbol{n} \cdot \nabla \boldsymbol{u}_{|\partial B_D} = \boldsymbol{d}\} \subset [H^2(B)]^3. \tag{15}$$

Functions from $\boldsymbol{W}$ satisfy the homogeneous Dirichlet boundary conditions:

$$\boldsymbol{W} = \{\boldsymbol{w} \in [H^2(B)]^3 \mid \boldsymbol{w}_{|\partial B_u} = 0,\ \boldsymbol{w}_{|C_u} = 0,\ \boldsymbol{n} \cdot \nabla \boldsymbol{w}_{|\partial B_D} = 0\} \subset [H^2(B)]^3. \tag{16}$$

Galerkin's method suggests to represent an approximate solution (trial function) and test function by using a finite number of basis functions

$$\boldsymbol{u}^h = \sum_{i=1}^{n} \boldsymbol{u}_i \varphi_i, \qquad \boldsymbol{w}^h = \sum_{i=1}^{n} \boldsymbol{w}_i \varphi_i, \tag{17}$$

at that $\boldsymbol{u}^h \in \boldsymbol{U}^h$ and $\boldsymbol{w}^h \in \boldsymbol{W}^h$, where $\boldsymbol{U}^h$ and $\boldsymbol{W}^h$ are finite-dimensional approximations of $\boldsymbol{U}$ and $\boldsymbol{W}$:

$$\boldsymbol{U}^h \subset \boldsymbol{U},\ \ \boldsymbol{W}^h \subset \boldsymbol{W}. \tag{18}$$

An important fact arises from (18) for a conforming method: the solution space must be a subspace of Sobolev space $[H^2(B)]^3$. It means that basis functions φ_i must provide at least C^1 continuity from element to element. Classical Lagrange basis functions can provide only C^0 continuity and consequently they are not suitable for solving the gradient elasticity problems. There are a lot of different modifications of Galerkin's methods but probably the most universal and advanced one is Isogeometric Analysis introduced by Hughes et al. (2005).

3 Isogeometric Analysis

Isogeometric Analysis (IGA) can be considered as the "next generation" of the finite-element methods family. It has been under development at a quick rate during last 10 years. The main idea of IGA is to use the non-uniform rational B-splines (NURBS) as basis functions φ_i. This peculiarity causes a lot of advantages of IGA methods such as the exact representation of the problem geometry. Interested readers advised to reach for the book devoted to Isogeometric Analysis (Cottrell et al. 2009).

In the context of the gradient elasticity theory the most important property of IGA is the C^{p-1} continuity provided across the elements boundaries in each parametric direction. Here p is an order of the NURBS basis functions in one of the directions. The NURBS basis is constructed by using 1D B-spline basis functions which can be defined by using Cox-de Boor recursion formula:

$$N_{i,p}(\xi) = \frac{\xi - \xi_i}{\xi_{i+p} - \xi_i} N_{i,p-1}(\xi) + \frac{\xi_{i+p+1} - \xi}{\xi_{i+p+1} - \xi_{i+1}} N_{i+1,p-1}(\xi) \text{ for } p = 1, 2, 3, \ldots$$
$$N_{i,0}(\xi) = \begin{cases} 1 & \text{if } \xi_i \leq \xi \leq \xi_{i+1}; \\ 0 & \text{otherwise.} \end{cases} \tag{19}$$

Definition of 3D NURBS basis functions is presented below:

$$R^{p,q,r}_{i,j,k}(\xi, \eta, \zeta) = \frac{N_{i,p}(\xi) N_{j,q}(\eta) N_{k,r}(\zeta) w_{i,j,k}}{\sum_{\hat{i}=1}^{n} \sum_{\hat{j}=1}^{m} \sum_{\hat{k}=1}^{l} N_{\hat{i},p}(\xi) N_{\hat{j},q}(\eta) N_{\hat{k},r}(\zeta) w_{\hat{i},\hat{j},\hat{k}}}, \tag{20}$$

where $w_{i,j,k}$ are weight coefficients.

4 Numerical Results

This section is devoted to results of numerical solutions of some benchmark problems which were obtained from (10)–(11) by dimension reduction. For more results the interested reader can look at Niiranen et al. (2015a, b).

4.1 Static Rod in Tension

Consider a straight prismatic rod of constant cross section area A and length L. The displacement along the longitudinal axis x are denoted by u. Material of the rod is isotropic with Young's modulus E and gradient coefficient l_s. According to Papargyri-Beskou et al. (2010), the governing equilibrium equation of this rod in terms of displacements can be written as follows (body force is set to be zero):

$$AE(u'' - l_s^2 u'''') = 0 \text{ in } [0, L]. \tag{21}$$

Boundary conditions for the gradient rod under tension are assumed to be:

$$\begin{aligned} u|_{x=0} &= 0, \quad P|_{x=L} \equiv AE(u' - l_s^2 u''')|_{x=L} = P_0, \\ u'|_{x=L} &= 0, \quad R|_{x=0} \equiv AEl_s^2 u''|_{x=0} = 0. \end{aligned} \tag{22}$$

Solution for the problem (21)–(22) can be found analytically. It means that this problem can be used for the examination of the IGA applicability for solving of 1D problems of the gradient elasticity theory.

Convergence curves for different orders p of the NURBS basis functions presented on Fig. 1 seem to follow the next formula:

$$||u - u^h||_m \leq ch^\beta ||u||_{p+1}, \tag{23}$$

where $||\cdot||_m$ denotes the norm corresponding to the Sobolev space $H^m(L)$, u^h is the approximate solution, c is an unknown constant and h is the element size. Convergence rate is denoted by β and expression for it depends on the degree of the Sobolev norm m of the solution error. In context of the gradient elasticity H^2 norm is the energy norm and its convergence rate equals to $\beta = p - 1$. For L^2 norm (or H^0 norm) the convergence rate equals to $\beta = min\{p + 1,\ 2(p - 1)\}$ and for H^1 norm it is equal to $\beta = p$. These results are in agreement with the theoretical analysis in Niiranen et al. (2015b) and Cottrell et al. (2009).

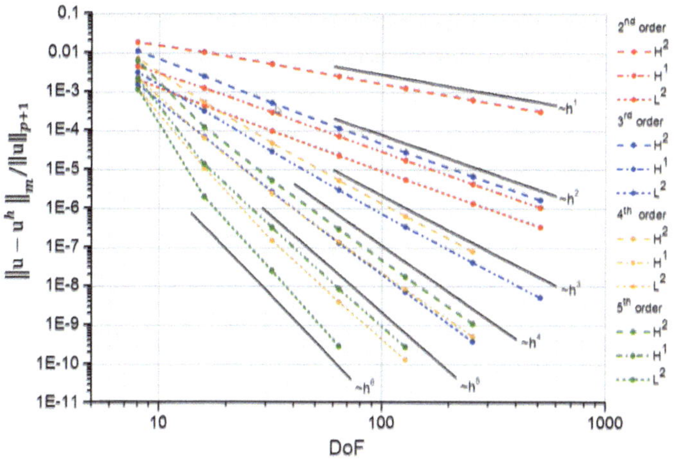

Fig. 1 L^2, H^1 and H^2 norms of error of the stretched rod displacements for different order of the NURBS basis functions

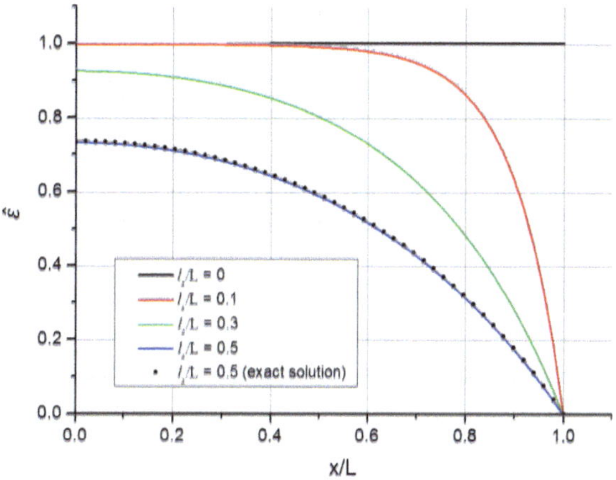

Fig. 2 Dimensionless axial strains of the stretched rod for different values of the gradient elasticity parameter

Figure 2 represents the dimensionless strain of stretched rod along axis x for the different values of the gradient parameter l_s:

$$\hat{\varepsilon} = \frac{\varepsilon_{\text{gr}}}{\varepsilon_{\text{c}}(x/L = 1)}. \tag{24}$$

In distinction from the classical theory for which strain ε_{c} is uniform along the rod axis, the gradient theory gives strain ε_{gr} which is not uniform. It means that 1D gradient elasticity can explain the local elongation and further fracture in the middle

Table 1 Eigenfrequencies of a 2D square domain

Frequency number	Frequency value, Hz		Gradient/Classical frequencies ratio
	Classical elasticity	Gradient elasticity	
1	1.60E+5	1.66E+5	1.03
3	2.27E+5	2.40E+5	1.06
10	4.81E+5	5.98E+5	1.24

of a tension specimen: primarily plastic deformations will most likely occur in a zone of the maximum elastic strain.

4.2 2D Dynamic Problem

Consider a square domain Ω with the side length $L = 10$ mm. Material properties are defined by Lame parameters λ and μ, mass density ρ (they are assumed to be equal to the parameters of standard steel), and gradient coefficients $l_s = 1$ mm and $l_d = 0.5$ mm. The governing equation of motion of 2D domain in plane strain/stress state can be written as follows (body force is set to be zero):

$$(1 - l_s^2\Delta)(\mu\Delta\boldsymbol{u} + (\lambda + \mu)\nabla\nabla\cdot\boldsymbol{u}) = \rho\ddot{\boldsymbol{u}} - \rho l_d^2\Delta\ddot{\boldsymbol{u}} \text{ in } \Omega. \tag{25}$$

Boundary conditions are assumed to be

$$\boldsymbol{u}\cdot\boldsymbol{s}|_{\partial\Omega} = 0; \quad \boldsymbol{P}\cdot\boldsymbol{n}|_{\partial\Omega} = 0; \quad \boldsymbol{R}|_{\partial\Omega} = 0. \tag{26}$$

Numerical solution for the spectral problem (25) and (26) is presented in Table 1. As one can see, gradient elasticity theory changes the body eigen frequencies and difference between results of classical and gradient elastic theories rises with increasing of the frequency number.

References

Aifantis E (1992) On the role of gradients in the localization of deformation and fracture. Int J Eng Sci 30:1279–1299

Altan B, Aifanis E (1997) On some aspects in the special theory of gradient elasticity. J Mech Behav Mater 8:231–282

Cottrell J, Hughes T, Bazilevs Y (2009) Isogeometric analysis: toward integration of CAD and FEA. Wiley, Chichester

Hughes T, Cottrell J, Bazilevs Y (2005) Isogeometric analysis: CAD, finite elements, nurbs, exact geometry and mesh refinement. Comput Method Appl Mech Eng 194:4135–4195

Mindlin R (1964) Micro-structure in linear elasticity. Arch Ration Mech Anal 16(1):51–78
Niiranen J, Balobanov V, Kiendl J, Hosseini B (2015a) Variational formulation and isogeometric analysis of six-order boundary value problems of gradient Bernoulli beams (submitted)
Niiranen J, Khakalo S, Balobanov V, Niemi A (2015b) Variational formulation and isogeometric analysis for fourth-order boundary value problems of gradient-elastic bar and plane strain/stress problems (submitted)
Papargyri-Beskou S, Giannakopoulos A, Beskos D (2010) Static analysis of gradient elastic bars, beams, plates and shells. Open Mech J 4:65–73
Polizzotto C (2012) A gradient elasticity theory for second-grade materials and higher order inertia. Int J Solids Struct 49:2121–2137

A Fast Fourier Transform-Based Approach for Generalized Disclination Mechanics Within a Couple Stress Theory

Stéphane Berbenni, Vincent Taupin, Claude Fressengeas and Laurent Capolungo

Abstract Recently, a small-distortion theory of coupled plasticity and phase transformation accounting for the kinematics and thermodynamics of generalized defects called generalized disclinations (abbreviated g-disclinations) has been proposed by Acharya and Fressengeas (2012, 2015). Then, a first numerical spectral approach has been developed to solve the elasto-static equations of field dislocation and g-disclination mechanics set out in this theory for periodic media and for linear elastic media using the classic Hooke's law within a Cauchy stress theory (Berbenni et al. 2014). Here, given a spatial distribution of generalized disclination density tensors in a homogenous linear higher order elastic media, a couple stress theory with elastic incompatibilities of first and second orders is developed. The incompatible and compatible elastic second and first distortions are obtained from the solution of Poisson and Navier-type equations in the Fourier space. The efficient Fast Fourier Transform (FFT) algorithm is used based on intrinsic Discrete Fourier Transforms (DFT) that are well adapted to the discrete grid to compute higher order partial derivatives in the Fourier space. Therefore, stress and couple stress fields can be calculated using the inverse FFT. The numerical examples are given for straight wedge disclinations and associated wedge disclination dipoles which are of importance to geometrically describe tilt grain boundaries at fine scales in polycrystalline solids.

Keywords Generalized disclinations · Couple stress theory · Fast fourier transform

S. Berbenni (✉) · V. Taupin · C. Fressengeas
Laboratoire d'Etude des Microstructures et de Mécanique des Matériaux, LEM3, UMR CNRS 7239, University of Lorraine, Ile du Saulcy, 57045 Metz, France
e-mail: stephane.berbenni@univ-lorraine.fr

V. Taupin
e-mail: vincent.taupin@univ-lorraine.fr

C. Fressengeas
e-mail: claude.fressengeas@univ-lorraine.fr

L. Capolungo
George Woodruff School of Mechanical Engineering, Georgia Institute of Technology, UMI 2958 Georgia Tech-CNRS, 57070 Metz, France
e-mail: laurent.capolungo@me.gatech.edu

H. Altenbach and S. Forest (eds.), *Generalized Continua as Models for Classical and Advanced Materials*, Advanced Structured Materials 42,
DOI 10.1007/978-3-319-31721-2_4

1 Introduction

In crystalline media, the internal stresses and couple stresses result from an incompatible process where crystal defects - dislocations, disclinations or "generalized disclinations" (abbreviated "g-disclinations")—induce the discontinuity of (elastic, i.e. lattice) displacement or distortion across surfaces in the body. Dislocations and disclinations were mathematically introduced by Volterra (1907). In the sole presence of dislocations, incompatibility fields were smoothly described in the continuum theory of dislocations initiated by Kröner (1958, 1981) and many others (Bilby et al. 1955; Mura 1963; Willis 1967; Kosevich 1979) by using Nye's dislocation density tensor (Nye 1953). The continuum theory of dislocations was recently revisited by Acharya (2001), Roy and Acharya (2005), Acharya and Roy (2006). One of the key features of the revisited continuum dislocation theory resides in the Stokes-Helmholtz decomposition of the elastic distortion and the associated side conditions yielding a unique solution for the incompatible part associated with a prescribed dislocation density field, while the compatible part is unambiguously determined from the satisfaction of the balance of linear momentum together with boundary conditions. When disclinations are present in the body in addition to dislocations, the displacement and rotation vectors are both multi-valued functions. Such a situation typically occurs in solids exhibiting kink bands, grain and subgrain boundaries and triple junctions. In this case, the elastic curvature tensor has an incompatible part complementing the compatible gradient component (deWit 1970; Fressengeas et al. 2011). Beyond Volterra's construct, the entire distortion tensor including the strain tensor in addition to the rotation tensor may be multivalued along some surface. Such situations are commonplace in materials science. They include terminating twinning and phase boundaries, terminating shear bands, sharp corners of inclusions in a matrix of dissimilar media, in addition to grain boundaries and triple junctions. As recently discussed by Acharya and Fressengeas (2012), the discontinuity of the distortion field is reflected by the incompatibility of the elastic 2-distortion, (i.e. the second gradient of displacement in strain gradient elasticity theory) in the presence of a non-vanishing g-disclination density tensor field. A Weingarten theorem and a finite strain framework were developed in Acharya and Fressengeas (2015).

An increasingly attractive alternative to the finite-element method is a computationally efficient scheme based on the Fast Fourier Transform (FFT) for the solution of periodic boundary-value problems in continuum mechanics. Pioneering works in this field can be found in Moulinec and Suquet (1994, 1998), Müller (1996), Dreyer et al. (1999), Eyre and Milton (1999), Lebensohn (2001), Michel et al. (2001), Neumann et al. (2002), Vinogradov and Milton (2008). This numerical approach solves the Lippmann-Schwinger integral equation of the periodic boundary-value problems by means of the Green's function of a chosen reference medium. It has been applied so far to elastic and elasto-plastic composites and polycrystals in the absence of crystal defects. The main interest of the FFT approach relies on its computational efficiency (Moulinec and Suquet 1998; Prakash and Lebensohn 2012). Its main drawbacks are the need for a periodic representative volume element and the possible occurrence of spurious Gibbs oscillations arising from the presence of strong spatial gradients. The elasto-static equations of FDM, which provide the long-range internal elastic fields associated with a prescribed distribution of dislocation densities in a body, were recently solved within the FFT framework (Brenner et al. 2014). In the latter, the equations for the incompatible elastic distortions and the balance of momentum are solved in the Fourier space, while the resulting elastic fields are obtained in the real space by using the inverse Fourier transforms. Independently, an extension of this spectral approach to field dislocation and generalized-disclination mechanics (FDGDM) was first proposed in Berbenni et al. (2014) using classic Cauchy stress theory, with additional features including a different discretization treatment of FFT-induced Gibbs oscillations in comparison with Brenner et al. (2014). Extensive 2D simulations showed that the numerical spectral approach is as accurate as optimized finite element approximations, but computationally much more efficient (Berbenni et al. 2014).

Motivated by the accuracy and the speed of such spectral approaches for the solution of classic elasto-static problems, we extend in the present contribution the theory developed in Berbenni et al. (2014) to account for the second order couple stress tensor (which is related to the skew-symmetric part of the third order hyperstress tensor) and the second order elastic curvature (which is related to the third order elastic 2-distortion tensor). Among various higher order theories, the Cosserat, couple-stress, micromorphic, strain-gradient theories are mostly documented, see e.g. Kröner (1968), Nowacki (1986), Eringen (2002), Forest (2006). The couple-stress theory originally developed by Mindlin and Tiersten (1962), Koiter (1964) contains the least material parameters in the constitutive equations compared with other non-conventional theories involving multiple materials length scale dependent elastic constants which may be difficult to identify at fine scales. Analytical elastic fields of straight dislocations and disclinations were obtained in a couple stress theory by Kröner (1963), Lubarda (2003), Anthony (1970), Gourgiotis and Georgiadis (2008). Furthermore, in a different context dedicated to homogenization and composites, the Green's function technique for isotropic centrosymmetric couple stress materials was derived in Smyshlyaev and Fleck (1994), Zheng and Zhao (2004) and a DFT-based approach was proposed for both Cosserat and couple stress linear elastic materials in Kassbohm (2006), Kassbohm et al. (2006). Recently, a general free energy density functional for crystalline materials with third order hyperstress tensor undergoing incompatible fields due to dislocations, disclinations and g-disclinations was proposed in Upadhyay et al. (2013), Upadhyay (2014). Here, the constitutive model will be built up starting from this general free energy density functional but will be simplified to only consider the second order deviatoric elastic curvature within a couple stress theory with incompatibilities.

The paper is organized as follows:

In Sect. 2, the notations are introduced. The kinematics for generalized disclination (abbreviated g-disclination in the sequel) mechanics is reviewed in Sect. 3 and the solutions for incompatible fields are given in the same section. Then, the generalized constitutive and equilibrium equations within a couple stress theory are introduced in Sect. 4. General three-dimensional solutions for incompatible elastic fields of g-disclinations are derived in Sect. 5 to compute their stress and couple stress fields in the Fourier space. In Sect. 6, the DFT method is introduced in the case of two-dimensional (2D) problems, and the FFT algorithm (Frigo and Johnson 2005) will be used to solve the Poisson and Navier-type equations with microstructural length scale in the case of infinite straight g-disclinations. In Sect. 7, g-disclination densities are distributed on 2D FFT pixelized grids for different configurations: single wedge disclination and wedge disclination dipole. The incompatible and compatible elastic fields are obtained in the discrete Fourier space and then used to derive the stresses and couple stresses by using the inverse FFT for an isotropic centrosymmetric elastic solid. The present numerical method is validated by comparisons with existing analytical expressions (deWit 1973; Anthony 1970; Romanov and Vladimirov 1992).

2 Notations

A bold symbol denotes a tensor or a vector. The symmetric part of tensor $\mathbf{A}$ is denoted $\mathbf{A}^{sym}$. Its skew-symmetric part is $\mathbf{A}^{skew}$ and its transpose is denoted by $\mathbf{A}^t$. The tensor $\mathbf{A}\cdot\mathbf{B}$, with rectangular Cartesian components $A_{ik}B_{kj}$, results from the dot product of tensors $\mathbf{A}$ and $\mathbf{B}$, and $\mathbf{A}\otimes\mathbf{B}$ is their tensorial product, with components $A_{ij}B_{kl}$. The vector $\mathbf{A}\cdot\mathbf{V}$, with rectangular Cartesian components $A_{ij}V_j$, results from the dot product of tensor $\mathbf{A}$ and vector $\mathbf{V}$. A ":" represents the trace inner product of the two second order tensors

$$\mathbf{A}:\mathbf{B}=A_{ij}B_{ij},$$

in rectangular Cartesian components, or the product of a higher order tensor with a second order tensor, e.g., $\mathbf{A}:\mathbf{B}=A_{ijkl}B_{kl}$. A "$\dot{:}$" represents the trace inner product of the two third order tensors $\mathbf{A}\dot{:}\mathbf{B}=A_{ijk}B_{ijk}$, in rectangular Cartesian components, or, it denotes the product of a higher order tensor with a third order tensor, e.g.,

$$\mathbf{A}\dot{:}\mathbf{B}=A_{ijklm}B_{klm}.$$

The cross product of a second-order tensor $\mathbf{A}$ and a vector $\mathbf{V}$, the **div** and **curl** operations for second/third-order tensors are defined row by row, in analogy with the vectorial case. For any base vector $\mathbf{e}_i$ of the reference frame:

$$(\mathbf{A}\times\mathbf{V})^t\cdot\mathbf{e}_i=(\mathbf{A}^t\cdot\mathbf{e}_i)\times\mathbf{V}, \tag{1}$$

$$(\mathbf{div}\ \mathbf{A})^t\cdot\mathbf{e}_i=\mathbf{div}(\mathbf{A}^t\cdot\mathbf{e}_i), \tag{2}$$

$$(\mathbf{curl}\ \mathbf{A})^t\cdot\mathbf{e}_i=\mathbf{curl}(\mathbf{A}^t\cdot\mathbf{e}_i). \tag{3}$$

In rectangular Cartesian components:

$$(\mathbf{A} \times \mathbf{V})_{ij} = e_{jkl} A_{ik} V_l, \tag{4}$$
$$(\mathbf{A} \times \mathbf{V})_{ijk} = e_{klm} A_{ijl} V_m, \tag{5}$$
$$(\mathbf{div} \mathbf{A})_i = A_{ij,j}, \tag{6}$$
$$(\mathbf{div} \mathbf{A})_{ij} = A_{ijk,k}, \tag{7}$$
$$(\mathbf{curl}\, \mathbf{A})_{ij} = e_{jkl} A_{il,k} = -(\mathbf{grad}\, \mathbf{A} : \mathbf{X})_{ij}, \tag{8}$$
$$(\mathbf{curl} \mathbf{A})_{ijk} = e_{klm} A_{ijm,l}, \tag{9}$$

where e_{jkl} is a component of the third-order alternating Levi-Civita tensor $\mathbf{X}$ and the spatial derivative with respect to a Cartesian coordinate is indicated by a comma followed by the component index.

3 Kinematics of Generalized Crystal Defects and Incompatibilities

3.1 Linear Theory

The analysis is developed in the small distortion framework (linear theory). The body V, with boundary ∂V, is assumed to be a continuum, with smooth displacement and rotation vector fields ($\mathbf{u}$, $\vec{\omega} = 1/2\, \mathbf{curl}\, \mathbf{u}$). The total 1-distortion (first distortion) tensor field $\mathbf{U} = \mathbf{grad}\, \mathbf{u}$, the curvature tensor field, $\kappa = \mathbf{grad}\, \vec{\omega}$, and the 2-distortion (second distortion) tensor field, $\mathbf{G} = \mathbf{grad}\, \mathbf{U}$, are therefore assumed to be integrable (compatible, or curl free). Under such assumptions, the possibility of developing cracks in the body is discarded. The total 1-distortion writes as the sum of the elastic distortion, $\mathbf{U}_e$, and plastic distortion, $\mathbf{U}_p$:

$$\mathbf{U} = \mathbf{U}_e + \mathbf{U}_p. \tag{10}$$

Similarly, the 2-distortion tensor can be decomposed into elastic and plastic 2-distortion tensors:

$$\mathbf{G} = \mathbf{G}_e + \mathbf{G}_p. \tag{11}$$

In a compatible body in the absence of polarized crystal defect density, the elastic/plastic distortions and 2-distortions are curl-free gradient tensors. However, they will contain incompatible, non-gradient parts, in the presence of a polarized crystal defect density, while total 1 and 2-distortions remain compatible. Such general incompatibilities are now discussed in terms of crystal defects.

3.2 Volterra's Crystal Translation and Rotation Line Defects

Volterra (1907) introduced six types of crystal line defects. Three of them, known as dislocations, are translational defects, and the other three, referred to as disclinations, are rotational defects. Like disclinations, dislocations have a smooth elastic distortion field $\mathbf{U}_e$ in a non-simply-connected domain excluding their core. However, their (elastic) displacement field features a discontinuity denoted $[[\mathbf{u}_e]]$ across a (non-unique) smooth surface in this domain. The geometry of any such surface is arbitrary except that, in a discrete modeling framework, it terminates along the dislocation line. A line integral of the elastic distortion field along any curve encircling the dislocation line, i.e. a Burgers circuit, is constant and is equal to the discontinuity of the elastic displacement. This constant $\mathbf{b} = [[\mathbf{u}_e]]$ is referred to as the Burgers vector of the dislocation. It represents the strength of the dislocation. In contrast with Volterra's discrete representation of crystal defects, we presently choose a continuous setting, in order to regularize this classical description. We consider smooth elastic distortion fields in simply connected domains, in which they are point-wise irrotational outside the core region, whereas their non-vanishing curl defines a smooth dislocation density tensor field inside the core (of non-zero volume):

$$\boldsymbol{\alpha} = \mathbf{curl}\,\mathbf{U}_e. \tag{12}$$

The Burgers vector is then obtained by integrating the dislocation density tensor field, referred to as Nye's tensor field, along appropriate surface patches S with unit normal $\mathbf{n}$:

$$\mathbf{b} = \int_S \boldsymbol{\alpha}.\mathbf{n}\mathrm{d}S. \tag{13}$$

Similarly, disclinations result from a discontinuity denoted $[[\vec{\boldsymbol{\omega}}_e]]$ in the rotation field over a surface terminating on the disclination line in a discrete setting, even though a smooth elastic curvature field $\boldsymbol{\kappa}_e$ exists in this region. The strength of disclinations is characterized by their Frank vector $\boldsymbol{\Omega}$, which represents the magnitude and direction of the rotational discontinuity $\boldsymbol{\Omega} = [[\vec{\boldsymbol{\omega}}_e]]$ over a closed circuit encircling the disclination line. In deWit's continuous setting (deWit 1970), also adopted in the present paper, the smooth elastic curvature field is irrotational outside the disclination core region, and the disclination density tensor is defined as the curl of this field inside the core, of non-zero volume:

$$\boldsymbol{\theta} = \mathbf{curl}\,\boldsymbol{\kappa}_e. \tag{14}$$

The Frank vector is then obtained by integrating the disclination density tensor field along appropriate surface patches S:

$$\boldsymbol{\Omega} = \int_S \boldsymbol{\theta}.\mathbf{n}\mathrm{d}S. \tag{15}$$

3.3 *Generalized Disclination (G-Disclination) Kinematics*

Acharya and Fressengeas (2012) adopted a similar approach in introducing generalized disclinations , as shown in Fig. 1. The g-disclination concept goes beyond the Volterra construct, in the sense that the distortion field now has a discontinuity denoted $[[\mathbf{U}_e]]$ along a surface terminating at the g-disclination line (Fig. 1), whereas the elastic 2-distortion tensor field $\mathbf{G}_e$ is still smooth in the non-simply connected region excluding the g-disclination line. As already mentioned, this surface of discontinuity is referred to as a phase or grain boundary. The strength $\boldsymbol{\Pi}$ of the g-disclination is defined as the jump in the elastic distortion tensor field across the interphase: $\boldsymbol{\Pi} = [[\mathbf{U}_e]]$. In a continuous setting, the elastic 2-distortion field is pointwise irrotational in the defect free volume of the body. Its curl in the defected part provides for the definition of the third order g-disclination density tensor field $\boldsymbol{\pi}$:

$$\pi = \mathbf{curl}\,\mathbf{G}_e, \tag{16}$$

and the integration of the latter over an appropriate surface patch yields the jump of the elastic distortion tensor field:

$$\boldsymbol{\Pi} = \int_S \pi.\mathbf{n}\mathrm{d}S. \tag{17}$$

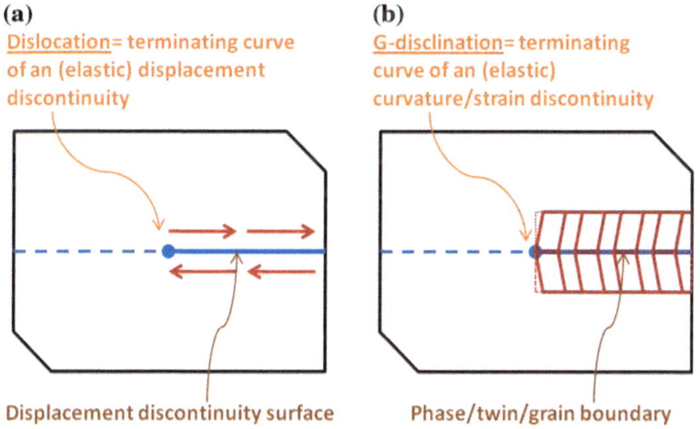

Fig. 1 Cross sectional view of two different types of straight line defects: dislocation **a** seen as the terminating curve of the surface of displacement discontinuity (the *arrows* with reverse directions along the displacement discontinuity surface describe different displacement directions), g-disclination **b** seen as the terminating curve of the surface of curvature/strain discontinuity (the differently inclined parallel lines in the vicinity of the distortion discontinuity surface describe different strains like different shears for example)

When, as a special case, strain continuity, while a rotation discontinuity $[[\vec{\omega}_e]]$ is persisting, the g-disclinations reduce to standard disclinations. In the context of g-disclinations, the dislocation density tensor $\boldsymbol{\alpha}$ needs to be redefined by alternating of the elastic 2-distortion tensor (Acharya and Fressengeas 2012):

$$\alpha = -\mathbf{G}_e : \mathbf{X} \tag{18}$$

instead of Eq. (12).

3.4 Incompatible Field Equations

Invoking the Stokes-Helmholtz orthogonal decomposition of the square-integrable elastic 2-distortion tensor field $\mathbf{G}_e$ (see for example (Jiang 1998)), there exist unique tensor fields $\boldsymbol{\chi}$ and $\mathbf{Z}$ such that $\mathbf{G}_e$ writes as the sum:

$$\mathbf{G}_e = \mathbf{G}_e^{\perp} + \mathbf{G}_e^{\parallel} = \mathbf{curl}\,\chi + \mathbf{grad}\,\mathbf{Z} \tag{19}$$

with the orthogonality condition $\int_V \mathbf{curl}\,\chi : \mathbf{grad}\,\mathbf{Z}\,dv = 0$. Thus, taking the curl of $\mathbf{G}_e$ in Eq. (19) extracts $\mathbf{curl}\,\boldsymbol{\chi}$ and discards $\mathbf{grad}\,\mathbf{Z}$, whereas taking its divergence extracts $\mathbf{grad}\,\mathbf{Z}$ and eliminates $\mathbf{curl}\,\boldsymbol{\chi}$. Therefore, Eq. (16) involves only $\mathbf{curl}\,\boldsymbol{\chi}$, which we will identify below as the incompatible part $\mathbf{G}_e^{\perp}$ of $\mathbf{G}_e$:

$$\mathbf{curl}\,\mathbf{G}_e^{\perp} = \mathbf{curl}\,\mathbf{curl}\,\chi = \pi. \tag{20}$$

Similarly, $\mathbf{grad}\,\mathbf{Z}$ will be the compatible part $\mathbf{G}_e^{\parallel}$ of the elastic 2-distortion $\mathbf{G}_e$, and $\mathbf{Z}$ will be the elastic distortion $\mathbf{U}_e$, up to a constant. To ensure correctness of this identification, $\mathbf{G}_e^{\perp}$ must vanish identically throughout the body when $\pi = 0$. In this aim, following Jiang (1998), Eq. (20) is augmented with the side conditions:

$$\mathbf{div}\,\mathbf{G}_e^{\perp} = 0 \text{ in } V, \tag{21}$$

$$\mathbf{G}_e^{\perp}.\mathbf{n} = 0 \text{ on } \partial V \tag{22}$$

with unit normal $\mathbf{n}$ on ∂V. Then taking the curl of Eq. (16) and using the side condition (21), it follows that:

$$\mathbf{curl}\,\mathbf{curl}\,\mathbf{G}_e^{\perp} = \mathbf{grad}\,\mathbf{div}\,\mathbf{G}_e^{\perp} - \mathbf{div}\,\mathbf{grad}\,\mathbf{G}_e^{\perp} = -\mathbf{div}\,\mathbf{grad}\,\mathbf{G}_e^{\perp} = \mathbf{curl}\,\pi \tag{23}$$

Hence, $\mathbf{G}_e^{\perp}$ satisfies a first Poisson-type equation

$$\mathbf{div}\,\mathbf{grad}\,\mathbf{G}_e^{\perp} = -\mathbf{curl}\,\pi \text{ in } V, \tag{24}$$

$$\mathbf{G}_e^{\perp}.\mathbf{n} = 0 \text{ on } \partial V. \tag{25}$$

In component form, Eq. (24) reads

$$G^{e,\perp}_{ijk,ll} = -e_{klm}\pi_{ijm,l} \tag{26}$$

As a consequence, the field of incompatible elastic 2-distortion $\mathbf{G}_e^{\perp}$ is uniquely determined once the g-disclination density field $\boldsymbol{\pi}$ is prescribed. In particular, it vanishes uniformly when $\boldsymbol{\pi} = 0$. Using Eqs. (8) and (18), the dislocation density tensor can therefore be written as:

$$\boldsymbol{\alpha} = -\mathbf{G}_e^{\perp} : \mathbf{X} - \mathbf{grad}\,\mathbf{U}_e : \mathbf{X} = \mathbf{curl}\,\mathbf{U}_e - \mathbf{G}_e^{\perp} : \mathbf{X}. \tag{27}$$

In turn, the Stokes-Helmholtz decomposition of the elastic distortion $\mathbf{U}_e$ can be used to separate its compatible part, $\mathbf{U}_e^{\parallel}$, from its incompatible part, $\mathbf{U}_e^{\perp}$:

$$\mathbf{U}_e = \mathbf{U}_e^{\perp} + \mathbf{U}_e^{\parallel} = \mathbf{curl}\,\boldsymbol{\psi} + \mathbf{grad}\,\mathbf{w}, \tag{28}$$

and to ensure uniqueness of the latter through the solution of a Poisson-type equation. In the decomposition (28), $\mathbf{U}_e^{\parallel} = \mathbf{grad}\,\mathbf{w}$ again belongs to the null-space of the curl operator since $\mathbf{curl}\,\mathbf{grad}\,\mathbf{w} = 0$, while $\mathbf{U}_e^{\perp} = \mathbf{curl}\,\boldsymbol{\psi}$ must additionally satisfy the side conditions:

$$\mathbf{div}\mathbf{U}_e^{\perp} = 0 \text{ in } V, \tag{29}$$

$$\mathbf{U}_e^{\perp}.\mathbf{n} = 0 \text{ on } \partial V. \tag{30}$$

Invoking the identity $\mathbf{curl}\,\mathbf{curl}\,\mathbf{U}_e^{\perp} = \mathbf{grad}\,\mathbf{div}\,\mathbf{U}_e^{\perp} - \mathbf{div}\,\mathbf{grad}\,\mathbf{U}_e^{\perp}$, taking the curl of Eq. (27) and using Eq. (29) then leads to a second Poisson-type equation:

$$\mathbf{div}\,\mathbf{grad}\,\mathbf{U}_e^{\perp} = -\mathbf{curl}\,(\boldsymbol{\alpha} + \mathbf{G}_e^{\perp} : \mathbf{X}) \text{ in } V, \tag{31}$$

$$\mathbf{U}_e^{\perp}.\mathbf{n} = 0 \text{ on } \partial V. \tag{32}$$

In component form, Eq. (31) reads

$$U^{e,\perp}_{ij,kk} = -e_{jkl}\alpha_{il,k} - (G^{e,\perp}_{ijk} - G^{e,\perp}_{ikj})_{,k}. \tag{33}$$

Hence, $\mathbf{U}_e^{\perp}$ is uniquely determined once the dislocation and g-disclination density fields $(\boldsymbol{\alpha}, \boldsymbol{\pi})$ are prescribed. In particular, it vanishes uniformly in V when $\boldsymbol{\alpha} = 0$ and $\boldsymbol{\pi} = 0$. Eqs. (31) and (33) will be transformed in the Fourier space in what follows (Sect. 5) and solved using the discrete Fourier transform (DFT) method (Sect. 6).

4 Constitutive and Equilibrium Equations

4.1 Constitutive Relationships

Recently, a general free energy density functional $\Psi\,(\boldsymbol{\epsilon}_e, \mathbf{G}_e)$ for crystalline materials undergoing incompatible fields due to dislocations, disclinations and g-disclinations was reported in Upadhyay et al. (2013). Here, in order to reduce this general form to a couple stress theory with incompatible fields, Ψ is supposed to only depend on the elastic strain $\boldsymbol{\epsilon}_e$ (i.e. the symmetric part of $\mathbf{U}_e$) and the skew-symmetric part of the elastic 2-distortion tensor $\mathbf{G}_e$ denoted $\mathbf{G}_e^{skew}$ as follows:

$$\Psi = \frac{1}{2}\boldsymbol{\epsilon}_e : \mathbf{C} : \boldsymbol{\epsilon}_e + \boldsymbol{\epsilon}_e : \mathbf{B} \vdots \mathbf{G}_e^{skew} + \mathbf{G}_e^{skew} \vdots \mathbf{D} : \boldsymbol{\epsilon}_e + \frac{1}{2}\mathbf{G}_e^{skew} \vdots \mathbf{E} \vdots \mathbf{G}_e^{skew}, \tag{34}$$

where $\mathbf{C}$ is the fourth order tensor of linear elastic moduli with the classic symmetry properties

$$C_{ijkl} = C_{jikl} = C_{ijlk} = C_{klij},$$

$\mathbf{B}$ is a fifth order tensor with

$$B_{ijklm} = B_{jiklm} = -B_{ijlkm} = -B_{jilkm},$$

$\mathbf{D}$ is a fifth order tensor with

$$D_{ijklm} = -D_{jiklm} = D_{ijkml} = -D_{jikml},$$

and $\mathbf{E}$ is a sixth order tensor with

$$E_{ijklmn} = -E_{jiklmn} = -E_{ijkmln} = E_{jikmln} = E_{lmnijk}.$$

The constitutive relationships are obtained by taking the partial derivatives of Ψ with respect to $\boldsymbol{\epsilon}_e$ and to $\mathbf{G}_e^{skew}$ in order to find the symmetric second order stress tensor $\mathbf{T}^{sym}$ (i.e. $T_{ij}^{sym} = T_{ji}^{sym}$) and the skew-symmetric third order hyperstress tensor denoted $\mathbf{M}^{skew}$ (i.e. $M_{ijk}^{skew} = -M_{jik}^{skew}$), respectively:

$$\mathbf{T}^{sym} = \mathbf{C} : \boldsymbol{\epsilon}_e + \mathbf{B} \vdots \mathbf{G}_e^{skew} + \mathbf{G}_e^{skew} \vdots \mathbf{D}, \tag{35}$$

$$\mathbf{M}^{skew} = \boldsymbol{\epsilon}_e : \mathbf{B} + \mathbf{D} : \boldsymbol{\epsilon}_e + \mathbf{E} \vdots \mathbf{G}_e^{skew}. \tag{36}$$

For isotropic and centrosymmetric materials as considered here, the free energy density functional Ψ further reduces to the one of a couple stress material (Mindlin and Tiersten 1962; Koiter 1964) with elastic incompatibilities under the quadratic form:

$$\Psi = \frac{1}{2}\boldsymbol{\epsilon}_e : \mathbf{C} : \boldsymbol{\epsilon}_e + \frac{1}{2}\boldsymbol{\kappa}_e^D : \mathbf{A} : \boldsymbol{\kappa}_e^D, \tag{37}$$

where $\boldsymbol{\kappa}_e^D$ is the second order deviatoric elastic curvature tensor, and, **C**, **A** read:

$$C_{ijkl} = \lambda\delta_{ij}\delta_{kl} + \mu\left(\delta_{ik}\delta_{jl} + \delta_{jk}\delta_{il}\right), \tag{38}$$

$$A_{ijkl} = A_1\delta_{ik}\delta_{jl} - A_2\delta_{jk}\delta_{il} \tag{39}$$

where μ and λ are respectively the classic shear modulus and Lamé constant of the material and A_1, A_2 are couple stress elastic constants that are length scale dependent. Let us note that the second order elastic curvature tensor $\boldsymbol{\kappa}_e$ is related to the skew-symmetric part of the third order elastic 2-distortion tensor $\mathbf{G}_e^{skew}$ as follows:

$$\boldsymbol{\kappa}_e = -\frac{1}{2}\mathbf{X} : \mathbf{G}_e^{skew}, \tag{40}$$

$$\mathbf{G}_e^{skew} = -\mathbf{X} \cdot \boldsymbol{\kappa}_e. \tag{41}$$

Taking now the thermodynamic conjugate of $\boldsymbol{\kappa}_e^D$ as the second order deviatoric couple stress tensor $\mathbf{m}^D$, we obtain from Eq. (37) together with Eq. (38) the constitutive relationships for homogeneous isotropic centro-symmetric materials:

$$T_{ij}^{sym} = C_{ijkl}\epsilon_{kl}^e = \lambda\epsilon_{kk}^e\delta_{ij} + 2\mu\epsilon_{ij}^e, \tag{42}$$

$$m_{ij}^D = A_{ijkl}\kappa_{kl}^{eD} = A_1\kappa_{ij}^{eD} - A_2\kappa_{ji}^{eD} \tag{43}$$

Let us note that only the deviatoric parts of second order elastic curvature and couple stress tensors are constitutively determined like in the so-called "undeterminate" couple stress theory originally derived by Mindlin and Tiersten (1962). The second order couple stress tensor **m** is related to the skew-symmetric part of the third order hyperstress tensor $\mathbf{M}^{skew}$ by the following operations:

$$\mathbf{m} = -\mathbf{X} : \mathbf{M}^{skew}, \tag{44}$$

$$\mathbf{M}^{skew} = -\frac{1}{2}\mathbf{X} \cdot \mathbf{m} \tag{45}$$

This initial couple stress theory is still very controversial, see for instance Neff et al. (2009), who proposed from homogenization theory with micro-randomness a couple stress theory with symmetric couple stress tensor. In the present study, the material constant A_1 is defined as function of a microstructural length sale (or characteristic size) l such that $A_1 = 4\mu l^2$ (Mindlin and Tiersten 1962). The constant A_2 is bounded in the Mindlin-Tiersten theory (Mindlin and Tiersten 1962).

4.2 Equilibrium Equations

In a couple stress theory, the linear and angular momentum balance equations for the elasto-static problem without body force and body couple force densities take the forms:

$$\mathbf{div}\,\mathbf{T} = 0 \text{ in } V, \tag{46}$$

$$\mathbf{div}\,\mathbf{m}^D - \mathbf{X} : \mathbf{T} = 0 \text{ in } V \tag{47}$$

where $\mathbf{T}$ is the force-stress tensor which splits into symmetric $\mathbf{T}^{sym}$ and skew-symmetric $\mathbf{T}^{skew}$ parts: $\mathbf{T} = \mathbf{T}^{sym} + \mathbf{T}^{skew}$ and where the second order tensor $\mathbf{m}^D$ is the deviatoric couple-stress tensor as defined earlier.

The equilibrium equations (46) and (47) are appended with force-stress vector field $\mathbf{t}^d$ and couple-stress vector $\mathbf{m}^d$ applied to a part of the boundary ∂V_t as detailed in (Mindlin and Tiersten 1962) (see also Koiter 1964) and the other part ∂V_u is subjected to the prescribed displacements $\mathbf{u}^d$.

From $\mathbf{T} = \mathbf{T}^{sym} + \mathbf{T}^{skew}$ with $T_{ij}{}^{skew} = \frac{1}{2} e_{ijk} e_{mnk} T_{mn}$ and using Eqs. (46) and (47), a single equilibrium equation involving $\mathbf{T}^{skew}$ and $\mathbf{m}^D$ can be written as follows:

$$\mathbf{div}\,\mathbf{T}^{sym} + \frac{1}{2}\mathbf{curl}\left(\mathbf{div}\,\mathbf{m}^D\right) = 0 \text{ in } V. \tag{48}$$

Using the Stokes-Helmholtz decomposition (Eqs. (19) and (28)), the equilibrium equation (48) together with the constitutive relationships can be rewritten in the form of a partial differential equation of Navier-type in V:

$$\mathbf{div}\,\mathbf{C} : \boldsymbol{\epsilon}_e^{\parallel} + \frac{1}{2}\mathbf{curl}\left(\mathbf{div}\left(\mathbf{A} : \boldsymbol{\kappa}_e^{D\parallel}\right)\right) + \mathbf{f}^{\perp} = 0, \tag{49}$$

where $\boldsymbol{\epsilon}_e^{\parallel}$, $\boldsymbol{\kappa}_e^{D\parallel}$ are respectively given by:

$$\epsilon_{ij}^{e,\parallel} = \frac{1}{2}\left(w_{i,j} + w_{j,i}\right), \tag{50}$$

$$\kappa_{ij}^{eD,\parallel} = \frac{1}{2} e_{ikl} w_{l,kj}, \tag{51}$$

and where the incompatible fictive body force density arising from the generalized defects is given by:

$$\mathbf{f}^{\perp} = \mathbf{div}\,\mathbf{C} : \boldsymbol{\epsilon}_e^{\perp} + \frac{1}{2}\mathbf{curl}\left(\mathbf{div}\left(\mathbf{A} : \boldsymbol{\kappa}_e^{D,\perp}\right)\right) \tag{52}$$

with $\epsilon_{ij}^{e,\perp} = \frac{1}{2}\left(U_{ij}^{e,\perp} + U_{ji}^{e,\perp}\right)$ and with $\kappa_{ij}^{e,\perp} = -\frac{1}{2} e_{mni} G_{mnj}^{e,skew,\perp}$. Equations (49) to (52) together with adequate boundary conditions on displacements, force and couple traction vectors (Mindlin and Tiersten 1962; Koiter 1964) set a couple stress elasticity problem for the unknown fields $\mathbf{w}$, $\boldsymbol{\epsilon}_e^{\parallel}$ and $\boldsymbol{\kappa}_e^{D,\parallel}$ which can therefore be determined uniquely. The incompatible volume fictive body force $\mathbf{f}^{\perp}$ is first determined by solving the Poisson-type Eq. (24) and (31) for $\mathbf{G}_e^{\perp}$, $\mathbf{U}_e^{\perp}$ to give $\boldsymbol{\kappa}_e^{D,\perp}$ and $\boldsymbol{\epsilon}_e^{\perp}$ after defect density fields $(\boldsymbol{\alpha}, \boldsymbol{\pi})$ have been initially prescribed.

Assuming a homogeneous reference medium with uniform elastic moduli and couple stress moduli C^0_{ijkl} and A^0_{ijkl}, such that $C_{ijkl} = C^0_{ijkl}$ and $A_{ijkl} = A^0_{ijkl}$, no "polarization tensor" fields are accounted for (i.e. no iterative scheme is needed). Therefore, Eqs. (49)–(52) yield, in component form:

$$C^0_{ijkl} w_{k,lj} + \frac{1}{4} e_{ikl} e_{prs} A^0_{lmpq} w_{s,rqmk} + C^0_{ijkl} \epsilon^{e,\perp}_{kl,j} + \frac{1}{2} e_{ikl} A^0_{lmpq} \kappa^{eD,\perp}_{pq,mk} = 0. \tag{53}$$

Using Eqs. (42) and (43) in Eq. (53) yields:

$$\begin{aligned} &\mu w_{i,kk} + (\lambda + \mu)\, w_{k,ki} + \mu l^2 \left(w_{k,ikmm} - w_{i,kkmm} \right) \\ &+ \lambda \epsilon^{e,\perp}_{kk,i} + 2\mu \epsilon^{e,\perp}_{ik,k} + 2\mu l^2 e_{ikl} \kappa^{eD,\perp}_{lm,mk} - \frac{A_2}{2} e_{ikl} \kappa^{eD,\perp}_{ml,mk} = 0. \end{aligned} \tag{54}$$

It is noteworthy that the classic size-independent theory with incompatibilities described in Berbenni et al. (2014) for instance is found with $l = 0$ and $A_2 = 0$. In this case, the term containing the incompatible elastic curvature fields due to g-disclinations as well as that with compatible elastic curvatures vanish and the classic Navier-type operator including only second order partial spatial derivatives with classic isotropic linear elasticity is retrieved.

5 Fourier Method

5.1 Solution of Poisson-Type Equations in Fourier Space

The previous Poisson and Navier-type equations can be solved using the Fourier Transform method. Indeed, the unknown vector field $\mathbf{w}(\boldsymbol{x})$ can be obtained by using the spectral method based on Fourier transforms to derive later on the stresses, elastic rotations etc. in the Fourier space. Then, the elastic fields are estimated in the real space using the inverse Fourier Transform. The FFT algorithm is well suited for periodic media. This one will be developed in Sect. 6 to estimate the discrete Fourier transforms on FFT grids.

In the Fourier space, let $\boldsymbol{\xi}$ be the Fourier vector of magnitude $\xi = \sqrt{\boldsymbol{\xi} \cdot \boldsymbol{\xi}}$ and components ξ_i in a cartesian coordinate system in a general three-dimensional setting. The complex imaginary number is denoted i and defined as $\mathrm{i} = \sqrt{-1}$.

Let $\widetilde{\boldsymbol{\alpha}}(\boldsymbol{\xi})$, $\widetilde{\mathbf{U}}_e^{\perp}(\boldsymbol{\xi})$, $\widetilde{\boldsymbol{\epsilon}}_e^{\perp}(\boldsymbol{\xi})$, $\widetilde{\boldsymbol{\pi}}(\boldsymbol{\xi})$, $\widetilde{\mathbf{G}}_e^{\perp}(\boldsymbol{\xi})$ and $\widetilde{\boldsymbol{\kappa}}_e^{D,\perp}(\boldsymbol{\xi})$ be the continuous Fourier transforms of $\boldsymbol{\alpha}(\boldsymbol{x})$, $\mathbf{U}_e^{\perp}(\boldsymbol{x})$, $\boldsymbol{\epsilon}_e^{\perp}(\boldsymbol{x})$, $\boldsymbol{\pi}(\boldsymbol{x})$,$\mathbf{G}_e^{\perp}(\boldsymbol{x})$ and $\boldsymbol{\kappa}_e^{D,\perp}(\boldsymbol{x})$. Then, the Poisson-type equations (Eqs. (24) and (31)) are solved using the differentiation theorem in Fourier space. Using component notations, Eq. (26) writes in the Fourier space

$$\begin{aligned}\widetilde{G}^{e,\perp}_{ijk}(\boldsymbol{\xi}) &= \tfrac{\mathrm{i}}{\xi^2}\xi_l e_{klm}\widetilde{\pi}_{ijm}(\boldsymbol{\xi})\ \forall\boldsymbol{\xi}\neq\mathbf{0},\\ \widetilde{G}^{e,\perp}_{ijk}(\mathbf{0}) &= \mathbf{0},\end{aligned} \tag{55}$$

and Eq. (33) yields in the Fourier space

$$\begin{aligned}\widetilde{U}^{e,\perp}_{ij}(\boldsymbol{\xi}) &= \tfrac{\mathrm{i}}{\xi^2}\xi_k\left(e_{jkl}\widetilde{\alpha}_{il}(\boldsymbol{\xi}) + \widetilde{G}^{e,\perp}_{ijk}(\boldsymbol{\xi}) - \widetilde{G}^{e,\perp}_{ikj}(\boldsymbol{\xi})\right)\ \forall\boldsymbol{\xi}\neq\mathbf{0},\\ \widetilde{U}^{e,\perp}_{ij}(\mathbf{0}) &= \mathbf{0}.\end{aligned} \tag{56}$$

Therefore, $\widetilde{\boldsymbol{\epsilon}}^{\perp}_e(\boldsymbol{\xi})$ is derived from the symmetric part of $\widetilde{\mathbf{U}}^{\perp}_e(\boldsymbol{\xi})$ and $\widetilde{\boldsymbol{\kappa}}^{D,\perp}_e(\boldsymbol{\xi})$ is obtained from the skew-symmetric part of $\widetilde{\mathbf{G}}^{\perp}_e(\boldsymbol{\xi})$ (see Eq. (40)).

5.2 Solution of Navier-Type Equation in Fourier Space

Let $\widetilde{\mathbf{w}}(\boldsymbol{\xi})$, $\widetilde{\boldsymbol{\epsilon}}^{\parallel}_e(\boldsymbol{\xi})$ and $\widetilde{\boldsymbol{\kappa}}^{D,\parallel}_e(\boldsymbol{\xi})$ be the continous Fourier transform of $\mathbf{w}(\boldsymbol{x})$, $\boldsymbol{\epsilon}^{\parallel}_e(\boldsymbol{x})$ and $\boldsymbol{\kappa}^{D,\parallel}_e(\boldsymbol{x})$. Then, the Fourier transform of the Navier-type equation (53) yields

$$\begin{aligned}C^0_{ijkl}\xi_l\xi_j\widetilde{w}_k(\boldsymbol{\xi}) &- \frac{1}{4}e_{ikl}e_{prs}A^0_{lmpq}\xi_r\xi_q\xi_m\xi_k\widetilde{w}_s(\boldsymbol{\xi})\\ &= \mathrm{i}C^0_{ijkl}\xi_j\widetilde{\epsilon}^{e,\perp}_{kl}(\boldsymbol{\xi}) + \frac{1}{2}e_{ikl}A^0_{lmpq}\xi_m\xi_k\widetilde{\kappa}^{eD,\perp}_{pq}(\boldsymbol{\xi}).\end{aligned} \tag{57}$$

The solution $\widetilde{\mathbf{w}}(\boldsymbol{\xi})$ can be obtained with the introduction of the Green tensor $\widetilde{G}_{ik}(\boldsymbol{\xi})$ in the Fourier space:

$$\widetilde{w}_i(\boldsymbol{\xi}) = \widetilde{G}_{ik}(\boldsymbol{\xi})\widetilde{f}^{\perp}_k(\boldsymbol{\xi}), \tag{58}$$

where:

$$\widetilde{G}_{ik}(\boldsymbol{\xi}) = \left(C^0_{ijkl}\xi_l\xi_j - \frac{1}{4}e_{ijl}A^0_{lmpq}e_{prk}\xi_r\xi_q\xi_m\xi_j\right)^{-1} \tag{59}$$

and:

$$\widetilde{f}^{\perp}_i(\boldsymbol{\xi}) = \mathrm{i}C^0_{ijkl}\xi_j\widetilde{\epsilon}^{e,\perp}_{kl}(\boldsymbol{\xi}) + \frac{1}{2}e_{ikl}A^0_{lmpq}\xi_m\xi_k\widetilde{\kappa}^{eD,\perp}_{pq}(\boldsymbol{\xi}). \tag{60}$$

It is noteworthy that Eqs. (58)–(60) can be applied to any centrosymmetric anisotropic couple stress materials.

For isotropic centrosymmetric couple stress materials, C^0_{ijkl} and A^0_{ijkl} are defined by Eqs. (38) and (39). In this case, the expression of the non local Green tensor can be found in Smyshlyaev and Fleck (1994), Zheng and Zhao (2004). Thus, the expression of $\widetilde{G}_{ik}(\boldsymbol{\xi})$ is given by:

$$\widetilde{G}_{ik}(\boldsymbol{\xi}) = \frac{1}{\mu\xi^2}\left[\frac{1}{1+l^2\xi^2}\left(\delta_{ik} - \frac{\xi_i\xi_k}{\xi^2}\right) + \frac{\mu}{\lambda+2\mu}\frac{\xi_i\xi_k}{\xi^2}\right] \forall\boldsymbol{\xi} \neq \mathbf{0}, \qquad \widetilde{G}_{ik}(\mathbf{0}) = \mathbf{0}. \tag{61}$$

The compatible elastic strain $\widetilde{\boldsymbol{\epsilon}}_e^{\parallel}(\boldsymbol{\xi})$ is obtained in the Fourier space from the differentiation rule:

$$\widetilde{\epsilon}_{ij}^{e,\parallel}(\boldsymbol{\xi}) = \frac{1}{2}\mathrm{i}\left(\xi_j\widetilde{w}_i(\boldsymbol{\xi}) + \xi_i\widetilde{w}_j(\boldsymbol{\xi})\right), \qquad \widetilde{\kappa}_{ij}^{eD,\parallel}(\boldsymbol{\xi}) = -\frac{1}{2}e_{ikl}\xi_j\xi_k\widetilde{w}_l(\boldsymbol{\xi}). \tag{62}$$

5.3 Stress and Couple Stress Fields

Knowing $\widetilde{\mathbf{U}}_e^{\perp}(\boldsymbol{\xi})$ and $\widetilde{\mathbf{U}}_e^{\parallel}(\boldsymbol{\xi})$, the (total) elastic distortion in the Fourier space yields

$$\widetilde{U}_{ij}^{e} = \widetilde{U}_{ij}^{e,\perp} + \widetilde{U}_{ij}^{e,\parallel}. \tag{63}$$

The stress $\widetilde{\mathbf{T}}(\boldsymbol{\xi})$ in Fourier space is obtained in component form as:

$$\widetilde{T}_{ij}(\boldsymbol{\xi}) = C_{ijkl}^{0}\widetilde{\epsilon}_{kl}^{e}(\boldsymbol{\xi}) \ \forall\boldsymbol{\xi} \neq \mathbf{0}, \qquad \widetilde{T}_{ij}(\mathbf{0}) = 0, \tag{64}$$

where $\widetilde{\epsilon}_{ij}^{e} = \frac{1}{2}\left(\widetilde{U}_{ij}^{e} + \widetilde{U}_{ji}^{e}\right)$ and where the far-field (overall) stress which is the spatial average of T_{ij} over the periodic unit cell is set to zero. Here, only the internal stress field will be computed in Sect. 7.

The couple stress $\widetilde{\mathbf{m}^D}(\boldsymbol{\xi})$ in Fourier space is obtained in component form as:

$$\widetilde{m}_{ij}^{D}(\boldsymbol{\xi}) = A_{ijkl}^{0}\widetilde{\kappa}_{kl}^{eD}(\boldsymbol{\xi}) \ \forall\boldsymbol{\xi} \neq \mathbf{0}, \qquad \widetilde{m}_{ij}^{D}(\mathbf{0}) = 0, \tag{65}$$

where the far-field (overall) deviatoric couple stress which is taken as the spatial average of m_{ij}^D over the periodic unit cell is set to zero. Here only the internal couple stress field due to generalized defects will be computed in Sect. 7.

The elastic stress and couple stress moduli C_{ijkl}^0 and A_{ijkl}^0 in Eqs. (64) and (65) are defined by Eqs. (38) and (39). Then, the inverse Fourier transforms of $\widetilde{\mathbf{T}}(\boldsymbol{\xi})$ and $\widetilde{\mathbf{m}^D}(\boldsymbol{\xi})$ are numerically computed using the FFT algorithm and inverse FFT allows finding $\mathbf{T}$ and $\mathbf{m}^D$ on the discretized periodic unit cell.

6 Fast Fourier Transform Numerical Implementation

6.1 Discrete and Fast Fourier Transforms

The field equations derived in the Fourier space are now solved by 2D discrete Fourier transforms with the Fast Fourier Transform (FFT) algorithm. Here, periodicity is assumed for the distribution of g-disclination densities (i.e. $\boldsymbol{\pi}$), with spatial periods T_1 and T_2 in the x_1 and x_2 directions, respectively. The periodic representative volume element (RVE) or unit cell is discretized by a regular rectangular grid with $N_1 \times N_2$ pixels with position vector $\boldsymbol{x} = ((i-1)\delta_1, (j-1)\delta_2)$, where $i = 1 \rightarrow N_1, j = 1 \rightarrow N_2$ and δ_1, δ_2 are the pixel sizes in the x_1 and x_2 directions with $\delta_1 = \delta_2 = \delta$. The total number of FFT grid points is $N_{tot} = N_1 \times N_2$. Here, the FFTW package of Matlab is used to compute discrete Fourier transforms (Frigo and Johnson 2005). The discrete FFT of a given spatial function f is $\widehat{f} = \mathrm{FFT}(f)$. Its inverse Fourier transform is $f = \mathrm{FFT}^{-1}\left(\widehat{f}\right)$. They write with the Matlab FFT convention:

$$\widehat{f}(k,l) = \sum_{i=1}^{N1}\sum_{j=1}^{N2} f(i,j)\exp\left(-2\pi\mathrm{i}\left(\frac{(i-1)(k-1)}{N1} + \frac{(j-1)(l-1)}{N2}\right)\right) \tag{66}$$

and

$$f(i,j) = \frac{1}{N_{tot}}\sum_{k=1}^{N1}\sum_{l=1}^{N2} \widehat{f}(k,l)\exp\left(+2\pi\mathrm{i}\left(\frac{(i-1)(k-1)}{N1} + \frac{(j-1)(l-1)}{N2}\right)\right) \tag{67}$$

It should be pointed out that Eqs. (66) and (67) are finite sums which can be determined exactly by FFT for periodic unit cells.

6.2 Discrete Fourier Transform Differentiation Rules Based on Centered Finite Difference Approximation

Here, the following differentiation rules are used for first-, second- and fourth- order partial derivatives calculated on the discrete grid based on a 9-pixel centered finite difference approximation (Press et al. 2002):

$$\begin{aligned}
\frac{\partial f(i,j)}{\partial x_1} &= \frac{f(i+1,j) - f(i-1,j)}{2\delta_1}, \\
\frac{\partial f(i,j)}{\partial x_2} &= \frac{f(i,j+1) - f(i,j-1)}{2\delta_2}, \\
\frac{\partial^2 f(i,j)}{\partial x_1^2} &= \frac{f(i+1,j) - 2f(i,j) + f(i-1,j)}{\delta_1^2},
\end{aligned} \tag{68}$$

$$\begin{aligned}
\frac{\partial^2 f(i,j)}{\partial x_2^2} &= \frac{f(i,j+1) - 2f(i,j) + f(i,j-1)}{\delta_2^2}, \\
\frac{\partial^2 f(i,j)}{\partial x_1 \partial x_2} &= \frac{f(i+1,j+1) + f(i-1,j-1)}{4\delta_1\delta_2}, \\
&- \frac{f(i+1,j-1) + f(i-1,j+1)}{4\delta_1\delta_2}, \\
\frac{\partial^4 f(i,j)}{\partial x_1^4} &= \frac{f(i-2,j) - 4f(i-1,j) + 6f(i,j) - 4f(i+1,j) + f(i+2,j)}{\delta_1^4}, \\
\frac{\partial^4 f(i,j)}{\partial x_2^4} &= \frac{f(i,j-2) - 4f(i,j-1) + 6f(i,j) - 4f(i,j+1) + f(i,j+2)}{\delta_2^4}, \\
\frac{\partial^4 f(i,j)}{\partial x_1^2 \partial x_2^2} &= \frac{4f(i,j)}{\delta_1^2\delta_2^2} \\
&- 2\frac{f(i-1,j) + f(i+1,j) + f(i,j-1) + f(i,j+1)}{\delta_1^2\delta_2^2} \\
&+ \frac{f(i+1,j-1) + f(i-1,j-1)}{\delta_1^2\delta_2^2} \\
&+ \frac{f(i+1,j+1) + f(i-1,j+1)}{\delta_1^2\delta_2^2}, \\
\frac{\partial^4 f(i,j)}{\partial x_2 \partial x_1^3} &= \frac{f(i+2,j+1) + 2f(i-1,j+1)}{4\delta_1^3\delta_2} \\
&- \frac{f(i-2,j+1) + 2f(i+1,j+1)}{4\delta_1^3\delta_2} \\
&+ \frac{f(i-2,j-1) + 2f(i+1,j-1)}{4\delta_1^3\delta_2} \\
&- \frac{f(i+2,j-1) + 2f(i-1,j-1)}{4\delta_1^3\delta_2}, \\
\frac{\partial^4 f(i,j)}{\partial x_1 \partial x_2^3} &= \frac{f(i+1,j+2) + 2f(i+1,j-1)}{4\delta_1\delta_2^3} \\
&- \frac{f(i-1,j+2) + 2f(i+1,j+1)}{4\delta_1\delta_2^3} \\
&+ \frac{f(i-1,j-2) + 2f(i-1,j+1)}{4\delta_1\delta_2^3} \\
&- \frac{f(i-1,j+2) + 2f(i-1,j-1)}{4\delta_1\delta_2^3}.
\end{aligned} \tag{69}$$

Using Eqs. (66)–(69), the substitutions due to correspondence between continuous and discrete Fourier transform derivatives are the following:

$$\mathrm{i}\xi_1 \leftrightarrow \frac{\mathrm{i}}{\delta_1}\sin\left(\frac{2\pi(k-1)}{N1}\right), \tag{70}$$

$$\mathrm{i}\xi_2 \leftrightarrow \frac{\mathrm{i}}{\delta_2}\sin\left(\frac{2\pi(l-1)}{N2}\right), \tag{71}$$

$$-\xi_1^2 \leftrightarrow \frac{2}{\delta_1^2}\left(\cos\left(\frac{2\pi(k-1)}{N1}\right)-1\right), \tag{72}$$

$$-\xi_2^2 \leftrightarrow \frac{2}{\delta_2^2}\left(\cos\left(\frac{2\pi(l-1)}{N2}\right)-1\right), \tag{73}$$

$$\begin{aligned} -\xi_1\xi_2 \leftrightarrow & \frac{1}{2\delta_1\delta_2}\cos\left(2\pi\left(\frac{(k-1)}{N1}+\frac{(l-1)}{N2}\right)\right) \\ & -\frac{1}{2\delta_1\delta_2}\cos\left(2\pi\left(\frac{(k-1)}{N1}-\frac{(l-1)}{N2}\right)\right), \end{aligned} \tag{74}$$

$$\xi_1^4 \leftrightarrow \frac{4}{\delta_1^4}\left(\cos\left(\frac{2\pi(k-1)}{N1}\right)-1\right)^2, \tag{75}$$

$$\xi_2^4 \leftrightarrow \frac{4}{\delta_2^4}\left(\cos\left(\frac{2\pi(l-1)}{N2}\right)-1\right)^2, \tag{76}$$

$$\xi_1^2\xi_2^2 \leftrightarrow \frac{4}{\delta_1^2\delta_2^2}\left(\cos\left(\frac{2\pi(k-1)}{N1}\right)-1\right)\left(\cos\left(\frac{2\pi(l-1)}{N2}\right)-1\right), \tag{77}$$

$$\begin{aligned} \xi_1^3\xi_2 \leftrightarrow & \frac{2}{\delta_1^3\delta_2}\sin\left(2\pi\left(\frac{(k-1)}{N1}\right)\right)\sin\left(2\pi\left(\frac{(l-1)}{N2}\right)\right) \\ & \times\left(1-\cos\left(\frac{2\pi(k-1)}{N1}\right)\right), \end{aligned} \tag{78}$$

$$\begin{aligned} \xi_1\xi_2^3 \leftrightarrow & \frac{2}{\delta_1\delta_2^3}\sin\left(2\pi\left(\frac{(k-1)}{N1}\right)\right)\sin\left(2\pi\left(\frac{(l-1)}{N2}\right)\right) \\ & \times\left(1-\cos\left(\frac{2\pi(l-1)}{N2}\right)\right). \end{aligned} \tag{79}$$

In Sect. 7 of the present paper, the 9-pixel approximation is sufficient to give accurate enough results for strong gradients of stress/couple stress fields of g-disclinations in comparison with existing analytical solutions . Higher order pixel approximations may also be developed (Neumann et al. 2002) to further refine the FFT analysis. In Berbenni et al. (2014), it was shown that the present FFT method with centered-difference based-DFT avoids spurious Gibbs oscillations occurring with classic FFT techniques, especially when defect densities are prescribed to a single pixel.

7 Application to Infinite Straight Wedge Disclinations

7.1 *Materials and Numerical Data*

In the forthcoming applications, the g-disclination densities are prescribed using a regular Gaussian function. 2D FFT $N \times N$ square grids with $\delta_1 = \delta_2 = \delta$, $N = N_1 = N_2$ and $N_{tot} = N^2$ are considered. In this section, the FFT grid will be set to 1024×1024 pixels with pixel size: $\delta = 0.1b$ where b is the magnitude of the Burgers vector. Here, a face-centered cubic metal like Copper is studied, for which the lattice parameter is $a_0 = 0.36151$ nm. The isotropic elastic constants of Copper (Cu) will be used for the simulations: $\mu = 47800$ MPa, $\nu = 0.34$. The magnitude of the Burgers vector of Cu is $b = \sqrt{2}a_0/2$, i.e. $b = 0.25563$ nm. Following Lubarda (2003) for dislocations, Upadhyay et al. (2011), Taupin et al. (2013), Fressengeas et al. (2014) for disclinations the length scale l is set to $b/2$ to make $A_1 \approx \mu b^2$. For wedge disclinations, it will be seen that the term containing $A2$ in Eq. (54) vanishes.

In Berbenni et al. (2014), the stresses for both pure screw and edge dislocations were already computed to assess the present numerical spectral method by comparing the FFT solutions to analytical expressions (Hirth and Lothe 1982; Acharya 2001) and finite element results. Here, numerical FFT results for disclinations, disclination dipole and walls will be compared to analytical results reported by Anthony (1970), deWit (1973), Romanov and Vladimirov (1992).

7.2 *Two-Dimensional Equations for G-Disclinations*

In the following applications, we consider straight g-disclinations such that the defect line lies along the $\mathbf{e}_3$ axis. Thus, the elastic fields are invariant with respect to x_3. Here, the defect is based on elastic distortion discontinuities described by non zero $[[U^e_{12}]]$ and $[[U^e_{21}]]$. Thus, in this case, Eqs. (16) and (17) simplify into

$$\int_S \pi_{123} dS = \int_S \left(G^{e,\perp}_{122,1} - G^{e,\perp}_{121,2} \right) dS = [[U^e_{12}]], \tag{80}$$

$$\int_S \pi_{213} dS = \int_S \left(G^{e,\perp}_{212,1} - G^{e,\perp}_{211,2} \right) dS = [[U^e_{21}]]. \tag{81}$$

Consequently, given $\pi_{123}(\boldsymbol{x})$ and $\pi_{213}(\boldsymbol{x})$, the incompatible elastic 2-distortions are solutions of the four following Poisson-type equations (see Eq. (26))

$$G^{e,\perp}_{122,11} + G^{e,\perp}_{122,22} = \pi_{123,1}, \tag{82}$$

$$G^{e,\perp}_{121,11} + G^{e,\perp}_{121,22} = -\pi_{123,2}, \tag{83}$$

$$G^{e,\perp}_{212,11} + G^{e,\perp}_{212,22} = \pi_{213,1}, \tag{84}$$

$$G^{e,\perp}_{211,11} + G^{e,\perp}_{211,22} = -\pi_{213,2}. \tag{85}$$

Once $G^{e,\perp}_{122}$, $G^{e,\perp}_{121}$, $G^{e,\perp}_{212}$ and $G^{e,\perp}_{211}$ are obtained, four other Poisson-type equations are needed to find in turn the incompatible elastic 1-distortions $U^{e,\perp}_{11}$, $U^{e,\perp}_{22}$, $U^{e,\perp}_{12}$ and $U^{e,\perp}_{21}$ using Eq. (33) (without dislocation densities)

$$U^{e,\perp}_{12,11} + U^{e,\perp}_{12,22} = -G^{e,\perp}_{121,1}, \tag{86}$$

$$U^{e,\perp}_{11,11} + U^{e,\perp}_{11,22} = G^{e,\perp}_{121,2}, \tag{87}$$

$$U^{e,\perp}_{21,11} + U^{e,\perp}_{21,22} = -G^{e,\perp}_{212,2}, \tag{88}$$

$$U^{e,\perp}_{22,11} + U^{e,\perp}_{22,22} = G^{e,\perp}_{212,1}. \tag{89}$$

All the previous equations is solved successively in the Fourier space for the particular cases of straight wedge disclination and wedge disclination dipole. Then, the generalized Navier-type equation is solved in the Fourier space for the compatible 1- and 2-elastic distortions (compatible elastic strain and elastic curvature tensors).

7.3 Single Straight Wedge Disclination

First, the case of a pure straight wedge disclination is considered. This corresponds to an elastic distortion discontinuity in the negative half-plane ($x_1 = 0, x_2 \leq 0$). The only non zero discontinuities are $[[U^e_{12}]] = [[\omega^e_{12}]] = -[[\Omega^e_3]]$ and $[[U^e_{21}]] = [[\omega^e_{21}]] = +[[\Omega^e_3]]$, where $[[\Omega^e_3]] = \Omega^e_{3+} - \Omega^e_{3-}$ is the elastic rotation discontinuity along the $\mathbf{e}_3$ axis (the domains (+) and (-) respectively correspond to $x_1 > 0$ and $x_1 < 0$). This g-disclination is equivalent to a pure disclination with the positive Frank vector component along the $\mathbf{e}_3$ axis (Romanov and Vladimirov 1992) with $[[\Omega^e_3]] = \omega$. Thus, the discontinuity in the elastic distortion is such that

$$[[\Omega^e_3]] = \omega = \int_S \pi_{213} \mathrm{d}S, \tag{90}$$

where $\pi_{213} = -\pi_{123}$. Thus, only two Poisson equations containing π_{213} are considered and solved in the Fourier space. For the simulations, the g-disclination density follows a Gaussian distribution:

$$\pi_{213}(\boldsymbol{x}) = \frac{\omega}{2\pi\sigma^2}\exp\left(-\frac{r^2}{2\sigma^2}\right) \text{ if } r \le r_0, \qquad \pi_{213}(\boldsymbol{x}) = 0 \text{ if } r > r_0 \tag{91}$$

with $r = \sqrt{{x_1}^2 + {x_2}^2}$ and $\sigma = 0.1 r_0$.

Then, the non zero incompatible elastic curvature components (equivalent to incompatible elastic 2-distortions) are obtained as

$$\kappa_{31}^{e,\perp} = -G_{121}^{e,\perp}, \tag{92}$$

$$\kappa_{32}^{e,\perp} = -G_{122}^{e,\perp}. \tag{93}$$

By solving the non zero incompatible and compatible elastic curvatures, the elastic curvatures $\kappa_{32}^e = \kappa_{32}^{e,\perp} + \kappa_{32}^{e,\parallel}$ and $\kappa_{31}^e = \kappa_{31}^{e,\perp} + \kappa_{31}^{e,\parallel}$ are computed by inverse FFT on a 2D grid with $r_0 = 0.5b$ and $\omega = 5/6$ rad. Therefore, the $A2$ term in Eq. 54 vanishes because of invariance of incompatible elastic curvature along x_3 together with only non zero $\kappa_{32}^{e,\perp}$ and $\kappa_{31}^{e,\perp}$ for this particular 2D case. The numerical results are reported in Fig. 2. The results show that the respective variations of κ_{32}^e and κ_{31}^e along x_1 and x_2 match exactly the analytical solutions of (deWit 1973) or (Anthony 1970):

$$\kappa_{31}^e = -\frac{[[\Omega_3^e]]}{2\pi}\frac{x_2}{r^2}, \tag{94}$$

$$\kappa_{32}^e = \frac{[[\Omega_3^e]]}{2\pi}\frac{x_1}{r^2}. \tag{95}$$

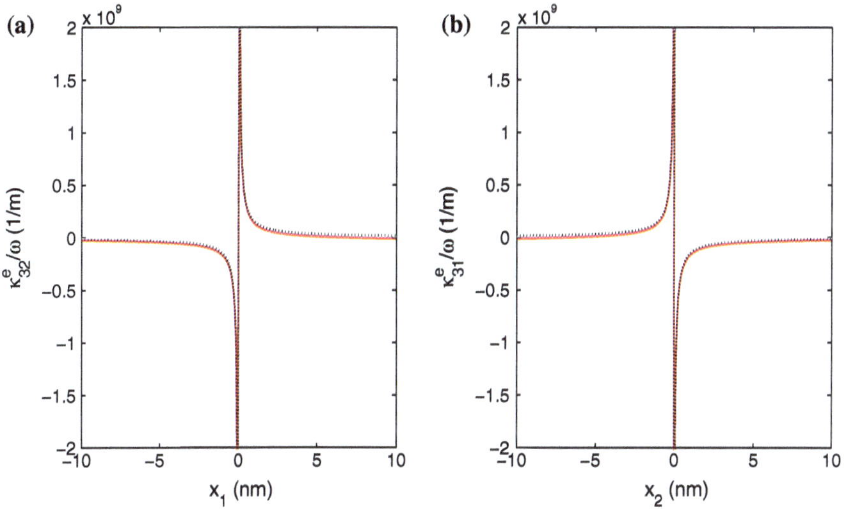

Fig. 2 Elastic curvatures for a single straight wedge disclination: **a** κ_{32}^e, (b) κ_{31}^e normalized by $\omega = [[\Omega_3^e]]$. Comparisons with the solutions given by Anthony (1970), deWit (1973) (*dashed lines*)

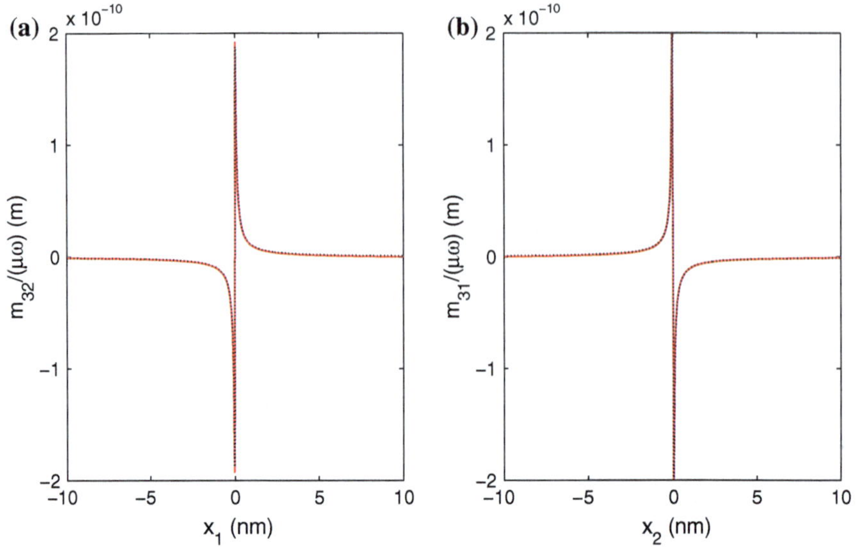

Fig. 3 Couple stresses for a single straight wedge disclination: m_{32}^D (a), m_{31}^D **b** non zero couple stress components normalized by $\mu\omega$. Comparisons with the solutions given by Anthony (1970) (*dashed lines*)

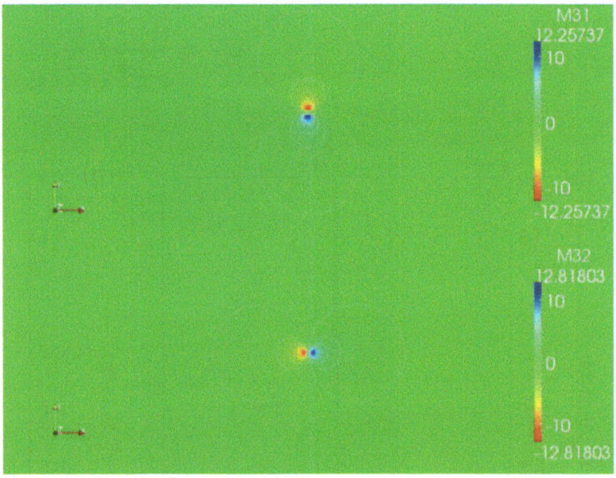

Fig. 4 FFT simulations of couple stresses for a single straight wedge disclination: m_{31}^D (top), m_{32}^D (*bottom*) non zero couple stress components normalized by $\omega = [[\Omega_3^e]]$

Once the elastic curvatures are calculated, the non zero couple stress components m_{31}^D and m_{32}^D are obtained using the constitutive relationship (Eq. (43)), see Fig. 3. The results are consistent with the analytical solutions obtained with a couple stress theory with disclinations (Anthony 1970). Couple stress contours for m_{31}^D and m_{32}^D are also reported in Fig. 4.

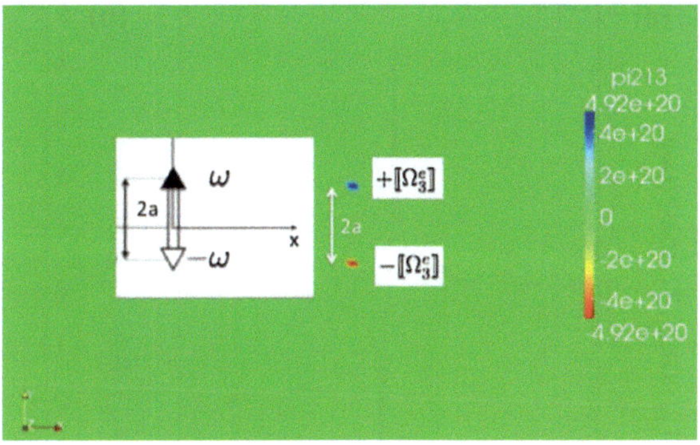

Fig. 5 Spatial distribution of G-disclination density in rad.m^{-2} for a straight wedge disclination dipole with associated positive and negative rotation jumps: $+[[\Omega^e_3]]$ and $-[[\Omega^e_3]]$

7.4 Straight Wedge Disclination Dipole

A second application is the case of a wedge disclination dipole as described in Fig. 5 which is commonly used to describe tilt grain boundaries as originally proposed by Li (1972) and later improved by Gertsman et al. (1989), Nazarov et al. (2000). In this representation, the grain boundary is represented in the form of a complex arrangement of periodic disclination dipole walls associated with the minority structural units (see Fig. 6). For a pure wedge disclination dipole, the positive (resp. negative) pole is distributed by using the same Gaussian distribution as in Eq. (91) at location $(x_1 = 0, x_2 = +a)$ (resp. $(x_1 = 0, x_2 = -a)$) with disclination strength $[[\Omega^e_3]]$ (resp. $-[[\Omega^e_3]]$).

In order to get stress fields similar to that of an equivalent straight edge dislocation, the semi-length of the dipole is set to $a = b/(2[[\Omega^e_3]])$ (see e.g. Romanov and Vladimirov 1992) with $[[\Omega^e_3]] = 5/6$ rad. For the FFT simulations, the stress components are obtained by inverse FFT on the 2D grid after successively computing

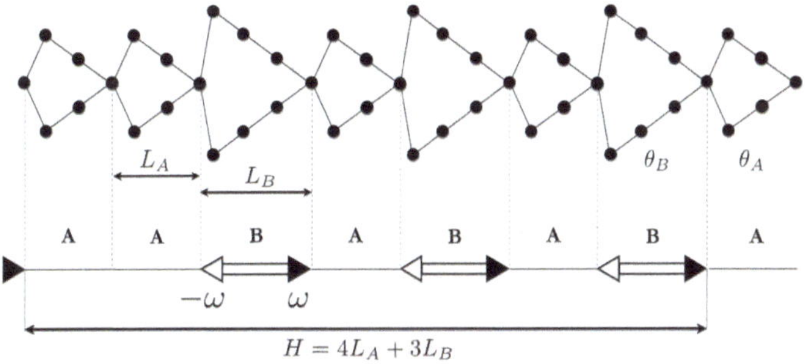

Fig. 6 Example of a disclination Structural Unit Model (DSUM) representation of the [001] $\Sigma 149$ (10 7 0) $\theta = 20.02°$ symmetric tilt grain boundary. Its structural decomposition is |AABABAB.AABABAB|, with B being the minority structural unit. The elastic fields of this grain boundary can be constructed as the superposition of three offsetted periodic walls of disclination dipoles B

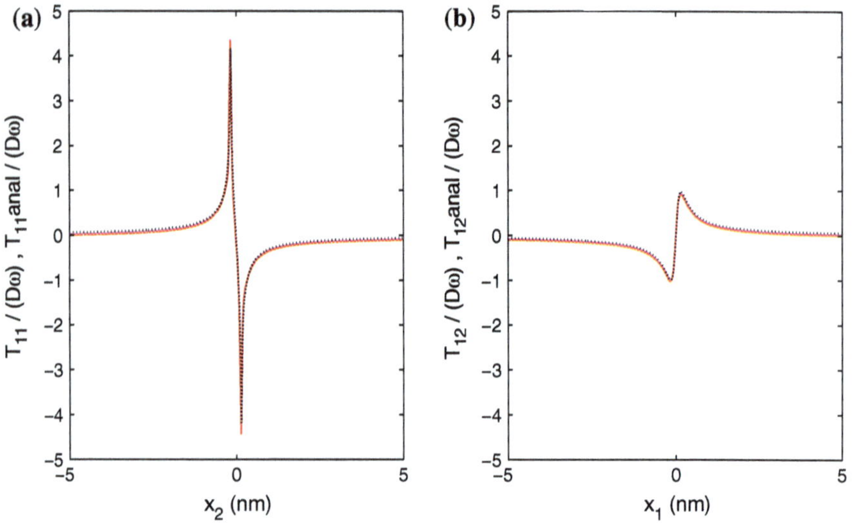

Fig. 7 Profiles of stress components: **a** T_{11}, (b) T_{12}, for a straight wedge disclination dipole (see Fig. 5) obtained by FFT and normalized by $D\omega$ where $D = \mu/(2\pi(1-\nu))$ (*solid lines*). Comparisons with the solutions given by Anthony (1970), deWit (1973) (*dashed lines*)

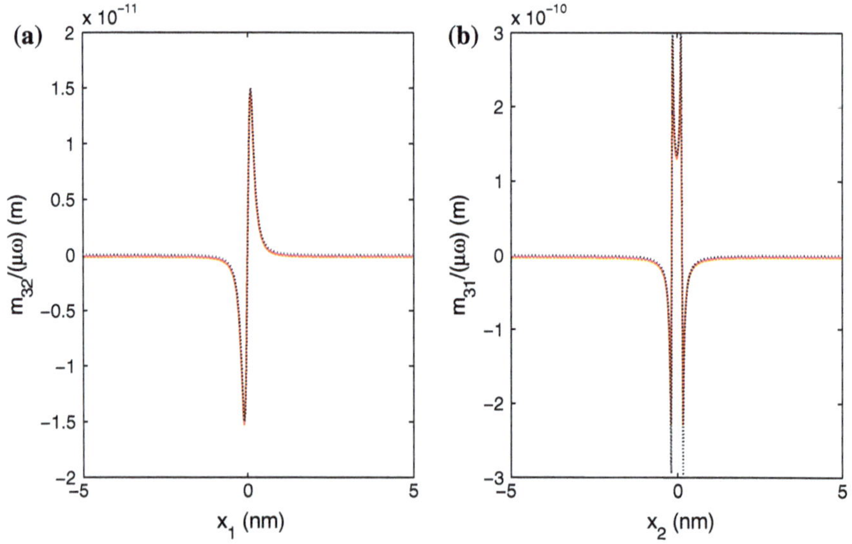

Fig. 8 Profiles of couple stress components: **a** m^D_{32}, (b) m^D_{31}, for a straight wedge disclination dipole (see Fig. 5) obtained by FFT and normalized by $\mu\omega$ (*solid lines*). Comparisons with the solutions given by Anthony (1970), deWit (1973) (*dashed lines*)

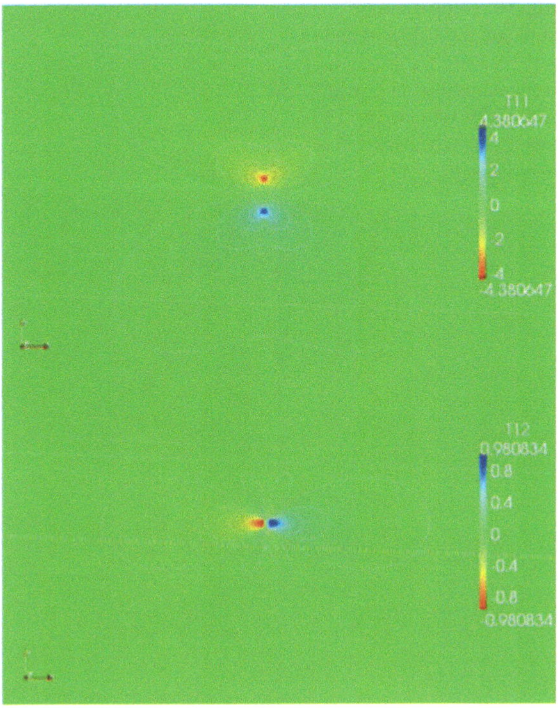

Fig. 9 Stress contours for T_{11} (*top*) and T_{12} (*bottom*) for a straight wedge disclination dipole (see Fig. 5) obtained by FFT and normalized by $D\omega$ where $\omega = [[\Omega_3^e]]$ and $D = \mu/(2\pi(1-\nu))$

in the discrete Fourier space the incompatible elastic curvatures, the incompatible elastic 1-distortions, the compatible elastic distortions and the stresses and couple stresses using the constitutive relationships.

Figure 7 displays the stress components T_{11}, T_{12} obtained by FFT and normalized by $D\omega$ where $\omega = [[\Omega_3^e]]$ and $D = \mu/(2\pi(1-\nu))$. Figure 8 describes the couple stress components m_{31}^D and m_{32}^D obtained by FFT and normalized by $\mu\omega$. Excellent agreement is found with the analytical stress components given by deWit (1973), Romanov and Vladimirov (1992), Anthony (1970) for wedge disclination dipoles. Normal and shear stresses contour plots normalized by $D\omega$ ($D = \mu/(2\pi(1-\nu))$) are reported in Fig. 9 and couple stress plots normalized by ω are reported in Fig. 10. It is shown that the present FFT results confirm that the stress contour reported in Fig. 9 respectively, are similar to that of an edge dislocation (see for instance Romanov and Vladimirov 1992; Upadhyay et al. 2011).

8 Concluding Remarks

A spectral approach for periodic media was developed to solve the elasto-static field equations of g-disclination mechanics (Acharya and Fressengeas 2012, 2015) in the framework of a couple stress theory as an extension of the work described in

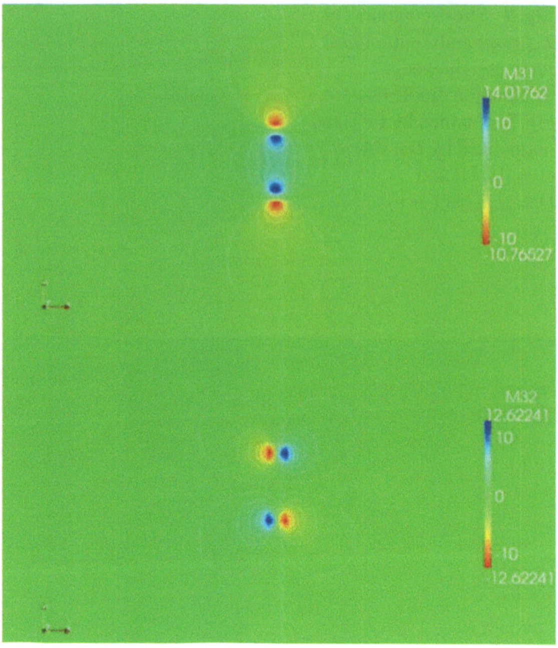

Fig. 10 Couple stress contours for m_{31}^D (*top*) and m_{32}^D (*bottom*) for a straight wedge disclination dipole (see Fig. 5) obtained by FFT and normalized by ω where $\omega = [[\Omega_3^e]]$

Berbenni et al. (2014), which was only dedicated to a Cauchy stress theory. Various results are obtained such as the solutions of Poisson-type equations in the Fourier space for a medium containing both dislocation and g-disclination densities. These solutions capture the incompatible part of elastic fields induced by g-disclinations in the Fourier space. The compatible elastic fields needed to retrieve the stress and couple stress fields are also solved with a non local character due to a characteristic internal length scale l involved in the equilibrium equations and associated with couple stresses. The latter is chosen to be related to the Burgers vector ($l = 0.5b$) of the material to give realistic and physical elastic fields near the defect core. The present discrete Fourier transform method uses the FFT algorithm and has been adapted to 2D periodic unit cells containing straight wedge disclinations in isotropic elasticity where defect line is parallel to the third dimension (i.e. x_3 axis). The discrete Fourier transform method is based on differentiation rules up to fourth-order partial derivatives allowing accurate calculations of stress and couple stress fields in comparison with analytical solutions for single disclinations or associated dipoles. The present theory and FFT implementation may be useful to derive the internal stress and couple stress fields of grain boundaries seen as DSUM (Disclination Structural Unit Model) (Gertsman et al. 1989) or more generalized defects, which may include a combination of stretching and rotation-discontinuities at the nanoscale. Furthermore, the numerical framework can be easily adapted to elastic anisotropy (see Eqs. (58) to (60)).

Acknowledgments SB would like to thank Professors P. Neff, W. H. Müller and S. Forest for fruitful discussions during the conference. This work is supported by the French State through the National Research Agency (ANR) under the program Investment in the future (LabEx DAMAS referenced as ANR-11-LABX-0008-01).

References

Acharya A (2001) A model of crystal plasticity based on the theory of continuously distributed dislocations. J Mech Phys Solids 49:761–785

Acharya A, Fressengeas C (2012) Coupled phase transformations and plasticity as a field theory of deformation incompatibility. Int J Fract 174:87–94

Acharya A, Fressengeas C (2015) Continuum mechanics of the interaction of phase boundaries and dislocations in solids. In: Chen GQ, Grinfeld M, Knops RJ (eds) Proceedings in mathematics and statistics for workshop on differential geometry and continuum mechanics, vol 137, Springer, Edinburgh, pp 125–168

Acharya A, Roy A (2006) Size effects and idealized dislocation microstructure at small scales: predictions of a Phenomenological model of Mesoscopic Field Dislocation Mechanics: Part I. J Mech Phys Solids 54:1687–1710

Anthony KH (1970) Die Theorie der Disklinationen. Arch Rat Mech Anal 39:43–88

Berbenni S, Taupin V, Djaka KS, Fressengeas C (2014) A numerical spectral approach for solving elasto-static field dislocation and g-disclination mechanics. Int J Solids Struct. 51:4157–4175

Bilby BA, Bullough R, Smith E (1955) Continuous distributions of dislocations: a new application of the methods of non-Riemannian geometry. Proc Roy Soc Lond A 231:263–273

Brenner R, Beaudoin AJ, Suquet P, Acharya A (2014) Numerical implementation of static field dislocation mechanics theory for periodic media. Philos Mag 94:1764–1787

deWit R (1970) Linear theory of static disclinations. In: Simmons JA, de Wit R, Bullough R (eds) Fundamental aspects of dislocation theory, NBS Spec Publ 317, vol 1, National Bureau of Standards, Washington, pp 651–680

deWit R (1973) Theory of disclinations: IV. Straight disclinations. J Res Nat Bur Stand, Phys Chem 77A(5):607–658

Dreyer W, Müller WH, Olschewski J (1999) An approximate analytical 2D-solution for the stresses and strains in eigenstrained cubic materials. Acta Mech 136(3–4):171–192

Eringen AC (2002) Non local continuum field theories. Springer, New York

Eyre DJ, Milton GW (1999) A fast numerical scheme for computing the response of composites using grid refinement. Eur Phys J Appl Phys 6:41–47

Forest S (2006) Milieux continus généralisés et matériaux hétérogènes. Presses de l'Ecole des Mines, Paris

Fressengeas C, Taupin V, Capolungo L (2011) An elasto-plastic theory of dislocation and disclination fields. Int J Solids Struct 48:3499–3509

Fressengeas C, Taupin V, Capolungo L (2014) Continuous modeling of the structure of symmetric tilt boundaries. Int J Solids Struct 51(6):1434–1441

Frigo M, Johnson SG (2005) The design and implementation of FFTW3. Proc IEEE 93(2):216–231

Gertsman VY, Nazarov AA, Romanov AE, Valiev RZ, Vladimirov VI (1989) Disclination-structural unit model of grain boundaries. Philos Mag A 59(5):1113–1118

Gourgiotis PA, Georgiadis HG (2008) An approach based on distributed dislocations and disclinations for crack problems in couple-stress theory. Int J Solids Struct 45:5521–5539

Hirth JP, Lothe J (1982) Theory of dislocations, 2nd edn. Wiley, New York

Jiang B (1998) Theory and applications in computational fluid dynamics and electromagnetics. The least-squares finite element method. Scientific Computation, Springer, Berlin

Kassbohm S (2006) Fourierreihen zur Berechnung repräsentativer Volumenelemente mit Mikrostruktur. Ph.D. thesis, Fakultät V der TU Berlin, Germany
Kassbohm S, Müller WH, Silber G, Fessler R (2006) Fourier series for continua with microstructure. PAMM—Proc Appl Math Mech 6:487–488
Koiter WT (1964) Couple stresses in the theory of elasticity, I and II. Proceedings of the Koninklijke Nederlandse Akademie van Wetenschappen, Series B 67:17–44
Kosevich AM (1979) Crystal dislocations and the theory of elasticity (Chap 1). In: Nabarro FRN (ed) Dislocations in solids, vol 1. Amsterdam, North-Holland, pp 33–141
Kröner E (1958) Kontinuumstheorie der Versetzungen und Eigenspannungen. In: Collatz L, Loesch F (eds) Ergebnisse der Angewewandte Mathematik 5. Springer, Berlin
Kröner E (1963) On the physical reality of torque stresses in continuum mechanics. Int J Eng Sci 1:261–278
Kröner E (1968) Mechanics of generalized media. In: Kröner E (ed) Proceedings of the IUTAM symposium on the generalized cosserat continuum and the continuum theory of dislocations with applications. Springer, Berlin
Kröner E (1981) Continuum theory of defects. In: Balian R et al. (eds) Physics of defects, Les Houches, Session 35. North Holland, New York, pp 215–315
Lebensohn RA (2001) N-site modeling of a 3D viscoplastic polycrystal using fast fourier transform. Acta Mater 49:2723–2737
Li JCM (1972) Disclination model of high angle grain boundaries. Surf Sci 31:12–26
Lubarda VA (2003) The effects of couple stresses on dislocation strain energy. Int J Solids Struct 40:3807–3826
Michel JC, Moulinec H, Suquet P (2001) A computational scheme for linear and non-linear composites with arbitrary phase contrast. Int J Num Method Eng 52:139–160
Mindlin RD, Tiersten HF (1962) Effects of couple-stresses in linear elasticity. Arch Rat Mech Anal 11:415–488
Moulinec H, Suquet P (1994) A fast numerical method for computing the linear and non linear properties of composites. C R Acad Sci Paris II 318:1417–1423
Moulinec H, Suquet P (1998) A numerical method for computing the overall response of nonlinear composites with complex microstructure. Comput Method Appl Mech Eng 157:69–94
Müller WH (1996) Mathematical versus experimental stress analysis of inhomogeneities in solids. J Phys IV 6(C1):139–148
Mura T (1963) Continuous distribution of moving dislocations. Philos Mag 89:843–857
Nazarov AA, Shenderova OA, Brenner DW (2000) On the disclination-structural unit model of grain boundaries. Mater Sci Eng A 281(1):148–155
Neff P, Jeong J, Ramidreza H (2009) Subgrid interaction and micro-randomness—Novel invariance requirements in infinitesimal gradient elasticity. Int J Solids Struct 46:4261–4276
Neumann S, Herrmann KP, Müller WH (2002) Stress/strain computation in heterogeneous bodies with discrete fourier transforms—different approaches. Comput Mater Sci 25:151–158
Nowacki W (1986) Theory of asymmetric elasticity. Pergamon Press, Oxford
Nye JF (1953) Some geometrical relations in dislocated crystals. Acta Metall 1:153–162
Prakash A, Lebensohn RA (2012) Simulation of micro mechanical behavior of polycrystals: Finite Elements versus Fast Fourier Transforms. Model Simul Mater Sci Eng 17:64,010–64,016
Press WH, Teukolsky SA, Vetterling WT, Flannery BP (2002) Numerical recipes in C++. The art of scientific computing, 2nd edn. Cambridge University Press, USA
Romanov AE, Vladimirov VI (1992) Disclinations in crystalline solids. In: Nabarro FRN (ed) Dislocations in solids, vol 9. North-Holland, Amsterdam, pp 191–402
Roy A, Acharya A (2005) Finite element approximation of field dislocation mechanics. J Mech Phys Solids 53:143–170
Smyshlyaev VP, Fleck NA (1994) Bounds and estimates for linear composites with strain gradient effects. J Mech Phys Solids 42:1851–1882
Taupin V, Capolungo L, Fressengeas C, Das A, Upadhyay M (2013) Grain boundary modeling using an elasto-plastic theory of dislocation and disclination fields. J Mech Phys Solids 61:370–384

Upadhyay MV (2014) On the role of defect incompatibilities on mechanical properties of polycrystalline aggregates: a multi-scale study. Ph.D. thesis, School of Mechanical Engineering, Georgia Institute of Technology, Altanta, USA
Upadhyay MV, Capolungo L, Taupin V, Fressengeas C (2011) Grain boundary and triple junction energies in crystalline media: a disclination based approach. Int J Solids Struct 48(22):3176–3193
Upadhyay MV, Capolungo L, Taupin V, Fressengeas C (2013) Elastic constitutive laws for incompatible crystalline media: the contributions of dislocations, disclinations and g-disclinations. Philos Mag 93:794–832
Vinogradov V, Milton GW (2008) An accelerated FFT algorithm for thermoelastic and non-linear composites. Int J Num Meth Eng 76:1678–1695
Volterra S (1907) Sur l'équilibre des corps élastiques multiplement connexes. Ann Sci Ecol Norm Sup III 24:401–517
Willis JR (1967) Second-order effects of dislocations in anisotropic crystals. Int J Eng Sci 5:171–190
Zheng QS, Zhao Z (2004) Green's function and Eshelby's fields in couple-stress elasticity. Int J Multiscale Comput Eng 2:15–27

Some Cases of Unrecognized Transmission of Scientific Knowledge: From Antiquity to Gabrio Piola's Peridynamics and Generalized Continuum Theories

Francesco dell'Isola, Alessandro Della Corte, Raffaele Esposito and Lucio Russo

"Pluralitas non est ponenda sine necessitate"
Plurality is not to be posited without necessity

(Duns Scotus)

Abstract The aim of this paper is to show some typical mechanisms in the transmission of scientific knowledge through the study of some examples. We will start by considering some ancient examples concerning Democritus, Heron, Galileo and the history of the theory of tides. Then we will mainly focus on the works of the Italian scientist Gabrio Piola (1794–1850). In particular: 1. we show clear similarities between Noll's postulation of mechanics and the 'ancient' presentation by Piola of the ideas needed to found Analytical Continuum Mechanics; 2. we prove that non-local and higher gradient continuum mechanics were conceived (and clearly formulated) already in Piola's works; 3. we explain the reasons of the unfortunate circumstances which caused the (temporary) erasure of the memory of many among

F. dell'Isola (✉)
Department of Structural and Gentechnical Engineering, University La Sapienza of Rome, Rome, Italy
e-mail: francesco.dellisola@uniroma1.it

F. dell'Isola · R. Esposito
International Research Center on Mathematics and Mechanics of Complex Systems (M&MoCS), University of L'Aquila, L'Aquila, Italy

R. Esposito
e-mail: raff.esposito@gmail.com

A. Della Corte
Department of Mechanical and Aerospace Engineering, University La Sapienza of Rome, Rome, Italy
e-mail: alessandro.dellacorte@uniroma1.it

L. Russo
Department of Mathematics, University Tor Vergata of Rome, Rome, Italy
e-mail: russo@mat.uniroma2.it

H. Altenbach and S. Forest (eds.), *Generalized Continua as Models for Classical and Advanced Materials*, Advanced Structured Materials 42,
DOI 10.1007/978-3-319-31721-2_5

Piola's contributions to mechanical sciences. Moreover, we discuss how the theory which has recently been called peridynamics, i.e. a mechanical theory which assumes that the force applied on a material particle of a continuum depends on the deformation state of a neighbourhood of the particle, was first formulated in Piola's works. In this way we argue that in the passage from one a cultural tradition to another the content of scientific texts may often be lost, and it is possible to find more recent sources which are scientifically more primitive than some more ancient ones.

Keywords Transmission of scientific knowledge · Principle of virtual work · Generalized continua · Peridynamics

1 Introduction

Recently the role of the ancient Hellenistic koiné (cultural and linguistic community speaking a *lingua franca* derived by Greek dialects) in the process leading to the modern illuminists scientific description of nature and consequent technology development has been re-examined in Russo (2004). The thesis presented in that work has sometimes been considered controversial. Indeed, it is there shown that Hellenistic science, and in particular Hellenistic mechanics, was much more developed, general, rigorous and technology-oriented than what is often believed. The opponents of this vision base their criticism on a series of (often unconscious) prejudices, such as:

1. every text or theory or body of doctrines which is more recent than another one is necessarily also more sophisticated and advanced;
2. when a modern scholar discovers in an ancient text some theories and mathematical theorems which are more advanced than those found in subsequent texts, then this scholar is 'forcing' a non-existing-in-reality intelligence into primitive sources, so distorting their meaning with his 'modernistic' lenses;
3. scientific and in particular mathematical knowledge cannot be lost, and increases in quality and scope as time is passing.

This vision does not take into account many phenomena that actually do occur in the transmission of scientific knowledge. In particular, it does not account for (re-) elaboration, (mis)understanding, biased selection and (in)voluntary neglect of scientific sources. Indeed:

- Scientific knowledge is difficult to transmit and to learn: only after years of study a young apprentice may start to understand the true content of more and more sophisticated theories. It sometimes happens that the elaboration and re-elaboration of precedent texts by subsequent scholars produces texts whose quality is worse than that of their sources simply because for any reason (decadence of scientific tradition, massive emigration of scholars, or lower interest by leading classes in funding scientific research) the successors are not able to understand their predecessors.

- When a scientific tradition is interrupted, the capability of understanding scientific treatises becomes impaired because of the nature itself of the mathematical reasoning, which is based on linguistic conventionalism and on the mastering of technical capabilities. As a consequence, sometimes in few generations very complicated and sophisticated theories are transformed into very naive or incomprehensible ones. The informative content of scientific theories can be in this way lost, or at least can become blurred.
- When successors cannot understand what written by their predecessors, they generally start to operate a **biased selection and involuntary or voluntary neglect** of the sources which they use.

The effects of these phenomena on the progress of knowledge cannot be underestimated. Indeed, many results seem to be rediscovered periodically and to be lost with the same periodicity, and very often the quality of re-discoveries is worse than that of the primary sources. It is very often impossible to determine the true scientific context where one novel method, theory and technique was first elaborated and very often the 'epic' vision of advancement of science prevails: indeed it is often believed that single discoverers were able to invent enormous bodies of doctrine, while they were simply elaborating results they were reading in their sources. Finally, the role of education institutions in forming creative scientists by transmitting the most advanced knowledge in a given field becomes more difficult when primary sources become blurred because of the described mechanisms.

2 Some Ancient Examples of Not Recognized Transmission of Knowledge

2.1 Galileo and Heron

After the development of rigorous philological methods in the middle of the XIX century, and the subsequent flourishing of critical editions, the study of the transmission of written knowledge has been based on solid evidence provided by a wide documentary basis. While this certainly entails great advantages in terms of soundness and consistence of produced results, it may lead to an excessive sternness in the interpretation of cases of logically and historically plausible cultural inheritance. In this way, indeed, one may be led to give up any investigation which lacks direct material evidence in the available sources. One case of this kind is the problem of the diffusion of Heron's Mechanics in the Modern Era. Heron's Mechanics (Reference editions are Heron 1988, 1899, 1900; de Vaux 1893) is generally believed to have been written in the first century (AD), even though there is no absolute agreement about the dating of the activity period of its author. Up to the end of the 19th century, the only parts of the Mechanics whose transmission is directly documented in the sources are:

- A discussion about the duplication of the cube reported in the book III of the Collection of Pappus (around 300 AD) and in the comment by Eutocius to the second book of Archimedes' *On the Sphere and Cylinder* (4th century AD).
- Some excerpts and summaries provided by Pappus (possibly interpolations, see Heron 1899, 1900, pp. 224–226) in the book VIII of the Collection about various mechanical arguments, among which there are simple machines, gears and centers of gravity (all Greek fragments are reported in Heron 1899, 1900, pp. 255–300).

Among Latin authors, other passages related to the content of Heron's *Mechanics* are generally thought to be found in Pliny, Cato and Vitruvius (see Heron 1899, 1900, pp. 374–393). The passages by Pappus were published in 1588 and were surely read, among others, by Galileo (see Galilei 1890–1909, reprinted 1964–1966, vol. II, p. 181). Only in 1893, Carra de Vaux published (and translated into French) an Arabic code of the *Mechanics* which he found in Leyda (de Vaux 1893), and in 1900 another edition appeared, based on the previous one and three new Arabic codes (Heron 1899, 1900). Summarizing, we have no direct evidence of the fact that Galileo knew parts of the Mechanics different from the ones he read in Pappus. However, following a remark proposed in Russo (2004) and some arguments provided in Voicu (1999), we propose here an analysis of two passages from which strong arguments can be made in favor of this conjecture. The passages are taken from Galileo's work *Le mecaniche*, which was published in 1629, but was most probably written several years before (see Carugo and Crombie 1983; Drake 2003, for a discussion), and treats several mechanical topics, from the balance and the simple machines to motion. In the following, we give in **bold** the translation of Galileo's words, while the original text is given in the footnotes.

1. Concerning the equilibrium configuration of a balance, Galileo exposes the need to measure the distances horizontally (Galilei 1890–1909, reprinted 1964–1966, vol. II, pp. 164–165, the English translation from the Italian text are by the authors):
   Another thing, before going ahead, should be considered; and it is about distances at which weights should be suspended: because it is very important to know how to figure out whether the distances are equal or not, and thus in which way one has to measure them. […] And finally one has to take care to measure the distances with lines which are perpendicular to the ones along which the weights hang down and would move if they were free to descend.[1]
   About the same subject, Heron wrote (Galilei 1890–1909, reprinted 1964–1966, pp. 109–110, the English translation from the French text are by the authors):
   […] and Archimedes has proven that, also in this case, the ratio between the two weights equals the inverse ratio of the respective distances. What those distances are in case of irregular and sloped beam, one can imagine

[1] *Un'altra cosa, prima che più oltre si proceda, bisogna che sia considerata; e questa è intorno alle distanze, nelle quali i gravi vengono appesi: per ciò che molto importa il sapere come s'intendano distanze eguali e diseguali, ed in somma in qual maniera devono misurarsi. […] E finalmente si deve aver avvertenza di misurare le distanze con linee, che ad angoli retti caschino sopra quelle nelle quali i gravi stanno pendenti, e si moveriano quando liberamente scendessero.*

considering a chord descending from the point G towards the point Z. Let us consider a line originating from the point Z which is the line HZQ; it should then be chosen in such a way to intersect the chord forming right angles.

It is worth noting, besides the fact that the two scientists treat the same question, the similarity in the way the question is posed, as both Heron and Galileo, rather than directly formulate the law corresponding to the general case, prefer to modify the notion of distance so as to obtain the law 'even' in case of an oblique balance. Moreover, they both use the result in the subsequent reasoning made to reduce other machines to the balance (in particular, Galileo exploits it when considering balances with mobile arms centered on a fixed point). An example of the previously mentioned sternness among the scholars can be observed, in our opinion, by Clagett, who about this passage by Galileo conjectured (Clagett 1959) that he could have been influenced by the tradition headed by the *Liber de ratione ponderis*, which was published by Tartaglia in 1565, denying the possibility of an influence by Heron **because of** the lack of direct evidence. In the authors' opinion the probability that such a complex problem could be solved independently in exactly the same way is totally negligible.

2. Galileo states on various occasions that it is equivalent to balance a weight and to lift it, because the additional force needed in the second case can be as small as one wants. For example, he writes (Galilei 1890–1909, reprinted 1964–1966, vol. II, p. 164):

   To move down the weight B, any minimal increased graveness is sufficient, and therefore, ignoring this imperceptible difference, we will not consider different for a weight to be able to balance another one or to move it.[2]

   Heron uses the same concept (Galilei 1890–1909, reprinted 1964–1966, vol. II):

   When we want to lift a weight, we need a force which equals it. [...] Thus when the weight receives an increasing however small, the other weight is led upward.

   Moreover, concerning a body in motion over an inclined plane, Galileo writes (Galilei 1890–1909, reprinted 1964–1966, vol. II, p. 183):

   It is sufficient that the force which has to move the weight imperceptibly exceeds the one which sustains it.[3]

   On the same subject, Heron wrote (Galilei 1890–1909, reprinted 1964–1966, p. 92):

   One thus needs a power to balance a weight and, when one adds to this power the smallest excess, it will prevail over the weight.

Both Heron and Galileo use arguments based on inclined planes with decreasing slopes to study the motion of a particle over a horizontal plane. In their reasoning, both authors pay attention to the practical problems caused by friction, and both use the example of descending water, which is set in motion by any slope (however

[2]*Per fare descendere il peso B, ogni minima gravità accresciutagli è bastante, però, non tenendo noi conto di questo insensibile, non faremo differenza dal potere un peso sostenere un altro al poterlo movere.*

[3]*La forza per muover il peso basta che insensibilmente superi quella che lo sostiene.*

slight), to argue that the reason for which the same conclusion does not hold for solid bodies is connected to (sliding) friction. Moreover, they both conclude stating a germinal form of what will be called the first law of dynamics. Galileo indeed writes (Galilei 1890–1909, reprinted 1964–1966, vol II, p. 180):
From this we can assume, as an undoubted axiom, this conclusion: that heavy bodies, once removed all the external and occasional obstacles, can be set in motion over a horizontal plane by a force which is however small.[4]
On the other hand, Heron writes (Galilei 1890–1909, reprinted 1964–1966, p. 89):
Thus the weight [on a horizontal plan] **is moved by any force, however small it is.**

Let us now summarize our findings. The similarities between the two texts in this second example concern the following features:

1. the key conclusions;
2. the approximating method to reach them;
3. the way in which the statements are formulated (especially noticeable because of the distance between the languages in which the works were originally written), and in particular the frequent reference to a quantity that is 'however small';
4. the problems connected with the applicability to reality of the mathematical model considered by the authors.
5. The example chosen to persuade the reader of the 'occasional' character of the observable exceptions to the last quoted statements.

As said before, all these features strongly support the idea that Galileo knew parts of Heron's *Mechanics* whose transmission at his age is not directly documented by philological facts.

Of note, a few historically sensible considerations are possible about the plausibility of this conclusion. A conjecture, proposed in Russo (2004) (p. 353, footnote 84), concerns the content of the mechanics courses thought by Cristoph Clavius at the Collegio Romano in 1579 (or 1580). In the notes by Clavius (probably concerning his lectures in mechanics) indeed, one can find 'mechanical questions of Heron, Pappus and Aristotle' (see Baldini 1992, p.175). Since Pappus is explicitly mentioned, one can conjecture that the reference to Heron did not mean the passages of Heron included in Pappus. Carra de Vaux (Galilei 1890–1909, reprinted 1964–1966, pp. 25–27) found that in the catalogs of several libraries (in particular located in Rome) a few manuscripts are mentioned which contained Heron's *Mechanics*. While the codex in Venice simply resulted a copy of Heron's *Pneumatica* with the wrong title, no further progress has ever been made about the identification of the other cited codices, whose tracks have been lost.

Let us briefly return to the opinion expressed by Clagett about the passage on inclined balances previously cited. His way of proceeding, which could seem just sensibly guided by prudence, implies the very strong and unjustified assumption that all the manuscripts which were available in Galileo's times are still accessible

[4] *Dal che possiamo prendere, come per assioma indubitato, questa conclusione: che i corpi gravi, rimossi tutti l'impedimenti esterni ed adventizii, possono esser mossi nel piano dell'orizonte da qualunque minima forza.*

today. This assumption is based on the understandable but rather shortsighted and exclusive preference towards 'material' proofs over arguments based on chains of logical connections. This procedure has the paradoxical consequence that in particular cases (i.e. when indeed strong arguments based on the content and on historical considerations are possible) one may be led to consider what is the most unlikely possibility as the 'sounder' one. In our opinion, a change of paradigm in this kind of questions is by now simply unavoidable.

2.2 Galileo and Democritus

The tendency to disregard the importance of the transmission of scientific knowledge over the centuries is observable even when the transmission itself is fully documented. To stay within the universe of Galileo, let us consider now the following passage from The *Assayer* (Chap. 48):

To excite in us tastes, odors, and sounds I believe that nothing is required in external bodies except magnitudes, shapes, quantities, and slow or rapid movements. I think that if ears, tongues, and noses were removed, shapes and quantities and motions would remain, but not odors or tastes or sounds. The latter, I believe, are nothing more than names when separated from living beings.[5]

Because of passages like this last, many scholars attributed to Galileo the distinction among primary and secondary qualities which will be very important for the subsequent history of science and philosophy, while others recognized that the origin of this idea was much more ancient, dating back to Democritus, who is actually often presented as a 'precursor'. Both the strict dependence of Galileo's ideas from the sources about Democritus and the way in which the transmission took place are generally ignored.

About the first point, we can observe that even if in many ancient sources (posterior to Democritus) there is the idea that warmth is caused by the velocity of atomic motion (an idea which will be recovered, among others, by Boyle in the 17th century), Galileo in this context was stuck to the old Democritus' idea of 'atoms of fire', which he calls *ignicoli*. As for the second point, we may recall that Galileo started his academic studies in the faculty of Medicine at the University of Pisa, where among the textbooks at use there was Galenus' treatise *De elementis secundum Hippocratem*, in which in one of the first pages one could read:

"Conventional is color, conventional is what is sweet or bitter, while true are the atoms and the void", states Democritus, considering all the sensitive appearances which we can perceive as originating from the encounter between the atoms, since all of these qualities are imagined by us, while he does not believe

---

[5]*Ma che ne' corpi esterni, per eccitare in noi i sapori, gli odori e i suoni, si richiegga altro che grandezze, figure, moltitudini e movimenti tardi o veloci, io non lo credo; e stimo che, tolti via gli orecchi le lingue e i nasi, restino bene le figure, i numeri e i moti, ma non già gli odori né i sapori né i suoni, li quali fuor dell'animal vivente non credo che sieno altro che nomi.*

that, in nature, white or black or yellow or red or bitter or sweet exist.[6] The transmission of ideas between Democritus and Galileo was not, thus, vague or indirect as one may think reading S. Drake, who in Drake (2003) refers to Lucretius as a possible intermediary between Democritus and Galileo. It was based, instead, on a direct use of easily accessible sources.

2.3 The Transmission of the Scientific Explanation of Tides

An extremely important case of unrecognized transmission of scientific knowledge is provided (as proven in Russo 2003) by the theory of tides. It is generally believed that Newton was the first one to scientifically explain tides in his *Philosophiae Naturalis Principia Mathematica* (1687). The theory by Newton is indeed a successful synthesis of three ideas:

1. the tide's cycles (daily, monthly and annual) can be explained by combining the actions of the sun and the moon, each one of which entails a lift of the water towards the luminary and in the opposite direction;
2. the lifting of the water can be explained as the combined effect of gravity and centrifugal force;
3. the application of 2. for the explanation of the tides.

Concerning 1., it is extremely probable that Newton took the idea from the work *Euripus, sive de fluxu et refluxu maris sententiae* (1624) by the Archbishop Marcantonio De Dominis, where the aforementioned explanation for tides is clearly exposed. Indeed, De Dominis taught in Cambridge, and Newton, in his *Opticks*, in quoting his theory about the rainbow, cites him as "the famous Archbishop De Dominis".

De Dominis' theory of tides, in turn, was not new at all. We can indeed follow backwards its footsteps, in a series of works by authors related to the University of Padua (the main of them being Jacopo Dondi and Federico Chrisogono), up to the beginning of the 14th Century. The theory was actually much more ancient, being exposed by Posidonius (1st century BC) in his lost work on tides, as we can reconstruct through the testimonies by Strabo, Pliny the Elder and mainly the Byzantine author Priscianus Lidius (6th Century AD). Most probably, therefore, the idea was transmitted from Constantinople to the Venetian State, which used to monopolize the relationship between the Byzantine Empire and the Western world.

As for 2., the idea of the equilibrium between gravity and centrifugal force is clearly explained by Plutarch in his *De facie quae in orbe Lunae apparet* referring

---

6 Νόμῳ γὰρ χροιὴ νόμῳ γλυκὺ νόμῳ πικρόν, ἐτεῇ δ' ἄτομα καὶ κενόν ὁ Δημόκριτός φησιν ἐκ τῆς συνόδου τῶν ἀτόμων γίγνεσθαι νομίζων ἁπάσας τὰς αἰσθητὰς ποιότητας ὡς πρὸς ἡμᾶς τοὺς αἰσθανομένους αὐτῶν, φύσει δ' οὐδὲν εἶναι λευκὸν ἢ μέλαν ἢ ξανθὸν ἢ ἐρυθρὸν ἢ γλυκὺ ἢ πικρόν (Galenus, De elementis secundum Hippocratem, ed. Kuhn, 417, 9-14).

to the motion of the moon around the Earth. It is also mentioned by Seneca (in the seventh book of his *Naturales Quaestiones*) in connection with the motion of the planets around the sun, and was recovered in modern times by Giovanni Alfonso Borelli in his work *Theoricae mediceorum planetarum ex causis physicis deductae* (1666).

Finally, concerning 3., the idea of using the equilibrium between gravity and centrifugal force to explain the lunar tides had become natural after the *Essay* about tides by John Wallis (1666), in which the idea of a monthly motion of the Earth around the barycenter of the system Earth-moon was introduced. Wallis, in turn, elaborated his theory modifying previous ideas of Paolo Sarpi, Galileo Galilei and Giovanni Battista Baliani, who tried to explain the tides as a consequence of the motion of the Earth. Wallis, indeed, in his *Essay* directly cites both Galileo and Baliani. As in the previous case 1., in Russo (2003), the origin of this last idea is recognized to be a very ancient one, dating back to the work of Seleucus of Babylon (2nd Century BC), who was probably among the sources of the aforementioned work by Posidonius.

It is important to notice that the transmission we are considering was mostly an unconscious one. When Galileo and De Dominis were disputing, they had no idea of the depth of the roots they were following, but still contributed to their recovery in modern science. The fact that the theory of tides is generally attributed to Newton alone (even if all the mentioned sources have always been available!) is certainly linked to this unconscious character of the transmission, and provides an example of a general tendency in the history of science: that of attributing to few "geniuses" results which were actually obtained thanks to the efforts of many scientists from different ages. This feature links this example to the previous ones and to the following.

3 Pristine Formulations of the Principle of Virtual Powers (or Work) as a Basic Postulate for Mechanics

The Principle of Virtual Work (PVW) is one of the most important conceptual tools in mechanics and, generally, in physics. The fact that its correct formulation for continuum mechanics has been erased from the awareness of the majority of scholars (and only subsequently rediscovered) deserves to be considered carefully. In this work we do not want to establish the detailed and historically correct discovery process which led to the formulation of the PVW. What we try here is rather to fix a 'stronghold': actually we want to determine a precise moment and some well-determined authors since when a 'complete' formulation of the PVW has to be considered well-established as the fundamental postulate of (Continuum) Mechanics. We will refrain from delving into complex scholarly studies about absolute historical priority, as we do not aim to find the first certain occurrence—in mechanics textbooks—of an exact and sufficiently complete version of the PVW. To cite simply one among the most careful studies, already in the work of Vailati (1987) it is attempted a first modern reconstruction of some mechanics text authored by Greek scientists (among which the pseudo-Aristotle, Archimedes and Heron of Alexandria) which are dealing with several problems involving the use of the PVW. The thesis of Vailati is in line with what is claimed in Russo (2003, 2004). In the text Mechanical Problems belonging to the Aristotelian corpus and attributed by Winter (2007) to Archytas of Tarentum, one can find a first formulation[7] of the PVW. Moreover in some text of Heron of Alexandria this principle is extensively used. It is still debated if Archimedes studied the equilibrium of the lever having in mind a form of the PVW (see e.g. Paipetis and Ceccarelli 2010). As said, however, we do not want here to be distracted by controversial issues. It is sufficient for our aims to establish that already in the celebrated textbooks by D'Alembert (*Traité de Dynamique* 1758) and by Lagrange (*Méchanique Analytique*, 1788) this principle is systematically used in order to deduce all other laws of Mechanics. In particular, we will focus on the version of this principle applied by Lagrange in fluid dynamics.

3.1 The Traité de Dynamique by D'Alembert

Let us start by reading a fragment of the *Traité de Dynamique* (1758) by D'Alembert which we translate in English (in **bold**) nearly word by word. The passage could indeed be very useful to provide a sort of methodological introduction to the technical content of the Mechanics in the view of D'Alembert. The Principle which is in the mind of the author, as clearly stated in the rest of the text (as it is also recognized by Lagrange 1788) is the Principle of Virtual Velocities (the name given to the PVW by D'Alembert and Lagrange).

The certainty of mathematics is an advantage which these sciences owe to the simplicity of their object. [...] the most abstract notions, those which the layman regards as the most inaccessible, are often those which carry with them the greatest light: [...] in order to treat following the best possible method [...] any Science whatsoever it is necessary [...] to imagine, in the most abstract and simple way possible,the particular object of this Science, (it is necessary) to suppose and admit in this subject anything else, than the properties which this same Science treats and supposes. From this standing two advantages result: the principles receive all clarity to which they are susceptible: (and these principles) are finally reduced to the smallest number possible [...] as the object of a Science is necessarily determined, the principles will be more fecunds if they will be less numerous [...].[8]

[7] See Aristotle's Mechanics 3, 850 a-b as translated on p. 431 by Thomas (1939).

[8] *La certitude des Mathématiques est un avantage que ces Sciences doivent principalement à la simplicité de leur objet. [...] les notions les plus abstraites, cellesque le commun des hommesregarde comme les plus inaccessibles, sontsouventcelles qui portent avec elles une plus grande lumiere: [...] pour traiter suivant la meilleure Méthode possible [...] quelque Science que cepuisse être il est nécessaire [...] d'envisager, de la maniere la plus abstraite et la plus simple qu'il se puisse, l'objet particulier de cette Science; de ne rien supposer, ne rien admettre dans cet objet, que les propriétés que la Science même qu'on traite y suppose. Delà résultent deux avantages: les principes reçoivent*

In the following, D'Alembert refers more specifically to Mechanics, claiming its special need, among all exact sciences, for a clear and solid foundation:
Mechanics, above all, seems to be (the Science) which has been more neglected from this point of view: also the great majority of his principles either obscure by them-selves, or enunciated and demonstrated in an obscure way have given place to several spiny problems [...] I proposed to my-self to move back the limits of Mechanics and to make its approach easier, (I proposed to my-self) not only to deduce the principles of Mechanics from the most clear notions, but also to apply them to new uses, to make it clear at the same time both the inutility of the many and various principles which have been used up to now in Mechanics and the advantage which can be drawn by the combination of others (principles) in order to have the progress of this Science in one word (I want to make clear which is the advantage) of extending the principles by reducing them.[9]

We will not try here to choose some excerptions from the work of D'Alembert to present his vision about the range of applicability of the Principle of Virtual Velocities, as he uses there notations and a language which could lead to some controversies about their interpretation. Instead we will present in great detail the point of view of Lagrange, who openly and frequently credits D'Alembert for his fundamental contributions in the correct and more comprehensive formulation of the Principle of Virtual Velocities.

Here we simply want to recall that at the beginning of the *Traité de Dynamique* we find the following (very impressive) statements:

1. **I have proscribed completely the forces relative to the bodies in motion, entities obscure and metaphysical, which are capable only to throw darkness on a Science which is clear by itself.**
2. **I must warn [the reader] that in order to avoid circumlocutions, I have used often the obscure term 'force', & some other terms which are used commonly when treating the motion of bodies; but I never wanted to attach to this term any other idea different from those which are resulting from the Principles**

(Footnote 8 continued)
toute la clarté dont ils sont susceptibles: ils se trouvent d'ailleurs réduits au plus petit nombre possible [...] puisque l'objet d'une Science étant nécessairement déterminé, les principes en sont d'autant plus féconds, qu'ils sont en plus petit nombre.

[9] *La Méchanique surtout, est celle qu'il paroit qu'on a négligée le plus à cet égard: aussi la plûpart de ses principes, ou obscurs par eux-mêmes, ou énoncés et démontrés d'une maniere obscure, ont-ils donné lieu à plusieurs questions épineuses. [...] Je me suisproposé [...] de reculer les limites de la Méchanique et d'en applanir l'abord [...] non seulement de déduire les principes de la Méchanique des notions les plus claires, mais de les appliquer aussi à de nouveaux usages; de faire voir tout à la fois, et l'inutilité de plusieurs principes qu'on avoit employés jusqu'ici dans la Méchanique et l'avantage qu'on peut tirer de la combinaison des autres pour le progrès de cette Science; en un mot, d'étendre les principes en les réduisant.*

which I have established both in the Preface and in the first part of this treatise.[10]

3.2 The Treatise Méchanique Analytique by Lagrange

For our aims it is sufficient to read just some well-chosen parts of Lagrange's *Méchanique Analytique* (1788) (Lagrange 1788). Lagrange presentation is very elegant, precise and rigorous: every scholar interested in mechanics will read it with great pleasure, as even nowadays it is an exciting and fruitful experience. As Lagrange's textbook is easily available, because of its recent reprinting, we often present in what follows, only our English translations of some chosen excerptions, indicating the pages from Lagrange textbook from which they are taken. Words in **bold** are the translation of Lagrange's original French text. Our comments are in *italic*, while some relevant excerpts from the original text are in the footnotes.

From p. 1:

One uses in general the word 'force' or 'power' *[puissance]* **for denoting the cause, whatever it will be, which is impressing or tends to impress motion to the bodies to which it is assumed to be applied.**

The reader is warned: Lagrange uses the word force as a synonym of the word power. This circumstance, carefully discussed by Lagrange and based on a choice of nomenclature intended to parallel the nomenclature previously introduced by Galileo, has been misleading for many scholars who seem to believe that Lagrange was not able to distinguish between the concept of force and our concept of power. Actually Lagrange uses the word 'moment' for meaning (using modern nomenclature)

---

[10]The complete original passage reads indeed:

PREFACE

A l'égard des démonstrations de ces Principes en eux-mêmes, le plan que j'ai suivi pour leur donner toute la clarté & la simplicité dont elles m'ont paru susceptibles, a été de les déduire toujours de la considération seule du Mouvement, envisagé de la maniére la plus simple & la plus claire. Tout ce que nous voyons bien distinctement dans le Mouvement d'un Corps, c'est qu'il parcourt un certain espace, & qu'il employe un certain tems à le parcourir. C'est donc de cette feule idée qu'on doit tirer tous les Principes de la Méchanique, quand on veut les démonstrer d'une maniére nette & précise ; ainsi on ne fera point surpris qu'en conséquence de cette réflexion, j'ai, pour ainsi dire, détourné la vûe de dessus les causes motrices, pour n'envisager uniquement que le Mouvement qu'elles produisent; que j'aie entiérement proscrit les forces inhérentes au Corps en Mouvement, être obscurs & Métaphysiques, qui ne font capables que de répandre les ténèbres sur une Science claire par elle-même. [...]
Au reste, comme cette feconde Partie est destinée principalement à ceux, qui déja instruits du calcul différentiel & intégral, le seront rendus familiers les Principes établis dans la premiére, ou seront déja exercés à la solution des Problêmes connus & ordinaires de la Méchanique; je dois avertir que pour éviter les circonlocutions, je me suis souvent servi du terme obscur de force, & de quelques autres qu'on employe communément quand on traite du Mouvement des Corps; mais je n'ai jamais prétendu attacher à ces termes d'autres idées que celles qui résultent des Principes que j'ai établis, soit dans cette Préface, soit dans la premiére Partie de ce Traité.

'power'. It is asthonishing that some modern mathematicians -who were educated to the most formal nominalism ever developed in the history of science- could not follow Lagrange in his use of his own nominalistic choice. Indeed:

From p. 8:

Galileo uses the word 'moment' of a weight or a power applied to a machine the effort, the action, the energy, the 'impetus' of this power for moving this machine [...] and he proves that the moment is always proportional to the power times the virtual velocity, depending on the way in which the power acts.

From p. 9:

Nowadays one uses more commonly the word 'moment' for the product of a power times the distance along its direction to a point or a line, that is the lever arm by which it acts [...], but it seems to me that the notion of moment given by Galileo and Wallis is much more natural and general, and I do not see why it was abandoned for replacing it by another which expresses only the value of the moment in certain cases.[11]

From pp. 10–11:

The Principle of virtual velocities can be formulated in a very general way, as follows:

If a system whatsoever constituted by bodies or points each of which is pulled by powers whatsoever is in equilibrium and if one impresses to this system a small motion whatsoever, in virtue of which every point will cover an infinitesimally small distance which will express its virtual velocity, then it will be equal to zero the sum of the powers each multiplied times the distance covered by the points where it is applied along the line of application of this same power, when considering as positive the small distances covered in the same direction as the power and as negative the distances covered in the opposite direction.

One cannot see in this statement any limit for its applicability: the mechanical system is assumed to be constituted by points and bodies and the powers applied are whatsoever. This Principle is applied by Lagrange also to the equilibrium of continuous systems, as undoubtedly among them there are all incompressible and compressible fuids.

From p. 11:

And in general I believe to be able to state that all general Principles which will be possibly discovered in the science of equilibrium will reduce themselves to a form, differently conceived, of the Principle of Virtual Velocities, from which they will differ simply because of their expression. Moreover this Principle not only is by itself very simple and general, it has also the really precious and unique advantage of being able to be formulated by means of a general formula which includes all problems which can be proposed about the equilibrium of bodies.

It is astonishing how deeply founded this conjecture appears more than two centuries after it was formulated, nothwithstanding the efforts made by some 'modern'

---

[11] *It is interesting that Germain (see e.g.* Germain 1973b*) seems to share the same position as Lagrange in a very similar nominalistic issue.*

mechanicians to find more general Principles. Actually the only successful effort was that of changing the name of the Principle (which is nowadays called the Principle of Virtual Work or Virtual Powers). Someone tried to formulate a nonstandard form for this principle: but actually this 'nonstandard' form[12] *was actually very standard, as it was applied by Lagrange himself some centuries before (see infra the excerption from p. 195).*

From pp. 15–16:

One finally obtains in general for the equilibrium of a number whatsoever of powers P, Q, R etc., directed following the lines $p, q, r,$ & c, and applied to a system whatsoever of bodies or points disposed one respect the others in a generic manner, an equation having this form

$$P\mathrm{d}p + Q\mathrm{d}q + R\mathrm{d}r + \cdots = 0.$$

This is the general formula of the equilibrium of a whatsoever system of powers. We will call each term of this formula, as for instance $P\mathrm{d}p$, the moment of the force P, taking for the word moment the meaning which Galileo gives to it, that is, the product of the force times its virtual velocity. In this way the general formula of equilibrium will consist into the equality to zero of the sum of the moments of all forces.[13]

From p. 16:

In order to use this formula *(i.e. the formula appearing before)* **the difficulty will be reduced to determine, following the nature of considered system, the values of the differentials $\mathrm{d}p, \mathrm{d}q, \mathrm{d}r,$ etc. One will consider therefore the system in two different positions, and infinitesimally close, and he will look for the most general expressions for the differences which are to be considered, by introducing in the expressions as many determined quantities as many arbitrary elements one can distinguish in the variation of the position of the system. One will replace then these expressions of $\mathrm{d}p, \mathrm{d}q, \mathrm{d}r,$ etc. in the proposed equation and it will be required that this equation be varied, independently of all the indetermined variables, so that the equilibrium of the system may in general subsist and in all directions.**

In the following, Lagrange observes that the problem one gets in the way above described is always a well-posed one:

---

[12] *'Nonstandard' is actually the word used by Gurtin himself for this form of the Principle.*

[13] *On a donc en général pour l'équilibre d'un nombre quelconque de puissances $P, Q, R,$ & c, dirigées suivant les lignes $p, q, r,$ & c & appliquées à un système quelconque de corps ou points disposés entr'eux d'une maniere quelconque, une équation de cette forme,*

$$P\mathrm{d}p + Q\mathrm{d}q + R\mathrm{d}r + \&c = 0.$$

C'est la formule générale de l'équilibre d'un systême quelconque de puissances.

Nous nommerons chaque terme de cette formule, tel que $P\mathrm{d}p$, le moment de la force P, en prenant le mot de moment dans le sens que Galilée lui a donné, c'est-à-dire, pour le produit de la force par la vitesse virtuelle. De sorte que la formule générale de l'équilibre consistera dans l'égalité à zero, de la somme des moments de toutes les forces.

One will then equate to zero the sum of the terms influenced by each and the same of the indetermined quantities and he will get, in this way, as many particular equations as many are these indetermined quantities. Now it is not difficult to be persuaded that their number must always be equal to the number of the unknown quantities determining the position of the system; therefore one will have, by means of this method, as many equations as many are necessary for determining the equilibrium state of the system.[14]

Lagrange states now that the Principle of Virtual Velocity includes as a particular case the Principle of Stationary Energy.

From pp. 36–37:

We will now consider the maxima and minima which can occur at equilibrium; and to this aim we recall the general formula

$$P\mathrm{d}p + Q\mathrm{d}q + R\mathrm{d}r + \cdots = 0,$$

stating the equilibrium among the forces $P, Q, R,$ **etc., applied along the lines** $p, q, r,$ **etc. One can assume that these forces could be in such a way that the quantity** $P\mathrm{d}p + Q\mathrm{d}q + R\mathrm{d}r + \ldots$, **be an exact differential of a function of** $p, q, r,$ **etc., function which will be denoted** Φ**, in such a way that one have**

$$\mathrm{d}\Phi = P\mathrm{d}q + Q\mathrm{d}q + R\mathrm{d}r + \ldots.$$

Then one will have as equilibrium condition $\mathrm{d}\Phi = 0$**, which shows that the system must be placed in such a way that the function** Φ **be generally speaking a maximum or a minimum. I say generally speaking, as it is known that the equality of a differential to zero is not always indicating a maximum or a minimum, as one knows from the theory of curves. The previous hypothesis is verified when the forces** $P, Q, R,$ **etc., attract really either to some fixed points or to some bodies of the same system and are proportional to some functions of the mutual distance, which is actually the case of nature. Therefore in this hypothesis about the forces, the system will be at equilibrium when the function**

[14]From p. 16:

3. Pour faire usage de cette formule, la difficulté se réduira à déterminer, conformément à la nature du systême donné, les valeurs des différentielles $\mathrm{d}p, \mathrm{d}q, \mathrm{d}r$, *& c. On considérera donc le systême dans deux positions différentes, & infiniment voisines, & on cherchera les expressions les plus générales dont il s'agit, en introduisant dans ces expressions autant de quantités déterminées, qu'il y aura d'élémens arbitraires dans la variation de position du systême. On substituera en suite ces expressions de* $\mathrm{d}p, \mathrm{d}q, \mathrm{d}r$, *& c., dans l'équation proposée, & il faudra que cette équation ait lieu, indépendamment de toutes les indéterminées, afin que l'équilibre du systême subsiste en général & dans tous les sens. On égalera donc séparément à 0, la somme des termes affectés de chacune des mêmes indéterminées; & l'on aura, par ce moyen, autant d'équations particulieres, qu'il y aura de ces indéterminées; or il n'est pas difficile de se convaincre que leur nombre doit toujours être égal à celui des quantités inconnues dans la position du systême; donc on aura par cette méthode, autant d'équations qu'il en faudra pour déterminer l'état d'équilibre du systême.*

Φ will be a maximum or a minimum; this is in what consists the Principle which M. de Maupertuis has proposed under the name of *law of rest*.[15]

In the following passage, Lagrange clearly states that continuous mechanical systems (in the sense used in modern literature) can be studied by means of the method he is presenting.

From p. 52:

I remark now that instead of considering the given mass as an assembly of an infinity of contiguous points it will be needed, following the spirit of infinitesimal calculus, to consider it rather as composed by infinitesimally small elements, which will have the same dimensions of the whole mass; [it will be needed] similarly that in order to have forces impressing motion to each of these elements, one must multiply times this same elements the forces P, Q, R, etc. (*here Lagrange introduces the density of force per mass unity*) **which are assumed to be applied to each point of these elements, and which will be regarded as analog to those which are due to the action of the gravity.**

If therefore one calls m the total mass, and dm one of its generic elements (*it is difficult here to deny that Lagrange considers the generic sub-body of the considered body*) **one will have $P\mathrm{d}m$, $Q\mathrm{d}m$, $R\mathrm{d}m$, etc., for the forces which pull the element $\mathrm{d}m$, along the directions of the lines p, q, r, etc. Therefore multiplying these forces times the variations $\delta p, \delta q, \delta r$, etc., one will get their moments whose sum for every element dm will be represented by the formula $(P\delta p + Q\delta q + R\delta r + \ldots)\, dm$; and for having the sum of the moments of all forces of the system, one will need simply to calculate the integral of this formula with respect to all given mass. We will denote these total integrals, that is relative to**

---

[15]*Nous allons considérer maintenant les maxima & minima qui peuvent avoir lieu dans l'équilibre; & pour cela nous reprendrons la formule générale.*

$$P\mathrm{d}q + Q\mathrm{d}q + R\mathrm{d}r + \&c, = 0,$$

de l'équilibre entre les forces P, Q, R, & c, dirigées suivant les lignes p, q, r, & c. (Sect. 2, art. 2).

On peut supposer que ces forces soient exprimées de maniere que la quantité Pdq + Qdq + Rdr+ & c, soit une différentielle exacte d'une fonction de p, q, r, & c, la quelle soit représentée par ϕ, ensorte que l'on ait

$$\mathrm{d}\phi = P\mathrm{d}q + Q\mathrm{d}q + R\mathrm{d}r + \&c.$$

Alors on aura pour l'équilibre cette équation $d\phi = 0$, laquelle fait voir que le systême doit être disposé de maniere que la fonction ϕ y soit généralement parlant un maximum ou un minimum.

Je dis généralement parlant; car on fait que l'égalité d'une différentielle à zéro, n'indique pas toujours un maximum ou un minimum, comme on le voit par la théorie des courbes.

La supposition précédente a lieu en général lorsque les forces P, Q, R, &c, tendent réellement ou à des points fixes ou à des corps du même systême, & sont proportionnelles à des fonctions quelconques des distances (Sect. 2, art. 4); ce qui est proprement le cas de la nature.

Ainsi dans cette hypothèse de forces le systême sera en équilibre lorsque la fonction ϕ sera un maximum ou un minimum; c'est en quoi consiste le principe que M. de Maupertuis avoit proposé sous le nom de loi de repos.

the extension of all [considered] mass, by the distinctive symbol *S*, and we will reserve the usual distinctive $\int$ to designate the definite or indefinite integrals.[16]

In the following, Lagrange teaches us how to perform the integration for continuous systems, integrating by parts (eventually in presence of integrals in which higher gradients of virtual displacements appear). Lagrange includes also a general expression for boundary conditions which can be deduced from the Principle of Virtual Velocities. Actually on p. 89 Lagrange starts to deal with the study of the equilibrium of wires; on p. 122 he studies the equilibrium of fluids and on p. 156 he considers, together with the moment of external forces, also the first variation of internal deformation energy (the moment of internal forces).[17]

---

[16]From p. 52: *11. Je remarque ensuite qu'au lieu de considérer la masse donnée comme un assemblage d'une infinité de points contigus, il faudra, suivant l'esprit du calcul infinitésimal, la considérer plutôt comme composée d'élémens infiniment petits, qui soient du même ordre de dimension que la masse entiere; qu'ainsi pour avoir les forces qui animent chacun de ces élémens, il faudra multiplier par ces mêmes élémens, les forces P, Q, R, & c., qu'on regardera comme analogues à celles qui proviennent de l'action de la gravité. 12. Si donc on nomme m la masse totale, et* d*m un de ces élémens quelconque, on aura P*d*m, Q*d*m, R*d*m, & c., pour les forces qui tirent l'élément* d*m, suivant les directions des lignes p, q, r, & c. Donc multipliant respectivement ces forces par les variations δp, δq, δr, & c., on aura leurs momens, dont la somme pour chaque élément dm, sera représentée par la formule* $(P\delta p + Q\delta q + R\delta r + \ \&\ c.)\mathrm{d}m$*; & pour avoir la somme des momens de toutes les forces du systême, il n'y aura qu'à prendre l'intégrale de cette formule par rapport à toute la masse donnée. Nous dénoterons ces intégrales totales, c'est-à-dire, relatives à l'étendue de toute la masse, par la caractéristique majuscule S, en conservant la caractéristique ordinaire $\int$ pour désigner les intégrales partielles ou indéfinies.*

[17]From pp. 55–57:

Or les différentielle $\mathrm{d}\delta x, \mathrm{d}\delta y, \mathrm{d}\delta z, \mathrm{d}^2\delta x$, & c, qui se trouvent sous le signe S, peuvent être éliminées par l'opération connue des intégrations par parties. Car en général

$\int \Omega \mathrm{d}\delta x = \Omega\delta x - \int \delta x \mathrm{d}\Omega$, $\int \Omega \mathrm{d}^2\delta x = \Omega \mathrm{d}\delta x - \mathrm{d}\Omega\delta x + \int \delta x \mathrm{d}^2\Omega$, & ainsi des autres, ou il faut observer que les quantités hors du signe $\mathcal{S}$ se rapportent naturellement aux derniers points des intégrales, mais que pour rendre ces intégrales complettes, il faut nécessairement en retrancher les valeurs des même quantité hors du signe, lesquelles répondent aux premiers points des intégrales, afin que tout s'évanouisse dans ce point; ce qui est évident par la théorie des intégrations.

Ainsi en marquant par un trait les quantités qui se rapportent au commencement des intégrales totales désignées par $\mathcal{S}$, & par deux traits celles qui se rapportent à la fin de ces intégrales, on aura les réductions suivantes,

$$\mathcal{S}\Omega \mathrm{d}\delta x = \Omega'' \delta x'' - \Omega' \delta x' - \mathcal{S}\delta x \mathrm{d}\Omega,$$

$$\mathcal{S}\Omega \mathrm{d}^2\delta x = \Omega'' \mathrm{d}\delta x'' - \mathrm{d}\Omega'' \delta x'' - \Omega' \mathrm{d}\delta x' + \mathrm{d}\Omega' \delta x' + \mathcal{S}\delta x \mathrm{d}^2\Omega,$$

&c.

lesquelles serviront à faire disparoître toutes les différentielles des variations qui pourront se trouver sous le signe $\mathcal{S}$. Ces réductions constituent le second principe fondamental du calcul des variations.

De cette maniere donc l'équation générale de l'équilibre se réduira à la forme suivant,

$$\mathcal{S}\,(\Pi\delta x + \Sigma\delta y + \Psi\delta z) + \Delta = 0,$$

dans laquelle Π, Σ, Ψ seront des fonctions de x, y, z, & de leurs différentielles, & Δ contiendra les termes affectés des variations $\delta x'$, $\delta y'$, $\delta z'$, $\delta x''$, $\delta y''$, & c, & de leurs différentielles.

Donc pour que cette équation ait lieu, indépendamment des variations des différentes cordonnées, il faudra que l'on ait, I°. Π, Σ, Ψ, nuls dans toute l'étendue de l'intégrale $\mathcal{S}$, c'est-à-dire, dans chaque point de la masse, 2°. chaque terme de Δ aussi égal à zéro].

On p. 158, starts the Lagrangian study of Dynamics. Our apologia of the work by Lagrange must be suspended: it is clear that Lagrange believes that Greek scientists had not obtained any result in dynamics, which is in our opinion false (see Russo 2004*).*

However Lagrange cannot be blamed too much as he credits all the results obtained by his predecessors whose works are known to him: and he needs more than 20 printed pages for accounting his bibliographical researches!

The Dynamics is the Science of accelerating forces and of the varied motions which forces can produce. This Science is entirely due to the Moderns and Galileo is the one who has laid its first foundations.[18]

On p. 179, in particular, Lagrange credits D'Alembert, as being the first to have found a Principle being able to generally found Dynamics.

The treatise of Dynamics by M. D'Alembert, printed in 1743, finally ended all these challenges, by offering a direct and general method able to solve, or at least to supply the set of equations [needed to solve], all the problems in Dynamics which one can imagine. This method reduces all laws governing the motion of bodies to the equations governing their equilibrium and therefore reduces dynamics to statics.[19]

On p. 195, Lagrange perfectly formulated the Principle of Virtual Works in its most 'modern' and complete scope (calling it the Principle of Virtual Velocities, circumstance for which he cannot be blamed: he could not comply to the preferences of his future readers), as the reader will be easily persuaded by carefully considering the passage:

Now the general formula of equilibrium consists in this exact statement: that the sum of the moments of all forces of the system must be vanishing [...] Therefore we will get the searched formula by equating to zero the sum of all quantities

$$m\left(\frac{\mathrm{d}^2x}{\mathrm{d}t^2}\delta x+\frac{\mathrm{d}^2y}{\mathrm{d}t^2}\delta y+\frac{\mathrm{d}^2z}{\mathrm{d}t^2}\delta z\right)+m\left(P\delta p+Q\delta q+R\delta r+\&c\right),$$

relative to each body of the proposed system. Therefore if one denotes this formula by means of the integral sign $\int$, which must include all bodies of the system, we will get

$$\int\left(\frac{\mathrm{d}^2x}{\mathrm{d}t^2}\delta x+\frac{\mathrm{d}^2y}{\mathrm{d}t^2}\delta y+\frac{\mathrm{d}^2z}{\mathrm{d}t^2}\delta z+P\delta p+Q\delta q+R\delta r+\&c.\right)m=0,$$

---

[18]From p. 158:

La Dynamique est la Science des forces accélératrices ou retardatrices, & des mouvemens variés qu'elles peuvent produire. Cette Science est due entiérement aux Modernes, & Galilée est celui qui en a jetté les primiers fondemens.

[19]From p. 179:

Le traité de Dynamique de M. d'Alembert qui parut en 1743, mit fin à ces especes de défis, en offrant une méthode directe & générale pour résoudre, ou du moins pour mettre en équations tous les problêmes de Dynamique que l'on peut imaginer. Cette méthode réduit toutes les loix du mouvement des corps à celles de leur équilibre, et ramene ainsi la Dynamique à la Statique.

for the general formula of the motion of a whatsoever system of bodies, regarded as points and subjected to accelerating forces whatsoever $P, Q, R,$ & $c.$[20]

The reader will remember—when we will discuss the works by Noll—that Lagrange ALREADY treats inertial forces exactly on the same ground as the other externally applied forces. Remark that Lagrange uses a different signs convention for the virtual displacements when considering inertial forces or externally applied forces (see p. 193), as he seems to like formulas without the minus sign, in which an equality appears and one term of the equality is zero.

3.3 *Attested Lagrange's Version of the Principle of Virtual Work*

The careful reading of some relevant parts of the *Méchanique Analytique* have allowed us to establish that—in easily accessible bibliographical sources—it is attested a version of the Principle of Virtual Velocities dating back to the 18th century which is equivalent to the most modern and general versions of the PVW. Summarizing what found in the previous pages, one can state that in the *Méchanique Analytique:*

1. The Principle is formulated for a generic continuous system, and the sum of moments (powers in modern language) is postulated to be zero for every body.
2. The Principle is first formulated for characterizing the equilibrium and then simply generalized (introducing inertia) to dynamics.
3. It is clearly stated that an integration by parts of the expression of virtual moments is needed in order to consider the differential conditions characterizing motion, which include also boundary conditions.
4. Lagrange explicitly considers the possibility of integrating by parts expressions for the moments of forces calculated on virtual displacements in which second and higher gradients of these displacements appear.

---

[20] *Or la formule générale de l'équilibre consiste en ce que la somme des momens de toutes les forces du systême doit être nulle (Part. I, Sect. 2, art. 2); donc on aura la formule cherchée en égalant à zéro la somme de toutes les quantités*

$$m\left(\frac{\mathrm{d}^2x}{\mathrm{d}t^2}\delta x+\frac{\mathrm{d}^2y}{\mathrm{d}t^2}\delta y+\frac{\mathrm{d}^2z}{\mathrm{d}t^2}\delta z\right)+m\left(P\delta p+Q\delta q+R\delta r+\&c\right),$$

relatives à chacun des corps du systême proposé.

7. Donc si on dénote cette somme par la ligne intégral $\int$, *qui doit embrasser tous les corps du systême, on aura*

$$\int\left(\frac{\mathrm{d}^2x}{\mathrm{d}t^2}\delta x+\frac{\mathrm{d}^2y}{\mathrm{d}t^2}\delta y+\frac{\mathrm{d}^2z}{\mathrm{d}t^2}\delta z+P\delta p+Q\delta q+R\delta r+\&c.\right)m=0,$$

pour la formule générale du mouvement d'un systême quelconque de corps, regardés comme des points, & animés par des forces accélératrices quelconques $P, Q, R,$ & $c.$

5. Lagrange presents several examples of the application of the Principle to infinite-dimensional systems corresponding to important continuous systems: e.g. wires and compressible or incompressible fluids.
6. Lagrange is aware of the more general scope of the Principle of Virtual Velocities when compared to the the Principle of Stationary Action: indeed, calculating the first variation of the Action, by identifying the variations of motions with D'Alembert virtual displacements one gets a version of the Principle of Virtual Velocities.

Although the treatise is written in French, it can be easily read nowadays, as it is clear, rigorous (a notion which of course has to be intended in a historical sense) and precise. The only limit it shows is shared by many textbooks which were written more recently: it is not using Levi-Civita absolute calculus, for the very obvious reason that Levi-Civita developed it about one hundred and fifty years later.[21] The agreement about the listed points seems widely spread (see e.g. Fraser 1983) and Truesdell himself seems in some of his works to be ready to credit to Lagrange the first formulation of the PVW for continua (Truesdell 1968).

3.4 Gabrio Piola: An Italian Follower of Lagrange, One of the Founders of Modern Continuum Mechanics

Gabrio Piola was the author of relatively few works (we have a list of 13 works complexively, see Piola 1825, 1833, 1835, 1845–1846, 1856). Five of them can be regarded as a unique work, aiming to give a Lagrangian basis to Continuum Mechanics (i.e. the mechanics '*di corpi qualsivogliono considerati secondo la naturale loro forma e costituzione*', of whatsoever bodies, considered following their own natural shape and constitution). The first (Piola 1825) was assuring to the author a prize given by the R. Istituto di Scienze di Milano, the last (Piola 1856, *in* dell'Isola et al. 2015) was published posthumous under the supervision of Prof. Francesco Brioschi, the founder of the Politecnico di Milano. The other works by Piola deal either with the mathematical tools which he uses and develops for his investigations in Mechanics, or with applications of his theoretical results to particular mechanical systems. Remark that in Piola (1845–1846) (*in* dell'Isola et al. 2015) continua whose deformation energy depends on nth gradients of displacement field are introduced: one can find there already the bulk equations governing their motion (without, however, the associated boundary conditions).

Piola's works—written in a very elegant Italian—were recognized in their full scientific value in Truesdell and Toupin's *Classical Field Theories* (Truesdell and Toupin 1960), where it was named after him (and Kirchhoff) the Lagrangian dual of the velocity gradient in the expression of internal work for first order continua. The rediscovery of the value of Gabrio Piola continued more recently with the works by

[21] In the opinion of the authors, Lebedev et al. (2010) is a very good technical reference on the subject for the inexperienced reader.

Capecchi and Ruta (2007, 2015). Piola decided to write his works in Italian even if, presumably, he could have written them in French, a choice which, in our opinion, would have given a greater audience to them (this was the choice made by Lagrange, even when Lagrange was still working and living in Turin). This linguistic choice was related to the political situation in which Piola operated: Italian Risorgimento (Resurgence) ideology required a re-affirmation of national identity, also through the choice of using Italian language for scientific writings. Therefore, in recent times, very few specialists can appreciate directly the content of his works.

3.5 "Quel Principio Uno, di Dove Emanano Tutte le Equazioni Che Comprendono Innumerabili Verità"

Piola is persuaded that the PVW can be used as a basic Principle also in Continuum Mechanics. He claims that if Lagrange were still alive he would easily have completed his works by extending his methods to continuum deformable bodies. In Piola (1845–1846) (dell'Isola et al. 2015, pp. 100–111) one can read:

I will invite the reader to consider **that fundamental principle from which are emanating all those equations which include innumerable truths** (*this is our translation of Piola's words in the subsection title*). Such a principle consists in the simultaneous reference of a system whatsoever to two triples of orthogonal axes: it can be used in two manners and in both of them it produces grandiose effects. It can be used in a first manner to make clearer what was already said about the minimizing motions compatible with the equations of conditions in order to demonstrate the Principle of Virtual Velocities together with the other ones i.e. the Conservation Principles of the motion of the centre of mass and of the areas. In this first manner, instead of conceiving the variations δx, δy, δz, of the different points of the system as virtual velocities or very small infinitesimal displacements covered during that fictitious motion (which was called after Carnot also a geometric motion) it is much more natural, and there is noting of mysterious in doing so, to regard them as the variations which are imposed to the coordinates of the aforementioned points when the system is referred to three other orthogonal axes very close to the first reference axes, as if these last were undergoing a very small displacement.

In the previous excerption one has to read the expression 'equations of conditions' as equivalent to its modern countepart 'constitutive equation'. The concept of 'equation of a constraint' as conceived by Lagrange is generalized by Piola to include the concept of 'constitutive equation', i.e. that equation which allows for the representation -in terms of the kinematical descriptors- of the dual in power of their time variations. In the following, Piola justifies his last statement, and then proceeds with the exposition of his method. Piola follows:

Everybody knows that we perceive the idea of motion when observing the relationships among distances: the said coordinates may vary either because of a motion of the system, remaining the axes fixed, or because of a motion of the axes, remaining fixed the system. When the relationship among distances is intended in this last second way, one can avoid the consideration of so-called geometric motions, and then it is possible to understand clearly as the variations of the coordinates take place without any alteration of reciprocal actions of one part of the system on the others. This way of reasoning is induced, without any effort, when one considers that it is arbitrary in the space the position of the axes to which one refers a system, which may be at rest or in motion: [I claim that] it was right to consider the consequences of such arbitrariness, which once transformed into calculations had necessarily to lead to some results which are different from those obtained when said arbitrariness is not considered. Because of such motion of the reference axes the variations $\delta x, \delta y, \delta z$ of the different points assume the values given by the equations no. 42 which are those particular values which satisfy all the equations of conditions which express the effects of internal forces as we have seen in the no. 48. The simultaneous reference of the system to two triples of orthogonal axes can be also exploited in another manner, as there are actually two methods with which one can treat the equations of conditions, exactly as shown in the no. 17. Cap. II. Here we refer to that method which leaves to the variations $\delta x, \delta y, \delta z$ all their generality, and treats the equations of conditions by introducing some indeterminate multipliers. In such case the consideration of the two triples of axes is very useful to establish the nature of said equations of conditions, which otherwise could not be assigned in general: in them -through the indication of partial derivatives- do appear those variables p; q; r; which, when the operations are concluded, will disappear from the calculations. Such point of view -in my opinion- seems to have been neglected by Lagrange and by other Geometers: to this point of view it has to be referred when one wants to underline which part of this Memoir deserves more attention. Finally I refer to the to the general considerations developed in the Prologue for clarifying how the aforementioned six equations of conditions can describe the effects of internal forces.

Remark 5.1 The statements which follow the sentence 'The simultaneous reference of the system to two triples of orthogonal axes can be also exploited in another manner' refer to the objectivity requirement which Piola imposes, and that in modern terms is called 'the invariance under change of observer' of the equations of conditions.

3.6 Truesdell and Toupin in Their Classical Field Theories cite Piola's Works

It has to be recognized that, notwithstanding his irreducible contrariety to Lagrangian Postulation, Truesdell gave (together with Toupin, in Truesdell and Toupin 1960, p. 597 and following) a comprehensive description of Piola's point of view. The elegance of Piola's writing style (we can say that Piola was a true man of letters) may have contributed to induce in Truesdell a form of respect for such a famous member of the Accademia dei Lincei. Due, probably, to Toupin's predilection for the

Principle of Least Action, Trusdell actually managed, for once, to partially balance his aversion towards the PVW, aversion which he has always shown in all his other works. Here we start quoting Truesdell and Toupin footnote 3 at p. 596 in Truesdell and Toupin (1960).

The pioneer work of PIOLA [1833, 3][22] [1848, 2, 34.38, 46.50] is somewhat involved. First, Piola used the material variables, and his condition of rigidity is $\partial C_{MK} = 0$ or $\partial C^{-1}_{MK} = 0$ [...] Second, he seemed loth to confess that his principle employed rigid virtual displacements; instead, he claimed to establish it first for rigid bodies only. In the former work, he promised to remove the restriction in a later memoir; in the latter, he claimed to do so by use of an intermediate reference state. He was also the first to derive the stress boundary conditions from a variational principle [1848, 2, Par. 52], and he formulated an analogous-variational principle for one-dimensional and two-dimensional systems [1848, 2, Chap. VIII].

We reported this quote for two reasons:

1. Truesdell's authority agrees, at least this time, with our opinion, for what concerns the content of Par. 52 of Piola (1845–1846) (in dell'Isola et al. 2015). It should be noted, by the way, that for some reasons Truesdell, increasing the possibility of misunderstanding, calls it "Piola (1845–1846)" though in Truesdell's references it is clearly written that the work was printed in 1845.
2. It proves that Truesdell actually misunderstood one part of Piola's argument: Truesdell does not understand that Piola is using the intermediate reference state to impose what later Noll will call 'frame indifference'. From the careful reading of the previous passages we can conclude that:
   in Piola (1845–1846) (dell'Isola et al. 2015) *Par. 52, the Cauchy formulas expressing contact actions are intended as valid at the boundary of every continuous sub-body.*

First we need to confute the opinion by Truesdell when he tries to prove that Piola limits his analysis to rigid bodies. Indeed, let us consider what can be read at the beginning of Par. 43, where the reasoning culminating in the following Par. 53 is started.

Del moto e dell'equilibrio di un corpo qualunque.

Dico qualunque quel corpo che può mutare di forma, cangiandosi per effetto di moti intestini le posizioni relative delle sue molecole. Lagrange trattò nella sua M. A. varie questioni che si riferivano a sistemi variabili di simil natura: trattò dell' equilibrio di fili e di superficie estensibili e contrattili, trattò dell' equilibrio e del moto de' liquidi e de' fluidi elastici.

---

[22] See Piola (1833).

In English the previous sentences read as follows:

On the motion and equilibrium of a body whatsoever

I call whatsoever that body which can change its shape, this changing being caused by the internal motions of the relative positions of its molecules. Lagrange treated in his analytical mechanics various questions which were referred to systems which were undergoing similar changes: he treated the equilibrium of wires and surfaces, extensible and contractible, treated the motion and the equilibrium of liquids and elastic liquids (see our previous sections).

We are now ready to discuss the content of Par. 53. Our aim is the following: **To assess that Piola intended that the PVW (he names the statement 'Virtual Work equal zero' with the expression 'formula generalissima', (i.e. the most general formula) to be valid for every sub-body of a given continuous body.**

To prove the previous statement we can use an argument based on plain logics. Indeed Piola assumes his 'formula generalissima' for every deformable body. Then, a sub-body $\mathscr{S}$ of a given body $\mathscr{B}$ is itself a body, whose external world is composed by the external world of $\mathscr{B}$ union the complement of $\mathscr{S}$ with respect to $\mathscr{B}$. So externally applied forces to $\mathscr{S}$ include the forces exerted by this complement onto $\mathscr{S}$.

Perhaps some reader may argue that Piola was not aware of the set-theoretic arguments by Noll on universes of bodies and will accuse us to 'assume a modern maturity and depth of knowledge' to ancient 'primitive scientists'. However, the previous argument does not require actually any technicality of the set theory, is based on a very ancient idea[23] and was clearly stated by Piola himself as can be seen in the following.

From pp. 94–96 of Piola (1845–1846) (in dell'Isola et al. 2015):

Prima di lasciare queste considerazioni sulle quantità ai limiti, dirò che da esse può facilmente cavarsi tutta quella dottrina che diede argomento a varie *Memorie* del Sig. Cauchy inserite ne' suoi primi *Esercizj di Matematica*. Ci è lecito in fatti immaginare per entro alla massa del corpo e per la durata di un solo istante di tempo (quando trattasi di moto) un parallelepipedo rettangolo grande o piccolo come più piace, e restringerci a riguardare il moto o l'equilibrio di esso solo, astraendolo col pensiero dall' equilibrio o dal moto di tutto il resto del corpo, e intendendo supplito l' effetto di tutta la materia circostante col mezzo di pressioni esercitate sulle sei facce di quel parallelepipedo. Allora in virtù delle tre equazioni che sul fine del num.° precedente insegnammo a dedurre e che in tale particolare supposizione diventano assai più semplici, troveremo tre equazioni fra le componenti λ, μ, ν, parallele agli assi, della pressione per un punto qualunque di una faccia, e le sei quantità Λ, Ξ, Π, Σ, Φ, Ψ nelle quali le variabili x, y, z abbiano assunti i valori proprj delle coordinate di quel punto.

Our English translation of aforementioned exceptions (we insert some comments in italic):

[23] Let us recall that Archimedes, in the treatise *On Floating Bodies* (in which, among other things, he demonstrates the spherical shape of the ocean and determines the conditions for the stability of the equilibrium for a floating segment of a paraboloid of revolution), bases his hydrostatics on a postulate concerning the interactions between **any given** contiguous portions of fluid.

Before leaving the reasonings about the boundary quantities, I will say that from them it is easy to draw all that doctrine which was object of several Memoirs by Mr. Cauchy, inserted in his first Mathematical Exercises. Indeed we are allowed to imagine INSIDE the mass of the body and for the duration of only a time instant (when dealing with bodies in motion) a rectangular parallelepiped big or small as we prefer better, and to restrict ourselves to consider the motion or the equilibrium of it alone, by abstracting it -with our mind- from the equilibrium or the motion of all the rest of the body (*it is difficult here to state that Piola was not considering sub-bodies of a given body, however the critical reader could state that Piola is simply considering here sub-bodies with the shape of parallelepiped: to this objection we can answer referring him to the works in which Piola deals with the theory of integration, where he proves to be able to reconstruct, via limits, integral over generic regions as sums of integrals over unions of parallelepipeds*) **and considering that the effect of all surrounding matter can be replaced by means of pressures exerted on the six facets of that parallelepiped. Then, by means of the three equations which at the end of the previous number we have taught to deduce** (*here clearly Piola shows that he intends to deduce from his 'generalissima formula' the correct boundary conditions at the boundary of every sub-body of the given body*) **[equations] which in such a particular case become much simpler, we will find three equations relating the three components** λ, μ, ν (*which are the three components of 'externally applied contact forces', in this case the contact forces applied by the remaining part of the body on the parallelepiped which our mind abstracted from the whole considered body: remark that Piola is considering only dead loads in the commented work*) **parallels to the three axes of the pressure at a generic point of the facet, and the six quantities** $\boldsymbol{\Lambda}$, $\boldsymbol{\Xi}$, $\boldsymbol{\Pi}$, $\boldsymbol{\Sigma}$, $\boldsymbol{\Phi}$, $\boldsymbol{\Psi}$ (*which are obtained by means of several transformations from the duals of deformation gradients of the body and which correspond to six independent the components of the Cauchy stress tensor*) **in which the coordinates** x, y, z **have assumed the values relatives to the coordinates of the considered point.**

We consider that the previous words by Piola prove without any doubt that he intended to apply the PVW for every virtual displacement of every sub-body of a considered deformable body. Maybe the only reason for which Piola was nearly never cited until Truesdell and Toupin's *Classical Field Theories* has to be determined in his choice of writing in Italian language his works. We believe nevertheless that his influence in the works of the subsequent writers in Continuum Mechanics has been enormous. Indeed, as observed in the introduction of the present work, not being cited does not mean not being known, even via secondary sources.

Let us consider, now, the following passages from Truesdell (1977) (Vol. I, pp. 62–63).

NOLL'S Axiom For every assignment of forces to bodies, the working of a system of forces acting on each body is frame-indifferent, no matter what be the motion.

Formally, in the notations (I.8-7) and (I.11-1),

Axiom A3.

$$W^* = W \qquad \forall \mathscr{B}, \in \bar{\Omega}, \forall \chi .$$

On the assumption that A2 is satisfied, we can demonstrate that (3) expresses a necessary and sufficient condition for the resultant force and torque on each body β to vanish. Indeed, by applying (I.9–13) to the definition (I.8-7) we see that, for given $\mathscr{B}$ and χ,

$$\begin{aligned} W^* - W &= \int_{\mathscr{B}} \left[(\dot{\chi}^*) \cdot \mathrm{d}f^*_{\mathscr{B}^e} - \dot{\chi} \cdot \mathrm{d}f_{\mathscr{B}^e} \right] \\ &= \int_{\mathscr{B}} \left[(\dot{\chi}^*_0) + \dot{Q}\,(\chi - \chi_0) + Q\dot{\chi} \right] \cdot Q\mathrm{d}f_{\mathscr{B}^e} - \int_{\mathscr{B}} \dot{\chi} \cdot \mathrm{d}f_{\mathscr{B}^e} \\ &= Q^T\,(\dot{\chi}^*_0) \cdot \int_{\mathscr{B}} \mathrm{d}f_{\mathscr{B}^e} - Q^T\dot{Q} \cdot \int_{\mathscr{B}} (\chi - \chi_0) \otimes \mathrm{d}f_{\mathscr{B}^e} \\ &= Q^T\,(\dot{\chi}^*_0) \cdot f\,(\mathscr{B}, \mathscr{B}^e) - \frac{1}{2} Q^T\dot{Q} \cdot F\,(\mathscr{B}; \mathscr{B}^e)_{x_0}\,. \end{aligned}$$

By axiom A3 the right-hand side of this equation must vanish for all choices of the functions Q and χ^*_0. We consider a particular time t and choose Q such that $\dot{Q}\,(t) = 0$. Since $Q\,(t)^T\,(\dot{\chi}^*_0)\,(t)$ may be any vector whatever, Axiom A3 requires that

$$f\,(\mathscr{B}, \mathscr{B}^e) = 0.$$

This being so, Axiom A3 again applied to (4) shows that in the space of skew tensors $F\,(\mathscr{B}, \mathscr{B}^e)_{x_0}$ must be perpendicular to every tensor of the form $Q\,(t)^T\,\dot{Q}\,(t)$, the values of $Q\,(t)^T$ being orthogonal tensors. If W is a constant skew tensor, and if $Q\,(t) := \mathrm{e}^{(t-t_0)W}$, then $Q\,(t_0) = 1$ and $\dot{Q}\,(t_0) = W$, and so $Q\,(t_0)^T\,\dot{Q}\,(t_0) = W$. Thus the skew tensor $F\,(\mathscr{B}, \mathscr{B}^e)_{x_0}$ must be perpendicular to every skew tensor. Therefore

$$F\,(\mathscr{B}, \mathscr{B}^e)_{x_0} = 0.$$

Theorem 5.1 (NOLL) *The working of a system of forces is frame-indifferent if and only if that system and its associated in system of torques are both balanced.*

We will show in the following that the previous attribution to Noll was not correct, and will try to reconstruct the missing links between Piola and Truesdell.

4 The Reconstruction of the Transmission Line of Piola's Ideas and Results

It is very likely that the ideas of Piola reached—in a way or another- the Cosserat Brothers, as we have seen in a previous section. It does not absolutely mean that Cosserat Brothers are to be considered to be a sort of plagiarists of Piola works: instead they were influenced by Piola's ideas and stream of cultural tradition via Melittas (see the section dedicated to them) or via a series of passages in written form in which some of the transmitters wanted to erase the original source. In our opinion, as observed, the same process occurred many times in the history of science. Indeed very often some very specific examples, theorems, mathematical procedures, formulas or arguments have reappeared in written form after a long period of 'karstic' flow in underground riverbeds (i.e. after a period in which the transmission occurred in a not-written form). And even more often the majority of scholars do believe that there was no transmission at all, as we illustrated before discussing the resurfacing of the content of the works of Heron of Alexandria and Democritus in those by Galileo Galilei.

The works of Piola are cited by Hellinger (1913) who, however, clearly underestimated the main part of their content. The information transmitted by Hellinger is already corrupted, although the corruption does not consist in a wrong statement but in a drastic reduction of the original content of the message. The source of Hellinger seems to be Müller et al. (1907) which is cited often in Hellinger's work. Indeed at the beginning of p. 20 of Hellinger (1913) we read[24]:

In close connection with these facts there is a different point of view in the formulation of the principle of virtual work (displacements),[25] which includes in its formulation only the internal forces, the forces per unit mass X; Y; Z and the surface forces $\overline{X}$, $\overline{Y}$, $\overline{Z}$, considered as given; here it is (with slight modifications) the statement found in the formulation of G. Piola:
For the balance it is necessary that the virtual work of the forces mentioned above

$$\int\int\int_{(V)} (X\delta x + Y\delta y + Z\delta z)\,\mathrm{d}V + \int\int_{(S)} \left(\bar{X}\delta x + \bar{Y}\delta y + \bar{Z}\delta z\right)\mathrm{d}S$$

vanishes for all pure virtual translational displacements of the considered region V.

The reader will remark that Hellinger cites a small part of the original statement by Piola (1845–1846) (dell'Isola et al. 2015, p. 86): indeed Piola states that the balance of forces and torques can be deduced[26] from the Principle of Virtual Velocities for every body (rigid, elastic, solid and fluid). Moreover Piola adds the proof of the validity of the 'conservation of the areas' which is a nomenclature clearly reminiscent of Kepler's law on the motion of planets. However the formula (16) on p. 86 (Piola (1845–1846), in dell'Isola et al. 2015) cannot be misunderstood: it is the global balance of angular momentum. Hellinger ignores this result from Piola.

Then Hellinger continues, loosing the contact with the real statements which actually can be found in Piola (1845–1846) (dell'Isola et al. 2015):

---

[24]The authors thank here Prof. Victor Eremeyev for his help in translating and interpreting the German text.

[25]Remark that Hellinger still considers a Lagrangian version of the name for the Principle: The Principle of Virtual Displacements, which is closer to the Lagrangian name, i.e. Principle of Virtual Velocities.

[26]The deduction presented in Piola (1845–1846) (see dell'Isola et al. 2015, Chap. 1 p. 86, in particular Eqs. (14), (15) and (16)) is clearly the proof of an equivalence.

Expressing this constraint for the displacements, namely the 9 partial differential equations:

$$\frac{\partial \delta x}{\delta x} = 0, \qquad \frac{\partial \delta x}{\delta y} = 0, \,......., \frac{\partial \delta x}{\delta z} = 0,$$

and using the known calculus of variations one can introduce 9 associated Lagrangian factors $-X_x, -Y_x, \ldots, -Z_z$ and then one gets exactly the Eq.(4) of the old principle, in which, therefore, the components of the stress dyad as Lagrangian factors are to be determined from to constraint conditions, those of rigidity. These of course will not be determined by this variational principle, rather they are playing exactly the same role as the internal stresses in statically indeterminate problems of rigid body mechanics.

Here Hellinger ignores that in Piola (1845–1846), *in* dell'Isola et al. (2015) the reader is slowly accompanied to more and more general formulations, passing gradually through simpler ones. Indeed Hellinger describes the content of Capo IV, completely ignoring the content of CAPO VI, starting from p. 146 (Piola 1845–1846, *in* dell'Isola et al. 2015) where the general case of deformable bodies (with even nonlocal deformation energies) is carefully treated. The observations of Hellinger are correct, when referred to the Piola's approach in Capo IV based on the application of rigid body constraint. However Piola proposes a much more general family of continuum models in the subsequent Capo VI. Then Hellinger adds:
If one imposes the same requirements for all rigid motions of V 1 (rather than just for the translations), he obtains exactly the IV in 23, p. 23 (in Müller's and Gimpe's paper) which reproduces Piola's approach, in which appear only six constraints and therefore only 6 Lagrangian multipliers and thus provide a symmetric stress tensor.

Hellinger forgot that the global invariance under rotations does not neglect to talk about the consequence of such invariance property on stress tensor. As discussed again in Capo VI by Piola (1845–1846) (dell'Isola et al. 2015) the nonlocal deformation energy densities may be approximated by expanding in series the placement field: after replacing these series and after integrations in which a nonlocal kernel weights the placement gradients, Piola gets local deformation energies depending on its nth order gradients (eventually truncating the series). To our knowledge Piola is the first author in which such a general setting for continuum mechanics is proposed and used. Also in Piola (1845–1846) (dell'Isola et al. 2015) are treated bodies having bidimensional or onedimensional extension. This results is echoed in Hellinger (1913), although the reference to Piola is lost. Indeed on p. 622 (end) and p. 623 (beginning) we read:

4. *Extensions of the Principle of virtual work.*
4a. Higher-order derivatives of displacements.
One can also add to those formulated in No. 3, some statements of the principle of virtual work containing a number of enhancements which enable at first to include all laws occurring in the mechanics of continua. The next generalization considers in the density of virtual work per volume a linear form of the 18 second derivatives of the

virtual displacements $\frac{\partial^2 \delta x}{\partial x^2}$. *In fact, one has the problem in which the energy function depends also on the second derivatives of placement functions which leads to such expression. Primarily this comes into consideration for the one- and two-dimensional continua (wires and plates).*

As we discuss in the section dedicated to the interpretation of Piola's works given by Truesdell, it is clear that this last author accepts Hellinger's misinterpretation of Piola's ideas and results. Reading the previous excerptions from Hellinger this is not surprising. It is also likely that Noll, who studied in German Universities (see Walter Noll's web page), had to study the work by Hellinger or some textbooks based on it: in any case Noll co-authors with Truesdell many works discussing Variational Principles in Mechanics (see e.g. Truesdell and Noll 2004*) and cites Piola.*[27]

When considering also the fact that Piola's Italian writing style is rather complex, and very elegant, so that many Italians today can find very difficult to read it, it becomes reasonable to conjecture that Truesdell was judging Piola's scientific quality without fully possessing the required linguistic capability. On the other hand the huge publishing activity which can be attributed to Truesdell (Including more than 500 reviews for Mathrev[28]) implies, for a simple consideration of the time spent on every written page, that his judgements were obtained devoting to them, in average, very little time. Of course it is not absolutely impossible to keep a very high scientific level also in this case, but many examples can supply evidence in the opposite direction, suggesting that in average there is an inverse correlation between production pace and quality. We can for instance recall that among ancient philologists one of the less interesting (using a polite expression) was Didymus Chalcenterus (which means "bronze-guts"), who according to Seneca authored about 4000 books, and since antiquity he was taken as a model case of empty and useless erudition. Indeed, he was also named *Bibliolatas* (book-forgetter) because he used to contradict in successive works what he himself had written before.

[27]We take the opportunity to recall the enormous importance of Variational Methods for today science in general (much beyond purely mechanical universe), first of all for rich and multidirectional theoretical developments (among those closer to the research lines of the authors, the reader can see e.g. Courant (1943), Schröder et al. (2005), Milton et al. (2009), dell'Isola and Seppecher (1997), Placidi (2015), Piccardo et al. (2014), Steigmann and Ogden (1999, 1997)), and also (a point which is sometimes missed by theoreticians and historians) for the birth of the most powerful tools for computation today available in continuum mechanics, which are based on the application of Finite Element Method (FEM); these method, indeed, could only see the light as a consequence of the development of rigorous variational theories. Powerful variants of FEM are now available to attack a large class of problems (see e.g. Cuomo et al. 2014; Hughes 2012, for some recent results which we found very interesting), and they are considered by now simply indispensable in practical computation. See below for further considerations on this point.

[28]See the data at: http://www.lib.utexas.edu/taro/utcah/00308/cah-00308.html.

5 Non-local Continuum Theories in Piola's Works

The homogenized theory which is deduced in Piola (1845–1846) on the basis of the identification of powers in the discrete micro-model and in the continuous macro-model is (in the language used by Eringen 1999; Eringen and Edelen 1972) a non-local theory. In dell'Isola et al. (2015) the parts of Piola's work which are most relevant in the present context are translated. Here we transform into modern symbols the formulas which the interested reader can find there in their original form.

It is our opinion that some of Piola's arguments can compete in depth and generality, even nowadays, with the most advanced modern presentations. We describe here the continuum model that he deduces from the Principle of Virtual Velocities for a discrete mechanical system constituted by a finite set of molecules, which he considers to be the most fundamental Principle in his Postulation process.[29]

In Piola (1845–1846) (Capo I, p. 8) the reference configuration of the considered deformable body is introduced by labeling each material particle with the three Cartesian coordinates (it is suggestive to remark that the same notation is used in Hellinger 1913, see e.g., p. 605). We denote by the symbol X the position occupied by each of the considered material particles in the reference configuration. The placement of the body is then described by the set of three scalar functions (Capo I, p. 8 and then pp. 11–14)

$$x\,(a, b, c)\,, y\,(a, b, c)\,, z\,(a, b, c)$$

which, by using a compact notation, we will denote with the symbol χ mapping any point in the reference configuration into its position in the actual one.

5.1 *Piola's Non-local Internal Interactions*

In Capo VI, on p. 149 of Piola (1845–1846) Piola introduces:
"The quantity ρ Eqs. (3), (5) and (6) has the value given by the equation

$$\begin{aligned}\rho^2 &= \left[x\,(a+f, b+g, c+k) - x\,(a, b, c)\right]^2 \\ &+ \left[y\,(a+f, b+g, c+k) - y\,(a, b, c)\right]^2 \\ &+ \left[z\,(a+f, b+g, c+k) - z\,(a, b, c)\right]^2 .\text{''}\end{aligned} \tag{1}$$

So by denoting with the symbol $\bar{X}$ the particle labeled by Piola with the coordinates $(a+f, b+g, c+k)$ we have, in modern notation, that

[29] We also remark that this kind of approach, starting from a discrete system with a very large number of degrees of freedom and then proceeding by means of heuristic homogenization, is today so vital that entire chapters in modern theoretical and computational mechanics closely follow it, as for instance molecular dynamics and granular mechanics (see e.g. Misra and Chang 1993; Misra and Singh 2013; Misra and Yang 2010).

$$\rho^2(X, \bar{X}) = \left\| \chi(\bar{X}) - \chi(X) \right\|^2. \tag{2}$$

In Capo VI on p. 150 we read the following expression for the internal work, relative to a virtual displacement $\delta\chi$, followed by a very clear remark:

$$\int da \int db \int dc \int df \int dg \int dk \cdot \frac{1}{2} K \delta\rho \tag{3}$$

"[...] In it the integration limits for the variables f, g, k will depend on the surfaces which bound the body in the antecedent configuration, and also on the position of the molecule m, which is kept constant, that is by the variables a, b, c which after the first three will also vary in the same domain".

Here the scalar quantity K is introduced as "the mass intensity of the force" exerted by the particle $\bar{X}$ on the particle X and the factor $\frac{1}{2}$ is present as the action reaction principle holds. The quantity K is assumed to depend on $\bar{X}, X$ and ρ and manifestly it is measured in $\mathrm{Nm}^{-6}\,\mathrm{kg}^{-1}$ (SI Units).[30] In the number 72 starting on p. 150 of Piola (1845–1846) it is discussed the physical meaning of this scalar quantity and consequently some restrictions on the constitutive equations which have to be assigned to it.

Indeed he refrains from any effort to obtain for it an expression in terms of microscopic quantities and limits himself to require its objectivity by assuming its dependence on ρ, an assumption which will produce in the sequel some important consequences. Moreover he argues that if one wants to deal with continua more general than fluids (for a discussion of this point one can have a look on the recent paper Auffray et al. 2015) then it may depend (in a symmetric way) also on the Lagrangian coordinates of both $\bar{X}$ and X : therefore

$$K(\bar{X}, X, \rho) = K(X, \bar{X}, \rho).$$

On pp. 151–152 in Piola (1845–1846) we then read some statements which cannot be rendered clearer:

"As a consequence of what was were said up to now we can, by adding up the two integrals (1), (10), and by replacing the obtained sum in the first two parts of the general Eq. (1) n. 16., formulate the equation which includes the whole molecular mechanics. Before doing so we will remark that it is convenient to introduce the following definition

$$\bar{\Lambda} = \frac{1}{4}\frac{K}{\rho} \tag{4}$$

[30]We remark that, in the original work, when passing from the discrete to the continuous formulation, Piola replaces the elementary mass of the molecules with elementary volumes. In so doing he changes implicitly the dimension of K. In dell'Isola et al. (2015) this point was left unmentioned, while here we will introduce some new symbol (specifically $\bar{\Lambda}$) in order to better highlight it and to use a form which the modern reader is more familiar with.

by means of which it will be possible to introduce the quantity $\bar{\Lambda}\delta\rho^2$ instead of the quantity $\frac{1}{2}K\delta\rho$ in the sextuple integral (1); and that inside this sextuple integral it is suitable to isolate the part relative to the triple integral relative to the variables f, g, k, placing it under the same sign of triple integral with respect to the variables a, b, c which includes the first part of the equation: which is manifestly allowed. In this way the aforementioned general equation becomes

$$\int \mathrm{d}a \int \mathrm{d}b \int \mathrm{d}c \cdot \left\{ \left(X - \frac{\mathrm{d}^2 x}{\mathrm{d}t^2}\right) \delta x + \left(Y - \frac{\mathrm{d}^2 y}{\mathrm{d}t^2}\right) \delta y + \left(Z - \frac{\mathrm{d}^2 z}{\mathrm{d}t^2}\right) \delta z \right.$$
$$\left. + \int \mathrm{d}f \int \mathrm{d}g \int \mathrm{d}k \cdot \Lambda\delta\rho^2 \right\} + \Omega = 0, \qquad (5)$$

where it is intended that (as mentioned at the beginning of the n. 71.) it is included in the Ω the whole part which may be introduced because of the forces applied to surfaces, lines or well-determined points and also because of particular conditions which may oblige some points to belong to some given curve or surface."

Piola is aware of the technical difficulty to calculate the first variation of a square root: as he knows that these difficulties have no physical counterparts, instead of K he introduces another constitutive quantity Λ which is the dual in work of the variation $\delta\rho^2$.

Remark 5.2 (*Boundedness and attenuation assumptions on K and* $\bar{\Lambda}$) Note that Piola explicitly assumes the summability of the function $\bar{\Lambda}\delta\rho^2 = \frac{1}{4}\frac{K}{\rho}\delta\rho^2 = \frac{1}{2}K\delta\rho$ and the boundedness of the function K. As a consequence, when ρ is increasing then Λ decreases.

Remark 5.3 (*Objectivity of Virtual Work*) Note that $\delta\rho^2$ and $\bar{\Lambda}(X, \bar{X}, \rho)$ are invariant (see Steigmann 2003) under any change of observer and as Piola had repeatedly remarked (see e.g. Capo IV, n. 48, pp. 86–87) the expression for virtual work has to verify this condition. We remark also that, as the work is a scalar, in this point Piola's reasoning is made difficult by his ignorance of Levi-Civita's tensor calculus (Ricci-Curbastro and Levi-Civita 1900; Levi-Civita 1927). In another formalism the previous formula can be written as follows

$$\int_{\mathscr{B}} \left[(b_m(X) - a(X))\,\delta\chi(X) + \left(\int_{\mathscr{B}} \Lambda(X, \bar{X}, \rho)\delta\rho^2 \mu(\bar{X})\mathrm{d}\bar{X} \right) \right] \mu(X)\mathrm{d}X + \delta W(\partial\mathscr{B}) = 0, \qquad (6)$$

where $\mathscr{B}$ is the considered body, $\partial\mathscr{B}$ its boundary,

$$\Lambda(X, \bar{X}, \rho) := \bar{\Lambda}(X, \bar{X}, \rho)(\mu(X)\mu(\bar{X}))^{-1}$$

($\mu(X)$ being the volume mass density), $b_m(X)$ is the (volumic) mass specific externally applied density of force, $a(X)$ the acceleration of material point X, and $\delta W(\partial\mathscr{B})$

the work expended on the virtual displacement by actions on the boundary $\partial\mathscr{B}$ and eventually the first variations of the equations expressing the applied constraints on that boundary times the corresponding Lagrange multipliers.

In Eringen and Edelen (1972), Eringen (1999, 2002), the non-local continuum mechanics is founded on a Postulation based on Principles of balance of mass, linear and angular momenta, energy and entropy. However in Eringen (2002) a chapter on variational principles is presented.

One can easily recognize by comparing, for example, the presentation in Eringen (2002) with (6) that in the works by Piola the functional

$$\left(\int_{\mathscr{B}} \Lambda(X, \bar{X}, \rho)\delta\rho^2\mu(\bar{X})\mathrm{d}\bar{X}\right) \tag{7}$$

is assumed to satisfy a slightly generalized version of what in Eringen (2002) p. 34 is called the

Smooth Neighborhood Hypothesis,

which reads (in Eringen's work the symbol V is used with the same meaning as our symbol $\mathscr{B}$, X' instead of $\bar{X}$, x instead of χ, t' denotes a time instant, the symbol $()_{,K_i}$ denotes the partial derivatives with respect to K_ith coordinate of X, and is assumed the convention of sums over repeated indices) as follows:

"Suppose that in a region $V_0 \subset V$, appropriate to each material body, the independent variables admit Taylor series expansions in $X' - X$ in V_0 [...] terminating with gradients of order P, Q, etc.,

$$\begin{aligned} x(X', t') &= x(t') + \left(X'_{K_1} - X_{K_1}\right) x_{,K_1}(t') \\ &\quad + \cdots + \frac{1}{P!}\left(X'_{K_1} - X_{K_1}\right) \ldots \left(X'_{K_P} - X_{K_P}\right) x_{,K_1 \ldots K_P}(t'), \end{aligned}$$

and [...]. If the response functionals are sufficiently smooth so that they can be approximated by the functionals in the field of real functions

$$\begin{aligned} &x(t'), x_{,K_1}(t'), \ldots, x_{,K_1 \ldots K_P}(t'), \\ &[\ldots] \end{aligned} \tag{8}$$

we say that the material at X [...] satisfies a *smooth neighborhood hypothesis. Materials of this type, for* $P > 1, Q > 1$ are called *nonsimple materials of gradient type.*"

Actually Piola is not truncating the series and keeps calculating the integrals on the whole body $\mathscr{B}$. Although no explicit mention can be found in the text of Piola, because of the arguments presented in Remark 5.2, it is clear that he uses a weaker form of the *Attenuating Neighborhood Hypotheses* stated on p. 34 of Eringen (2002).

The idea of an internal interaction which does not fall in the framework of Cauchy continuum mechanics is nowadays attracting the attention of many researchers.

Following Piola's original ideas, modern peridynamics[31] assumes that the force applied on a material particle of a continuum actually depends on the deformation state of a whole neighborhood of the particle. We will see more on this later on.

5.2 An Explicit Calculation of the Strong Form of the Variational Principle (6)

In this section we compute explicitly the Euler-Lagrange equation corresponding to the Variational Principle (6). To this end we need to treat algebraically the expression

$$\int_{\mathscr{B}} \left(\int_{\mathscr{B}} \Lambda(X, \bar{X}, \rho) \delta\rho^2 \mu(\bar{X}) d\bar{X} \right) \mu(X) \mathrm{d}X \tag{9}$$

by calculating explicitly

$$\delta\rho^2 = \delta \left(\sum_{i=1}^{3} \left(\chi_i(\bar{X}) - \chi_i(X) \right) \left(\chi_i(\bar{X}) - \chi_i(X) \right) \right).$$

With simple calculations we obtain that (Einstein convention is applied from now on)

$$\delta\rho^2 = \left(2 \left(\chi^i(\bar{X}) - \chi^i(X) \right) \left(\delta\chi_i(\bar{X}) - \delta\chi_i(X) \right) \right),$$

which once placed in (9) produces

$$\int_{\mathscr{B}} \int_{\mathscr{B}} \left(2\Lambda(X, \bar{X}, \rho) \mu(\bar{X}) \mu(X) \left(\chi^i(\bar{X}) - \chi^i(X) \right) \right) \left(\delta\chi_i(\bar{X}) - \delta\chi_i(X) \right) \mathrm{d}\bar{X} \mathrm{d}X$$

$$= \frac{1}{2} \left(\int_{\mathscr{B}} f^i(\bar{X}) \delta\chi_i(\bar{X}) \mathrm{d}\bar{X} + \int_{\mathscr{B}} f^i(X) \delta\chi_i(X) \mathrm{d}X, \right)$$

where we have introduced the internal interaction force (recall that Piola assumes that $\Lambda(X, \bar{X}, \rho) = \Lambda(\bar{X}, X, \rho)$) by means of the definition

$$f^i(\bar{X}) := \int_{\mathscr{B}} \left(4\Lambda(X, \bar{X}, \rho) \mu(\bar{X}) \mu(X) \left(\chi^i(\bar{X}) - \chi^i(X) \right) \right) \mathrm{d}X$$

[31] We remark that (luckily!) the habit of inventing new names (although sometimes the related concepts are not so novel) is not lost in modern science (see Russo 2004, for a discussion of the importance of this attitude in science) and that the tradition of using Greek roots for inventing new names is still alive.

By a standard localization argument one easily gets that (6) implies

$$a^i(X) = b^i_m(X) + f^i(X) \tag{10}$$

This is exactly the starting point of modern peridynamics.

5.3 *Modern Perydinamics: A New/Old Model for Deformable Bodies*

Many non-local continuum theories were formulated since the first formulation by Piola seen before. We cite here, for instance, Eringen (1999, 2002), Eringen and Edelen (1972), Soubestre and Boutin (2012). Remarkable also are the following more modern papers: dell'Isola et al. (1995, 2000), dell'Isola and Seppecher (1995), Demmie and Silling (2007), Du et al. (2013), Emmrich et al. (2013), Lehoucq and Silling (2008), Seleson et al. (2013), Silling et al. (2007), Steinmann (2008), Steinmann et al. (2007), Sunyk and Steinmann (2003). The non-local interaction described by the integral operators introduced in the present subsections are not to be considered exclusively as interactions of a mechanical nature: indeed recently a model of biologically driven tissue growth has been introduced (see e.g. Andreaus et al. 2014a, b; Madeo et al. 2011) where such a non-local operator is conceived to model the biological stimulus to growth.

Starting from a balance law of the form (10) for instance in Di Paola et al. (2010a, b), Silling (2000) (but many other similar treatments are available in the literature) one finds a formulation of Continuum Mechanics which relaxes the standard one and seems suitable (see the few comments below) to describe many and interesting phenomena e.g. in crack formation and growth.

However even those scientists whose native language is Italian actually seem unaware of the contribution due to Gabrio Piola in this field: this loss of memory and this lack of credit to the major sources of our knowledge, even in those cases in which their value is still topical, is very dangerous, as proven in detail by the analysis developed in Russo (2013, 2004).

In Silling (2000) the analysis started by Piola is continued, seemingly as if the author, Silling, were one of his closer pupils: the arguments are very similar and also a variational formulation of the presented theories is found and discussed. In Lehoucq and Silling (2008), Silling and Lehoucq (2008) it is stated in the Abstract that:

"The peridynamic model is a framework for continuum mechanics based on the idea that pairs of particles exert forces on each other across a finite distance. The equation of motion in the peridynamic model is an integro- differential equation. In this paper, a notion of a peridynamic stress tensor derived from nonlocal interactions is defined."

"The peridynamic model of solid mechanics is a nonlocal theory containing a length scale. It is based on direct interactions between points in a continuum separated from each other by a finite distance. The maximum interaction distance provides a length scale for the material model. This paper addresses the question of whether the peridynamic model for an elastic material reproduces the classical local model as this length scale goes to zero. We show that if the motion, constitutive model, and any nonhomogeneities are sufficiently smooth, then the peridynamic stress tensor converges in this limit to a Piola-Kirchhoff stress tensor that is a function only of the local deformation gradient tensor, as in the classical theory. This limiting Piola-Kirchhoff stress tensor field is differentiable, and its divergence represents the force density due to internal forces."

The reader is invited to compare these statements with those which can be found in the original works by Piola.

It is very interesting to see how fruitful can be the ideas formulated more than 150 years ago by Piola. It is also useful to read the abstract of Askari et al. (2008): "The paper presents an overview of peridynamics, a continuum theory that employs a nonlocal model of force interaction. Specifically, the stress/strain relationship of classical elasticity is replaced by an integral operator that sums internal forces separated by a finite distance. This integral operator is not a function of the deformation gradient, allowing for a more general notion of deformation than in classical elasticity that is well aligned with the kinematic assumptions of molecular dynamics. Peridynamics' effectiveness has been demonstrated in several applications, including fracture and failure of composites, nanofiber networks, and polycrystal fracture. These suggest that peridynamics is a viable multiscale material model for length scales ranging from molecular dynamics to those of classical elasticity."

Or also the abstract of the paper by Parks et al. (2008):
"Peridynamics (PD) is a continuum theory that employs a nonlocal model to describe material properties. In this context, nonlocal means that continuum points separated by a finite distance may exert force upon each other. A meshless method results when PD is discretized with material behavior approximated as a collection of interacting particles. This paper describes how PD can be implemented within a molecular dynamics (MD) framework, and provides details of an efficient implementation. This adds a computational mechanics capability to an MD code enabling simulations at mesoscopic or even macroscopic length and time scales".

It is remarkable how strictly related are non-local continuum theories with the discrete theories of particles bound to the nodes of a lattice. How deep was the insight of Piola can be understood by looking at the literature about the subject which includes, for instance, Demmie and Silling (2007), Di Paola et al. (2010a, b), Du et al. (2013), Emmrich et al. (2013), Lehoucq and Silling (2008), Seleson et al. (2013), Silling (2000), Silling et al. (2007), Silling and Lehoucq (2008).

5.4 *Piola's Higher Gradient Continua*

The state of deformation of a continuum in the neighborhood of one of its material points can be approximated by means of the Green deformation measure and of all its derivatives with respect to Lagrangian referential coordinates. Piola never considers the particular case of linearized deformation measures (which is physically rather unnatural): his spirit has been recovered in many modern works, among which we cite Silling and Lehoucq (2008), Steigmann (2002).

Indeed in Capo VI, on p. 152, Piola develops in Taylor series $\delta\rho^2$ (also by using his regularity assumptions about the function $\Lambda(X, \bar{X}, \rho)$ and the definition (11)) and replaces the obtained development in (7).

In a more modern notation (see dell'Isola et al. 2015, for the word by word translation) starting from

$$\chi_i(\bar{X}) - \chi_i(X) = \sum_{N=1}^{\infty} \frac{1}{N!} \left(\frac{\partial^N \chi_i(X)}{\partial X_{i_1} \ldots . \partial X_{i_N}} (\bar{X}_{i_1} - X_{i_1}) \ldots . (\bar{X}_{i_N} - X_{i_N}) \right)$$

Piola gets an expression for the Taylor expansion with respect to the variable $\bar{X}$ of center X for the function,

$$\rho^2(\bar{X}, X) = \left(\chi^i(\bar{X}) - \chi^i(X)\right) \left(\chi_i(\bar{X}) - \chi_i(X)\right).$$

He estimates and explicitly writes first, second and third derivatives of ρ^2 with respect to the variable $\bar{X}$. This is what we will do in the sequel, repeating his algebraic procedure with the only difference consisting in the use of Levi-Civita tensor notation.

We start with the first derivative

$$\frac{1}{2} \frac{\partial \rho^2(\bar{X}, X)}{\partial \bar{X}_\alpha} = \left(\chi^i(\bar{X}) - \chi^i(X)\right) \frac{\partial \chi_i(\bar{X})}{\partial \bar{X}_\alpha}. \tag{11}$$

We remark that when $\bar{X} = X$ this derivative vanishes. Therefore the first term of Taylor series for ρ^2 vanishes. We now proceed by calculating the second and third order derivatives:

$$\begin{aligned} \frac{1}{2} \frac{\partial^2 \rho^2(\bar{X}, X)}{\partial \bar{X}_\alpha \partial \bar{X}_\beta} &= \frac{\partial \chi^i(\bar{X})}{\partial \bar{X}_\beta} \frac{\partial \chi_i(\bar{X})}{\partial \bar{X}_\alpha} + \left(\chi^i(\bar{X}) - \chi^i(X)\right) \frac{\partial^2 \chi_i(\bar{X})}{\partial \bar{X}_\alpha \partial \bar{X}_\beta} \\ &=: C_{\alpha\beta}(\bar{X}) + \left(\chi^i(\bar{X}) - \chi^i(X)\right) \frac{\partial^2 \chi_i(\bar{X})}{\partial \bar{X}_\alpha \partial \bar{X}_\beta}; \end{aligned}$$

$$\frac{1}{2}\frac{\partial^3\rho^2(\bar{X},X)}{\partial\bar{X}_\alpha\partial\bar{X}_\beta\partial\bar{X}_\gamma} = \frac{\partial C_{\alpha\beta}(\bar{X})}{\partial\bar{X}_\gamma} + \frac{\partial\chi_i(\bar{X})}{\partial\bar{X}_\gamma}\frac{\partial^2\chi^i(\bar{X})}{\partial\bar{X}_\alpha\partial\bar{X}_\beta} + \left(\chi^i(\bar{X}) - \chi^i(X)\right)\frac{\partial^3\chi_i(\bar{X})}{\partial\bar{X}_\alpha\partial\bar{X}_\beta\partial\bar{X}_\gamma}. \tag{12}$$

The quantities of this last equation are exactly those described in Piola (1845–1846) on p. 157 concerning the quantities appearing in formulas (14) on p. 153.

We now introduce a fundamental analytical identity found by Piola and reformulated in Appendix D of dell'Isola et al. (2016) as follows

$$F_{i\gamma}\frac{\partial^2\chi^i}{\partial X^\alpha\partial X^\beta} = \frac{1}{2}\left(\frac{\partial C_{\alpha\gamma}}{\partial X^\beta} + \frac{\partial C_{\beta\gamma}}{\partial X^\alpha} - \frac{\partial C_{\beta\alpha}}{\partial X^\gamma}\right).$$

By replacing in (12) we get

$$\frac{1}{2}\frac{\partial^3\rho^2(\bar{X},X)}{\partial\bar{X}_\alpha\partial\bar{X}_\beta\partial\bar{X}_\gamma} = \frac{1}{2}\left(\frac{\partial C_{\alpha\gamma}}{\partial X^\beta} + \frac{\partial C_{\beta\gamma}}{\partial X^\alpha} + \frac{\partial C_{\beta\alpha}}{\partial X^\gamma}\right) + \left(\chi^i(\bar{X}) - \chi^i(X)\right)\frac{\partial^3\chi_i(\bar{X})}{\partial\bar{X}_\alpha\partial\bar{X}_\beta\partial\bar{X}_\gamma}, \tag{13}$$

so that when $\bar{X} = X$ we get that the third order derivatives of ρ^2 can be expressed in terms of the first derivatives of $C_{\gamma\beta}$.

Now we go back to read in Capo VI n.73 pp. 152–153:

"73. What remains to be done in order to deduce useful consequences from the Eq. (12) is simply a calculation process. Once recalled the Eq. (8), it is seen, transforming into series the functions in the brackets, so that one has

$$\rho^2 = \left(f\frac{dx}{da} + g\frac{dx}{db} + k\frac{dx}{dc} + \frac{f^2}{2}\frac{d^2x}{da^2} + ec.\right)^2 + \left(f\frac{dy}{da} + g\frac{dy}{db} + k\frac{dy}{dc} + \frac{f^2}{2}\frac{d^2y}{da^2} + ec.\right)^2 + \left(f\frac{dz}{da} + g\frac{dz}{db} + k\frac{dz}{dc} + \frac{f^2}{2}\frac{d^2z}{da^2} + ec.\right)^2;$$

and by calculating the squares and gathering the terms which have equal coefficients:

$$\rho^2 = f^2t_1 + g^2t_2 + k^2t_3 + 2fgt_4 + 2fkt_5 + 2gkt_6 + f^3T_1 + 2f^2gT_2 + 2f^2kT_3 + fg^2T_4 + etc., \tag{14}$$

in which expression the quantities $t_1, t_2, t_3, t_4, t_5, t_6$ represent the six trimonials which are alreay familiar to us, as we have adopted such denominations since the Eq. (6) in the num°.34.; and the quantities $T_1, T_2, T_3, T_4, ec.$ where the index goes to infinity, represent trinomials of the same nature in which derivatives of higher and higher order appear."

Then the presentation of Piola continues with the study of the algebraic structure of the trinomial constituting the quantities $T_1, T_2, T_3,$ as shown by the formulas appearing in Capo VI, n. 73 on pp. 153–160. The reader will painfully recognize that these huge component-wise formulas actually have the same structure which becomes easily evident in formula (13) and in all formulas deduced, with Levi-Civita Tensor Calculus, in Appendices D and E.

What Piola manages to recognize (also with a courageous conjecture, see Appendices D and E) is that in the expression of Virtual Work all the quantities which undergo infinitesimal variation (which are naturally to be chosen as measures of deformation) are indeed either components of the deformation measure C or components of one of its gradients.

Indeed in the n.74 p. 156 one reads:

"74. A new proposition, which the reader should pay much attention to, is that all the trinomials $T_1, T_2, T_3, etc.$ where the index goes to infinity, which appear in the previous Eq. (17), can be expresses by means of the only first six $t_1, t_2, t_3, t_4, t_5, t_6,$ and of their derivatives with respect to the variables a, b, c of all orders. I started to suspect this analytical truth because of the necessary correspondence which must hold between the results which are obtained with the way considered in this Capo and those results obtained with the way considered in the Capos III and IV."

In order to transform the integral expression (7)

$$\left(\int_{\mathscr{B}} \Lambda(X, \bar{X}, \rho)\delta\rho^2(X, \bar{X})\mu(\bar{X})\mathrm{d}\bar{X}\right)$$

Piola remarks that (pp. 155–156):

"When using the Eq. (13) to deduce the value of the variation $\delta\rho^2$, it is clear that the characteristic δ will need to be applied only to the trinomials we have discussed up to now, so that we will have:

$$\begin{aligned}\delta\rho^2 &= f^2\delta t_1 + g^2\delta t_2 + k^2\delta t_3 + 2fg\,\delta t_4 + 2fk\,\delta t_5 + 2gk\,\delta t_6\\ &\quad + f^3\delta T_1 + 2f^2g\delta T_2 + 2f^2k\delta T_3 + f\,g^2\delta T_4 + etc.\end{aligned} \tag{15}$$

Indeed the coefficients $f^{\,2}, g^{\,2}, k^{\,2}, 2fg$, etc. are always of the same form as the functions giving the variables x, y, z in terms of the variables a, b, c, and therefore cannot be affected by that operation whose aim is simply to change the form of these functions. Vice versa, by multiplying the previous Eq. (16) times Λ and then integrating with respect to the variables f, g, k in order to deduce from such calculation the value to be given to the forth term under the triple integral of the Eq. (12), such an operation is affecting only the quantities $\Lambda f^{\,2}, \Lambda g^{\,2}$, etc. and the variations $\delta t_1, \delta t_2, \delta t_3 .... \delta T_1, \delta T_2, ec.$ cannot be affected by it, as the trinomials $t_1, t_2, t_3 .... T_1, T_2, ec.$ (one has to consider carefully which is their origin) do not contain the variables f, g, k: therefore such variations result to be constant factors, times which are to be multiplied the integrals to be calculated in the subsequent terms of the series."

Using a modern notation we have that

$$\rho^2(\bar{X}, X) = \sum_{N=1}^{\infty} \frac{1}{N!} \left. \frac{\partial^N \rho^2(\bar{X}, X)}{\partial \bar{X}_{i_1} \ldots . \partial \bar{X}_{i_N}} \right|_{X=\bar{X}} (\bar{X}_{i_1} - X_{i_1}) \ldots . (\bar{X}_{i_N} - X_{i_N}),$$

and therefore that

$$\delta\rho^2(\bar{X}, X) = \sum_{N=1}^{\infty} \frac{1}{N!} \left(\delta \left. \frac{\partial^N \rho^2(\bar{X}, X)}{\partial \bar{X}_{i_1} \ldots . \partial \bar{X}_{i_N}} \right|_{X=\bar{X}} \right) (\bar{X}_{i_1} - X_{i_1}) \ldots . (\bar{X}_{i_N} - X_{i_N}).$$

As a consequence

$$\int_{\mathscr{B}} \Lambda(X, \bar{X}, \rho) \delta\rho^2(\bar{X}, X) \mu(\bar{X}) d\bar{X} = \sum_{N=1}^{\infty} \frac{1}{N!} \left(\delta \left. \frac{\partial^N \rho^2(\bar{X}, X)}{\partial \bar{X}^{i_1} \ldots . \partial \bar{X}^{i_N}} \right|_{X=\bar{X}} \right)$$
$$\times \left(\int_{\mathscr{B}} \Lambda(X, \bar{X}, \rho) \left((\bar{X}^{i_1} - X^{i_1}) \ldots . (\bar{X}^{i_N} - X^{i_N}) \right) \mu(\bar{X}) \mathrm{d}\bar{X} \right).$$

If we introduce the tensors

$$T_{\cdot}^{i_1 \ldots i_N}(X) := \left(\int_{\mathscr{B}} \Lambda(X, \bar{X}, \rho) \left((\bar{X}^{i_1} - X^{i_1}) \ldots . (\bar{X}^{i_N} - X^{i_N}) \right) \mu(\bar{X}) \mathrm{d}\bar{X} \right).$$

we get:

$$\int_{\mathscr{B}} \Lambda(X, \bar{X}, \rho) \delta\rho^2(\bar{X}, X) \mu(\bar{X}) \mathrm{d}\bar{X}$$
$$= \sum_{N=1}^{\infty} \frac{1}{N!} \left(\delta L_{\alpha_1 \ldots . \alpha_n} \left(C(X), .., \nabla^{n-2} C(X) \right) \right) T_{\cdot}^{i_1 \ldots i_N}(X).$$

Piola states that:

"After these considerations it is manifest the truth of the equation:

$$\int df \int dg \int dk \cdot \Lambda\delta\rho^2 = (1)\ \delta t_1 + (2)\ \delta t_2 + (3)\ \delta t_3 + (4)\ \delta t_4 + (5)\ \delta t_5$$
$$+ (6)\ \delta t_6 + (7)\ \delta T_1 + (8)\ \delta T_2 + (9)\ \delta T_3 + (10)\ \delta T_4$$
$$+\ etc., \tag{16}$$

where the coefficients (1), (2), etc. indicated by means of numbers in between brackets, must be regarded to be each a function of the variables a, b, c as obtained after having performed the said definite integrals."

In order to establish the correct identification between Piola's notation and the more modern notation which we have introduced, the reader may simply consider the following table ($i = 1, 2, \ldots .n, \ldots .$)

$$T_{\cdot}^{i_1 \ldots i_N} \leftrightarrows (1), (2), etc.\ \delta L_{\alpha_1 \ldots . \alpha_n} \left(C, \ldots, \nabla^{n-2} C\right) \leftrightarrows \delta T_i \,.$$

After having accepted Piola's assumptions the identity (6) becomes

$$\int_{\mathscr{B}} \left((b_m(X) - a(X))\, \delta\chi(X) + \sum_{N=1}^{\infty} \frac{1}{N!} \left(\delta L_{\alpha_1 \ldots . \alpha_n} \left(C(X), .., \nabla^{n-2} C(X) \right) \right) T_{\cdot}^{i_1 \ldots i_N}(X) \right) \mu(X) \mathrm{d}X$$
$$+ \delta W(\partial \mathscr{B}) = 0.$$

By a simple re-arrangement and by introducing a suitable notation the last formula becomes

$$\int_{\mathscr{B}} \left((b_m(X) - a(X))\, \delta\chi(X) + \sum_{N=1}^{\infty} \langle \nabla^N \delta C(X) | S_{\cdot}(X) \rangle \right) \mu(X) \mathrm{d}X$$
$$+ \delta W(\partial \mathscr{B}) = 0, \tag{17}$$

where S is a Nth order contravariant totally symmetric tensor[32] and the symbol $\langle | \rangle$ denotes the total saturation (inner product) of a pair of totally symmetric contravariant and covariant tensors.

Indeed on pp. 159–160 of Piola (1845–1846) we read:
"75. Once the proposition of the previous num. has been admitted, it is manifest that the Eq. (17) can assume the following other form

[32]The constitutive equations for such tensors must verify the condition of frame invariance. When these tensors are defined in terms of a deformation energy (that is when the Principle of Virtual Work is obtained as the first variation of a Least Action Principle) the objectivity becomes a restriction on such an energy. The generalization of the results in Steigmann (2003) to the Nth gradient continua still needs to be found.

$$\int \mathrm{d}f \int \mathrm{d}g \int \mathrm{d}k \cdot \Lambda\delta\rho^2 = (\alpha)\ \delta t_1 + (\beta)\ \delta t_2 + (\gamma)\ \delta t_3 + \cdots + (\epsilon)\ \frac{\delta \mathrm{d}t_1}{\mathrm{d}a} + (\zeta)\ \frac{\delta \mathrm{d}t_1}{\mathrm{d}b}$$
$$+ (\eta)\ \frac{\delta \mathrm{d}t_1}{\mathrm{d}c} + (\vartheta)\ \frac{\delta \mathrm{d}t_2}{\mathrm{d}a} + \cdots + (\lambda)\ \frac{\delta \mathrm{d}^2 t_1}{\mathrm{d}a^2}$$
$$+ (\mu)\ \frac{\delta \mathrm{d}^2 t_1}{\mathrm{d}a\mathrm{d}b} + \cdots + (\xi)\ \frac{\delta \mathrm{d}^2 t_2}{\mathrm{d}a^2} + (o)\ \frac{\delta \mathrm{d}^2 t_2}{\mathrm{d}a\mathrm{d}b} + etc., \tag{18}$$

in which the coefficients (α), (β) .... (ϵ) .... (λ) ....$ec.$ are suitable quantities given in terms of the coefficients (1), (2)....(7), (8).... of the Eq. (17): they depend on the quantities $t_1, t_2....t_6$, and on all order derivatives of these trinomials with respect to the variables a, b, c. Then the variables δt_1, δt_2.... (with the index varying up to infinity) and the variations of all their derivatives of all orders $\frac{\delta \mathrm{d}t_1}{\mathrm{d}a}$, $\frac{\delta \mathrm{d}t_1}{\mathrm{d}b}$, $etc.$ appear in the (18) only linearly."

Nowadays, higher order continua are commonly met in the literature as the homogenized limit of various types of mechanical systems, among which a noticeable example is constituted by reticular structures (see e.g. Atai and Steigmann 1997; Boutin et al. 2011; Boutin and Hans 2003; Chesnais et al. 2012; Haseganu and Steigmann 1996; Nadler et al. 2006; Rinaldi and Lai 2007; Rinaldi et al. 2008; Steigmann 1992, 1996). The development of new technical possibility of controlling and manufacturing objects at the micro- and nano-scale makes this research line one of the most vital in today's mechanics.

6 Weak and Strong Evolution Equations for Piola Continua

We shortly comment here about the relative role of Weak and Strong formulations, framing it in a historical perspective.

Since at least the pioneering works by Lagrange the Postulation process for Mechanical Theories was based on the Least Action Principle or on the Principle of Virtual Work.

One can call Variational both these Principles as the Stationarity Condition for Least Action requires that for all admissible variations of motion the first variation of Action must vanish, a statement which, as already recognized by Lagrange himself, implies a form of the Principle of Virtual Work.

However in order to compute the motion relative to given initial data the initiators of Physical Theories needed to integrate by parts the Stationarity Condition which they had to handle.

In this way they derived some PDEs with some boundary conditions which sometimes were solved by using analytical or semi-analytical methods.

From the mathematical point of view this procedure is applicable when the searched solution have a stronger regularity than the one strictly needed to formulate the basic variational principle.

It is a rather ironic circumstance that very often nowadays those mathematicians who want to prove well-posedness theorems for PDEs (which originally were obtained by means of an integration by part procedure) start their reasonings by applying in the reverse direction the same integration by parts process: indeed very often the originating variational principle of all PDEs is forgotten. Some examples of mathematical results which exploit in an efficient way the power of variational methods are those presented for instance by Neff (2007, 2002, 2006c).

Actually, even if one refuses to accept the idea of basing all physical theories on variational principles, he is indeed obliged, in order to find the correct mathematical frame for his models, to prove the validity of a weak form applicable to his painfully formulated balance laws. In reality (see dell'Isola and Placidi 2011) his model will not be acceptable until he has been able to reformulate it in a weak form. This seems what occurred sometimes in Continuum Mechanics: the Euler-Lagrange equations, obtained by means of a process of integration by parts, were originally written, starting from a variational principle, to supply a calculation tool to applied scientists. They soon became (for simplifying) the bulk of the theories and often the originating variational principles were forgotten (or despised as too mathematical). For a period balance equations were (with some difficulties which are discussed e.g. in dell'Isola and Placidi 2011) postulated on physical grounds. The vitality of variational methods is nowadays shown by many relevant result, most of which cannot be obtained without the generality and the rigorousness provided by the variational framework. Among the general works on variational methods, we have to cite Germain (1973a), Daher and Maugin (1986), Maugin (2011), Polizzotto (2001), Berdichevsky (2009), Epstein and Segev (1980), Steigmann and Faulkner (1993), Eremeyev and Lebedev (2013), Germain (1973b). Moreover, well established results show that even non-conservative systems can be described by means of suitable variational formulations (see e.g. the systems considered in Carcaterra and Akay 2007, 2011; Carcaterra et al. 2006; Carcaterra 2005.

When the need of proving rigorous existence and uniqueness theorems met the need of developing suitable numerical methods, and when many failures of the finite difference schemes became evident, the variational principles were re-discovered *starting from the balance equations*. Moreover, they then have been recovered as a computational tool, via finite element analysis or other numerical optimization methods (see, for instance, Contrafatto and Cuomo 2002, 2005, 2006; Cuomo and Contrafatto 2000; Cuomo and Ventura 1998; Cazzani et al. 2014a, b).

One question needs to be answered: why in the modern paper (dell'Isola et al. 2012) a strong formulation was searched for the evolution equation for Nth gradient continua? The answer is simple: because of the need of finding for those theories the most suitable boundary conditions.

This point is discussed also in Piola (1845–1846) as remarked already in Auffray et al. (2015).

Piola (1845–1846) on pp. 160–161 claims indeed that:

"Now it is a fundamental principle of the calculus of variations (and we used it also in this Memoir in the num.° 36. and elsewhere) that one series as the previous one, where the variations of some quantities and the variations of their derivatives with respect to the fundamental variables a, b, c appear linearly can be always be transformed into one expression which contains the first quantities without any sign of derivation, with the addition of other terms which are exact derivatives with respect to one of the three simple independent variables. As a consequence of such a principle, the expression which follows to the Eq. (18) can be given

$$\int \mathrm{d}f \int \mathrm{d}g \int \mathrm{d}k \cdot \Lambda\delta\rho^2 = A\delta t_1 + B\,\delta t_2 + C\,\delta t_3 + D\delta t_4 + E\delta t_5 + F\,\delta t_6 + \frac{\mathrm{d}\Delta}{\mathrm{d}a} + \frac{\mathrm{d}\Theta}{\mathrm{d}b} + \frac{\mathrm{d}\Upsilon}{\mathrm{d}c}. \tag{19}$$

The values of the six coefficients A, B, C, D, E, F are series constructed with the coefficients

$$(\alpha)\,,(\beta)\,,(\gamma)\,....\,(\epsilon)\,,(\zeta)\,....\,(\lambda)\,,ec.$$

of the Eq. (18) which appear linearly, with alternating signs and affected by derivations of increasingly higher order when we move ahead in the terms of said series: the quantities Δ, Θ, Υ are series of the same form of the terms which are transformed, in which the coefficients of the variations have a composition similar to the one which we have described for the six coefficients A, B, C, D, E, F.

Once -instead of the quantity under the integral sign in the left hand side of the Eq. (12)-one introduces the quantity which is on the right hand side of the Eq. (19), it is clear to everybody that an integration is possible for each of the last three addends appearing in it and that as a consequence these terms only give quantities which supply boundary conditions. What remains under the triple integral is the only sestinomial which is absolutely similar to the sestinomial already used in the Eq. (10) num.° 35. for rigid systems. Therefore after having remarked the aforementioned similarity the analytical procedure to be used here will result perfectly equal to the one used in the num.° 35, procedure which led to the Eqs. (26) and (29) in the num.° 38 and it will become possible the demonstration of the extension of the said equations to every kind of bodies which do not respect the constraint of rigidity, as it was mentioned at the end of the num.° 38. It will also be visible the coincidence of the obtained results with those which are expressed in the Eq. (23) of the num.°50. which hold for every kind of systems and which were shown in the Capo IV by means of those intermediate coordinates p, q, r, whose consideration, when using the approach used in this Capo, will not be needed."

The works (nowadays considered fundamental) by Mindlin (1964, 1965), Mindlin and Eshel (1968), Sedov (1972, 1968), and Toupin (1962, 1964) have developed a more complete study of Piola Continua, at least up to those whose deformation energy depends on the Third Gradient, completely characterizing the nature of contact actions in these cases, or for continua having a kinematics richer than that considered by Piola, including micro-deformations and micro-rotations. Moreover, a deep understanding of the geometric features involved in the mathematical formulation of generalized continuum theories has also proven fundamental (see e.g. Epstein and Segev 1980; Segev 1986, 2000).

Many important results has been obtained for higher gradient materials, as shown by the theoretical investigations performed in Alibert et al. (2003), dell'Isola et al.

(2000, 2009), Gatignol and Seppecher (1986), Seppecher (1987, 1989, 1996a, 2001), Eremeyev and Altenbach (2014), and the applications described for instance in Madeo and Gavrilyuk (2010), Madeo et al. (2012), Seppecher (2002, 1988, 1993, 1996b), Yang et al. (2011), Yang and Misra (2010), Placidi et al. (2014), Forest et al. (2011), Forest (2009), Steeb and Diebels (2004), McBride et al. (2011, 2012).

A further generalization of higher gradient continua is represented by those models in which additional independent kinematical descriptors are considered, i.e. micromorphic continua. This line was ideally started in the works of the already cited (Cosserat and Cosserat 1909), and developed later by Green and Rivlin (1964a, b, c, 1965), Eringen (1999). This research field, as well, is receiving increasing attention because of the links it has with the newly arisen (especially computer-aided) manufacturing possibilities (for recent interesting results in the subject see e.g. Neff 2004, 2006a, b, c, 2007; Neff and Chelmiński 2007; Neff and Jeong 2009; Scerrato et al. 2014; Yeremeyev and Zubov 1999; Steigmann 2009).

7 A Half-Facetious Conclusion: Melittas or the Role of 'Ideas Spreader' in the Erasure of Authors and in the Diffusion of Ideas

There is a phenomenon which has a great influence in the process of diffusion of knowledge and progress of science, and which has been underestimated. We want here to attract the attention of the reader to it and to its consequences. We are talking about the existence of 'melittas'.[33] We define a scientific melitta a savant who hates writing works, textbooks or memoirs, but likes studying, understanding, discussing. When they are asked to write a work in which they expose their results they have frequent attacks of a disease which is characterized by the three Ps: Perfectionism, Procrastination, Paralysis.

Of note, they can be very deep thinkers: they, for instance, can find problems in other savants' reasoning and solve them with clever suggestions. They spend more time in thinking about other people's research than developing their own. They prefer to be victims of plagiarism than being obliged to sign a paper which they did not digest for weeks or months per page. They feel more or less like raped if their contribution in a research is recognized by the addition of their name in the list of the authors of a paper, they feel happy if their idea is published with somebody else's name, as they feel relieved by the duty of writing the paper, duty which costs them painfully (and generally useless) hours of impossible search for perfection. Melittas love mental activity and hate reordering ideas in written form, because what is written cannot

[33]Melitta is a word from Greek that indicates both the bee which is capable to produce honey and the mythological figure (nymph) who taught the bees to produce honey.

be changed, is crystallized in an immutable form. For a melitta the work of Piola is impossible: Piola wrote hundred pages of deep ideas, published them and then started again in rewriting them for answering to the objections or to his own (often very demanding) new requests of rigor and elegance. Melittas, moreover, use to talk with everybody about their deepest ideas, and often also about the ideas of those who believe that they are able to write a common paper and naively share with them their results, thus actually encouraging free appropriation of someone's ideas (i.e., plagiarism). The typical melitta can be personified in Paul Ehrenfest. Indeed (see Klein 1981).

'It is not by discoveries only, and the registration of them by learned societies, that science is advanced. The true seat of science is not in the volume of Transactions, but in the living mind, and the advancement of science consists in the direction of men's minds into a scientific channel; whether this is done by the announcement of a discovery, the assertion of a paradox, the invention of a scientific phrase, or the exposition of a system of doctrine'.

The words are James Clerk Maxwell's, and they are particularly appropriate in talking about Paul Ehrenfest, who was born a century ago. Ehrenfest did advance science in all the ways that Maxwell mentions. The activity of the true scientific 'melitta' personifies the metaphor of Bacon's bee. Melittas render the work of the historian of science the true hell which it is. Why Cosserat Brothers wrote something very similar to Piola's works without citing him? Why some authors write an amount of works which any human being could never formulate and write alone? It is clear that there is a hidden way for transmitting the information which is different from the one based on the written texts. And melittas do this job: propagate the ideas without leaving any detectable trace; plagiarians, not surprisingly, usually like melittas very much. However the enormous work in favor of the advancement of science made by melittas must be recognized.

Melittas erase authors but keep ideas and theories alive.

References

Alibert JJ, Seppecher P, dell'Isola F, (2003) Truss modular beams with deformation energy depending on higher displacement gradients. Math Mech Solids 8:51–73

Andreaus U, Giorgio I, Lekszycki T (2014a) A 2-d continuum model of a mixture of bone tissue and bio-resorbable material for simulating mass density redistribution under load slowly variable in time. Zeitschrift für Angewandte Mathematik und Mechanik (ZAMM) 94(12):978–1000

Andreaus U, Giorgio I, Madeo A (2014b) Modeling of the interaction between bone tissue and resorbable biomaterial as linear elastic materials with voids. Zeitschrift für angewandte Mathematik und Physik 66(1):209–237

Askari E, Bobaru F, Lehoucq RB, Parks ML, Silling SA, Weckner O (2008) Peridynamics for multiscale materials modeling. J Phys: Conf Ser 125:012078

Atai AA, Steigmann DJ (1997) On the nonlinear mechanics of discrete networks. Arch Appl Mech 67:303–319

Auffray N, dell'Isola F, Eremeyev V, Madeo A, Rosi G, (2015) Analytical continuum mechanics a la Hamilton-Piola least action principle for second gradinet continua and capillary fluids. Math Mech Solids 20(4):375–417

Baldini U (1992) Legem impone subactis. Studi su filosofia e scienza dei Gesuiti in Italia. Bulzoni Editore, Rome, pp 1540–1632

Berdichevsky V (2009) Variational principles of continuum mechanics, vols I–II. Springer, Berlin

Boutin C, Hans S (2003) Homogenisation of periodic discrete medium: application to dynamics of framed structures. Comput Geotech 30:303–320

Boutin C, Hans S, Chesnais C (2011) Generalized beams and continua. Dynamics of reticulated structures. In: Altenbach H, Erofeev V, Maugin G (eds) Mechanics of generalized continua. Springer, Heidelberg, pp 131–141

Capecchi D, Ruta G (2015) Strength of materials and theory of elasticity in 19th Century Italy. A brief account of the history of mechanics of solids and structures, advanced structured materials, vol 52. Springer, Berlin

Capecchi D, Ruta GC (2007) Piola's contribution to continuum mechanics. Arch Hist Exact Sci 61(4):303–342

Carcaterra A (2005) Ensemble energy average and energy flow relationships for nonstationary vibrating systems. J Sound Vib 288:751–790

Carcaterra A, Akay A (2007) Theoretical foundations of apparent-damping phenomena and nearly irreversible energy exchange in linear conservative systems. J Acoust Soc Am 12:1971–1982

Carcaterra A, Akay A (2011) Dissipation in a finite-size bath. Phys Rev E 84(011):121

Carcaterra A, Akay A, Ko IM (2006) Near-irreversibility in a conservative linear structure with singularity points in its modal density. J Acoust Soc Am 119:2141–2149

Carugo A, Crombie AC (1983) The jesuits and galileo's ideas of science of nature. Annali dell'Istituto e Museo di storia della scienza di Firenze pp 63–62

Cazzani A, Malagú M, Turco E (2014a) Isogeometric analysis: a powerful numerical tool for the elastic analysis of historical masonry arches. Continuum Mechanics and thermodynamics, pp 1–18

Cazzani A, Malagú M, Turco E (2014b) Isogeometric analysis of plane-curved beams. Math Mech Solids. doi:10.1177/1081286514531265

Chesnais C, Boutin C, Hans S (2012) Effects of the local resonance on the wave propagation in periodic frame structures: generalized newtonian mechanics. J Acoust Soc Am 132:2873–2886

Clagett M (1959) The science of mechanics in the middle ages. The University of Wisconsin Press, Madison

Contrafatto L, Cuomo M (2002) A new thermodynamically consistent continuum model for hardening plasticity coupled with damage. Int J Solids Struct 39:6241–6271

Contrafatto L, Cuomo M (2005) A globally convergent numerical algorithm for damaging elasto-plasticity based on the multiplier method. Int J Numer Methods Eng 63:1089–1125

Contrafatto L, Cuomo M (2006) A framework of elastic-plastic damaging model for concrete under multiaxial stress states. Int J Plast 22:2272–2300

Cosserat E, Cosserat F (1909) Théorie des Corps deformables. A. Hermann et Fils, Paris

Courant R (1943) Variational methods for the solution of problems of equilibrium and vibrations. Bull Am Math Soc 49(1):1–23

Cuomo M, Contrafatto L (2000) Stress rate formulation for elastoplastic models with internal variables based on augmented Lagrangian regularisation. Int J Solids Struct 37:3935–3964

Cuomo M, Ventura G (1998) Complementary energy approach to contact problems based on consistent augmented Lagrangian regularization. Math Comput Model 28:185–204

Cuomo M, Contrafatto L, Greco L (2014) A variational model based on isogeometric interpolation for the analysis of cracked bodies. Int J Eng Sci 80:173–188

Daher N, Maugin GA (1986) The method of virtual power in continuum mechanics. Application to media presenting singular surfaces and interfaces. Acta Mech 60:217–240

d'Alembert JBLR (1758) Traité de l'équilibre et du mouvement des fluides: pour servir de suite au Traité de dynamique. Paris, chez David, l'aîné

dell'Isola F, Placidi L (2011) Variational principles are a powerful tool also for formulating field theories. In: dell'Isola F, Gavrilyuk S (eds) Variational models and methods in solid and fluid mechanics, CISM course and lectures, vol 535. Springer, Vienna, pp 1–16

dell'Isola F, Seppecher P, (1995) The relationship between edge contact forces, double force and interstitial working allowed by the principle of virtual power. Comptes rendus de l'Académie des Sciences Série IIb 321:303–308
dell'Isola F, Seppecher P, (1997) Edge contact forces and quasi-balanced power. Meccanica 32(1):33–52
dell'Isola F, Gouin H, Seppecher P, (1995) Radius and surface tension of microscopic bubbles by second gradient theory. Comptes rendus de l'Académie des Sciences Série IIb 320:211–216
dell'Isola F, Guarascio M, Hutter K, (2000) A variational approach for the deformation of a saturated porous solid. A second-gradient theory extending Terzaghi's effective stress principle. Arch Appl Mech 70:323–337
dell'Isola F, Sciarra G, Vidoli S, (2009) Generalized Hooke's law for isotropic second gradient materials. Proc R Soc Lond Ser A: Math Phys Eng Sci 465:2177–2196
dell'Isola F, Seppecher P, Madeo A, (2012) How contact interactions may depend on the shape of Cauchy cuts in *N*th gradient continua: approach à la D'Alembert. Zeitschrift für Angewandte Mathematik und Physik (ZAMP) 63:1119–1141
dell'Isola F, Maier G, Perego U, Andreaus U, Esposito R, Forest S, Piola G, (ed) (2015) The complete works of Gabrio Piola: commented english translation, advanced structured materials, vol 38. Springer, Heidelberg
dell'Isola F, Andreaus U, Placidi U, (2016) At the origins and in the vanguard of peridynamics, non-local and higher-gradient continuum mechanics: an underestimated and still topical contribution of Gabrio Piola. Math Mech Solids 20(8):887–928
Demmie PN, Silling SA (2007) An approach to modeling extreme loading of structures using peridynamics. J Mech Mater Struct 2(10):1921–1945
Di Paola M, Failla G, Zingales M (2010a) The mechanically-based approach to 3D non-local linear elasticity theory: long-range central interactions. Int J Solids Struct 47(18):2347–2358
Di Paola M, Pirrotta A, Zingales M (2010b) Mechanically-based approach to non-local elasticity: variational principles. Int J Solids Struct 47(5):539–548
Drake S (2003) Galileo at work: his scientific biography. Courier Dover Publications, New York
Du Q, Gunzburger M, Lehoucq RB, Zhou K (2013) A nonlocal vector calculus, nonlocal volume-constrained problems, and nonlocal balance laws. Math Models Methods Appl Sci 23(3):493–540
Emmrich E, Lehoucq RB, Puhst D (2013) Peridynamics: nonlocal continuum theory. Lecture notes in computational science and engineering, vol 89. Springer, Berlin, pp 45–65
Epstein M, Segev R (1980) Differentiable manifolds and the principle of virtual work in continuum mechanics. J Math Phys 21(5):1243–1245
Eremeyev VA, Altenbach H (2014) Equilibrium of a second-gradient fluid and an elastic solid with surface stresses. Meccanica 49(11):2635–2643
Eremeyev VA, Lebedev LP (2013) Existence of weak solutions in elasticity. Math Mech Solids 18:204–217
Eringen AC (1999) microcontinuum field theories, vol I. Foundations and solids, Springer, New York
Eringen AC (2002) Nonlocal continuum field theories. Springer, New York
Eringen AC, Edelen DGB (1972) On nonlocal elasticity. Int J Eng Sci 10:233–248
Forest S (2009) Micromorphic approach for gradient elasticity, viscoplasticity, and damage. J Eng Mech 135:117–131
Forest S, Cordero NM, Busso EP (2011) First vs. second gradient of strain theory for capillarity effects in an elastic fluid at small length scales. Comput Mater Sci 50:1299–1304
Fraser CG (1983) J. L. Lagrange's early contributions to the principles and methods of mechanics. Arch Hist Exact Sci 28(3):197–241
Galilei G (1890–1909, reprinted 1964–1966) Le opere, Ed Naz, vol 20. G Barbera, Florence
Gatignol R, Seppecher P (1986) Modelisation of fluid-fluid interfaces with material properties. Journal de Mécanique Théorique et Appliquée 225–247
Germain P (1973a) La méthode des puissances virtuelles en mécanique des milieux continus. Premiére partie. Thèorie du second gradient. Journal de Mécanique 12:235–274

Germain P (1973b) The method of virtual power in continuum mechanics. Part 2: microstructure. SIAM J Appl Math 25(3):556–575
Green AE, Rivlin RS (1964a) Multipolar continuum mechanics. Arch Ration Mech Anal 17:113–147
Green AE, Rivlin RS (1964b) On Cauchy's equations of motion. Zeitschrift für Angewandte Mathematik und Physik (ZAMP) 15:290–292
Green AE, Rivlin RS (1964c) Simple force and stress multipoles. Arch Ration Mech Anal 16:325–353
Green AE, Rivlin RS (1965) Multipolar continuum mechanics: functional theory. Proc R Soc Lond Ser A Math Phys Eng Sci 284:303–324
Haseganu EM, Steigmann DJ (1996) Equilibrium analysis of finitely deformed elastic networks. Comput Mech 17:359–373
Hellinger E (1913) Die allgemeinen ansätze der mechanik der kontinua. In: und C Müller FK (ed) Encyklopädie der mathematischen Wissenschaften, vol IV-4, No. 5, Teubner, Leipzig, pp 601–694
Heron (1899, 1900) Opera quae supersunt omnia, (reprinted by B.G. Teubner, Stuttgart, 1976), vol I (ed. W. Schmidt) and II (ed. L. Nix and W. Schmidt). B.G. Teubner, Leipzig
Heron, (1988) Les Mécaniques ou l'èlèvateur des corps lourds. Les Belles Lettres. Commented by A.G, Drachmann
Hughes TJ (2012) The finite element method: linear static and dynamic finite element analysis. Courier Corporation, Mineola
Klein MJ (1981) Not by discoveries alone: the centennial of Paul Ehrenfest. Physica 106A:3–14
Lagrange JL (1788) Mécanique Analytique. Editions Jaques Gabay, Sceaux
Lebedev LP, Cloud MJ, Eremeyev VA (2010) Tensor analysis with applications in mechanics. World Scientific, New Jersey
Lehoucq RB, Silling SA (2008) Force flux and the peridynamic stress tensor. J Mech Phys Solids 56(4):1566–1577
Levi-Civita T (1927) The Absolute Differential Calculus (Calculus of Tensors) (edited by E. Persico and English translation by M. Long). Dover Editions
Madeo A, Gavrilyuk S (2010) Propagation of acoustic waves in porous media and their reflection and transmission at a pure-fluid/porous-medium permeable interface. Eur J Mech-A/Solids 29(5):897–910
Madeo A, Lekszycki T, dell'Isola F, (2011) A continuum model for the bio-mechanical interactions between living tissue and bio-resorbable graft after bone reconstructive surgery. Comptes Rendus Mécanique 339(10):625–640
Madeo A, George D, Lekszycki T, Nierenberger M, Rémond Y (2012) A second gradient continuum model accounting for some effects of micro-structure on reconstructed bone remodelling. Comptes Rendus Mécanique 340(8):575–589
Maugin GA (2011) The principle of virtual power: from eliminating metaphysical forces to providing an efficient modelling tool. Contin Mech Thermodyn 25:127–146
McBride AT, Javili A, Steinmann P, Bargmann S (2011) Geometrically nonlinear continuum thermomechanics with surface energies coupled to diffusion. J Mech Phys Solids 59:2116–2133
McBride AT, Mergheim J, Javili A, Steinmann P, Bargmann S (2012) Micro-to-macro transitions for heterogeneous material layers accounting for in-plane stretch. J Mech Phys Solids 60:1221–1239
Milton GW, Seppecher P, Bouchitté G (2009) Minimization variational principles for acoustics, elastodynamics and electromagnetism in lossy inhomogeneous bodies at fixed frequency. Proc R Soc A: Math Phys Eng Sci 465(2102):367–396
Mindlin RD (1964) Micro-structure in linear elasticity. Arch Ration Mech Anal 16:51–78
Mindlin RD (1965) Second gradient of strain and surface tension in linear elasticity. Int J Solids Struct 1:417–438
Mindlin RD, Eshel NN (1968) On first strain-gradient theories in linear elasticity. Int J Solids Struct 4:109–124

Misra A, Chang CS (1993) Effective elastic moduli of heterogeneous granular solids. Int J Solids Struct 30:2547–2566
Misra A, Singh V (2013) Micromechanical model for viscoelastic-materials undergoing damage. Contin Mech Thermody 25:1–16
Misra A, Yang Y (2010) Micromechanical model for cohesive materials based upon pseudo-granular structure. Int J Solids Struct 47(21):2970–2981
Müller CH, Timpe A, Tedone O (1907) Die grundgleichungen der mathematischen elastizitätstheorie - allgemeine theoreme der mathematischen elastizitätslehre (integrationstheorie). In: und C Müller FK (ed) Encyklopädie der mathematischen Wissenschaften mit Einschluss ihrer Anwendungen, vol IV-1, Teubner, Leipzig
Nadler B, Papadopoulos P, Steigmann DJ (2006) Multiscale constitutive modeling and numerical simulation of fabric material. Int J Solids Struct 43(2):206–221
Neff P (2002) On Korn's first inequality with non-constant coefficients. R Soc Edinb Proc A 132(1):221–243
Neff P (2004) A geometrically exact cosserat shell-model including size effects, avoiding degeneracy in the thin shell limit. part i: Formal dimensional reduction for elastic plates and existence of minimizers for positive cosserat couple modulus. Contin Mech Thermodyn 16(6):577–628
Neff P (2006a) Existence of minimizers for a finite-strain micromorphic elastic solid. R Soc Edinb Proc A 136(5):997–1012
Neff P (2006b) A finite-strain elastic-plastic cosserat theory for polycrystals with grain rotations. Int J Eng Sci 44(8–9):574–594
Neff P (2006c) The Cosserat couple modulus for continuous solids is zero viz the linearized Cauchy-stress tensor is symmetric. ZAMM - Journal of Applied Mathematics and Mechanics / Zeitschrift für Angewandte Mathematik und Mechanik 86(11):892–912
Neff P (2007) A geometrically exact planar Cosserat shell-model with microstructure: existence of minimizers for zero Cosserat couple modulus. Math Models Methods Appl Sci 17(3):363–392
Neff P, Chełmiński K (2007) A geometrically exact Cosserat shell model for defective elastic crystals. Justification via γ-convergence. Interfaces Free Bound 9(4):455–492
Neff P, Jeong J (2009) A new paradigm: the linear isotropic Cosserat model with conformally invariant curvature energy. ZAMM - Journal of Applied Mathematics and Mechanics / Zeitschrift für Angewandte Mathematik und Mechanik 89(2):107–122
Paipetis SA, Ceccarelli M (eds) (2010) The Genius of Archimedes – 23 Centuries of Influence on Mathematics, Science and Engineering. Proceedings of an international conference held at Syracuse, Italy, 8–10 June 2010. Springer, Berlin
Parks M, Lehoucq R, Plimpton S, Silling S (2008) Implementing peridynamics within a molecular dynamics code. Comput Phys Commun 179:777–783
Piccardo G, Ranzi G, Luongo A (2014) A complete dynamic approach to the Generalized Beam Theory cross-section analysis including extension and shear modes. Math Mech Solids 19(8):900–924
Piola G, (1825) Sull'applicazione de' principj della meccanica analitica del Lagrange ai principali problemi. Memoria di Gabrio Piola presentata al concorso del premio e coronata dall'I.R. Istituto di Scienze, ecc. nella solennita del giorno 4 ottobre, (1824) Imp. Regia stamperia, Modena
Piola G (1833) La meccanica de' corpi naturalmente estesi: trattata col calcolo delle variazioni. Giusti, Milano
Piola G (1835) Nuova analisi per tutte le questioni della meccanica molecolare. Tipografia camerale, Modena
Piola G (1845–1846) Memoria intorno alle equazioni fondamentali del movimento di corpi qualsivogliono considerati secondo la naturale loro forma e costituzione. Tipi del R.D. Camera, Modena
Piola G (1856) Di un principio controverso della Meccanica analitica di Lagrange e delle molteplici sue applicazioni (memoria postuma pubblicata per cura del prof. Bernardoni, Milano, Francesco Brioschi)

Placidi L (2015) A variational approach for a nonlinear 1-dimensional second gradient continuum damage model. Contin Mech Thermodyn 17(4):623–638

Placidi L, Rosi G, Giorgio I, Madeo A (2014) Reflection and transmission of plane waves at surfaces carrying material properties and embedded in second-gradient materials. Math Mech Solids 94(10):862–877

Polizzotto C (2001) Nonlocal elasticity and related variational principles. Int J Solids Struct 38(42–43):7359–7380

Ricci-Curbastro G, Levi-Civita T (1900) Méthodes de calcul différentiel absolu et leurs applications. Mathematische Annalen 54(1–2):125–201

Rinaldi A, Lai YC (2007) Statistical damage theory of 2D lattices: Energetics and physical foundations of damage parameter. Int J Plast 23(10–11):1796–1825 (in honor of Professor Dusan Krajcinovic)

Rinaldi A, Krajcinovic D, Peralta P, Lai YC (2008) Lattice models of polycrystalline microstructures: a quantitative approach. Mech Mater 40(1–2):17–36

Russo L (2003) Flussi e Riflussi. Feltrinelli, Milano

Russo L (2004) The Forgotten revolution. How Science Was Born in 300 BC and Why it Had to Be Reborn. Springer, Berlin

Russo L (2013) L'America dimenticata. Mondadori Università, I rapporti tra le civiltà e un errore di Tolomeo

Scerrato D, Giorgio I, Madeo A, Limam A, Darve F (2014) A simple non-linear model for internal friction in modified concrete. Int J Eng Sci 80:136–152 (special issue on Nonlinear and Nonlocal Problems. In occasion of 70th birthday of Prof Leonid Zubov)

Schröder J, Neff P, Balzani D (2005) A variational approach for materially stable anisotropic hyperelasticity. Int J Solids Struct 42(15):4352–4371

Sedov LI (1968) Variational methods of constructing models of continuous media. Irreversible aspects of continuum mechanics and transfer of physical characteristics in moving fluids. Springer, Vienna, pp 346–358

Sedov LI (1972) Models of continuous media with internal degrees of freedom. J Appl Math Mech 32:803–819

Segev R (1986) Forces and the existence of stresses in invariant continuum mechanics. J Math Phys 27(1):163–170

Segev R (2000) The geometry of cauchy's fluxes. Arch Ration Mech Anal 154(3):183–198

Seleson P, Beneddine S, Prudhomme S (2013) A force-based coupling scheme for peridynamics and classical elasticity. Comput Mater Sci 66:34–49

Seppecher P (1987) Etude d'une modelisation des zones capillaires fluides: Interfaces et lignes de contact. Université Paris VI, Paris, Thése

Seppecher P (1988) Thermodynamique des zones capillaires. Annales de Physique 13:13–22

Seppecher P (1989) Etude des conditions aux limites en théorie du second gradient: cas de la capillarité. Comptes rendus de l'Académie des Sciences 309:497–502

Seppecher P (1993) Equilibrium of a Cahn and Hilliard fluid on a wall: Influence of the wetting properties of the fluid upon the stability of a thin liquid film. Eur J Mech B/Fluids 12:69–84

Seppecher P (1996a) Les fluides de cahn-hilliard. Université du Sud Toulon Var, Mémoire d'habilitation à diriger des recherches

Seppecher P (1996b) A numerical study of a moving contact line in Cahn-Hilliard theory. Int J Eng Sci 34:977–992

Seppecher P (2001) Line tension effect upon static wetting. Oil Gas Sci Technol Rev IFP 56:77–81

Seppecher P (2002) Second-gradient theory: application to Cahn-Hilliard fluids. In: Maugin GA, Drouot R, Sidoroff F (eds) Continuum thermomechanics, solid mechanics and its applications, vol 76. Springer, Netherlands, pp 379–388

Silling S (2000) Reformulation of elasticity theory for discontinuities and long-range forces. J Mech Phys Solids 48(1):175–209

Silling S, Lehoucq R (2008) Convergence of peridynamics to classical elasticity theory. J Elast 93(1):13–37

Silling S, Epton M, Weckner O, Xu J, Askari E (2007) Peridynamic states and constitutive modeling. J Elast 88(2):151–184
Soubestre J, Boutin C (2012) Non-local dynamic behavior of linear fiber reinforced materials. Mech Mater 55:16–32
Steeb H, Diebels S (2004) Modeling thin films applying an extended continuum theory based on a scalar-valued order parameter.: Part i: isothermal case. Int J Solids Struct 41(18–19):5071–5085
Steigmann D (1996) The variational structure of a nonlinear theory for spatial lattices. Meccanica 31(4):441–455
Steigmann D, Faulkner M (1993) Variational theory for spatial rods. J Elast 33(1):1–26
Steigmann DJ (1992) Equilibrium of prestressed networks. IMA J Appl Math 48(2):195–215
Steigmann DJ (2002) Invariants of the stretch tensors and their application to finite elasticity theory. Math Mech Solids 7(4):393–404
Steigmann DJ (2003) Frame-invariant polyconvex strain-energy functions for some anisotropic solids. Math Mech Solids 8(5):497–506
Steigmann DJ (2009) A concise derivation of membrane theory from three-dimensional nonlinear elasticity. J Elast 97(1):97–101
Steigmann DJ, Ogden RW (1997) Plane deformations of elastic solids with intrinsic boundary elasticity. Proc R SocLond A: Math Phys Eng Sci 453(1959):853–877
Steigmann DJ, Ogden RW (1999) Elastic surface–substrate interactions. Proc R Soc Lond A: Math Phys Eng Sci 455(1982):437–474
Steinmann P (2008) On boundary potential energies in deformational and configurational mechanics. J Mech Phys Solids 56(3):772–800
Steinmann P, Elizondo A, Sunyk R (2007) Studies of validity of the Cauchy-Born rule by direct comparison of continuum and atomistic modelling. Model Simul Mater Sci Eng 15(1):S271–S281
Sunyk R, Steinmann P (2003) On higher gradients in continuum-atomistic modelling. Int J Solids Struct 40(24):6877–6896
Thomas I (1939) Greek Mathematical Works: Vol. I Thales to Euclid. No. 335 in Loeb Classical Library, Loeb
Toupin RA (1962) Elastic materials with couple-stresses. Arch Ration Mech Anal 11(1):385–414
Toupin RA (1964) Theories of elasticity with couple-stress. Arch Ration Mech Anal 17(2):85–112
Truesdell C (1968) Essays in the history of mechanics. Springer, Berlin
Truesdell C (1977) A first course in rational continuum mechanics. Academic Press, New York
Truesdell C, Noll W (2004) The non-linear field theories of mechanics. Springer, Berlin
Truesdell C, Toupin R (1960) Classical field theories of mechanics. In: Flügge S (ed) Handbuch der Physik, vol III/1. Springer, Berlin
Vailati G (1987) Scritti, a cura di M. Quaranta, Bologna, Forni, pp 3–17
de Vaux C (1893) Les Mécaniques ou l'Élévateur de Heron d'Alexandrie, sur la version arabe de Qostā Ibn Lūqā. Journal asiatique, neuvième série 1, 2:286–472, 152–269, 420–514
Voicu A (1999) La continuità di varie questioni meccaniche da Stratone a Newton. Master degree thesis in mathematics (advisor: Lucio Russo), University of Rome Tor Vergata
Wallis J (1666) An essay of Dr. John Wallis, exhibiting his hypothesis about the flux and reflux of the sea, taken from the consideration of the common center of gravity of the earth and moon; Together with an appendix of the same, containing an answer to some objections, made by several persons against that hypothesis. Philos Trans Roy Soc London 1(1–22):263–281
Winter TN (2007) The Mechanical Problems in the Corpus of Aristotle, digitalCommons@University of Nebraska - Lincoln
Yang Y, Misra A (2010) Higher-order stress-strain theory for damage modeling implemented in an element-free galerkin formulation. Comput Model Eng Sci 64:1–26
Yang Y, Ching WY, Misra A (2011) Higher-order continuum theory applied to fracture simulation of nanoscale intergranular glassy film. J Nanomechanics Micromechanics 1(2):60–71
Yeremeyev VA, Zubov LM (1999) The theory of elastic and viscoelastic micropolar liquids. J Appl Math Mech 63(5):755–767

Computational Analysis of the Size Effects Displayed in Beams with Lattice Microstructures

Martin A. Dunn and Marcus A. Wheel

Abstract The mechanical behaviour of finite element based computational representations of heterogeneous materials with regular or periodic cellular microstructure is compared to existing closed form analytical predictions of their constitutive behaviour available within the open literature. During the computational investigation, slender, geometrically similar rectangular beams of different sizes which are comprised of regular, repeating arrangements of square cellular microstructures were represented using the finite element analysis (FEA) software ANSYS. Flexural loading of the virtual samples reveals that the materials exhibit the theoretically forecast size effect from which the relevant material constitutive properties, notably the flexural modulus and characteristic length can be identified. Initial findings suggest that while there is agreement between the numerically determined and theoretically predicted moduli the characteristic lengths in bending, l_b, calculated from the numerical data appear to differ from the theoretical forecasts. Moreover, the computational representations indicate that finite sized material samples are capable of exhibiting size effects not predicted by the more general higher order constitutive theories. Results indicate that the nature of the size effect appears to depend on the prescription of the sample surfaces with respect to the specified microstructure of the material. While these unanticipated size effects show qualitative agreement with that forecast for a simple laminate material comprised of alternating stiff and compliant layers the consequences may be profound for experimental mechanical testing of such materials.

Keywords Micropolar continuum · Lattice microstrucutre · Negative size effects · Flexural loading

M.A. Dunn (✉) · M.A. Wheel
Department of Mechanical and Aerospace Engineering, University of Strathclyde, Glasgow G1 1XJ, UK
e-mail: martin.dunn@strath.ac.uk

M.A. Wheel
e-mail: marcus.wheel@strath.ac.uk

H. Altenbach and S. Forest (eds.), *Generalized Continua as Models for Classical and Advanced Materials*, Advanced Structured Materials 42,
DOI 10.1007/978-3-319-31721-2_6

1 Introduction

This body of work investigates the use of Finite Element Analysis, FEA, to determine the underlying material constitutive properties of idealised heterogeneous materials comprised of a regular array of repeating elements that form a lattice like microstructure. All of the materials being considered in this investigation are of a two dimensional composition whereby the material is uniformly extruded in the third dimension.

The mechanical properties of heterogeneous materials have been the focus of a considerable amount of research in recent years due to the naturally lower density such materials offer in comparison to the volume they occupy. Cellular materials are one such type of heterogeneous structure which is of great interest to the engineering community (Gibson and Ashby 1999). Materials which consist of a geometrically repeating microstructure are of particular interest due to the fact that it would allow for optimised structures to be generated which have predictable, deterministic mechanical properties.

Improvements in the capabilities of additive manufacturing, 3D printing and technologies such as selective laser melting and digital light processing have allowed for the creation of materials with geometrically exact and regular microstructures to be produced (Luxner et al 2005). This allows for the creation of materials which have mechanical or thermal properties which are tailored to the specific application in mind.

With this ever increasing demand for lightweight materials, there is a need to be able to better define the underlying mechanical properties of such materials. One way of characterising heterogeneous materials is through the use of generalised continuum theories. Higher order theories such as couple stress, micropolar and micromorphic elasticity have been developed to do this. Generalised continuum theories of this kind were first introduced in the early twentieth century by the Cosserat brothers (Cosserat and Cosserat 1909). However, further research in this area remained relatively dormant until the middle of the twentieth century when Eringen (1966), Eringen and Suhubi (1964) among others popularized micropolar and related elasticity theories.

It has been shown that in heterogeneous materials exhibiting generalized continuum behaviour there exists an inherent size effect in geometrically similar material samples of differing scales (Gauthier and Jahsman 1975). While this behaviour can be categorised well with theoretical solutions, it has proved troublesome to effectively show this behaviour through experimental means.

Gauthier and Jahsman (1975) attempted to create an idealised micropolar material using cylinders which had aluminium shot evenly dispersed throughout the material. Yang and Lakes (1981) also investigated the size effects displayed in testing cortical bone. More recently researchers like Beveridge (2010), Frame (2013) investigated the size effects that were displayed in idealised structures with regular arrays of voids. It was shown that the underlying micropolar material properties of these materials could be found by either experimental testing of real materials or computational analysis of their virtual counterparts.

While this more recent research shows that it is possible to model materials whereby the cell size is comparable to the overall macrostructure of the material, there is still a practical limit in the ability to both accurately model and test materials in which the cell walls are very thin in comparison to the cell size. Such limitations include the variation in deformation modes and sensitivity to loading effects that can potentially mask behaviour and thereby corrupt the identification of the underlying material properties.

It is the aim of this paper to highlight the potential difficulties that are exhibited by investigating computational models of very low mass density cellular materials in order to provide a better understanding of how such materials might behave when examined in practice.

2 Objective of Numerical Modelling

The aim of this work is to provide a direct comparison between the theoretically predicted behaviour for a material comprised of a repetitive planar lattice and its computationally predicted behaviour as identified through virtual experiment. The micropolar constitutive properties are derived using the homogenisation method seen in dos Reis and Ganghoffer (2011).

The relevant material constitutive parameters were derived by virtually or computationally testing the materials using the finite element analysis (FEA) software ANSYS and applying the size effect method in Eringen (1999).

3 Micropolar Elasticity

Micropolar or Cosserat elasticity is a higher order theory that incorporates an additional degree of freedom, a micro rotation, to describe deformation within the microstructure of the material. It is a more general theory which converges to either classical elasticity or couple stress theory (Eringen 1966) in the appropriate limit. In order to differentiate between classical, couple stress or genuinely micropolar behaviour, a series of experimental or virtual tests can be performed on the material to determine the constitutive properties.

Six elastic moduli are required to fully define 3-dimensional Cosserat elastic solids (Eringen 1966). These are λ, μ, α, β, γ and κ. These constants can be compared to their classical counterparts (Lakes 1995). In the two-dimensional case the six constants can be expressed in terms of just four parameters relevant to this investigation as summarised below.

$$\text{Young's Modulus,} \quad E_{\mathrm{m}} = \frac{(2\mu + \kappa)(3\lambda + 2\mu + \kappa)}{(2\lambda + 2\mu + \kappa)} \tag{1}$$

$$\text{Shear Modulus,} \quad G_m = \frac{(2\mu + \kappa)}{2} \tag{2}$$

$$\text{Poisson's Ratio,} \quad \nu_m = \frac{\lambda}{(2\lambda + 2\mu + \kappa)} \tag{3}$$

$$\text{Characteristic Length for Bending,} \quad l_b = \sqrt{\frac{\gamma}{2(2\mu + \kappa)}} \tag{4}$$

This paper concentrates on the constants which can be derived from uniaxial and flexural tests; the micropolar Young's Modulus, E_m and the characteristic length in bending, l_b.

3.1 *Micropolar Elasticity and Flexure*

The formulation for the generalised theory of a micropolar beam in 3 point bending has been derived in Beveridge (2010). Similar approaches have also been employed by dos Reis and Ganghoffer (2012). The maximum displacement for such a beam subjected to this loading is,

$$v_{max} = \frac{WL^3}{48(E_m I + \gamma A)} \tag{5}$$

where W (N) is the applied load, L (mm) is the length of the beam, E_m (Nmm^{-2}) is the micropolar Young's Modulus, I (mm^4) is the second moment of area, γ (N) is a length scale dependent micropolar couple modulus relating the couple stresses to the curvatures and A (mm^2) is the cross sectional area of the beam. This equation can be rearranged in terms of the beam stiffness, K (Nmm$^-$2) as a function of the beam depth, d (mm).

For a rectangular cross section the second moment of area, I and area, A are given as,

$$I = \frac{bd^3}{12} \tag{6}$$

$$A = bd \tag{7}$$

where b and d are the beam breadth and depth respectively.

The stiffness, K of the beam is then given as,

$$K = 4E_m b\left(\frac{d}{L}\right)^3\left[1 + \left(\frac{l_c}{d}\right)^2\right] \tag{8}$$

where l_c (mm) is the characteristic length for a rectangular cross section given by:

$$l_c = \sqrt{\frac{12\gamma}{E_m}} \tag{9}$$

Note this characteristic length measure differs from that of Eq. (4) by a factor of $\sqrt{24}$. Thus two different characteristic lengths have been defined in this paper; the characteristic length in bending, l_b, as conventionally defined for a micropolar material according to Eq. (4) and the characteristic length of a beam with a rectangular cross section, l_c, appearing in Eq. (8). The first parameter, l_b denotes the length scale dependent size effect that is observed in a heterogeneous material that is subjected to a bending load.

It should be noted that when l_c is equal to zero Eq. (8) reduces to the stiffness of a classically elastic beam subjected to 3 point loading,

$$K = 4E_m b \left(\frac{d}{L}\right)^3 \tag{10}$$

Thus it can be seen that in a classically elastic beam no size effect is expected to be observed between beams of different sizes but which have the same breadth, b, and the same length to depth aspect ratio, (L/d).

The stiffness, K of the beam in bending has been derived in terms of the linearly varying direct stresses, τ_{xx}, and uniform couple stresses, m_{xz}, acting on the beam cross section. It can be seen from Eq. (8) that the stiffness of the beam varies with the reciprocal of the square of the overall depth. When plotted on a graph, the characteristic length of the beam, l_c^2 can thus be calculated from the slope or gradient of the stiffness variation which is expected to vary linearly with the size measure $1/d^2$. Although Eq. (8) is derived in terms of a beam subjected to 3 point bending, similar derivations can be obtained for beams loaded in 4 point or pure bending.

3.2 *Constitutive Properties of a Regular Square Lattice*

The microstructure of the test material comprised of a regular array of square voids as seen in Fig. 1. Each unit cell has a width and height, a and a thickness, t. Where two cells join the wall thickness is thus $2t$.

The stiffness matrix for a material with this square microstructure has been derived by dos Reis and Ganghoffer (2011) and is given as follows;

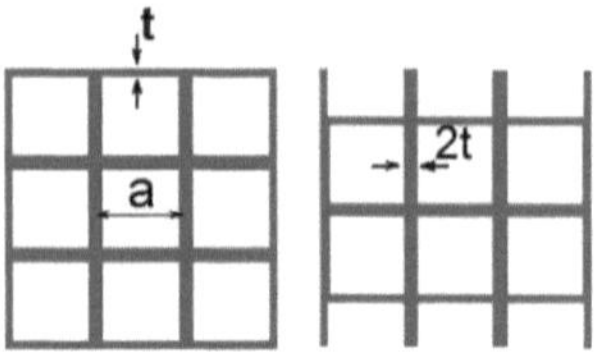

Fig. 1 Beam microstructure with closed and open outer surfaces

$$[K] = \begin{bmatrix} E_s\eta & 0 & 0 & 0 & 0 & 0 \\ 0 & E_s\eta & 0 & 0 & 0 & 0 \\ 0 & 0 & E_s\eta^3 & 0 & 0 & 0 \\ 0 & 0 & 0 & E_s\eta^3 & 0 & 0 \\ 0 & 0 & 0 & 0 & \frac{a^2E_s\eta^3}{12} & 0 \\ 0 & 0 & 0 & 0 & 0 & \frac{a^2E_s\eta^3}{12} \end{bmatrix} \tag{11}$$

where E_s is the modulus of the underlying material that makes up the matrix, a is the cell size and η is the internal cell wall aspect ratio, $2t/a$.

The characteristic length of a micropolar beam relates to the matrix as follows (dos Reis and Ganghoffer 2011),

$$l_c^2 = \frac{12\gamma}{E_m} = \frac{12K_{55}}{(K_{33}+K_{34})} = \frac{a^2E_s\eta^3}{(E_s\eta^3+0)} = a^2 \tag{12}$$

K_{33}, K_{34} and K_{55} are the relevant stiffness elements in the 6×6 stiffness matrix, Eq. (11). They have been derived in dos Reis and Ganghoffer (2011) in terms of the micropolar constants. K_{33} is the bending stiffness, K_{34} is the micropolar Poisson's ratio, μ_m and K_{55} is the micropolar constant γ which is the couple modulus or microbending stiffness.

The characteristic length in bending for a square lattice can therefore be shown to be a function of the cell size and independent of the cell wall thickness. Substituting Eq. (12) into Eq. (4) yields:

$$l_b^2 = \frac{a}{24} \tag{13}$$

It can be seen that the characteristic length of a material with a square microstructure is therefore independent of the cell wall thickness and is only dependent on the cell size.

4 Computational Representation of Cellular Material

Computational models were generated in the FEA software ANSYS using Beam189 elements. Beam189 elements are 3 node Timoshenko beam elements which have six degrees of freedom, three displacements and three rotations at each node. The models were tested under static loading conditions.

The models comprised of geometrically similar rectangular beams which had between 1 and 10 rows of cellular microstructure through the depth. The microstructure of the material consists of an array of square unit cells that forms a regular grid structure (Fig. 1). Each cell has a length, depth and breadth of 1 mm. The thickness, t of each cell wall is defined for each analysis. The cell centres have a horizontal and vertical pitch of 1 mm from their neighbours. Where two neighbouring cells join, the resulting wall thickness, $t_w = 2t$. Symmetry conditions were also employed at half the length in order to reduce the model size. An example of a typical beam loaded in 3 point bending is shown in Fig. 2. As half symmetry was evoked in this case, the loading was simulated with a cantilever load. The beam was constrained vertically at the nodes on the left side of the beam. Additionally, a horizontal constraint was applied to the centre node of this face to prevent any horizontal movement of the beam at this face. The applied load, W at the opposite face is half that required for a full model to keep continuity in the analysis.

In each model the matrix material was assigned a Young's Modulus, E of $70\,\text{KNmm}^{-2}$ and a Poisson's ratio, ν of 0. The breadth, b was assigned a unit value, 1 mm for all models. Beams with a variety of depths and a common aspect ratio were then virtually loaded as illustrated. Two different wall thicknesses, $t = 0.1$ and 0.01 mm were considered in the analysis to assess what effect this has on the behaviour of the beams in bending.

The stiffness of each beam was computed from the applied load and resulting deflection data was plotted in the graphical form of Eq. (8) as stiffness against the reciprocal of the depth squared.

Two different unit cells are defined in this analysis; an 'open' and 'closed' square cell (Fig. 3). When assembled, the underlying microstructure of the beams is identical for each case. The key difference being that in the case of the 'closed' unit cell, the top and bottom surfaces of the beam macrostructure do not intersect the square voids constituting the microstructure. In the case of the 'open' cell however; the top and bottom surfaces of the beam bisect the microstructure at half the cell depth thus opening or exposing those cells lying adjacent to the surfaces.

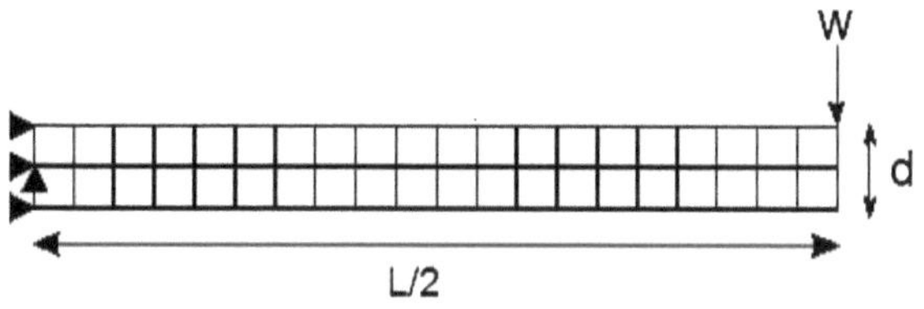

Fig. 2 Half symmetry model of a beam in 3-point bending

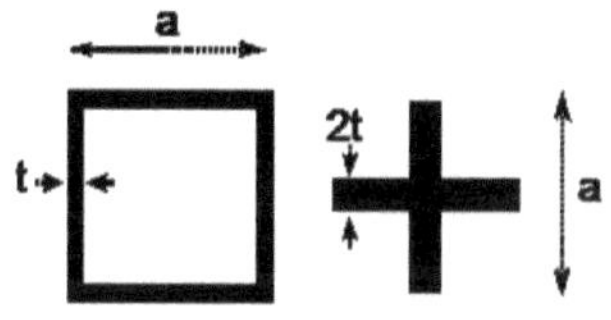

Fig. 3 Closed and open unit cell

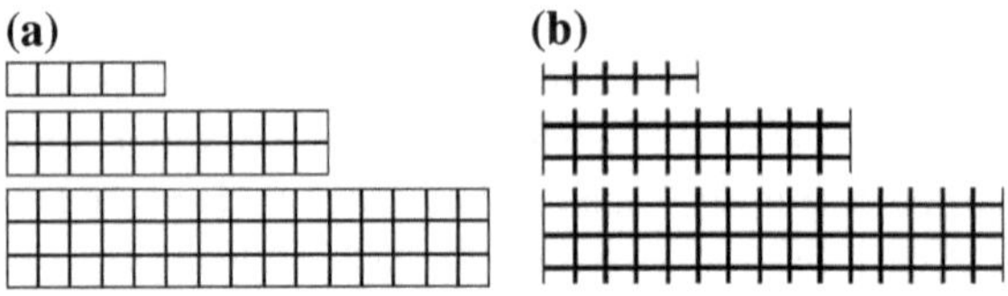

Fig. 4 Examples of beams created from closed and open unit cells with varying rows of microstructure

The prescription of the microstructure with regards to the surfaces of the macrostructure is significant as the second moment of area through the cross section of a beam with the same number of voids will vary. The idealised second moment of area is the same for each beam with the same number of cells through the depth of the beam. Figure 4 shows the difference in the model surface topology for beams with varying rows of voids through the depth.

In the 'open' cell case, the surface of the beam intersects the underlying microstructure half way through a cell. The cells were divided by two in the 'open' case in order to keep an even mass distribution through the depth of the samples. All beam samples with the same aspect ratio and depth have the same mass density regardless of the microstructure surface topology.

The simulations carried out were as follows:

- Uniaxial tests of beams with open and closed cell microstructures.
- 3 Point, 4 Point and constant moment bend tests at various aspect ratios for beams with closed cell microstructure.
- Constant moment or pure bending tests of beams with an open unit cell on the top and bottom surfaces.

By comparing the different bending modes, it is the aim to highlight the difficulties in loading low density structures such as these by standard mechanical tests, with aims to make recommendations as to what might be the best procedure for testing samples by physical experiment.

5 Results

5.1 *Uniaxial Tests*

Initially uniaxial tests were conducted to determine the micropolar modulus, E_m of the samples. The samples were subjected to a uniaxial load, W of 1×10^{-6} N and the beam extension recorded. The uniaxial relationship of stress, σ to strain, ε was used to find the overall modulus of the beam so that the modulus could be derived from:

$$E_m = \frac{\sigma}{\varepsilon} = \frac{WL}{A\Delta L} \tag{14}$$

Table 1 Calculated modulus of beams with open and closed cell microstructure from uniaxial tests

Cell wall thickness, t (mm)	E_m, Simulation ($KNmm^{-2}$)	E_m, Theory ($KNmm^{-2}$)
0.1	13.999	14
0.01	1.399	1.4

Tests were conducted using various numbers of rows of voids to check for any size dependency. The results are summarised in Table 1 and compared to the theoretical value; this being given by dos Reis and Ganghoffer (2011), Wang and McDowell (2004),

$$E_m = \frac{E_s t_w}{a} \tag{15}$$

where t_w is the thickness of two adjoining cells and a is the cell width.

It can be seen that the uniaxial tests accurately predict the micropolar modulus of the beams being tested. As anticipated, no variation in modulus was observed for a given cell wall thickness when the number of cell rows was altered. As there is no size effect displayed in uniaxial testing, a Cauchy continuum is sufficient to describe the material behaviour in this mode of loading.

5.2 *Flexural Testing*

Beams with a varying number of cells through the depth were tested in various flexural modes including three point, four point and pure bending. The resulting stiffness was then calculated for each case and plotted against the reciprocal of the depth squared in accordance with Eq. (8). It should be noted that from the resulting plots no indication of material stiffness can be derived for a sample smaller than a single unit cell ($1/\text{depth}^2 = 1$) as this would imply an infinite increase in stiffness with decreasing depth which would be unphysical.

Figure 5 shows the normalised results for beams with a closed cell micro-structure loaded in 3 point bending obtained by FEA. The tests were undertaken using centre symmetry and an applied transverse load of 1×10^{-6} N. It was found that the width of the unit cells and the aspect ratio of the beams had a significant effect on the overall stiffness of the beam. For the beams which had a cell wall thickness of $t = 0.01$ mm a 'negative' size effect is observed whereby the beam is seen to get stiffer as the number of cells through the depth is increased. The variation in stiffness was also seen to be non-linear. This effect is independent of the overall beam aspect ratio that was used in the test. Both the non-linear stiffness variation and the negative size effect are contradictory to Eq. (8) which predicts a linear decrease in stiffness with an increasing number of rows of material microstructure.

However, when the cell wall thickness was increased to t=0.1 mm a 'positive' size effect was observed whereby the stiffness of the beams decreased with an increasing

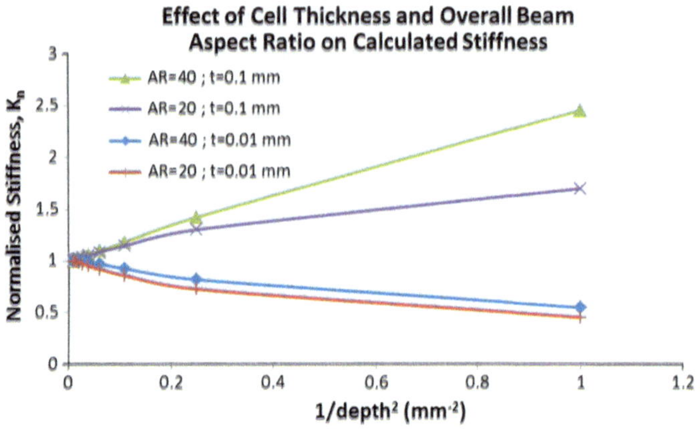

Fig. 5 3 Point bending analysis of beam with 1–10 rows of voids at various aspect ratios and internal wall thicknesses

number of voids through the depth. A linear variation in stiffness corresponding to Eq. (8) was not observed at an aspect ratio of 20 but becomes apparent when this is increased to 40. The resulting characteristic length derived from this linear variation using Eq. (8) has been summarised in Table 2.

Virtual 4 Point bend tests were then carried out on the beams. The beams were tested at a higher aspect ratio than previously in order to keep the aspect ratio of the section between the two inner loading points similar to the previous tests. Care was also taken to ensure that load and reaction points were located at joints in the microstructure. The vertical deflection in the analysis was calculated by taking the difference between the deflection at the centre of the beam, y_2 and a point inward of the load application point, y_1 as demonstrated in Fig. 6. The deflection of the beam was not recorded from the position of the applied load, W, as it was found in initial testing to give a nonlinear size effect. This was especially evident at lower aspect ratios and when there were only a few rows of microstructure through the depth of the beam. Two beam aspect ratios were considered in the analysis; an aspect ratio of 50 and one of 24. This correlated to an internal aspect ratio between the loading

Table 2 Calculated characteristic length derived from beams with a closed cell surface structure

	Calculated characteristic length (mm^2)		Theoretical characteristic length (mm^2)	
	l_c^2	l_b^2	l_c^2	l_b^2
3 Point bending	1.4685	0.08119	1	0.04167
4 Point bending	2.006	0.0836	1	0.04167
Constant moment	2.021	0.0842	1	0.04167

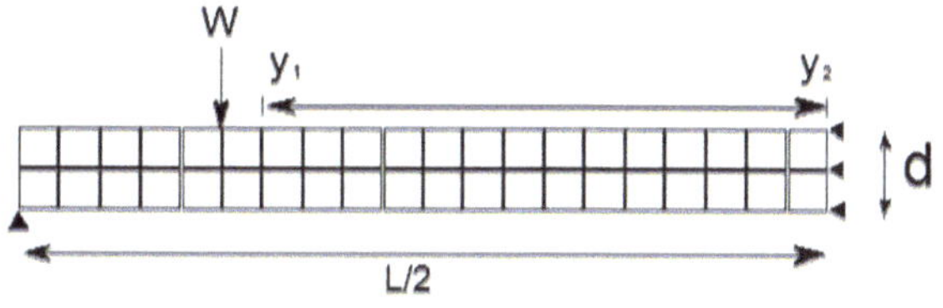

Fig. 6 4 Point beam loading configuration

points of 40 and 16 respectively. This aspect ratio was reduced by 1 in each case in order to avoid the loading effects that were described above.

From Fig. 7 it can be seen that a size effect approaching linearity is seen in all cases. At the lower aspect ratio the size effect is over estimated when compared to that which is seen in the beams with an aspect ratio of 50. It can also be seen that the size effect starts to become significantly nonlinear at the lower aspect ratio when the sample size increases above 4 rows of cells.

At the higher aspect ratio, the variation in the observed size effect is minimal between differing cell wall thicknesses. There is however a slight deviation from linearity in the material which had a cell wall thickness of $t = 0.01$ mm. While this deviation is minimal, it accounts for a 4 % increase in the calculated characteristic length. This again highlights the sensitivity of thin walled materials to the loading mode. The calculated characteristic length under 4 point loading was found to be $l_c^2 = 2.006\,\text{mm}^2$.

Figure 8 shows the normalised variation in beam stiffness for the beams that were subjected to pure bending or a constant moment. It can be seen that in this mode of flexure there is less sensitivity in the stiffness variation to the aspect ratio of the beam or the cell wall thickness. It is noted however that a slight non linearity is observed at an aspect ratio of 20 for the cell wall thickness of 0.01 mm. The characteristic length calculated from the size effect exhibited by the beams is very similar to that of the 4 point loading tests as noted in Table 2.

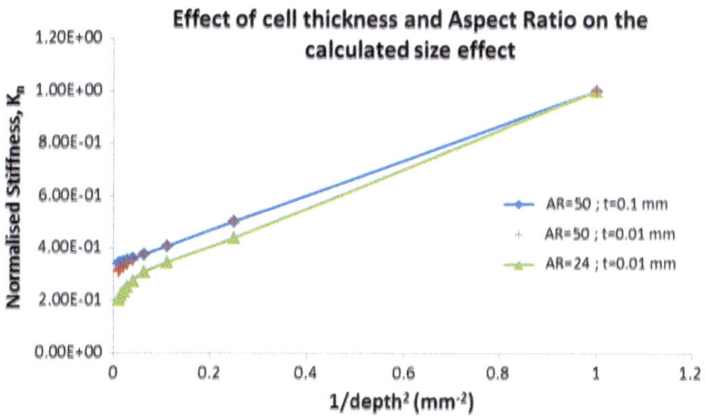

Fig. 7 Observed size effect in beams subjected to 4 point bending

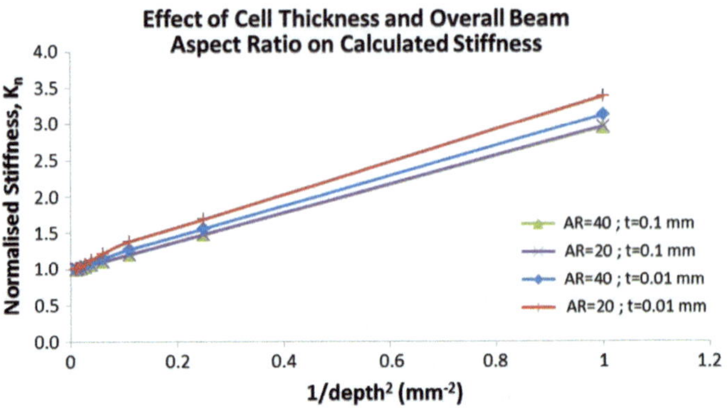

Fig. 8 Constant moment loading of beams with a closed cell microstructure

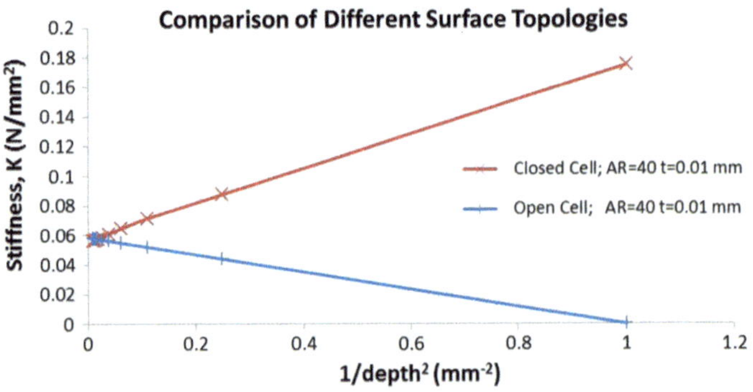

Fig. 9 Comparison between the size effects observed in beams which have different surface topologies from constant moment load tests in ANSYS

It is also useful to compare the characteristic length in bending, l_c and the characteristic length, l_b with the unit cell size of the microstructure. For the results seen in Table 2 a value of $l_c = 1.41$ mm and $l_b = 0.29$ mm were recorded for a unit cell size of 1 mm. The theoretical value of lb being 0.20 mm. It can be observed that the value of the characteristic length is of similar order to the microstructure of the beam which highlights that higher order effects need to be considered when modelling such structures.

The size effect observed in the material was also considered for when the cell microstructure consisted of open unit cells on the top and bottom surfaces of the beam. From Fig. 4b it can be seen that in this case the beam is comprised of rows of microstructure which is offset with regards to the beam surface by half a cell. The number of cells per beam is the same as the previous case, but the surface topology has been altered to assess the effect this has. The beams being tested had an aspect ratio of 40 and were subjected to a pure moment load of 2×10^{-5} N per row of voids. The modulus of the beam can be derived by substituting the y-intercept value

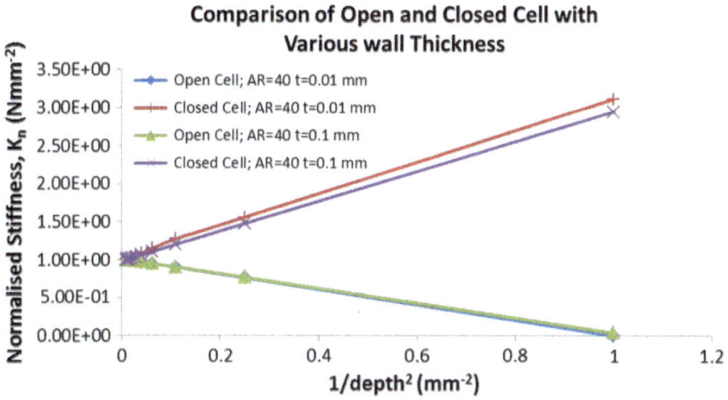

Fig. 10 Normalised comparison of the size effect observed in pure bending tests carried out in ANSYS between lattices with different wall thicknesses

found in Fig. 9 into Eq. (8). It was calculated that the modulus of the open cell beam was 1399.2 N/mm^2 and the closed cell, 1365.6 N/mm^2. This compares favourably to the theoretical modulus of a square lattice beam with an internal wall thickness t of 0.01 mm (1.4 KN/mm^2).

It can be seen from Fig. 9 that there is a negative size effect when the surface of the beam has an open cell structure at the surface. The variation in stiffness is linear and tends to a finite value as the sample depth increases.

While the negative size effect does not accord with the stiffness variation forecast by Eq. (8) it is nevertheless interesting to infer a characteristic length value from the gradient of the variation depicted in Fig. 8. The characteristic length extracted for the open cell microstructure beam was thus found to be $l_c^2 = 1\,\text{mm}^2$. This gives a characteristic length in bending for the square microstructure as $l_b^2 = 0.04165\,\text{mm}^2$. Although this size effect is negative, the value is approximately the same as the predicted characteristic length of a square lattice of the same cell size using Eq. (13) (dos Reis and Ganghoffer 2011) (0.04167 mm^2).

Figure 10 is a normalised comparison between the stiffness in beams of differing cell wall thickness. The samples are normalised by the stiffness of the largest sample for each given set of data. It can be seen that the normalized size effect is now independent of the cell wall thickness and only dependent on the surface topology.

6 Discussion

The uniaxial tests performed on the samples showed good correlation to the theoretical prediction. No size effect was observed when varying sample sizes were tested. This is in line with the theoretical predictions for a micropolar material.

The tests that were undertaken using 3 point bend loading were, in general, not in agreement with the predictions of this generalized continuum theory. A nonlinear size effect was observed in most tests and a negative effect was observed when the aspect ratio was reduced to 20:1. It has been concluded from this that shear deformation effects may well begin to dominate under these conditions. Three point bend tests may therefore be unsuitable for testing very low density materials with a lattice structure.

As the results of Fig. 7 indicate, 4 point bending provided a better means for testing the materials. The section of the beam which is located between the two loading points is subjected to a constant moment and is therefore free from any shear effects. The point at which the beam deflection was measured from was found to be significant. A substantial underestimate of the stiffness was observed when the displacement measurements were taken from the point of loading and the resulting size effect was nonlinear. It was especially evident in the cells with the smallest wall thickness. This is most likely due to the fact that the shear effects present in the outer section of the beam located between load points and adjacent support have a significant influence on the local beam deformation in the vicinity of the load point, thereby corrupting the resulting stiffness estimates. It is recommended that the measured beam deflection should be taken some way in towards the centre from the point of loading in order to minimise this local effect.

The calculated size effect for beams subjected to 4 point bending conformed well to the constant moment loading case. The beams subjected to the constant moment demonstrated the most consistent in this analysis. Although pure bending of a specimen is difficult to reproduce in a physical test, computationally it provides a good benchmark for the investigation as no shear deformation is induced in the beam sample.

From the tests that have been performed it has been established that the characteristic length for a square lattice derived from a rectangular beam is quantitatively different from the theoretical characteristic length. It was observed that l_b^2 calculated from the beams with the solid surface beams was twice as large as the theoretical prediction for the characteristic length for a square microstructure. The size effect in this case is positive which follows the theoretical predictions.

Interestingly, when the outer surface of the beam consists of open voids it was found that the characteristic length inferred from the magnitude of the size effect was in line with that of the theoretical predictions. However, a negative size effect is observed in this case whereby the beams are observed to get stiffer as the number of voids through depth increases. Since such an effect is not theoretically forecast, inferring a characteristic length value from it is certainly not rigorous and needs to be rated with caution. However, the modulus of the beams is in line with the theoretical predictions in both open and closed cell cases.

7 Conclusion

A size effect has been shown to exist for materials with an idealised microstructure. It can be seen from the results that this size effect is linear when external loading effects are removed. The magnitude and nature of the size effect observed was also shown to depend on the prescription of the surface topology. The resulting characteristic length that is obtained from testing these structures using FEA simulations is found to vary from the theoretically predicted value by a factor of $\sqrt{2}$. The characteristic length inferred when the microstructure is intersected by the surface of the beam is seen to match that of the predicted characteristic length in bending. This result suggests that it is necessary to consider the surface topology of actual structures in order to fully categorise the material properties of a heterogeneous material.

While this variation in predicted size effect is unexpected, it highlights the need for a better understanding of the relationship between the microstructure and macrostructure of physical materials in order to provide accurate predictions to the underlying material properties.

The work in this investigation only covers materials whose microstructure is 2 dimensional in nature. Further, 3 dimensional computational investigations are required to fully categorise the micropolar properties of these structures.

References

Beveridge AJ (2010) Novel computational methods to predict the deformation of macroscopic heterogeneous materials. Univerity of Strathclyde, Glasgow

Cosserat E, Cosserat F (1909) Theory of deformable bodies. Herman et fils, Paris

dos Reis F, Ganghoffer JF (2011) Construction of micropolar continua from the homogenization of repetitive planar lattices. In: Altenbach H, Eremeyev VA (eds) Mechanics of generalized continua, ASM, vol 7. Springer, Heidelberg, pp 193–226

dos Reis F, Ganghoffer JF (2012) Construction of micropolar continua from the asymptotic homogenization of beam lattices. Comput Struct 112–113:354–363

Eringen AC (1966) Linear theory of micropolar elasticity. Indiana Univ Math J 15(6):909–923

Eringen AC (1999) Microcontinuum field theories, vol I. Foundations and solids, Springer, New York

Eringen AC, Suhubi ES (1964) Nonlinear theory of simple micro-elastic solids - I. Int J Eng Sci 2(2):189–203

Frame JC (2013) A computational and experimental investigation into the micropolar elastic behaviour of cortical bone. Univerity of Strathclyde, Glasgow

Gauthier RD, Jahsman WE (1975) A quest for micropolar elastic constants. Trans ASME J Appl Mech 42(2):369–374

Gibson LJ, Ashby MF (1999) Cellular solids—structures and properties, 2nd edn. Cambridge University Press, Cambridge

Lakes R (1995) Experimental methods for study of cosserat elastic solids and other generalized elastic continua. In: Mühlhaus H (ed) Continuum models for materials with micro-structure. Wiley, New York, pp 1–22

Luxner M, Stampfl J, Pettermann H (2005) Finite element modeling concepts and linear analyses of 3d regular open cell structures. J Mater Sci 40(22):5859–5866

Wang AJ, McDowell DL (2004) In-plane stiffness and yield strength of periodic metal honeycombs. Trans ASME J Eng Mater Technol 126(2):137–156

Yang J, Lakes R (1981) Transient study of couple stress effects in compact bone: Torsion. J Biomech Eng 103:275–279

Inelastic Interaction and Splitting of Strain Solitons Propagating in a One-Dimensional Granular Medium with Internal Stress

Vladimir I. Erofeev, Vladimir V. Kazhaev and Igor S. Pavlov

Abstract A one-dimensional model of a granular medium with internal stress is considered that represents a chain consisting of elastically interacting ellipsoidal-shaped particles, which possesses translational and rotational degrees of freedom. By means of a long-wavelength approximation, nonlinear differential equations have been derived that describe the propagation of longitudinal, transverse and rotational waves in such a medium. Analytical dependencies of the velocities of elastic waves and the nonlinearity coefficients on the sizes of particles and the parameters of interactions between them have been found. If longitudinal waves are not excited in the medium and in the field of low frequencies, when the rotational degree of freedom of particles can be neglected, the obtained three-mode system reduces to one equation for the transverse mode. On the base of this equation containing cubic nonlinearity, numerical investigations of counter and passing interactions of strongly nonlinear soliton-like subsonic and supersonic waves have been performed. In particular, effects of splitting of supersonic solitary waves are demonstrated.

Keywords Granular medium with internal stresses · Ellipse-shaped particles · Cubic nonlinearity · Counter and passing interactions of strain solitons · Splitting of solitons

V.I. Erofeev (✉) · V.V. Kazhaev · I.S. Pavlov
Mechanical Engineering Research Institute of Russian Academy of Sciences,
Nizhny Novgorod Lobachevsky State University, 85 Belinskogo St.,
603024 Nizhny Novgorod, Russia
e-mail: erof.vi@yandex.ru; erf04@sinn.ru

V.V. Kazhaev
e-mail: ipmvvk@mail.ru

I.S. Pavlov
e-mail: ispavl@mts-nn.ru

H. Altenbach and S. Forest (eds.), *Generalized Continua as Models for Classical and Advanced Materials*, Advanced Structured Materials 42,
DOI 10.1007/978-3-319-31721-2_7

1 Introduction

Mathematical models describing propagation and interaction of nonlinear waves in distributed systems are usually divided into integrable by the inverse scattering problem and non-integrable by this method (Dodd et al. 1982). Such models are called, in the slang of experts, "integrable systems" and "non-integrable systems."

In many works (see, for example, Dodd et al. 1982; Scott et al. 1973), it is analytically and numerically shown that in the integrable systems the localized waves (solitons) behave like particles: they conserve their individuality under collision and receive a phase shift ("elastic" interaction). This fact was confirmed by experiments with nonlinear waves in plasma, in a liquid with gas bubbles, as well as with electromagnetic waves (Ostrovsky et al. 1972; Lonngren 1978).

Another scenario of interaction of localized waves, besides the "elastic" one, is also possible in non-integrable systems. If an overtaking (passing) interaction of soliton-like waves takes place, then they radiate a part of their energy in the form of quasi-linear wave packets ("inelastic" interaction) (Abdullow et al. 1976).

Both effects of inelastic interactions and effects of splitting of strongly nonlinear waves, when additional soliton-like waves are extracted from a wave packet after interaction, have been experimentally observed in Potapov and Vesnitsky (1994). Splitting effects have been obtained for the opposite interaction of strongly nonlinear waves propagating along a rubber band.

A set of nonlinear partial differential equations describing wave processes in a granular medium has been derived in this paper by the structural modeling method (Berglund 1982; Chunyu and Tsu-Wei 2003; Pavlov and Potapov 2008). Features of interaction of localized shear waves have been investigated in the framework of these equations. The obtained mathematical model belongs to the class of non-integrable systems.

Currently, there are no regular methods of obtaining analytical solutions describing both overtaking and contradirectional interaction of soliton-like waves, therefore only the results of numerical modeling are presented in this paper. Processes of interaction under head-on collision, strictly speaking, cannot be described by evolution equations of the single-wave approximation (Engelbrecht et al. 1988) and it is necessary to employ the complete equations of nonlinear dynamics, which take into account the waves moving in both directions. As a rule, such equations are non-integrable and their solutions describe solitary waves that are not solitons in the strict mathematical sense (Dodd et al. 1982), but for simplicity we will also refer to them as solitons.

2 A Mathematical Model of a One-Dimensional Granular Medium

We consider a chain consisting of homogeneous particles (grains or granules) with the mass M having the shape of an ellipse with the axes d_1 and d_2. In the initial state, they are concentrated in the lattice edges and the distance between the centers of gravity of the neighboring granules along the x axis is a (Fig. 1). When moving in the plane, each particle has three degrees of freedom: the displacement of the center of gravity of the particle with the number $N = N(i)$ along the axes x and y (translational degrees of freedom u_i and w_i) and the rotation with respect to the center of gravity (rotational degree of freedom φ_i) (Fig. 2). The kinetic energy of the particle with the number $N(i)$ is described by the formula

$$T_i = \frac{M}{2}\left(\dot{u}_i^2 + \dot{w}_i^2\right) + \frac{J}{2}\dot{\varphi}_i^2, \qquad (1)$$

where $J = M(d_1^2 + d_2^2)/16$ is the inertia of the particle with respect to the axis passing through the center of gravity. It is assumed that the particle N interacts only with the two nearest neighbors in the chain, the centers of gravity of which are located at the distance a along the axis x from the particle N (Fig. 2). The central and non-central interactions of the neighboring granules are simulated by elastic springs of three types: central (with rigidity K_0), non-central (with rigidity K_1), and diagonal (K_2) (Lisina et al. 2001; Pavlov 2010; Erofeev and Pavlov 2016). The points of junctions of the springs with the particles are in the apexes of the rectangle of the maximum area, $ABCE$, inscribed in the ellipse (Fig. 2). Each rectangle has the size $h_1 \times h_2$, where $h_1 = d_1/\sqrt{2}$, $h_2 = d_2/\sqrt{2}$. The elongations of the central springs are determined by the changes of the distances between the geometrical centers of the rectangles $ABCE$ (Fig. 2), and the tensions of other springs are characterized by the variations of the distances between the apexes of these rectangles.

It is supposed that in the initial state all the springs with rigidity K_0, K_1 and K_2 have been already deformed (stretched or compressed), accordingly, at quantities

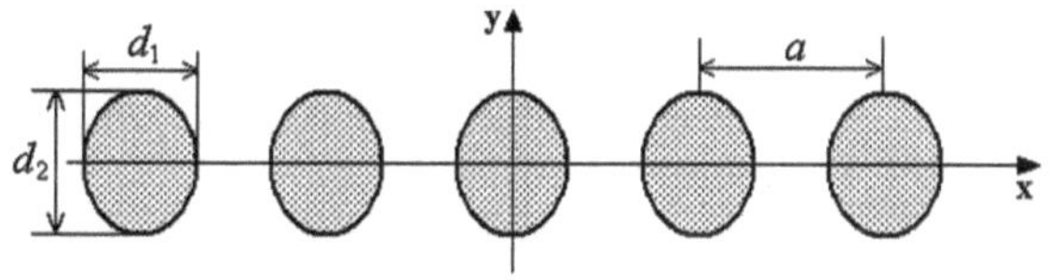

Fig. 1 Chain of ellipse-shaped particles

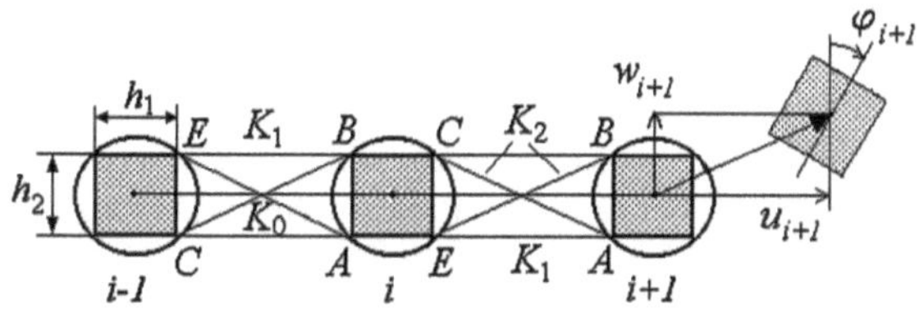

Fig. 2 Scheme of force interactions between the particles and kinematics

$\delta_0 \ll a$, $\delta_1 \ll a$, and $\delta_2 \ll a$ (Nikitina and Pavlov 2013). The preliminary deformations of the springs simulate internal stresses in the medium.

The displacements of the grains are supposed to be small in comparison with the period a of the considered one-dimensional lattice. The potential energy of a one particle is provided by the interaction of the particle N with two nearest neighbors in the chain and is described by the formula

$$U_N = \frac{1}{2}\left(\sum_{n=1}^{2}\frac{K_0}{2}D_{0n}^2 + \sum_{n=1}^{4}\frac{K_1}{2}D_{1n}^2 + \sum_{n=1}^{4}\frac{K_2}{2}D_{2n}^2\right), \tag{2}$$

where D_{ln} are the elongations of the springs ($l = 0, 1, 2$ is the type of the spring, respectively, K_0, K_1, or K_2; $n = i \pm 1$), which connect the central particle N with two nearest neighbors in the chain. This formula (2) contains an additional factor $1/2$, since the potential energy of each spring is equally divided between two particles connected by this spring.

In the approximation of smallness of the quantities

$$\begin{aligned} \Delta u_i &= (u_i - u_{i-1}) \sim a\varepsilon_0, \\ \Delta w_i &= (w_i - w_{i-1}) \sim a\varphi_i \sim a\varepsilon_0^{3/4}, \\ \Delta\varphi_i &= (\varphi_i - \varphi_{i-1}) \sim \varepsilon_0^{5/4}, \end{aligned} \tag{3}$$

where ε_0 is a deformation measure of the cell, and taking into account that

$$\Phi_i = (\varphi_{i-1,j} + \varphi_{i,j})/2 = \varphi_{i,j} - (\varphi_{i,j} - \varphi_{i-1,j})/2 = \varphi_{i,j} - \Delta\varphi_i/2$$

expressions for D_{ln} with accuracy up to quadratic terms take the form:

$$\begin{aligned} D_{0(i-1)} &= \delta_0 + \Delta u_i + \frac{(\Delta w_i)^2}{2a} \sim D_{0(i+1)}, \\ D_{1(i-1)}^{CB,EB} &= \delta_1 + \Delta u_i \pm \frac{h_2}{2}\Delta\varphi_i + \frac{(\Delta w_i + h_1\Phi_i)^2}{2(a-h_1)} \sim D_{1(i+1)}^{BC,AE}, \\ D_{2(i-1)}^{CA,EB} &= \delta_2 + \frac{1}{r}[(a-h_1)\Delta u_i \pm h_2\Delta w_i \pm ah_2\Phi_i] \\ &\quad + \frac{1}{2r^3}[h_2(\Delta u_i \pm h_2\Phi_i) \mp (a-h_1)(\Delta w_i + h_1\Phi_1)]^2 \sim D_{2(i+1)}^{AC,BE}. \end{aligned} \tag{4}$$

Here $r = \sqrt{(a+\delta_1-h_1)^2 + h_2^2}$ is the distance in the initial state between the neighboring particles (i.e. the initial length of the spring K_2) located along the axis x. In expressions (4) the elongations of all the springs, excepting the central ones with rigidity K_0, contain a third (upper) index. This index comprises apexes of rectangles $ABCE$, which are connected by the spring (Fig. 2). The apex of the central rectangle is given first. For the elongations of the springs with rigidity K_2 in formulas (4) there are signs $\pm$ and $\mp$; therefore, the third index of these formulas consists of two parts: first, the apexes of the rectangles connected by the first spring (the upper signs of the $\pm$ and $\mp$ symbols are taken for the elongations of such springs) are indicated and after the comma those of the second spring (in this case the bottom signs of such symbols are taken). The tensions of springs denoted by the equivalence signs have been obtained by the substitution of subscripts i by $i+1$. However, it is necessary to take into account that $\Phi_{i+1} = (\varphi_{i+1} + \varphi_i)/2 = \varphi_i + (\varphi_{i+1} - \varphi_i)/2 = \varphi_i + \Delta\varphi_{i+1}/2$.

By substituting the expressions (4) into Eq. (2) it is possible first to obtain a potential energy of the considered chain and then, taking into account (1), to make up the Lagrange function $L = T_i - U_i$ for the ith particle with accuracy up to the terms of ε_0^3-order (see estimates (3)). In this case in the continuum approximation the Lagrange function L of the considered medium consisting of anisotropic particles takes on the form:

$$\begin{aligned} L = {} & \frac{M}{2}\left(u_t^2 + w_t^2 + R^2\varphi_t^2\right) - \frac{M}{2}\left[c_1^2 u_x^2 + c_2^2 w_x^2 + R^2 c_3^2 \varphi_x^2 + 2\beta_1 w_x \varphi + 2\beta_2 \varphi^2\right. \\ & + \alpha_1 u_x^3 + \alpha_2 u_x w_x^2 + \alpha_3 u_x \varphi^2 + \alpha_4 u_x w_x \varphi + \alpha_5 w_x^4 + \alpha_6 w_x^3 \varphi + \alpha_7 w_x^2 \varphi^2 \\ & \left. + \alpha_8 w_x \varphi^3 + \alpha_9 \varphi^4\right]. \end{aligned} \tag{5}$$

A set of differential equations describing the nonlinear dynamic processes in the anisotropic crystalline medium is derived from the Lagrange function (5) in agreement with Hamilton's variational principle:

$$\begin{aligned} u_{tt} - c_1^2 u_{xx} = {} & \frac{1}{2}\frac{\partial}{\partial x}\left(3\alpha_1 u_x^2 + \alpha_2 w_x^2 + \alpha_3 \varphi^2 + \alpha_4 w_x \varphi\right), \\ w_{tt} - c_2^2 w_{xx} - \beta_1 \varphi = {} & \frac{1}{2}\frac{\partial}{\partial x}\left(2\alpha_2 u_x w_x + \alpha_4 u_x \varphi + 4\alpha_5 w_x^3 + 3\alpha_6 w_x^2 \varphi\right. \\ & \left. + 2\alpha_7 w_x \varphi^2 + \alpha_8 \varphi^3\right), \\ R^2\left(\varphi_{tt} - c_3^2 \varphi_{xx}\right) + \beta_1 w_x + 2\beta_2 \varphi = {} & -\alpha_3 u_x \varphi - \frac{\alpha_4}{2} u_x w_x - \frac{\alpha_6}{2} w_x^3 - \alpha_7 w_x^2 \varphi \\ & - \frac{3}{2}\alpha_8 w_x \varphi^2 - 2\alpha_9 \varphi^3. \end{aligned} \tag{6}$$

It should be noted that these equations are in analogy to the equations for the nonlinear one-dimensional Cosserat continuum (Erofeyev 2003). Here, the following notations are introduced: $c_i (1 = 1, 2, 3)$ are the propagation velocities, respectively, of the longitudinal, transverse and rotational waves, β_1 and β_2 are the dispersion parameters, $\alpha_i (i = 1, \ldots, 4)$ are the nonlinearity coefficients, $R = \sqrt{d_1^2 + d_2^2/4}$ is the inertia radius of the particle. Dependencies of the coefficients of equations (6) on the microstructure parameters (the force constants K_0, K_1, K_2, the lattice period a and the grain sizes $h_1 = d_1/\sqrt{2}$ and $h_2 = d_2/\sqrt{2}$) have the following form:

$$\rho c_1^2 = a\left[K_0 + 2K_1 + \left(\frac{2(a-h_1)^2}{r^2} + \frac{2\delta_2 h_2^2}{r^3}\right)K_2\right],$$
$$\rho c_2^2 = a\left[\frac{\delta_0}{a}K_0 + \frac{2\delta_1}{a-h_1}K_1 + \frac{2}{r^2}\left(h_2^2 + \frac{2\delta_2(a-h_1)^2}{r}\right)K_2\right],$$
$$\rho c_3^2 = \frac{a}{2R^2}\left[h_2^2K_1 + \frac{a^2h_2^2K_2}{r^2} + \frac{\delta_1 h_1^2}{a-h_1}K_1 + \frac{\delta_2}{r^3}K_2\left(h_2^4 + (a-h_1)^2h_1^2 - 2h_1h_2^2(a-h_1)\right)K_2\right], \tag{7a}$$

$$\rho\beta_1 = 2\left[\frac{ah_2^2}{r^2}K_2 + \frac{\delta_1 h_1}{a-h_1}K_1 + \frac{\delta_2(a-h_1)(ah_1 - h_1^2 - h_2^2)}{r^3}K_2\right],$$
$$\rho\beta_2 = \frac{ah_2^2}{r^2}K_2 + \frac{\delta_1 h_1^2}{a(a-h_1)}K_1 + \frac{\delta_2(h_2^4 + (a-h_1)^2h_1^2 - 2h_1h_2^2(a-h_1))}{ar^3}K_2, \tag{7b}$$

$$\rho\alpha_1 = \frac{K_2}{r^4}a^2(a-h_1)h_2^2,$$
$$\rho\alpha_2 = k_0a + K_1\frac{a^2}{a-h_1} + \frac{K_2}{r^4}a^2(a-h_1)[(a-h_1)^2 - 2h_2^2],$$
$$\rho\alpha_3 = K_1\frac{h_1^2}{a-h_1} + \frac{K_2}{r^4}(h_1^2 + h_2^2 - ah_1)[2ah_2^2 + (a-h_1)(h_1^2 + h_2^2 - ah_1)],$$
$$\rho\alpha_4 = K_1\frac{2ah_1}{a-h_1} + \frac{2a}{r^4}K_2[(h_1^2 + h_2^2 - ah_1)(h_2^2 - (a-h_1)^2) - ah_2^2(a-h_1)], \tag{7c}$$

$$\rho\alpha_5 = \frac{a}{4}\left[K_0 + K_1\frac{a^2}{(a-h_1)^2} + \frac{K_2}{r^6}a^2(a-h_1)^4\right],$$
$$\rho\alpha_6 = a\left[K_1\frac{ah_1}{(a-h_1)^2} - \frac{K_2}{r^6}ah_2^2(a-h_1)^4\right],$$
$$\rho\alpha_7 = a\left[K_1\frac{3h_1}{2(a-h_1)^2} + K_2\frac{3h_2^2(a-h_1)^2}{2r^6}(h_2^2 - 2h_1a + 2h_1^2)\right],$$
$$\rho\alpha_8 = K_1\frac{h_1^3}{(a-h_1)^2} + K_2\frac{h_2^2(a-h_1)}{r^6}[3h_1h_2^2(a-h_1) - h_2^4 - 3h_1^2(a-h_1)^2],$$
$$\rho\alpha_9 = K_1\frac{h_1^4}{4a(a-h_1)^2} + \frac{K_2}{4r^6}[h_2^8 - 4h_1h_2^6(a-h_1) + h_1^4(a-h_1)^4 + 6h_1^2h_2^4(a-h_1)^2 - 4h_1^3h_2^2(a-h_1)^3], \tag{7d}$$

where $\rho = M/a$ is the density of the one-dimensional medium per unit length. From (7b) the relationship $\beta_1 = 2\beta_2$ follows, if there are no preliminary deformations of the springs. It should be noted that in expressions (7a) there are no preliminary deformations of the springs, since they are not significant for the further investigations (Erofeev and Pavlov 2016).

3 The Low-Frequency Approximation

It is obvious that in Eq. (6) the transverse and rotational modes are interconnected even in the linear approximation, whereas the longitudinal mode is independent in the linear approximation. Therefore, further investigations will be performed for the case, that longitudinal waves do not propagate in the lattice, i.e. $u = 0$. In this case the set (6) degenerates into a two-mode system:

$$w_{tt} - c_2^2 w_{xx} - \beta_1 \varphi = \frac{1}{2}\frac{\partial}{\partial x}(4\alpha_5 w_x^3 + 3\alpha_6 w_x^2 \varphi + 2\alpha_7 w_x \varphi^2 + \alpha_8 \varphi^3),$$

$$R^2(\varphi_{tt} - c_3^2 \varphi_{xx}) + \beta_1 w_x + 2\beta_2 \varphi = -\frac{\alpha_6}{2} w_x^3 - \alpha_7 w_x^2 \varphi - \frac{3}{2}\alpha_8 w_x \varphi^2 - 2\alpha_9 \varphi^3. \quad (8)$$

In the low-frequency approximation, when no rotational waves propagate (Potapov et al. 2009, 2010; Erofeev et al. 2013), a relation between the microrotations φ and displacements w can be found from the linear part of the second Eq. (8) by a step-by-step approach. Since in this equation the term $\beta_1 w_x + 2\beta_2 \varphi$ plays the main role (according to estimates (3), it has $\varepsilon_0^{3/4}$-order of smallness), and the second term of the linear part of the equation, $R^2(\varphi_{tt} - c_3^2 \varphi_{xx})$, has the next order of smallness—$\varepsilon_0^{7/4}$, then in the first approximation

$$\varphi(x, t) \approx -\frac{\beta_1}{2\beta_2} w_x, \quad (9)$$

and in the second approximation the variable φ can be expressed in terms of w and its derivatives by the following way:

$$\varphi(x, t) \approx -\frac{\beta_1}{2\beta_2}\frac{\partial w}{\partial x} + \frac{R^2 \beta_1}{4\beta_2^2}\left(\frac{\partial^3 w}{\partial x \partial t^2} - c_3^2 \frac{\partial^3 w}{\partial x^3}\right).$$

Taking into account of relation (9) leads to “freezing” of the rotational degrees of freedom and, therefore, excluding φ from the system (8). As a result, in the first approximation the Lagrange function (5) degenerates into the following expression:

$$L = \frac{M}{2}\left(w_t^2 + \frac{R^2\beta_1^2}{4\beta_2^2} w_{xt}^2\right) - \frac{M}{2}\left((c_2^2 - \frac{\beta_1^2}{2\beta_2}) w_x^2 + \frac{R^2\beta_1^2}{4\beta_2^2} c_3^2 w_{xx}^2 + \gamma w_x^3\right), \quad (10)$$

where

$$\gamma = \alpha_5 - \frac{\alpha_6\beta_1}{2\beta_2} + \frac{\alpha_7\beta_1^2}{4\beta_2^2} - \frac{\alpha_8\beta_1^3}{8\beta_2^3} + \frac{\alpha_9\beta_1^4}{16\beta_2^4}.$$

This parameter depends on microstructure parameters according to (7b) and (7d). An analysis shows that γ can be positive, particularly, when $K_1 \ll K_0$ and $K_2 \ll K_0$. From Lagrange function (10) one can obtain the following equation containing terms of fourth-order derivatives:

$$w_{tt} - \tilde{c}_2^2 w_{xx} - \frac{R^2\beta_1^2}{4\beta_2^2}\frac{\partial}{\partial x}\left[\frac{\partial^2 w_x}{\partial t^2} - c_3^2\frac{\partial^2 w_x}{\partial x^2}\right] = \frac{\partial(\gamma w_x^3)}{\partial x}, \tag{11}$$

where $\tilde{c}_2^2 = c_2^2 - (\beta_1^2/2\beta_2)$. It should be noted that, taking into account (7a) and (7b),

$$\tilde{c}_2^2 \approx \frac{\delta_0}{\rho}K_0, \tag{12}$$

whereas $\beta_1 = 2\beta_2$ and $\tilde{c}_2^2 = 0$ for $\delta_0 = \delta_1 = \delta_2 = 0$. Thus, all further analysis is possible only for a medium with internal stress, which are simulated by the preliminary deformations of the springs.

4 Soliton Solutions

In order to write Eq. (11) in the dimensionless form, we introduce dimensionless displacement, coordinate and time: $\tilde{w} = w/w_0, \tilde{x} = x/X, \tilde{t} = t/T$. Then the following equation yields:

$$\tilde{w}_{\tilde{t}\tilde{t}} - \tilde{c}_2^2\frac{T^2}{X^2}\tilde{w}_{\tilde{x}\tilde{x}} - \frac{R^2T^2\beta_1^2}{4X^2\beta_2^2}\frac{\partial}{\partial\tilde{x}}\left[\frac{1}{T^2}\tilde{w}_{\tilde{t}\tilde{t}} - c_3^2\frac{1}{X^2}\tilde{w}_{\tilde{x}\tilde{x}}\right] = \frac{\gamma T^2 w_0^2}{X^4}\frac{\partial\tilde{w}_{\tilde{x}}^3}{\partial\tilde{x}}. \tag{13}$$

If to suppose $X^2/T^2 = \tilde{c}_2^2$ and $X^2 = R^2\beta_1^2/4\beta_2^2$, then Eq. (13) can be rewritten in the form:

$$\tilde{w}_{\tilde{t}\tilde{t}} - \tilde{w}_{\tilde{x}\tilde{x}} - \frac{\partial^2}{\partial\tilde{x}^2}\left(\tilde{w}_{\tilde{t}\tilde{t}} - \frac{c_3^2}{\tilde{c}_2^2}\tilde{w}_{\tilde{x}\tilde{x}}\right) = \frac{12\gamma}{\tilde{c}_2^2}\frac{w_0^2\beta_2^2}{R^2\beta_1^2}[\tilde{w}_{\tilde{x}}^2\tilde{w}_{\tilde{x}\tilde{x}}]. \tag{14}$$

Let $12\gamma w_0^2\beta_2^2/\tilde{c}_2^2R^2\beta_1^2 = 1$ (this equality is possible as γ can be positive—it was mentioned in item 3 below Eq. (10)), then $w_0^2 = \tilde{c}_2^2R^2\beta_1^2/12\gamma\beta_2^2$ and Eq. (14) takes on form:

$$\tilde{w}_{\tilde{t}\tilde{t}} - (1 + \tilde{w}_{\tilde{x}}^2)\tilde{w}_{\tilde{x}\tilde{x}} - \frac{\partial^2}{\partial\tilde{x}^2}\left(\tilde{w}_{\tilde{t}\tilde{t}} - c^2\tilde{w}_{\tilde{x}\tilde{x}}\right) = 0, \tag{15}$$

where $c = c_3^2/\tilde{c}_2^2$. From (12) and (7a) it follows that

$$c^2 \approx \frac{ah_2^2}{2\delta_0 R^2}\left(\frac{K_1}{K_0} + \frac{a^2}{r^2}\frac{K_2}{K_0}\right).$$

If, for instance, $\delta_0/a = 0.05$, then

$$c^2 \approx \frac{10h_2^2}{R^2}\left(\frac{K_1}{K_0} + \frac{a^2}{r^2}\frac{K_2}{K_0}\right) \approx \frac{80h_2^2}{h_1^2 + h_2^2}\left(\frac{K_1}{K_0} + \frac{a^2}{r^2}\frac{K_2}{K_0}\right).$$

It means that in a wide enough wide range of values of the microstructure parameters $c^2 > 1$. Particularly, at $h_1 < 2h_2$ and $K_0 < 16K_1$ this inequality is valid for any positive K_2.

The following notations are further introduced:

$$\tilde{w} = W, \tilde{t} = t, \tilde{x} = x.$$

Equation (15) allows for solutions describing solitons of displacements:

$$W(t,x) = \sqrt{24|v^2 - c^2|}\arctan\left(\exp\left(\frac{x - vt}{\Delta}\right)\right), \tag{16}$$

where $\Delta = (c^2 - v^2)/(1 - v^2)$ is the width of a soliton and v is its velocity.

The energy density of the displacement soliton has the form

$$E = \frac{1}{2}\left(W_t^2 + W_x^2 - \frac{1}{6}W_x^4 + W_{tx}^2 + c^2 W_{xx}^2\right), \tag{17}$$

and the complete energy is calculated by simply integrating

$$\int_{-\infty}^{+\infty} E\mathrm{d}(x - vt)$$

Its dependence on the velocity is described by the formula:

$$\int_{-\infty}^{+\infty} E\mathrm{d}(x - vt) = \frac{-(6v^4 - 5v^2c^2 + v^2 - c^2 - 1)\sqrt{v^2 - 1}}{\sqrt{v^2 - 1}}. \tag{18}$$

From the solution for displacements by ordinary differentiating equation (16), it is possible to obtain a strain soliton:

$$U(t,x) = \frac{\partial W(t,x)}{\partial x} = \frac{A}{\cosh\left(\frac{x - vt}{\Delta}\right)}, \tag{19}$$

where $A = \sqrt{6|v^2 - 1|}$ is the soliton amplitude. To the purpose of convenience of the graphical visualization and further interpretation of results, strain solitons are more preferable, since $w(t,x) \to 0$ at $x \pm vt \to \pm\infty$. An equation for solitons of displacements can also be obtained from Eq. (14) by ordinary differentiating:

$$\frac{\partial^2 U}{\partial t^2} - \frac{\partial^2}{\partial x^2}\left(U + \alpha\frac{U^3}{3}\right) - \frac{\partial^2}{\partial x^2}\left(\frac{\partial^2 U}{\partial t^2} - c^2\frac{\partial^2 U}{\partial x^2}\right). \qquad (20)$$

Solutions of Eq. (15) coincide completely with solutions of Eq. (20) with account of (19).

5 Subsonic Solitons

In the system there can be solitons corresponding to the solution (16) or (19) and propagating with the velocity v, which is smaller than the sound velocity. In this case the amplitude is calculated as follows: $A = \sqrt{6|1 - v^2|}$. The dependence of energy of subsonic solitons, satisfying (18), on their velocity v is plotted in Fig. 3 for various values of the parameter c^2. From Fig. 3 it is clear that the energy is maximum at $v = v^* = 0.78$. The formal solution (16) or (19) implies the existence of subsonic solitons with any, even infinitesimal, velocities, however, the numerical experiments show that solitons with velocities less than v^* are unstable. The graph of v^* versus c^2, having the horizontal asymptote $v^* = 0.78$, is plotted in Fig. 4.

Equation (15) and, accordingly, (20), unlike, for example, Korteweg–de Vries equation, have no infinite number of polynomial laws of conservation. Such systems are called *not completely integrated* and interaction (collision) of solitons is not elastic in them. It means that, as a result of collision, parameters of secondary solitons

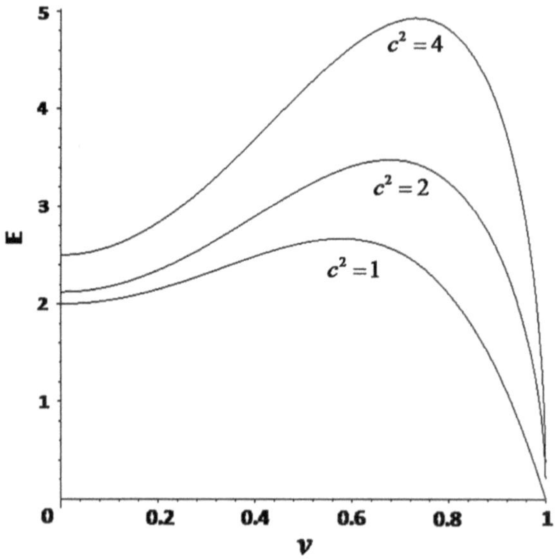

Fig. 3 Dependence of energy of subsonic solitons on their velocity

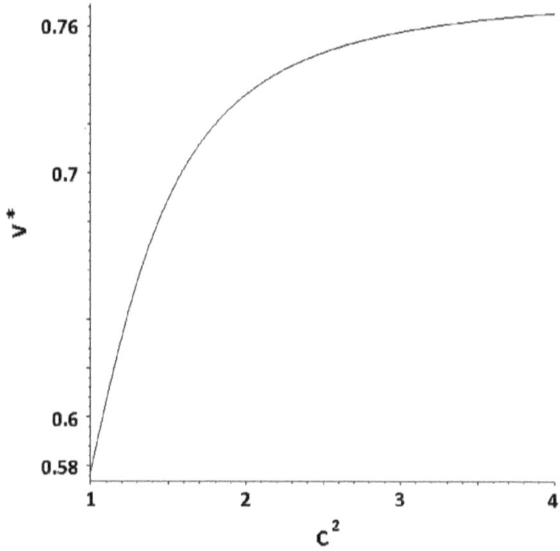

Fig. 4 Dependence of the maximal value of the soliton velocity v on c^2

change: they loose some energy, which can be realized in a quasiharmonic radiation and, if there is enough energy, in other solitons. So, it is interesting to investigate interactions of unipolar and bipolar solitons. But primarily it should be noted that, as the area of soliton interaction is restricted, for numerical modeling of these processes we assume that the boundary of the medium is at infinity and boundary conditions are natural: at infinity the function $W(t, x)$ is identical to zero together with all its derivatives.

Numerical experiments show that interaction of unipolar solitons with velocities, which are smaller than 0.77, is unstable. And if the velocity exceeds 0.77, solitons behave like linear waves: they are unified, their total amplitude grows by about 15–20 %, then the amplitude falls and the solitons move in opposite directions, being slightly deformed. The last fact means that their interaction is inelastic. "Classical" solitons, such as Korteweg–de Vries, Schrödinger and other solitons, are not unified, they approach to each other at a short distance, exchange by pulses and move in the opposite directions with a time delay (or acceleration), which is called a *phase shift*. But as a result, the "classical" solitons are not changed. In our case, the solitons also receive a time delay, however it is incorrect to call it a phase shift, since the interacting solitons change their shape—quasi-harmonic waves moving with the velocity close to 1 are added to them. But it is impossible to determine the exact time of interaction, it can only be concluded that this time is directly proportional to the width of interacting solitons.

Collision of bipolar solitons occurs according to another scheme: their joint amplitude reduces to zero, but the interaction is stable even from velocities v^* (see Fig. 4). As the amplitude of interacting solitons falls, their velocity increases, therefore time of interaction decreases. Assuming that the solitons located at a distance of 10Δ

do not interact, we offer the following formula for evaluation of "the factor of time distortion":

$$K_T = \frac{T_{ex} - T_{th}}{T_0}. \tag{21}$$

Here T_{ex} is the experimentally measured time of interaction, T_{th} is the theoretical time during which the soliton passes distance of 10Δ without interactions, T_0 is time within which the soliton passes its width Δ.

Figure 5 shows the dependence of the factor K_T (see (21)) on the velocity of interacting solitons for different values of c^2. From Fig. 5 it is clear that for the velocities close to 1 "the factor of time distortion" tends to 0, i.e. solitons degenerate into linear waves. The collision of bipolar solitons is shown in Fig. 6. First, the solitons are located at a distance of 10Δ and do not interact (Fig. 6a). After the collision, they again move away at a distance of 10Δ. As it is to be seen that the solitons are almost completely recovered after the collision, apart from a small deformation, which is a packet of quasi-harmonic waves moving with the velocity of sound. After a certain time, the soliton becomes "pure"—it remains behind the wave packet and moves in accordance with its own characteristics that are slightly changed after interaction. The wave packet evolution is further shown in an enlarged scale (in Fig. 6 the packet is allocated with an oval). Figure 6a has been performed for $c^2 = 1$, when no dispersion occurs in the linear approximation, but it to be seen that the low-frequency perturbation has a higher velocity than the high-frequency one. This fact indicates the presence of a nonlinear dispersion in the system. If after "purification" of the soliton to "switch off" the nonlinearity, then the wave packet will move with a constant shape for a very long time. Therefore, it can be concluded that, due to

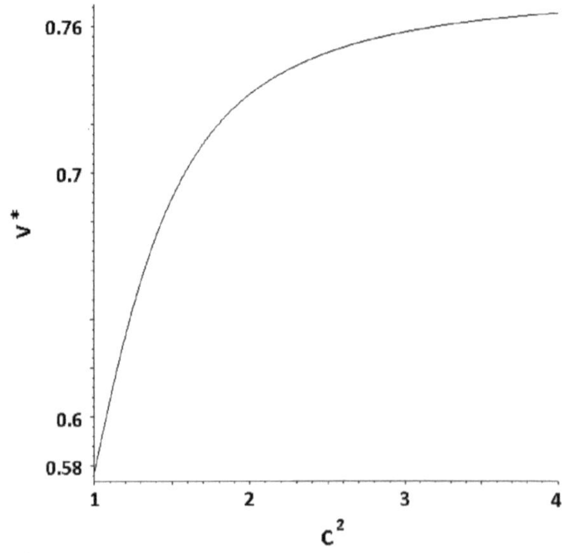

Fig. 5 Dependence of "the factor of time distortion" K_T on the velocity of interacting solitons for different values of c^2

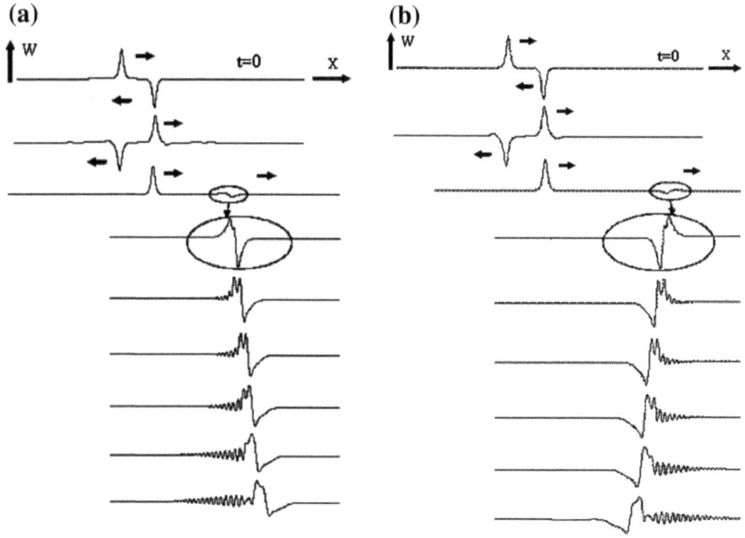

Fig. 6 Collision of bipolar solitons: **a** $c^2 = 1$, **b** $c^2 > 1$

discreteness of the calculation scheme, the dispersion is, at least, one order less than the nonlinearity.

Figure 6b shows the same process of collision of bipolar solitons at $c^2 > 1$, when dispersion of linear waves occurs. The process is actually identical to the previous one, except for the evolution of the quasi-linear packet—now the high-frequency components more with higher velocity than the low-frequency ones, and propagation of quasi-harmonic waves occurs like in a mirror image.

6 Supersonic Solitons

Supersonic solitons propagating with velocity $v > c^2$ are described by the same formulas (16) or (19) containing the amplitude $A = \sqrt{6|1 - v^2|}$. In this case, the energy is a monotonically increasing function of the velocity and the solitons are stable for all the velocities, so here it is possible to consider various types of interactions (passing or counter) of unipolar and bipolar solitons. For convenience of the graphical representation of the results, it is advisable to investigate the passing interaction of the unipolar solitons. Qualitatively different scenarios of passing interaction depend not only on the energy of the interacting solitons but on the relative velocity of collision.

If the relative velocity is small, the collision is similar to the scenario of the exchange interaction of the classical solitons. First, the soliton with the greater velocity overtakes the slower soliton, they approach to each other at a certain distance, but they are not unified. Then, the slow soliton amplitude increases, whereas the quick

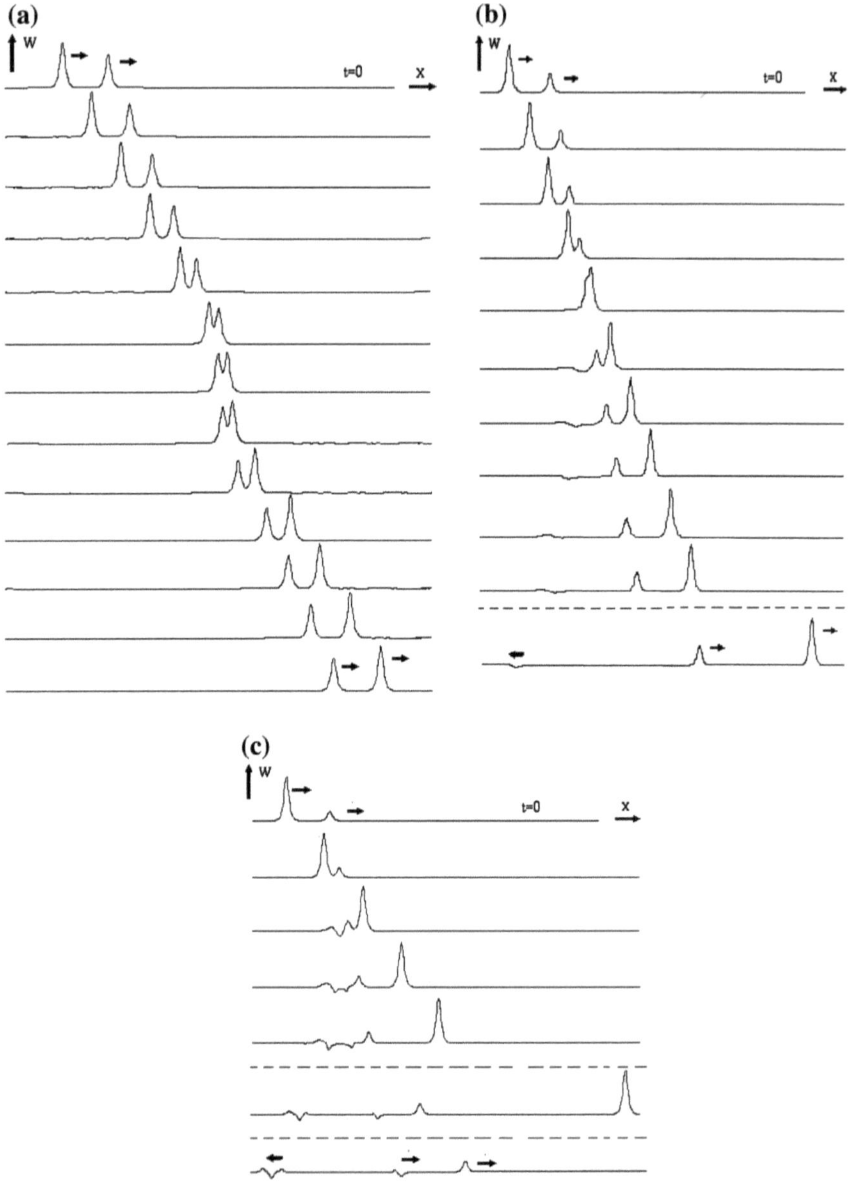

Fig. 7 Passing interaction of supersonic solitons with the collision velocities **a** $v_c \leq 1$, **b** $1 \leq v_c \leq 1.5$ and **c** $1.5 < v_c \leq 2$

soliton amplitude falls. In fact, the energy exchange occurs, after the solitons have moved away moving in accordance with their velocities as shown in Fig. 7a. It seems that the solitons interact elastically, as after the collision the soliton characteristics have not been changed.

When the velocity of interaction is higher, the solitons pass through each other as transparent, and the amplitude of the total perturbation does not increase. Due to collision, the solitons lose a relatively small amount of energy, which is implemented in the packet of quasi-harmonic waves as shown in Fig. 7b. It is interesting to note that the wave packet propagates in the opposite direction, i.e. if the interacting solitons move from left to the right, then the packet moves from right to the left.

Further increase of the interaction velocity leads to the formation of two packets of quasi-harmonic waves. One of which, like in Fig. 7b, moves in the opposite direction, whereas the other one moves in the same direction as the interacting solitons do (Fig. 7c). The packet of quasiharmonic waves is shown in an enlarged scale in Fig. 8a, its further evolution is similar to that shown in Fig. 6b. The passing packet is presented in Fig. 8b.

Figure 7a–c show the soliton interactions obtained for relative collision velocities equal to 0.8, 1.2, and 1.8, respectively. Rather high velocities of soliton collision, which lead to a splitting, are more convenient to explore if the solitons move in opposite directions (*head-on*, or *counter* interaction). For reasons of symmetry, the interacting solitons should be of the same type.

The most impressive is the process of splitting of bipolar solitons under head-on collision (see Fig. 9a). Splitting is considered as such an effect, if two interacting solitons with initial amplitude A_0 generate after the interaction a large number of secondary solitons with amplitudes $A_i (i \geq 1)$. Figure 9 shows the formation of two secondary solitons and, naturally, a non-stationary wave process. The amplitudes of the secondary solitons are distributed with respect to the initial amplitude A_0 by the following way: $A_1 = 0.96A_0$, $A_2 = 0.27A_0$ (a relative measurement accuracy is equal to 1.2 %). It is interesting to note that such a distribution of amplitudes of the secondary solitons does not depend on the value of the parameter c^2 in the range $1 < c^2 \leq 5$ and on the relative velocity of interaction in the range $5 \leq v_c \leq 16$. Unfortunately, it is impossible to demonstrate splitting under passing collision in one scale, as the velocities of the secondary solitons differ almost by 10 times and their amplitudes differ almost by 100 times. Therefore only some fragments of interaction are shown in Fig. 9b, where $c^2 = 64/25$, the great soliton velocity equals $v = 20$,

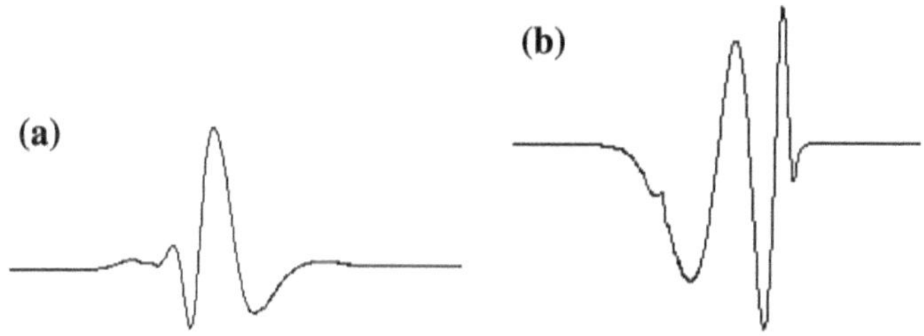

Fig. 8 Enlarged scale of the packet of quasi-harmonic waves (**a**) and the passing packet (**b**)

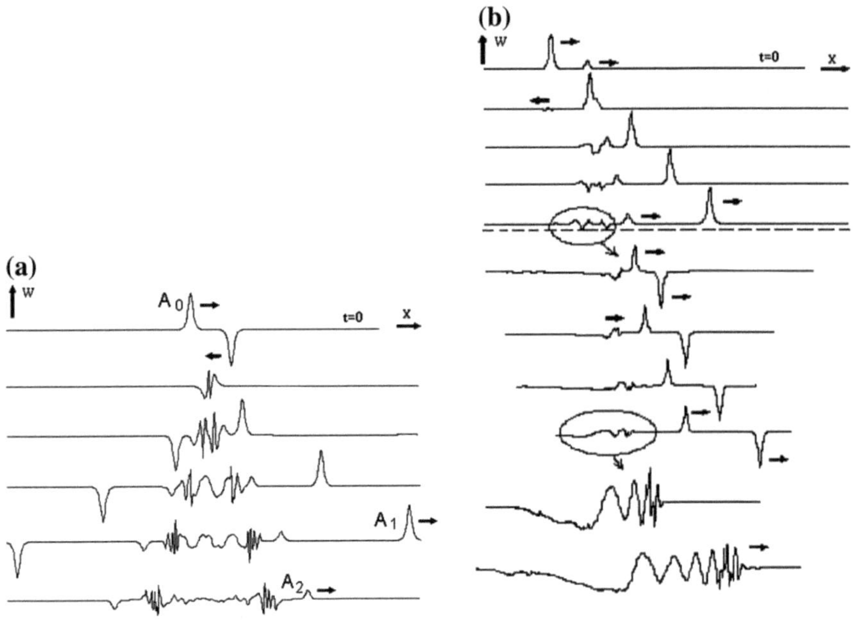

Fig. 9 Splitting of bipolar solitons under head-on collision (**a**) and of unipolar solitons in the passing interaction (**b**)

and the small soliton velocity is equal to $v = 5$; and some of these fragments are demonstrated in the enlarged scale. After the faster soliton overtakes the slower one and both solitons become "pure", the place of interaction is occupied by a nonstationary wave packet, from which two bipolar secondary solitons are extracted (see the upper oval and the solitons in an enlarged scale). After selecting of the last secondary solitons, the quasi-harmonic wave packet remains (in the second oval and below this packet is shown in an enlarged scale).

7 Conclusions

In this paper a nonlinear mathematical model of a one-dimensional granular medium with internal stress that consists of ellipsoidal-shaped particles with two translational and one rotational degrees of freedom has been introduced. The analytical dependencies of the elastic wave velocities and the nonlinearity coefficients of the particle size and the parameters of the interactions between them have been found. In the low-frequency range, when the rotational wave does not propagate, and if the longitudinal wave is not excited in the medium, the obtained equations in partial derivatives are reduced to one Eq. (11) for the transverse mode, which contains a cubic nonlinearity and is non-integrable. In the framework of this equation, numerical research of counter and passing interactions of strongly nonlinear soliton-like strain waves in a one-dimensional granular medium has been performed. It is shown that both subsonic and supersonic soliton-like waves can be realized in the medium.

The subsonic soliton-like waves steadily propagate in such a medium with some value of the velocity. Subsonic solitons interact inelastically and with mutual acceleration, i.e. a negative shift of phases is observed. The supersonic soliton-like waves interact inelastically, too, but the scheme of interaction depends on the relative velocity of collision. If this velocity is small, there is an exchange interaction. At passing collision with the supersonic velocity, one or two packets of quasiharmonic waves propagating in opposite directions can be formed. If the collision velocity is several times greater than the sound velocity, then splitting of the soliton into secondary solitons with generation of packets of quasiharmonic waves is observed both at counter and passing interactions.

Acknowledgments The research was carried out under the financial support of the Russian Scientific Foundation (project no. 14-19-01637).

References

Abdullow KO, Bogoloubsku IL, Makhankov VG (1976) One more example of inelastic soliton interaction. Phys Lett A 56:427–428

Berglund K (1982) Structural models of micropolar media. In: Brulin O, Hsieh RKT (eds) Mechanics of micropolar media. World Scientific, Singapore, pp 35–86

Chunyu L, Tsu-Wei C (2003) A structural mechanics approach for the analysis of carbon nanotubes. Int J Solids Struct 40:2487–2499

Dodd RK, Eilbeck JC, Gibbon JD, Morris HC (1982) Solitons and nonlinear wave equations. Academic Press, London

Engelbrecht JK, Fridman VE, Pelinovsky EN (1988) Nonlinear evolution equations. Pitman, London

Erofeyev VI (2003) Wave processes in solids with microstructure. World Scientific Publishing, Singapore

Erofeev VI, Pavlov IS (2016) Self-modulation of shear waves of deformation propagating in a one-dimensional granular medium with internal stresses. Math Mech Solids 21(1):60–72

Erofeev VI, Kazhaev VV, Pavlov IS (2013) Nonlinear localized strain waves in a 2d medium with microstructure. In: Altenbach H, Forest S, Krivtsov A (eds) Generalized continua as models for materials: with multi-scale effects or under multi-field actions, vol 22, Advanced structured materials. Springer, Heidelberg, pp 91–110

Lisina SA, Potapov AI, Nesterenko VF (2001) Nonlinear granular medium with rotations of the particles. One-dimensional model. Acoust Phys 47(5):666–674

Lonngren K (1978) Experimental investigations of solitons in nonlinear dispersive transmission lines. In: Lonngren K, Scott AC (eds) Solitons in actions. Academic Press, New York, pp 138–162

Nikitina NY, Pavlov IS (2013) Specificity of the phenomenon of acoustoelasticity in a two-dimensional internally structured medium. Acoust Phys 59(4):399–405

Ostrovsky LA, Papko VV, Pelinovsky EN (1972) Solitary electromagnetic waves in nonlinear lines. Radiophys Quantum Electron 15(4):438–446

Pavlov IS (2010) Acoustic identification of the anisotropic nanocrystalline medium with non-dense packing of particles. Acoust Phys 56(6):924–934

Pavlov IS, Potapov AI (2008) Structural models in mechanics of nanocrystalline media. Dokl Phys 53(7):408–412
Potapov AI, Vesnitsky AI (1994) Interaction of solitary waves under head-on collection. Experimental investigation. Wave Motion 19:29–35
Potapov AI, Pavlov IS, Lisina SA (2009) Acoustic identification of nanocrystalline media. J Sound Vib 322(3):564–580
Potapov AI, Pavlov IS, Lisina SA (2010) Identification of nanocrystalline media by acoustic spectroscopy methods. Acoust Phys 56(4):588–596
Scott AC, Chu FYF, McLaughlin D (1973) The soliton: a new concept in applied science. Proc IEEE 61:1443–1483

The Eigenmodes in Isotropic Strain Gradient Elasticity

Rainer Glüge, Jan Kalisch and Albrecht Bertram

Abstract We present the spectral decomposition of the isotropic stiffness hexadic that appears in Mindlin's strain gradient elasticity, where the kinematic variable is the second gradient of the displacement field. It turns out that four distinct eigenmodes appear, two of which are universal for all isotropic strain gradient materials, and two depend on an additional material parameter. With the aid of the harmonic decomposition, general interpretations of the eigenmodes can be given. Further, the material parameters are related to commonly employed special cases, namely the cases tabulated in Neff et al. (Int J Solids Struct 46(25–26):4261–4276, 2009) and isotropic gradient elasticity of Helmholtz type.

Keywords Strain gradient plasticity · Spectral decomposition · Stiffness hexadic

1 Introduction

It is well known that classical elasticity cannot account for size effects that are observed in very small structures (Liebold and Müller 2013). Mostly, the specific stiffness of fine structures is increased. It is also well known that one can overcome this shortcoming by including a strain gradient dependence in the elastic energy. The isotropic extension of linear elasticity has been given by Mindlin (1964). It involves

R. Glüge (✉) · A. Bertram
Lehrstuhl für Festigkeitslehre, Institut für Mechanik,
Fakultät für Maschinenbau, Otto-von-Guericke-Universität Magdeburg,
Universitätsplatz 2, 39106 Magdeburg, Germany
e-mail: gluege@ovgu.de

A. Bertram
e-mail: albrecht.bertram@ovgu.de

J. Kalisch
Graduiertenkolleg 1554 "Micro-Macro-Interactions in Structured
Media and Particle Systems", Otto-von-Guericke-Universität Magdeburg,
Universitätsplatz 2, 39106 Magdeburg, Germany
e-mail: j.kalisch@ovgu.de

H. Altenbach and S. Forest (eds.), *Generalized Continua as Models for Classical and Advanced Materials*, Advanced Structured Materials 42,
DOI 10.1007/978-3-319-31721-2_8

a sixth-order stiffness tensor with five independent parameters, which relates the strain gradient to the hyperstress tensor. The aim of the present work is to give a spectral decomposition of this hexadic. The eigenmodes of this hexadic may be interpreted geometrically, similar to the eigenmodes of the well-known Hooke-tetradic of classical isotropic linear elasticity, which are volume- and shape changing deformations. The eigenmodes can be interpreted in terms of displacement fields, curvature, volume- and shape changing deformations, local rotations, and so on. By this, the eigenvalues and hence the 5 independent parameters of the hexadic become interpretable. For convenience, we have included a conversion of some special cases from the literature to the five independent parameters in Mindlin's strain gradient elasticity. A more general account to strain gradient theories on this topic can be found in Bertram (2015).

The present article builds on isotropic strain gradient elasticity (Mindlin and Eshel 1968) and decomposition and representation theorems for isotropic tensors of arbitrary order, as found in Golubitsky et al. (1988), Zheng and Zou (2000), Olive and Auffray (2014), Auffray et al. (2013).

Notation

We prefer a direct notation, but make use of Einstein's summation convention (implicit summation over pairs of indices) whenever necessary. Scalars, vectors, second- and higher-order tensors are denoted by italic letters (like a or A), bold minuscules (like $\mathbf{a}$), bold majuscules (like $\mathbf{A}$), and blackboard bold majuscules (like $\mathbb{A}$), respectively. Moreover, $\{\mathbf{e}_i\}$ denotes an orthonormal basis. The single contraction and the dyadic product are denoted by $\cdot$ and $\otimes$, respectively. Multiple contractions act in the same sense on either tensor, e. g., $(\mathbf{a} \otimes \mathbf{b} \otimes \mathbf{c}) \cdot\cdot (\mathbf{d} \otimes \mathbf{e}) = (\mathbf{b} \cdot \mathbf{d})\,(\mathbf{c} \cdot \mathbf{e})\,\mathbf{a}$.

For groups and vector spaces, we use calligraphic letters, such as $\mathscr{H}$ for the space of right subsymmetric third-order tensors $\mathbb{H}$, for which $\mathbb{H} \cdot\cdot \mathbf{A} = \mathbb{H} \cdot\cdot \mathbf{A}^T$ holds. In particular, we denote harmonic tensor spaces of order i by $\mathscr{H}_i$. $\mathbf{I}$, $\mathbb{I}_S$, $\mathbb{I}$ and ε denote the identities on vectors, symmetric second-order tensors, subsymmetric third-order tensors and the third-order permutation tensor.

2 Isotropic Stiffness Hexadic

In general, the elastic energy of a strain gradient material is written in terms of the symmetric second-order strain tensor $\mathbf{E} = \mathrm{sym}(\mathbf{u} \otimes \nabla) = \mathrm{sym}(\mathbf{H})$ and a third-order tensor with one subsymmetry as the strain gradient variable $\mathbb{H}$. The latter may be the gradient of the strain $\mathbf{E} \otimes \nabla$, or the second gradient of the displacement $\mathbf{u}(\mathbf{x}_0, t)$. Mindlin (1964) refers to these two choices as strain gradient elasticity of form 1 and form 2. In any case, the third-order tensor $\mathbb{H}$ has one subsymmetry (left or right), and therefore only 18 independent components. It is interesting to note that these symmetries have different origins. In one case, the subsymmetry is a purely mathematical consequence (Schwartz' theorem), in the other case it comes from purging the rotations from the first gradient deformation measure.

Here, we will use the second gradient of the displacement $\mathbf{u}$ as the strain gradient variable, i.e., we take

$$\mathbb{H} = \mathbf{u} \otimes \nabla \otimes \nabla. \tag{1}$$

Approaching the elastic energy density as a quadratic form, we get

$$w = \frac{1}{2} C_{ijkl} E_{ij} E_{kl} + C_{ijklm} E_{ij} H_{klm} + \frac{1}{2} C_{ijklmn} H_{ijk} H_{lmn} \tag{2}$$

w.r.t. an ONB. Here appear the fourth-, fifth- and sixth-order stiffness tensors $\overset{\langle 4\rangle}{\mathbb{C}}$, $\overset{\langle 5\rangle}{\mathbb{C}}$ and $\overset{\langle 6\rangle}{\mathbb{C}}$, all of which are determined only up to some subsymmetric part that is due to the symmetries of the involved variables $\mathbf{E}$ and $\mathbb{H}$. Further, $\overset{\langle 4\rangle}{\mathbb{C}}$ and $\overset{\langle 6\rangle}{\mathbb{C}}$ have the principle symmetry, since they are multiplied twice with the same variable,

$$C_{ijkl} = C_{klij} = C_{jikl} = C_{ijlk}, \tag{3}$$

$$C_{ijklm} = C_{jiklm} = C_{ijkml}, \tag{4}$$

$$C_{ijklmn} = C_{lmnijk} = C_{ikjlmn} = C_{ijklnm}. \tag{5}$$

Presuming these index symmetries alone, $\overset{\langle 4\rangle}{\mathbb{C}}$, $\overset{\langle 5\rangle}{\mathbb{C}}$ and $\overset{\langle 6\rangle}{\mathbb{C}}$ have 21, 108, and 171 independent components, respectively. However, these numbers can be drastically reduced when material symmetries are taken into account. A particular case is isotropy. The components of any isotropic tensor can be given in terms of Kronecker- and Levi–Civita symbols w.r.t. an ONB. Due to the index symmetries of $\overset{\langle 4\rangle}{\mathbb{C}}$, $\overset{\langle 5\rangle}{\mathbb{C}}$ and $\overset{\langle 6\rangle}{\mathbb{C}}$ and the anti-symmetry of the Levi–Civita symbol, only Kronecker-deltas appear, which means that $\overset{\langle 5\rangle}{\mathbb{C}} = \mathbb{O}$ in case of centrosymmetric isotropy. For $\overset{\langle 4\rangle}{\mathbb{C}}$ and $\overset{\langle 6\rangle}{\mathbb{C}}$, we have the well-known representations (see, e.g., Mindlin 1964, 1965; Lazar and Maugin 2005; dell'Isola et al. 2009; Bertram and Forest 2014). In Mindlin's notation with $\eta_{ijk} := u_{k,ij}$, the strain gradient energy density is

$$w = 2c_2 \eta_{kii} \eta_{kjj} + c_4 \eta_{ijk} \eta_{ijk} + 2c_3 \eta_{ijk} \eta_{jki} + \frac{c_5}{2} \eta_{jji} \eta_{kki} + 2c_1 \eta_{iik} \eta_{kjj}. \tag{6}$$

This can be written as a quadratic form $u_{i,jk} C_{ijklmn} u_{l,mn}/2$ with a stiffness hexadic

$$\begin{aligned}
\overset{\langle 6\rangle}{\mathbb{C}} = \big[& c_1 (\delta_{jk}\delta_{im}\delta_{nl} + \delta_{jk}\delta_{in}\delta_{ml} + \delta_{ji}\delta_{kl}\delta_{mn} + \delta_{jl}\delta_{ik}\delta_{mn}) && (7)\\
+ & c_2 (\delta_{ji}\delta_{km}\delta_{nl} + \delta_{jm}\delta_{ki}\delta_{nl} + \delta_{ji}\delta_{kn}\delta_{ml} + \delta_{jn}\delta_{ik}\delta_{ml}) && (8)\\
+ & c_3 (\delta_{jm}\delta_{kl}\delta_{in} + \delta_{jl}\delta_{in}\delta_{km} + \delta_{jn}\delta_{im}\delta_{kl} + \delta_{jl}\delta_{im}\delta_{nk}) && (9)\\
+ & c_4 (\delta_{jn}\delta_{il}\delta_{km} + \delta_{jm}\delta_{kn}\delta_{il}) && (10)\\
+ & c_5 \delta_{jk}\delta_{il}\delta_{mn} \big] \mathbf{e}_i \otimes \mathbf{e}_j \otimes \mathbf{e}_k \otimes \mathbf{e}_l \otimes \mathbf{e}_m \otimes \mathbf{e}_n. && (11)
\end{aligned}$$

We summarize the different combinations of Kronecker symbols that belong to each parameter c_i with the basis $\{\mathbf{e}_i \otimes \mathbf{e}_j \otimes \mathbf{e}_k \otimes \mathbf{e}_l \otimes \mathbf{e}_m \otimes \mathbf{e}_n\}$ to five base hexadics $\{\mathbb{B}_i\}$, such that

$$\overset{\langle 6\rangle}{\mathbb{C}} = \sum_{i=1}^{5} c_i \mathbb{B}_i. \tag{12}$$

The metric of the basis $\{\mathbb{B}_i\}$ is

$$\mathbb{B}_i \cdots\cdots \mathbb{B}_j = \begin{bmatrix} 168 & 96 & 96 & 24 & 36 \\ 96 & 192 & 72 & 48 & 12 \\ 96 & 72 & 192 & 48 & 12 \\ 24 & 48 & 48 & 72 & 18 \\ 36 & 12 & 12 & 18 & 27 \end{bmatrix}. \tag{13}$$

We observe that $\frac{1}{2}\mathbb{B}_4$ maps every subsymmetric triadic onto itself, that $\frac{1}{3}\mathbb{B}_5$ maps every tensor of the form $\mathbf{v} \otimes \mathbf{I}$ onto itself, and that $\frac{1}{8}\mathbb{B}_2$ maps every tensor of the form $\mathbf{I} \otimes \mathbf{v}$ into its right subsymmetric part.

2.1 An Orthogonal Basis

Before turning to the spectral decomposition, a more suitable basis is introduced

$$\tilde{\mathbb{B}}_1 := -\frac{1}{15}(\mathbb{B}_1 + \mathbb{B}_2 + \mathbb{B}_5) + \frac{1}{6}(\mathbb{B}_3 + \mathbb{B}_4), \tag{14}$$

$$\tilde{\mathbb{B}}_2 := \frac{1}{12}(2\,\mathbb{B}_1 - \mathbb{B}_2 - 2\,\mathbb{B}_3 + 4\,\mathbb{B}_4 - 4\,\mathbb{B}_5), \tag{15}$$

$$\tilde{\mathbb{B}}_3 := \frac{1}{60}(6\,\mathbb{B}_1 - 9\,\mathbb{B}_2 + 16\,\mathbb{B}_5), \tag{16}$$

$$\tilde{\mathbb{B}}_4 := \frac{1}{6\sqrt{5}}(3\,\mathbb{B}_1 - 4\,\mathbb{B}_5), \tag{17}$$

$$\tilde{\mathbb{B}}_5 := \frac{1}{20}(-2\,\mathbb{B}_1 + 3\,\mathbb{B}_2 + 8\,\mathbb{B}_5). \tag{18}$$

The metric of this basis is diagonal with $\tilde{\mathbb{B}}_i \cdots\cdots \tilde{\mathbb{B}}_i = (7, 5, 6, 6, 6)$. The components of $\overset{\langle 6\rangle}{\mathbb{C}}$ with respect to this basis are

$$\tilde{c}_1 := 2\,(c_4 - c_3), \tag{19}$$

$$\tilde{c}_2 := 4\,c_3 + 2\,c_4, \tag{20}$$

$$\tilde{c}_3 := \frac{1}{6}(12\,c_1 - 16\,c_2 + 2\,c_3 + 9\,c_5), \tag{21}$$

$$\tilde{c}_4 := \frac{2\sqrt{5}}{3}\,(3\,c_1 + 2\,c_2 + 2\,c_3)\,, \tag{22}$$

$$\tilde{c}_5 := \frac{1}{2}(4\,c_1 + 8\,c_2 + 2\,c_3 + 4\,c_4 + 3\,c_5). \tag{23}$$

2.2 Eigenvalues and Projectors

In terms of the latter basis $\{\tilde{\mathbb{B}}_i\}$ and components $\tilde{c}_i$, the spectral decomposition of $\overset{\langle 6\rangle}{\mathbb{C}}$ is given by

$$\overset{\langle 6\rangle}{\mathbb{C}} = \sum_{i=1}^{4} \lambda_i \mathbb{P}_i \tag{24}$$

with the eigenvalues

$$\lambda_1 = \tilde{c}_1, \tag{25}$$

$$\lambda_2 = \tilde{c}_2, \tag{26}$$

$$\lambda_3 = \tilde{c}_5 + c_r, \tag{27}$$

$$\lambda_4 = \tilde{c}_5 - c_r \tag{28}$$

with

$$c_r = \sqrt{\tilde{c}_3^2 + \tilde{c}_4^2} \tag{29}$$

and the eigenprojectors

$$\mathbb{P}_1 = \tilde{\mathbb{B}}_1, \tag{30}$$

$$\mathbb{P}_2 = \tilde{\mathbb{B}}_2, \tag{31}$$

$$\mathbb{P}_3(\kappa) = \frac{1}{2}\left(\tilde{\mathbb{B}}_5 + \frac{\tilde{c}_3}{c_r}\tilde{\mathbb{B}}_3 + \frac{\tilde{c}_4}{\tilde{c}_r}\tilde{\mathbb{B}}_4\right), \tag{32}$$

$$\mathbb{P}_4(\kappa) = \frac{1}{2}\left(\tilde{\mathbb{B}}_5 - \frac{\tilde{c}_3}{c_r}\tilde{\mathbb{B}}_3 - \frac{\tilde{c}_4}{\tilde{c}_r}\tilde{\mathbb{B}}_4\right) \tag{33}$$

with

$$\cos\kappa = \frac{\tilde{c}_3}{c_r} \qquad \Leftrightarrow \qquad \sin\kappa = \frac{\tilde{c}_4}{c_r}. \tag{34}$$

For the spectral decomposition, the representation of $\overset{\langle 6\rangle}{\mathbb{C}}$ with the dimensionless parameter κ and the four eigenvalues is more convenient than with the parameters $\{c_1, c_2, c_3, c_4, c_5\}$ or $\{\tilde{c}_1, \tilde{c}_2, \tilde{c}_3, \tilde{c}_4, c_r\}$. One can check that

$$\mathbb{P}_3(\kappa) = \mathbb{P}_4(\kappa + \pi), \tag{35}$$
$$\lambda_3(\kappa) = \lambda_4(\kappa + \pi) \tag{36}$$

holds, i.e., it is reasonable to restrict κ to the interval $[0, \pi)$. The metric of the projectors is diagonal with $\mathbb{P}_i \cdot\cdot\cdot\cdot\cdot\cdot \mathbb{P}_i = (7, 5, 3, 3)$, thus the multiplicities of the eigenvalues are 7, 5, 3 and 3. Further, we have the projector properties

$$\mathbb{P}_i \cdots \mathbb{P}_j = \begin{cases} \mathbb{P}_i & \text{if } i = j, \\ \mathbb{O} & \text{if } i \neq j, \end{cases} \tag{37}$$

$$\sum_{i=1}^{4} \mathbb{P}_i = \mathbb{I}, \tag{38}$$

where $\mathbb{I}$ is the sixth-order identity tensor on triads with the right subsymmetry. These equations resemble those of the spectral decomposition of a transversely isotropic stiffness tetradic (see Appendix A of Kalisch and Glüge 2015), which also has in general five independent components and four distinct eigenvalues.

The above formulae are convenient when one knows the parameters $c_{1,2,3,4,5}$, and seeks the eigenvalues and the third and fourth eigenprojector. The other way around, the coefficients $c_{1,2,3,4,5}$ are given by

$$c_1 = (10\lambda_1 - 4\lambda_2 - 3(\lambda_3 + \lambda_4) + 3(\lambda_3 - \lambda_4)(\cos(\kappa) + \sqrt{5}\sin(\kappa)))/60, \tag{39}$$
$$c_2 = (-10\lambda_1 - 8\lambda_2 + 9(\lambda_3 + \lambda_4) + 9(-\lambda_3 + \lambda_4)\cos(\kappa))/120, \tag{40}$$
$$c_3 = (-\lambda_1 + \lambda_2)/6, \tag{41}$$
$$c_4 = (2\lambda_1 + \lambda_2)/6, \tag{42}$$
$$c_5 = (-5\lambda_1 - \lambda_2 + 3(\lambda_3 + \lambda_4) + (\lambda_3 - \lambda_4)(2\cos(\kappa) - \sqrt{5}\sin(\kappa)))/15 \tag{43}$$

in terms of $\{\lambda_{1,2,3,4}, \kappa\}$.

3 The Eigenmodes and the Harmonic Decomposition

The latter result becomes clearer from the point of view of the harmonic decomposition of a third-order tensor with one subsymmetry (Golubitsky et al. 1988; Zheng and Zou 2000; Olive and Auffray 2014). The third and fourth projector—more precisely: the parameter κ—distinguish a specific decomposition of the first-order harmonic contribution, which is discussed next.

The space of all second gradients $\mathbb{H} = \mathbf{u} \otimes \nabla \otimes \nabla$ is subsequently denoted by $\mathscr{H}$. By virtue of the harmonic decomposition a tensor is decomposed into a sum of mutually orthogonal tensors,

$$\mathbb{H} = \sum_{i=1}^{N} \mathbb{H}_i, \tag{44}$$

$$0 = \mathbb{H}_i \cdots \mathbb{H}_j, \qquad i \neq j. \tag{45}$$

These correspond to the eigentensors of $\overset{\langle 6 \rangle}{\mathbb{C}}$, where N is the number of different eigenvalues. Each $\mathbb{H}_i$ is related to a *harmonic tensor* $\overset{\langle n \rangle}{\mathbb{H}_i}$ by virtue of an isotropic linear mapping $\mathbb{L}_i$

$$\mathbb{H}_i = \overset{\langle 3+n \rangle}{\mathbb{L}}_i \underbrace{\cdots}_{n \text{ dots}} \overset{\langle n \rangle}{\mathbb{H}_i}. \tag{46}$$

The order n of the harmonic tensors does not exceed that of the decomposed tensor. The harmonic tensor spaces are denoted by $\mathscr{H}_i$. Their dimensions are

$$dim(\mathscr{H}_i) = 2i + 1, \tag{47}$$

which is due to the fact that all elements from $\mathscr{H}_i$ are completely symmetric, and all possible index contractions (like H_{ijj}) are zero.

On the whole, the tensor space $\mathscr{H}$ is decomposed into the direct sum ($\oplus$) of mutually orthogonal subspaces. These subspaces are closed under the action of the Rayleigh product with an orthogonal tensor $\mathbf{Q}$, which can be considered as a rotation of $\mathbb{H}$ by $\mathbf{Q}$. The Rayleigh product is defined as

$$\mathbf{Q} * (H_{ijk}\mathbf{e}_i \otimes \mathbf{e}_j \otimes \mathbf{e}_k) = H_{ijk}(\mathbf{Q} \cdot \mathbf{e}_i) \otimes (\mathbf{Q} \cdot \mathbf{e}_j) \otimes (\mathbf{Q} \cdot \mathbf{e}_k), \tag{48}$$

whereas the closedness under its action is

$$\mathbf{Q} * \mathbb{H} \in \mathscr{H}_i \Leftrightarrow \mathbb{H} \in \mathscr{H}_i \tag{49}$$

for all proper orthogonal tensors $\mathbf{Q}$. A further decomposition without loss of this property is not possible, which is why this decomposition is sometimes referred to as irreducible.

The harmonic decomposition can be thought of as the diagonalization of a matrix. The matrix originates from the action of the group of all proper orthogonal tensors on the tensor space (rotation of tensors by means of the Rayleigh product). Subspaces for harmonic spaces of equal order form block matrices on the main diagonal, the dimension of which corresponds to the number of subspaces of equal order. If we define additional orthogonal decompositions, we can diagonalize these block matrices as well. It is shown below that the angle κ parametrizes such an additional decomposition in the present case.

For sufficiently smooth fields $\mathbf{u}$, the respective tensor $\mathbb{H}$ can be represented by a linear combination of products of the form

$$\mathbb{H} = \sum_{i=1...3;j=1...6} C_{ij}\, \mathbf{e}_i \otimes \mathbf{E}_j, \tag{50}$$

where $\{\mathbf{e}_i\}$ and $\{\mathbf{E}_j\}$ are orthonormal bases in the three-dimensional Euclidean space and the space of symmetric second-order tensors, respectively. The harmonic decomposition of these spaces is given by $\mathscr{H}_1$ and $\mathscr{H}_0 \oplus \mathscr{H}_2$, respectively. The three-dimensional space cannot be decomposed into harmonic subspaces, hence it is represented by the three-dimensional space $\mathscr{H}_1$. The six-dimensional space of symmetric second-order tensors is decomposed into the well known spherical and deviatoric symmetric parts, the first is one-dimensional and corresponds to $\mathscr{H}_0$, and the second is five-dimensional and corresponds to $\mathscr{H}_2$.

Similar to the decomposition (50), the space $\mathscr{H}$ can be constructed as the dyadic product of the form

$$\mathscr{H} = \mathscr{H}_1 \otimes (\mathscr{H}_0 \oplus \mathscr{H}_2). \tag{51}$$

With the Clebsch–Gordan rule (Golubitsky et al. 1988)

$$\mathscr{H}_m \otimes \mathscr{H}_n = \bigoplus_{k=|m-n|}^{m+n} \mathscr{H}_k \tag{52}$$

we obtain

$$\begin{aligned}
\mathscr{H} &\cong \mathscr{H}_1 \otimes (\mathscr{H}_0 \oplus \mathscr{H}_2) && (53)\\
&= (\mathscr{H}_1 \otimes \mathscr{H}_0) \oplus (\mathscr{H}_1 \otimes \mathscr{H}_2) && (54)\\
&= \mathscr{H}_1 \oplus \mathscr{H}_1 \oplus \mathscr{H}_2 \oplus \mathscr{H}_3 && (55)\\
&= \mathscr{H}_3 \oplus \mathscr{H}_2 \oplus \mathscr{H}_1^{\oplus 2}. && (56)
\end{aligned}$$

Thus, we get two three-dimensional, one five-dimensional and one seven-dimensional subspace, altogether forming the 18-dimensional space of third-order tensors with one symmetry.

The harmonic decomposition is unique regarding the number and the dimensionality of the subspaces. However, when two equally-dimensioned subspaces appear, there is an arbitrariness in the isomorphisms that connect $\mathbb{H}_i$ and $\overset{\langle n \rangle}{\mathbb{H}}_i$. In our representation, this arbitrariness corresponds to the angle κ that determines the direction of the two eigenprojectors $\mathbb{P}_3$ and $\mathbb{P}_4$ of the eigenvalues λ_3 and λ_4, each having the multiplicity 3. The specifications of Eq. (46) are

$$\mathbb{H}_1 = \mathbb{H}_1, \tag{57}$$
$$\mathbb{H}_2 = \boldsymbol{\varepsilon} \cdot \mathbf{H}_2, \tag{58}$$
$$\mathbb{H}_3 = \mathbf{h}_3 \cdot (\cos(\kappa/2)\mathbb{P}_{4/1} + \sin(\kappa/2)\mathbb{P}_{4/2}/\sqrt{5}), \tag{59}$$
$$\mathbb{H}_4 = \mathbf{h}_4 \cdot (-\sin(\kappa/2)\mathbb{P}_{4/1} + \cos(\kappa/2)\mathbb{P}_{4/2}/\sqrt{5}). \tag{60}$$

On both sides of these equations, the index indicates the ordering of the eigenspaces. The $\mathbb{H}_i$ on the left side represent second displacement gradients that are eigentensors in the indexed eigenspaces. The $\mathbb{H}_1, \mathbf{H}_2, \mathbf{h}_3$ and $\mathbf{h}_4$ (denoted more general as $\overset{\langle n \rangle}{\mathbb{H}}_i$) on the right side are harmonic (fully symmetric and traceless) tensors of order 3, 2, 1 and 1, hence having 7, 5, 3 and 3 independent components. The number of these independent components corresponds to the dimension of the eigenspaces. Further, $\mathbb{P}_{4/1,2}$ are the isotropic projectors from the spectral decomposition of isotropic stiffness tetradics with the compression modulus K and the shear modulus G,

$$\overset{\langle 4 \rangle}{\mathbb{C}} = 3K \underbrace{\frac{1}{3}\mathbf{I} \otimes \mathbf{I}}_{\mathbb{P}_{4/1}} + 2G \underbrace{(\mathbb{I}_S - \frac{1}{3}\mathbf{I} \otimes \mathbf{I})}_{\mathbb{P}_{4/2}}. \tag{61}$$

$\mathbb{I}_S = \frac{1}{2}(\delta_{ik}\delta_{jl} + \delta_{il}\delta_{jk})\mathbf{e}_i \otimes \mathbf{e}_j \otimes \mathbf{e}_k \otimes \mathbf{e}_l$ is the identity on symmetric second-order tensors. With this symbolic representation of the eigenmodes, we can examine their properties by virtue of the traceless and symmetric properties of their corresponding harmonic tensors.

3.1 The 7-Dimensional Eigenspace $\mathscr{H}_3$

With $\mathbb{H}_1$ being harmonic, we find the traces and index symmetries

$$u_{i,jj} = 0, \tag{62}$$
$$u_{i,ik} = 0, \tag{63}$$
$$u_{i,jk} = u_{j,ik}. \tag{64}$$

Thus, $\mathbf{u}$ is a harmonic function, and the volumetric strain must be homogeneous. After Helmholtz' representation theorem (Helmholtz 1858), there exist a scalar field ϕ and a divergence free (solenoidal) vector field $\mathbf{a}$ (Coulomb's gauge) such that

$$\mathbf{u} = \nabla\phi + \nabla \times \mathbf{a}, \qquad \mathbf{a} \cdot \nabla = 0. \tag{65}$$

Using $u_{i,i} = \mathbf{u} \cdot \nabla = \Delta\phi$, we find with Eq. (64)

$$\nabla(\Delta\phi) = \mathbf{o}. \tag{66}$$

Equation (64) can be rewritten as

$$(\mathbf{u} \times \nabla) \otimes \nabla = \mathbf{O}, \tag{67}$$

i.e., the rotational part of **u** is homogeneous. Then, the Helmholtz representation and Coulomb's gauge imply

$$(\Delta\, \mathbf{a}) \otimes \nabla = \mathbf{O}. \tag{68}$$

Given sufficiently smooth fields, Laplacian and gradient commute. Thus, $\Delta\phi = u_{i,i}$ and $\Delta\mathbf{a}$ are homogeneous (Eqs. 66 and 68) and $\nabla\,\phi$ and $\mathbf{a} \otimes \nabla$ are harmonic functions.

In conclusion, the displacement fields that generate eigenstrain-gradients in $\mathscr{H}_3$

- are free from volumetric strain gradients,
- have zero mean curvature of the displacement components, and
- the gradient of the axial vector $\mathbf{u} \times \nabla$ vanishes everywhere, i.e., the rotational part of the displacement field is homogeneous.

3.2 The 5-Dimensional Eigenspace $\mathscr{H}_2$

For convenience, we drop the index at $\mathbf{H}_2$ in this paragraph. In index notation w.r.t. an ONB we get

$$u_{i,jk} = \frac{1}{2}(\varepsilon_{ijl}H_{lk} + \varepsilon_{ikl}H_{lj}), \tag{69}$$

where u_i is a displacement field that produces only strain gradients in the 5-dimensional eigenspace that is isomorphic to $\mathscr{H}_2$.

We cannot directly transfer the traceless and symmetric properties of **H** to the displacement gradient, since a summation index is involved in **H** but not in $u_{i,jk}$. Taking the two independent traces of $u_{i,jk}$ gives

$$u_{i,jj} = \frac{1}{2}(\varepsilon_{ijl}H_{lj} + \varepsilon_{ijl}H_{lj}) = \varepsilon_{ijl}H_{lj} = 0 \qquad \Leftrightarrow \qquad \mathrm{axi}(\mathrm{skw}(\mathbf{H})) = \mathbf{o}, \tag{70}$$

$$u_{j,jk} = \frac{1}{2}(\varepsilon_{jjl}H_{lk} + \varepsilon_{jkl}H_{lj}) = \frac{1}{2}\varepsilon_{ijl}H_{lj} = 0 \qquad \Leftrightarrow \qquad \frac{1}{2}\mathrm{axi}(\mathrm{skw}(\mathbf{H})) = \mathbf{o}, \tag{71}$$

i.e., they give the same information. The skew part of **H** (and hence the axial vector $\mathbf{w} = \mathrm{axi}(\mathrm{skw}(\mathbf{H}))$ implicitly defined by $\mathbf{w} \times \mathbf{x} = \mathrm{skw}(\mathbf{H}) \cdot \mathbf{x}$) is zero by definition. Thus, we find that the eigenstrain gradients of the 5-dimensional eigenspace belong to harmonic displacement fields without volumetric strain gradient, as in the case before. Now we consider

$$
\begin{aligned}
(\varepsilon_{nij}\, u_{i,j})_{,k} &= \varepsilon_{nij}\, u_{i,jk} && (72)\\
&= \varepsilon_{nij}\, (u_{i,jk} - u_{j,ik})/2 && (73)\\
&= \varepsilon_{nij}\, (2\varepsilon_{ijm}\, H_{mk} + \varepsilon_{ikm}\, H_{mj} - \varepsilon_{jkm}\, H_{mi})/4 && (74)\\
&= (2\varepsilon_{ijn}\varepsilon_{ijm} H_{mk} + \varepsilon_{ijn}\varepsilon_{ikm} H_{mj} - \varepsilon_{jni}\varepsilon_{jkm} H_{mi})/4 && (75)\\
&= \delta_{nm} H_{mk} + [(\delta_{jk}\, \delta_{nm} - \delta_{jm}\, \delta_{nk})\, H_{mj} && (76)\\
&\quad - (\delta_{nk}\, \delta_{im} - \delta_{nm}\, \delta_{ik}]\, H_{mi})/4 && (77)\\
&= H_{nk} + (H_{nk} - \delta_{nk}\, H_{mm} - \delta_{nk}\, H_{mm} + H_{nk})/4 && (78)\\
&= 3\, H_{nk}/2. && (79)
\end{aligned}
$$

In symbolic notation we thus have

$$
\begin{aligned}
\mathbf{H} &\propto (\mathbf{u} \times \nabla) \otimes \nabla && (80)\\
&= (-\Delta\, \mathbf{a}) \otimes \nabla && (81)\\
&= -\, \Delta\, (\mathbf{a} \otimes \nabla). && (82)
\end{aligned}
$$

$\mathbf{H}$ is symmetric and deviatoric. The latter property is in accordance with Coulomb's condition on $\mathbf{a}$. The symmetry of $\mathbf{H}$ implies another constraint on $\mathbf{a}$.

$$
\begin{aligned}
\Delta\, (\mathbf{a} \otimes \nabla) &= \Delta\, (\nabla \otimes \mathbf{a}), && (83)\\
\Leftrightarrow \quad \mathbf{O} &= \Delta\, (\mathbf{a} \otimes \nabla - \nabla \otimes \mathbf{a}) && (84)\\
\mathbf{O} &= \Delta\, \varepsilon \cdot (\mathbf{a} \times \nabla) && (85)\\
&= \varepsilon \cdot (\Delta\, (\mathbf{a} \times \nabla)), && (86)\\
\Leftrightarrow \quad \mathbf{o} &= \Delta\, (\mathbf{a} \times \nabla) && (87)\\
&= (\Delta\, \mathbf{a}) \times \nabla && (88)
\end{aligned}
$$

The divergence of Eq. (84) provides—by means of Coulomb's gauge

$$
\begin{aligned}
\mathbf{o} &= (\Delta\, (\mathbf{a} \otimes \nabla - \nabla \otimes \mathbf{a})) \cdot \nabla && (89)\\
&= \Delta\, [(\mathbf{a} \otimes \nabla) \cdot \nabla - (\nabla \otimes \mathbf{a}) \cdot \nabla] && (90)\\
&= \Delta\, (\Delta\, \mathbf{a} - \nabla\, (\mathbf{a} \cdot \nabla)) && (91)\\
&= \Delta\, \Delta\, \mathbf{a}. && (92)
\end{aligned}
$$

Thus, $\mathbf{a}$ must be a biharmonic function. In conclusion, the displacement fields that generate eigenstrain-gradients in $\mathscr{H}_2$

- are free from volumetric strain gradients,
- have zero mean curvature of the displacement components, and
- the divergence of the gradient of the axial vector $\mathbf{u} \times \nabla$ vanishes everywhere.

These restrictions are weaker (third bullet point) than in case of eigenstrain-gradients of $\mathbb{H}_1$. This is not surprising, as we have less constraints to exploit, namely only one zero trace and one index symmetry, due to $\mathbf{H}_2$ being a second order tensor.

3.3 The 3-Dimensional Eigenspaces

Unfortunately $\mathbf{h}_{3,4}$ have no symmetry or zero trace which could be exploited. The third and fourth eigenmode depend on the angle κ, which depends on the coefficients $c_{1,2,3,5}$ through Eq. (34). Thus, we can determine canonical angles κ by taking one of the c_i as infinite, or consider more general directional limits with fixed ratios between the c_i. In doing so, two special cases emerge, namely when c_2 or c_5 are taken to infinity. In both cases, the third eigenvalue λ_3 becomes infinite, and its eigenprojector $\mathbb{P}_3$ becomes $\frac{1}{8}\mathbb{B}_2$ or $\frac{1}{3}\mathbb{B}_5$, respectively. The angles κ that belong to these materials can be inferred from Eq. (34), and one finds

$$c_2 \to \infty: \qquad \cos\kappa \to -\frac{2}{3}, \qquad \mathbb{P}_3 = \frac{1}{8}\mathbb{B}_2, \qquad \lambda_3 \to \infty, \tag{93}$$

$$c_5 \to \infty: \qquad \cos\kappa \to 1, \qquad \mathbb{P}_3 = \frac{1}{3}\mathbb{B}_5, \qquad \lambda_3 \to \infty. \tag{94}$$

However, we can also adjust κ and the eigenvalues $\lambda_{1,2,3,4}$ independently.

3.3.1 The Case $\cos\kappa = -2/3$

The eigentensors of the third and fourth eigenvalue are related to the harmonic tensors $\mathbf{h}_3$ and $\mathbf{h}_4$ through

$$\mathbb{H}_3 = \mathbf{h}_3 \cdot \mathbb{I}_S = \mathrm{sym}_{23}\mathbf{I} \otimes \mathbf{h}_3, \tag{95}$$

$$\mathbb{H}_4 = \mathbf{h}_4 \cdot (\mathbb{I}_S - 6\mathbb{P}_{4/1})/\sqrt{5}. \tag{96}$$

This case is closest to the usual strain decomposition into dilatoric and deviatoric parts. The eigenmodes to the third eigenvalue are gradients of the volumetric strain. The fourth eigenmode does not correspond to a gradient of a deviatoric strain. By considering

$$\cos\kappa = \frac{\tilde{c}_3}{c_r} = -\frac{2}{3}, \tag{97}$$

$$\sin\kappa = \frac{\tilde{c}_4}{c_r} = \frac{\sqrt{5}}{3}, \tag{98}$$

(remember that $\kappa \in [0, \pi)$), eliminating c_r and summarizing, one finds that this case corresponds to

$$4c_1 + 2c_3 + c_5 = 0. \tag{99}$$

3.3.2 The Case $\cos\kappa = 1$

The eigentensors of the third and fourth eigenvalue are related to the harmonic tensors $\mathbf{h}_3$ and $\mathbf{h}_4$ through

$$\mathbb{H}_3 = \mathbf{h}_3 \cdot \mathbb{P}_{4/1}, \tag{100}$$

$$\mathbb{H}_4 = \mathbf{h}_4 \cdot \mathbb{P}_{4/2}. \tag{101}$$

A calculation similar to the symbolic examination of the 5- and 7-dimensional eigenspaces shows that both eigenstrain gradients $\mathbb{H}_3$ and $\mathbb{H}_4$ result from displacement fields with a biharmonic field ϕ in their Helmholtz representations. In terms of c_i, this case corresponds to

$$3c_1 + 2c_2 + 2c_3 = 0. \tag{102}$$

4 Relation to Other Forms of Strain Gradient Elasticity

For convenience, we summarize the conversion of parameters between the two forms of strain gradient elasticity and for special cases of the first form of strain gradient elasticity (Mindlin and Eshel 1968). We follow the list given in Neff et al. (2009) (Eq. 2.10) and Lazar's proposal of gradient elasticity of Helmholtz type (Po et al. 2014).

4.1 Mindlin's Second Form of Strain Gradient Elastictiy

The two forms of strain gradient elasticity (Mindlin and Eshel 1968) are

$$w_1 = \frac{1}{2}\mathbf{u} \otimes \nabla \otimes \nabla \cdots \mathbb{C} \cdots \mathbf{u} \otimes \nabla \otimes \nabla, \tag{103}$$

$$w_2 = \frac{1}{2}\nabla \otimes \operatorname{sym}(\mathbf{u} \otimes \nabla) \cdots \hat{\mathbb{C}} \cdots \nabla \otimes \operatorname{sym}(\mathbf{u} \otimes \nabla), \tag{104}$$

where we use the very same base tensors $\mathbb{B}_{1,2,3,4,5}$, but with the parameters $\hat{c}_{1,2,3,4,5}$. The conversion between the two variants is

$$c_1 = \hat{c}_1/2 + \hat{c}_2/2, \tag{105}$$

$$c_2 = \hat{c}_1/2 + \hat{c}_2/4 + \hat{c}_5/4, \tag{106}$$

$$c_3 = 3\hat{c}_3/4 + \hat{c}_4/4, \tag{107}$$

$$c_4 = \hat{c}_3/2 + \hat{c}_4/2, \tag{108}$$

$$c_5 = \hat{c}_2. \tag{109}$$

Our conversion differs from the one given in Mindlin and Eshel (1968) (Eq. 2.6), since here we considered the components of the stiffness hexadic w.r.t. the base tensors $\mathbb{B}_i$, whereas Mindlin considered the coefficients in the strain gradient energy. The differences are due to symmetrizations, see Eq. (6). Apart from that, the ordering is different.

4.2 Common Strain Gradient Extensions

We translate directly the forms in Table 2.10 from Neff et al. (2009) (Table 1).

4.3 Gradient Elasticity of Helmholtz Type

In order to reduce the number of elasticity constants, Lazar et al. (2006) recommend to use

$$C_{ijklmn} = l^2 C_{jkmn} \delta_{il}, \tag{110}$$

in the second form (Eq. 104), with the fourth-order stiffness tetradic and the additional material parameter l. In case of anisotropic elasticity, the second-order tensor that extends the stiffness tetradic is invariant under the action of the material symmetry group. In case of isotropy and cubic elasticity, this is a multiple of the identity tensor,

Table 1 Special cases of strain gradient elasticity translated into the parameter set $c_{1,2,3,4,5}$, where the left column contains the strain energy density, the center column the corresponding parameters c_i and the right column the eigenvalues λ_i and the angle κ

El. energy w	$c_{1,2,3,4,5}$	$\lambda_{1,2,3,4}, \kappa$
$\Vert\mathbf{u} \otimes \nabla \otimes \nabla\Vert^2$	0, 0, 0, 1, 0	2, 2, 2, 2, arbitrary
$\Vert\Delta \mathbf{u}\Vert^2$	0, 0, 0, 0, 2	0, 0, 6, 0, 0
$\Vert\text{sym}(\mathbf{u} \otimes \nabla) \otimes \nabla\Vert^2$	0, 0, 1/4, 1/2, 0	2, 1/2, 2, 1/2, arccos(1/9)
$\Vert\text{devsym}(\mathbf{u} \otimes \nabla) \otimes \nabla\Vert^2$	0, −1/6, 1/4, 1/2, 0	2, 1/2, 7/6, 0, arccos(19/21)
$\Vert\text{skw}((\mathbf{u} \times \nabla) \otimes \nabla)\Vert^2$	−1/2, 1/4, 0, 0, 1	0, 0, 3, 0, arccos(−1/9)
$\Vert(\mathbf{u} \times \nabla) \times \nabla\Vert^2$	−1, 1/2, 0, 0, 2	0, 0, 6, 0, arccos(−1/9)
$\Vert(\mathbf{u} \cdot \nabla)\nabla\Vert^2$	0, 1/2, 0, 0, 0	0, 0, 4, 0, arccos(−2/3)
$\Vert(\mathbf{u} \times \nabla) \otimes \nabla\Vert^2$	0, 0, −1/2, 1, 0	0, 3, 3, 0, arccos(−1/9)
$\Vert\text{dev}((\mathbf{u} \times \nabla) \otimes \nabla)\Vert^2$	0, 0, −1/2, 1, 0	0, 3, 3, 0, arccos(−1/9)
$\Vert\text{sym}((\mathbf{u} \times \nabla) \otimes \nabla)\Vert^2$	1/2, −1/4, −1/2, 1, −1	0, 3, 0, 0, arbitrary
$\Vert\text{devsym}((\mathbf{u} \times \nabla) \otimes \nabla)\Vert^2$	1/2, −1/4, −1/2, 1, −1	0, 3, 0, 0, arbitrary
$\Vert\text{sym}(\text{sym}(\mathbf{u} \otimes \nabla) \times \nabla)\Vert^2$	1/8, −1/16, −1/8, 1/4, −1/4	0, 3/4, 0, 0, arbitrary

with the parameter l^2. The conversion to $c_{1,2,3,4,5}$ is

$$c_1 = 0, \tag{111}$$

$$c_2 = l^2(K/4 - G/6), \tag{112}$$

$$c_3 = l^2 G/4, \tag{113}$$

$$c_4 = l^2 G/2, \tag{114}$$

$$c_5 = 0. \tag{115}$$

Thus, the third and fourth eigenmode depend via κ on the internal length parameter l and the compression and shear moduli K and G. In terms of Mindlin's second form of strain gradient elasticity (see Sect. 4.1), we have only two nonzero parameters, namely

$$\hat{c}_1 = 0, \tag{116}$$

$$\hat{c}_2 = 0, \tag{117}$$

$$\hat{c}_3 = 0, \tag{118}$$

$$\hat{c}_4 = Gl^2 = \mu l^2, \tag{119}$$

$$\hat{c}_5 = Kl^2 - 2Gl^2/3 = \lambda l^2, \tag{120}$$

where the inheritance from the classical isotropic stiffness tetradic with Lamé's constants λ and μ is more obvious.

Acknowledgments Valuable hints from Patrizio Neff and Markus Lazar are gratefully acknowledged.

References

Auffray N, Le Quang H, He Q (2013) Matrix representations for 3d strain-gradient elasticity. J Mech Phys Solids 61(5):1202–1223

Bertram A (2015) Compendium on gradient materials. http://www.ifme.ovgu.de/fl_veroeffentlichungen.html

Bertram A, Forest S (2014) The thermodynamics of gradient elastoplasticity. Contin Mech Thermodyn 26:269–286

dell'Isola F, Sciarra G, Vidoli S (2009) Generalized hooke's law for isotropic second gradient materials. Proc R Soc Lond A: Math Phys Eng Sci 465(2107):2177–2196

Golubitsky M, Stewart I, Schaeffer D (1988) Singularities and groups in bifurcation theory, vol II. Springer, New York

Helmholtz H (1858) Über Integrale der hydrodynamischen Gleichungen, welcher der Wirbelbewegungen entsprechen. Journal für die reine und angewandte Mathematik 55:25–55, p 38

Kalisch J, Glüge R (2015) Analytical homogenization of linear elasticity based on the interface orientation distribution – a complement to the self-consistent approach. Compos Struct 126:398–416

Lazar M, Maugin G (2005) Nonsingular stress and strain fields of dislocations and disclinations in first strain gradient elasticity. Int J Eng Sci 43(13):1157–1184

Lazar M, Maugin G, Aifantis E (2006) On a theory of nonlocal elasticity of bi-helmholtz type and some applications. Int J Solids Struct 43(6):1404–1421

Liebold C, Müller W (2013) Measuring material coefficients of higher gradient elasticity by using afm techniques and raman-spectroscopy. In: Altenbach H, Forest S, Krivtsov A (eds) Generalized continua as models for materials. Advanced structured materials, vol 22. Springer, Heidelberg, pp 255–271

Mindlin R (1964) Micro-structure in linear elasticity. Arch Ration Mech Anal 16(1):51–78

Mindlin R (1965) Second gradient of strain and surface-tension in linear elasticity. Int J Solids Struct 1:417–438

Mindlin R, Eshel N (1968) On first strain-gradient theories in linear elasticity. Int J Solids Struct 4:109–124

Neff P, Jeong J, Ramézani H (2009) Subgrid interaction and micro-randomness—novel invariance requirements in infinitesimal gradient elasticity. Int J Solids Struct 46(25–26):4261–4276

Olive M, Auffray N (2014) Symmetry classes for odd-order tensors. Zeitschrift für Angewandte Mathematik und Mechanik 94(5):421–447

Po G, Lazar M, Seif D, Ghoniem N (2014) Singularity-free dislocation dynamics with strain gradient elasticity. J Mech Phys Solids 68:161–178

Zheng QS, Zou WN (2000) Irreducible decompositions of physical tensors of high orders. J Eng Math 37(1–3):273–288

Limit Analysis of Lattices Based on the Asymptotic Homogenization Method and Prediction of Size Effects in Bone Plastic Collapse

Ibrahim Goda, Francisco Dos Reis and Jean-François Ganghoffer

Abstract Lattice structures possess a huge potential for energy absorbing applications, thus it is important to develop predictive tools for their mechanical response up to collapse. Yielding is generally premonitory of structural collapse for lattice structures, so a comprehensive and quantitative understanding of lattice yielding behavior is indispensable in engineering applications. In the present work, the overall plastic yield and brittle failure behaviors of three-dimensional lattices is investigated by a microstructural modeling approach based on the homogenization of the initially discrete microstructure. The multiaxial yield behavior of the lattice is analyzed to formulate a multiaxial plastic yield criterion. Furthermore, the brittle fracture of the lattice is modeled under triaxial stress states to construct the failure surfaces, defined in the tension–tension quadrant. In plastic yielding, the analyses are performed assuming an elastic perfectly plastic lattice, and a micromechanical model based on an homogenization scheme is applied to a representative unit cell to determine the macroscopic plastic yield surfaces in stress space. This general framework is applied to evaluate the yield and failure properties of trabecular bone, which are of key interest in understanding and predicting the fracture of bones and bone implant systems. The effective strength of trabecular bone is evaluated in the two situations of fully brittle (fracture with no tissue ductility) and fully ductile failure (yield with no tissue fracture) of the trabecular tissue. At high bone volume fraction, the real strut-level ductility is sufficiently high to effectively be fully ductile but at very low bone volume fraction, the real behavior of bone may fail in a brittle mode. An adaptation and extension of the discrete homogenization method towards

I. Goda · F. Dos Reis · J.-F. Ganghoffer (✉)
LEMTA, Université de Lorraine, 2, Avenue de la Forêt de Haye,
TSA 60604, 54518 Vandoeuvre-lès-Nancy Cedex, France
e-mail: jean-francois.ganghoffer@univ-lorraine.fr

F. Dos Reis
e-mail: francisco.dos-reis@hotmail.fr

I. Goda
Faculty of Engineering, Department of Industrial Engineering, Fayoum University,
Fayoum 63514, Egypt
e-mail: ibrahim.goda@univ-lorraine.fr

H. Altenbach and S. Forest (eds.), *Generalized Continua as Models for Classical and Advanced Materials*, Advanced Structured Materials 42,
DOI 10.1007/978-3-319-31721-2_9

a micropolar effective medium is introduced in order to construct the plastic yield surfaces for which the material point of the effective continuum supports couple stresses in addition to Cauchy-type stresses. The size effects in the ductile fracture mode are addressed by considering a micropolar behavior, reflecting the influence of additional degrees of freedom and internal bending length effects on the initiation of plasticity. It is observed that when the characteristic size of the microscale structure is comparable to the bending length, a significant difference is shown between the results based on the non-classical theory and those obtained by the classical theory.

Keywords Plastic yield · Elastoplasticity · Brittle failure · Discrete homogenization · Micropolar theory · Size effects

1 Introduction

Nowadays, cellular solids and network materials such as foams, lattice truss materials and grid materials are widely used in a variety of commercial and military applications such as automotive industry, aerospace, or marine engineering, due to their many advantages, including high mechanical properties and strength, energy absorption capacities, thermal and acoustic insulation properties, lightweight structural components.

Cellular lattice structures endowed with a specific mechanical behavior due to the presence of an inherent microstructure are encountered in many applications such as light-weight industrial components, trabecular bone and bone scaffolds. Such structures are a complex of cells made of an interconnected network of edges. They are usually categorized as 2D and 3D cellular solids, and coined honeycombs and foams respectively. Many man-made and biological structures present a discrete topology, such as fibrous materials (textiles, collagen fiber networks, biological membranes), with a more or less complex organization of the fibrous microstructure; the membrane of biological cells can be viewed as an assembly of filaments which are linked together as part of a network. Since those structures consist of many repetitive elements, a need arose to develop mechanical models for the prediction of their deformation behavior.

Among their notable mechanical properties, cellular lattice structures are of more importance, since the microstructure of the lattice can be adjusted so that desired mechanical properties can be achieved at the mesoscopic or macroscopic level. In view of exploiting the full potential of cellular solids, their constitutive mechanical response has to be entirely understood and modeled. Recently, researchers worldwide have widely used additive manufacturing methods to fabricate lattice structures. Although these fabrication methods are capable of producing lattice structures with tunable porosity and pore sizes and with high repeatability, they are costly and time consuming even for fabricating small sizes parts. Accordingly, developing numerical prediction models having the capability to predict the mechanical properties of lattice structures can reduce the required experimental measurements in addition to the manufacturing cost.

The material strength of architectured materials has been the topic of several studies related to either their elastic strength, or to their non-linear elastoplastic behavior. The choice of method of analysis of such lattices in the plastic range is especially guided by the nature and type of lattice under consideration. Several classification methods exist in the literature: Deshpande proposed to classify lattices in either stretching dominated or bending dominated lattices (Deshpande et al. 2001a), considering that beams are working either in tension-compression, also known as "direct action mechanism", or in bending (Christensen 2000; Mohr 2005). The elastic strength of beam lattices has been considered by various authors: Gibson and Ashby (1999) analyzed foams (bending dominated) with the relative density as the dominant criterion, Demiray et al. (2007), Sullivan et al. (2008), Kim and Al-Hassani (2002) with a more numerical or analytical approach, Florence and Sab (2006) with an energy homogenization method, Deshpande et al. (2001b), Wang and McDowell (2005) with various stretching dominated lattices, and Doyoyo and Hu (2006) analyzing the octet lattice. The stretching dominated lattices are proved to be much stronger than bending governed lattices, Deshpande et al. (2001a); this raises the interest of this material as a substitute for metallic foams in lightweight structures. More recently, the initial yield surface for 2D truss-lattice materials under biaxial loading was investigated by Alkhader and Vural (2009), based on FE analyses and analytical techniques relying on an energy criterion for orthotropic materials. The extended finite element method is used in Zhang et al. (2010) for the elastoplastic analysis of periodic truss materials in the small strain regime. Multiscale base functions are constructed to capture the small scale heterogeneities of the unit cells; this local information is then brought to the upper macroscopic scale to perform structural calculations. The mechanical properties of micro-lattice structures subjected to normal stresses are evaluated in Ushijima et al. (2013), based on an analytical method relying on classical beam theory. The yield surface is determined under an external biaxial loading state.

The investigation of the stress-strain relationships of beam lattices in the plastic range is more involved. The classical criteria of continuum mechanics do indeed not allow describing the nonlinearities in the plastic range (Fan et al. 2009). The effective behavior of three different lattice materials endowed with cubic symmetry has been studied by means of analytical and numerical techniques in Park et al. (2010). A multiscale finite element method has been developed by Zhang et al. (2010) to analyze the elastoplastic small strain behavior of 2D periodic lattices. A continuum mechanism based multi-surface plasticity model has been introduced by Mohr (2005) to simulate the mechanical behavior of 2D or 3D stretching dominated lattices. This method has been extended later by Fan et al. (2009); this model however relies on an underlying hypothesis of uniform deformation of the cells. This hypothesis is not necessary true in the case of lattice with internal nodes (in the unit cell), even if the lattice is stretching dominated.

Those lattices are endowed with a specific mechanical behavior because of the presence of an inherent microstructure. The prediction of the effective mechanical behavior of such lattices in relation to the geometrical and mechanical parameters at the micro level is an especially significant issue. The setting up of such predictive models allows to fully understand the microstructural origin of the mechanical behavior, and the lattice architecture required to achieve optimized properties at the structural level. The suitable size of the structural elements, the lattice topology and mechanical properties can be chosen relying on a quantitative understanding of the macroscopic impact of the microstructural parameters.

An essential category of such lattice materials consists of lattices such as trabecular bone having a discrete kinematics and topology, showing size effects at a macroscopic scale of description. Such size effects have been proven to be important when the sample dimensions are comparable with the size of the cell (Lakes 1986). In other words, the effective mechanical properties show a size-dependence if the dimension of a specimen or a structure is in a close order to the unit cell size. This behavior called "size effects" (Onck et al. 2001; Tekoglu and Onck 2008) designates the effect of the macroscopic sample size, relative to the unit cell size, on the mechanical behavior. However, these effects are not easily accessible from a direct analysis at the macroscopic structure scale. Consequently, the motivation of such micromechanical inspired constitutive models is to increase the understanding of the yielding behavior of those structures in different loading situations.

A key issue in bone biomechanics is the influence of the microarchitecture strut ductility on the overall strength of trabecular bone. With aging and disease, the individual trabeculae become more brittle, so the question arises as to how does this influence the strength of apparent trabecular bone? In the present work, we analyze the structural behavior of trabecular bone failure by investigating two extreme behaviors, known as fully ductile and fully brittle. For both strut level behaviors, we investigate the failed tissue amount and reactive strength evaluated at the trabecular bone scale at yield or ultimate point, respectively for ductile or brittle tissue level material behavior. The yield point corresponds to the stress at which a material begins to deform plastically, and the ultimate point corresponds to the highest point of the stress-strain curve. For the entirely ductile case, we consider tissue level failure by plastic yielding, whereas in the brittle case the yielding is not allowed (break without significant deformation or failed by fracture). Furthermore, when determining the plastic yield strength of trabecular bone structure, bone struts are assumed to be made of elastic perfectly plastic material (Gibson 1985; Keaveny et al. 1994; McDonald et al. 2010). Basically, once the ultimate yielding (plastic collapse) occurs, the structure cannot bear extra loads any more.

The objective of this work is then to develop an adequate three-dimensional model for describing the multiaxial yield and failure behavior of lattice-like structure, and to set up criteria for the brittle and ductile collapse based on micromechanical approaches. The discrete homogenization technique is presently developed as a convenient micromechanical approach to construct the plastic yield surfaces of 2D and 3D periodic lattices of articulated beams. The initial lattice is replaced by an effective Cauchy continuous medium at an intermediate scale, endowed with effective properties representative of an identified representative unit cell within the structure.

The effective elastoplastic response of general beam lattices is additionally obtained thanks to an adaptation and extension of the discrete asymptotic homogenization method. The stress-strain response accounting for ongoing hardening is then constructed. We determine the effective elastoplastic response for the case of stretching dominated lattices without considering bending effects. This methodology has been implemented in algorithmic format in a dedicated code as a user oriented subroutine in finite element calculations, allowing the analysis of a large variety of new 2D lattices. The proposed algorithm is applied to two lattices exhibiting a non uniform deformation: the asymmetric lattice and the square-star lattice. The obtained homogenized elastoplastic responses are validated by comparison with finite element simulations performed over entire lattices.

The 3D homogenization model is further applied to evaluate the yield and failure properties of trabecular bone, which are of key interest in understanding and predicting the fracture of bones and bone implant systems. The effective strength of trabecular bone is evaluated in the two situations of fully brittle (fracture with no tissue ductility) and fully ductile failure (yield with no tissue fracture) of the trabecular tissue. A size-dependent non-classical plastic yield criterion is finally developed relying on the Cosserat theory to capture the size-dependency of trabecular bone structures. Accounting for the moments will allow the homogenization towards the more complete micropolar framework, incorporating a microrotation in addition to the displacement as kinematic descriptors at the continuum level. Such extension and adaptation towards a micropolar effective medium leads to construct the plastic yield surfaces for which the material point supports couple stresses in addition to Cauchy-type stresses. As a consequence, the plastic yield surfaces of trabecular bone will be determined under the external applied couple stresses.

This work is organized as follows: in Sect. 2, we briefly recall the background behind discrete homogenization, and its adaptation in view of the construction of the plastic yield domain and brittle failure surfaces in 3D context. Issues related to the determination of plastic yield surfaces and brittle failures of the vertebral bone modeled as cellular solid model are additionally addressed in Sect. 2. The numerical results provide the plastic yield surface of the vertebral model under multi-axial loading, which is constructed together with the brittle fracture surface in stress space under tri-axial loading conditions. In Sect. 3 we expose the set of basic equations for the update of the plastic variables in presence of hardening of extensional lattices. The proposed algorithm is applied to two different 2D lattices exhibiting a non-uniform deformation: the asymmetric lattice and the square-star lattice. The obtained homogenized elastoplastic responses are validated by comparison with finite element simulations performed over entire lattices. In Sect. 4, on the basis of micropolar theory, the discrete homogenization method is extended to the plastic collapse of trabecular bone under the action of combined couple and Cauchy stresses in a micropolar framework. In addition, a non-classical size-dependent plastic yield criterion is developed relying on the non-classical continuum theory in order to capture the size dependency observed in the bone structures. Finally, a summary of the main achievements and perspectives are exposed in Sect. 5.

2 Lattice Homogenization Towards 3D Cauchy Continuum

2.1 Description of the Lattice Geometry

The lattice-like materials under consideration are described as a quasi 3D periodic lattice of beams and are fully defined by the positions of the nodes and their connectivity: we symbolize N_R and B_R, respectively as the set of nodes and beams within the reference unit cell. The cells are numbered by the triplet of integers $\nu^i = (\nu^1, \nu^2, \nu^3)$. Within the reference cell, one can select the origin node $O(b)$ of a beam b so that it belongs to the reference cell. Nevertheless, the end node $E(b)$ does not necessarily belong to the reference cell, but is necessarily included in an adjacent cell numbered globally with the set of integers $\lambda^i = (\nu^1 + \delta^{1b}, \nu^2 + \delta^{2b}, \nu^3 + \delta^{3b})$. We pass from an elementary cell to the adjacent cells by a translation, parameterized by the integers $\delta^{ib} \in \{-1, 0, 1\}$ as described in Fig. 1. Note that due to the assumption of periodicity, the infinite structure is built from the repetition of a reference cell by translation in 3D. As a consequence, we consider a family of lattices parameterized by a small parameter, defined as the ratio of a characteristic length of the basic cell to a characteristic length of the lattice; and occupying a fixed reference volume, versus which all static and kinematic variables will be expanded. By letting the small parameter tends to zero, we simulate a continuous increase of the number of cells within the fixed reference volume; as a result, a continuous medium is obtained in the limit. Accordingly, the topology of the lattice is completely described by the identification of the reference unit cell and its topology (in terms of its nodes and beams), and the three periodicity vectors in the case of 3D periodic lattices. The discrete sums over all lattice nodes (equilibrium of efforts) are finally converted in the limit of a continuous density of beams into Riemann integrals over the reference unit volume, thereby highlighting the effective continuum medium.

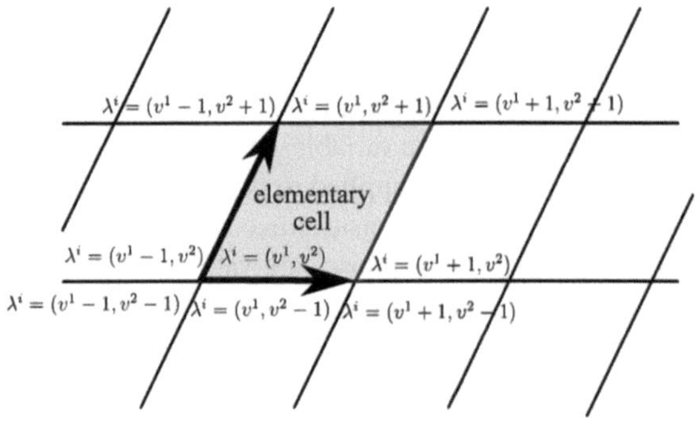

Fig. 1 Curvilinear coordinates λ^i of the lattice

2.2 3D Elementary Unit Cell

Results from Goda et al. (2014) provide the force equilibrium equations of any lattice written in virtual power format as

$$\sum_{b\in B_R} \boldsymbol{F}^b \cdot \boldsymbol{v}^{E(b)} - \boldsymbol{F}^b \cdot \boldsymbol{v}^{O(b)} = \sum_{b\in B_R} \boldsymbol{F}^b \cdot \left(\boldsymbol{v}^{E(b)} - \boldsymbol{v}^{O(b)}\right) = 0 \tag{1}$$

with $\boldsymbol{v}^n$ the virtual velocity field chosen nil on the edges of the considered domain, and $\boldsymbol{F}^b$ the force applied on each node of the beams. Since each beam within the unit cell is equilibrated, the force applied to the extremity $\boldsymbol{F}^b$ is equal and opposite to the force applied to the origin, vector $-\boldsymbol{F}^b$.

We write in addition to previous equilibrium of forces the discrete equilibrium of moments—useful for the resolution of the kinematic unknowns—expressed as the following sum over all unit cell nodes

$$\sum_{b\in B_R} \left(\boldsymbol{M}^{O(b)} \cdot \boldsymbol{w}^{O(b)} - \boldsymbol{M}^{E(b)} \cdot \boldsymbol{w}^{E(b)}\right) = 0 \tag{2}$$

with $\boldsymbol{w}^n$ is the virtual rotational velocity of node n.

The beam resultant $\boldsymbol{F}^b$ in Eq. (1) is the sum of the extensional force F_x^b and transversal forces, F_y^b and F_z^b, expressing successively as

$$F_x^b = \frac{E_s^b A^b}{L^b}\left(\boldsymbol{e}_x \cdot \left(\Delta \boldsymbol{U}_1^b\right)\right) = \frac{E_s^b A^b}{L^b}\left(\boldsymbol{e}_x \cdot \left(\boldsymbol{u}^{E(b)} - \boldsymbol{u}^{O(b)} + \frac{\partial \boldsymbol{u}}{\partial \lambda^i}\delta^{ib}\right)\right), \tag{3}$$

$$\begin{aligned} F_y^b &= \frac{12E_s^b I_z^b}{(L^b)^3(1+\Phi_y)}\left(\boldsymbol{e}_y \cdot \left(\Delta \boldsymbol{U}_1^b\right) - \frac{L_b}{2}\left(\boldsymbol{e}_z \cdot \left(\Phi^{O(b)} + \Phi^{E(b)}\right)\right)\right) \\ &= \frac{12E_s^b I_z^b}{(L^b)^3(1+\Phi_y)}\left(\boldsymbol{e}_y \cdot \left(\boldsymbol{u}^{E(b)} - \boldsymbol{u}^{O(b)} + \frac{\partial \boldsymbol{u}}{\partial \lambda^i}\delta^{ib}\right) - \frac{L_b}{2}\left(\boldsymbol{e}_z \cdot \left(\Phi^{O(b)} + \Phi^{E(b)}\right)\right)\right), \end{aligned} \tag{4}$$

$$\begin{aligned} F_z^b &= \frac{12E_s^b I_y^b}{(L^b)^3(1+\Phi_z)}\left(\boldsymbol{e}_z \cdot \left(\Delta \boldsymbol{U}_1^b\right) - \frac{L_b}{2}\left(\boldsymbol{e}_y \cdot \left(\Phi^{O(b)} + \Phi^{E(b)}\right)\right)\right) \\ &= \frac{12E_s^b I_y^b}{(L^b)^3(1+\Phi_z)}\left(\boldsymbol{e}_z \cdot \left(\boldsymbol{u}^{E(b)} - \boldsymbol{u}^{O(b)} + \frac{\partial \boldsymbol{u}}{\partial \lambda^i}\delta^{ib}\right) - \frac{L_b}{2}\left(\boldsymbol{e}_y \cdot \left(\Phi^{O(b)} + \Phi^{E(b)}\right)\right)\right), \end{aligned} \tag{5}$$

where $\boldsymbol{e}_x, \boldsymbol{e}_y$, and $\boldsymbol{e}_z$ are the direction cosines, describing the transformation between the local and global coordinate system in 3D. The coefficients $\Phi_y = \Phi_z = 12E_s^b I_z^b / G_s^b A^b ks(L^b)^2$ for a circular cross section of vertebrae therein vanish when transverse shear can be neglected, E_s^b and G_s^b are the tensile and shear modulus of the vertebrae material, and k_s is the Timoshenko shear correction coefficient. Note that the displacement difference $\Delta \boldsymbol{U}_1^b$ between the extremity and origin node of each beam is written as the first order expansion $\Delta \boldsymbol{U}_{b1} = \boldsymbol{u}^{E(b)} - \boldsymbol{u}^{O(b)} + \frac{\partial \boldsymbol{u}}{\partial \lambda^i}\delta^{ib}$, and the asymptotic expansion of the rotation is accordingly limited to the zeroth order. Additionally, the displacements and rotations unknowns, the variables $\boldsymbol{u}^n = [u_x^n, u_y^n, u_z^n]$ and $\boldsymbol{\phi}^n = [\phi_x^n, \phi_y^n, \phi_z^n]$ respectively, are determined for all nodes by solving the equilibrium equations in translation and rotation, (1) and (2), over the base unit cell.

The bending moments at both extremities $\boldsymbol{M}^{O(b)}$ and $\boldsymbol{M}^{E(b)}$ of the beam introduced in Eq. (2) can be expressed as $\boldsymbol{M}^{O(b)} = M_y^{O(b)}\boldsymbol{e}_y + M_z^{O(b)}\boldsymbol{e}_z$ and $\boldsymbol{M}^{E(b)} = M_y^{E(b)}\boldsymbol{e}_y + M_z^{E(b)}\boldsymbol{e}_z$, with the components given by

$$\begin{aligned}
M_y^{O(b)} &= \frac{6E_s^b I_y^b}{(L^b)^2(1+\Phi_z)}\left(\boldsymbol{e}_z \cdot \left(\boldsymbol{u}^{E(b)} - \boldsymbol{u}^{O(b)} + \frac{\partial \boldsymbol{u}}{\partial \lambda^i}\delta^{ib}\right)\right) \\
&\quad + \frac{E_s^b I_y^b}{L^b(1+\Phi_z)}\left(\boldsymbol{e}_y \cdot \left((4+\Phi_z)\boldsymbol{\phi}^{O(b)} + (2-\Phi_z)\boldsymbol{\phi}^{E(b)}\right)\right), \\
M_y^{E(b)} &= \frac{6E_s^b I_y^b}{(L^b)^2(1+\Phi_z)}\left(\boldsymbol{e}_z \cdot \left(\boldsymbol{u}^{E(b)} - \boldsymbol{u}^{O(b)} + \frac{\partial \boldsymbol{u}}{\partial \lambda^i}\delta^{ib}\right)\right) \\
&\quad + \frac{E_s^b I_y^b}{L^b(1+\Phi_z)}\left(\boldsymbol{e}_y \cdot \left((2-\Phi_z)\boldsymbol{\phi}^{O(b)} + (4+\Phi_z)\boldsymbol{\phi}^{E(b)}\right)\right), \\
M_z^{O(b)} &= \frac{6E_s^b I_z^b}{(L^b)^2(1+\Phi_y)}\left(-\boldsymbol{e}_y \cdot \left(\boldsymbol{u}^{E(b)} - \boldsymbol{u}^{O(b)} + \frac{\partial \boldsymbol{u}}{\partial \lambda^i}\delta^{ib}\right)\right) \\
&\quad + \frac{E_s^b I_z^b}{L^b(1+\Phi_y)}\left(\boldsymbol{e}_z \cdot \left((4+\Phi_y)\boldsymbol{\phi}^{O(b)} + (2-\Phi_y)\boldsymbol{\phi}^{E(b)}\right)\right), \\
M_z^{E(b)} &= \frac{6E_s^b I_z^b}{(L^b)^2(1+\Phi_y)}\left(-\boldsymbol{e}_y \cdot \left(\boldsymbol{u}^{E(b)} - \boldsymbol{u}^{O(b)} + \frac{\partial \boldsymbol{u}}{\partial \lambda^i}\delta^{ib}\right)\right) \\
&\quad + \frac{E_s^b I_z^b}{L^b(1+\Phi_y)}\left(\boldsymbol{e}_z \cdot \left((2-\Phi_y)\boldsymbol{\phi}^{O(b)} + (4+\Phi_y)\boldsymbol{\phi}^{E(b)}\right)\right).
\end{aligned} \tag{6}$$

The homogenization of the discrete equilibrium of efforts (1) leads to the following continuous self-equilibrium in virtual power form

$$\int_{\Omega} \boldsymbol{S}^i \cdot \frac{\partial \boldsymbol{v}}{\partial \lambda^i} \mathrm{d}\lambda = 0 \tag{7}$$

with the force vectors $\boldsymbol{S}^i$ expressed as the following sum over all struts of the reference unit cell

$$\boldsymbol{S}^i = \sum_{b \in B_R} \boldsymbol{F}^b \delta^{ib} \tag{8}$$

The equilibrium equation of the equivalent Cauchy continuum is then written in virtual power form, in order to highlight the stress tensor as the dyadic product of the force vector $\boldsymbol{S}^i$ with the gradient of the position vector $\boldsymbol{R}$ with respect to the curvilinear coordinates. The transformation from the Cartesian to the curvilinear coordinates λ^i is expressed as

$$\frac{\partial \boldsymbol{v}}{\partial \lambda^i} = \boldsymbol{\nabla}_x \boldsymbol{v} \cdot \frac{\partial \boldsymbol{R}}{\partial \lambda^i}. \tag{9}$$

This leads to the following expression of the equilibrium equation of the equivalent Cauchy continuum

$$\int_{\Omega} \boldsymbol{S}^i \cdot \frac{\partial \boldsymbol{v}}{\partial \lambda^i} \mathrm{d}\lambda = \int_{\Omega} \boldsymbol{S}^i \cdot \left(\boldsymbol{\nabla}_x \boldsymbol{v} \cdot \frac{\partial \boldsymbol{R}}{\partial \lambda^i} \right) = \int_{\Omega} \left(\boldsymbol{S}^i \otimes \frac{\partial \boldsymbol{R}}{\partial \lambda^i} \right) : \frac{1}{g} (\boldsymbol{\nabla}_x \boldsymbol{v}) \, \mathrm{d}x = \int_{\Omega} (\boldsymbol{\Sigma} \cdot \boldsymbol{\nabla}_x) \cdot \boldsymbol{v} \mathrm{d}x = 0. \tag{10}$$

From the comparison of the homogenized equilibrium to the equilibrium of a postulated continuum Cauchy medium, it is natural to set the following definition of the macroscopic (apparent) Cauchy stress

$$\boldsymbol{\Sigma} = \frac{1}{g} \boldsymbol{S}^i \otimes \frac{\partial \boldsymbol{R}}{\partial \lambda^i} \tag{11}$$

with g the Jacobian associated with the function transformation from Cartesian to curvilinear coordinates. Note that the scalar quantities λ^i are curvilinear Lagrangian coordinates suitable for a general parametrization of any material point.

It should be noticed that the terms $\frac{\partial \boldsymbol{u}}{\partial \lambda^i}$ appearing in Eqs. (3)–(6) are strain functions, and all displacements and rotations unknowns are also functions of the strain tensor $[\boldsymbol{E}]$, which are determined by solving the equilibrium equations in translation and rotation. Accordingly, the constitutive equation for the equivalent continuum writes in 3D matrix format as

$$\{\Sigma_x, \Sigma_y, \Sigma_z, \Sigma_{xy}, \Sigma_{yz}, \Sigma_{xz}\} = [K] \left\{ \frac{\partial u_x}{\partial x}, \frac{\partial u_y}{\partial y}, \frac{\partial u_z}{\partial z}, \frac{\partial u_y}{\partial x}, \frac{\partial u_z}{\partial y}, \frac{\partial u_z}{\partial x} \right\} \tag{12}$$

with $[K]$ the equivalent rigidly matrix; the corresponding compliance matrix can be evaluated as

$$[S] = [K]^{-1}. \tag{13}$$

2.3 *Determination of the Plastic Yield and Brittle Fracture Surfaces in Stress Space*

We have two different types of analyses to be performed in the so-called ductile and brittle modes in order to assess the effects of strut level material behavior on the apparent network behavior. For this purpose, we expose a sequence of equations written in stress space based on homogenization method in order to determine the onset of plastic yielding of lattice structure and the fracture surface under multiaxial loading conditions, relying on a micromechanical analysis.

2.3.1 Microscopic Stresses Versus the Macroscopic Deformation Tensor

In this subsection, the homogenization method previously described is used to analytically determine the matrix relating the microscopic stress tensor to the apparent (effective) stress tensor. We first construct a vector of forces that gather all resultants (axial and transverse) and moments acting on each beam b of the reference unit cell.

Let $(F_x^b, F_{y,z}^b, M_{y,z}^{O(b)}, M_{y,z}^{E(b)})$ denote the microscopic stress components of beam b, with $b = 1, 2, 3, \ldots$. Recall that F_x^b is the axial force, $F_{y,z}^b$ the shearing forces and $M_{y,z}^{O(b)}, M_{y,z}^{E(b)}$ the bending moments with respect to y and z at both beam extremities. Therefore, the vector of the forces is written lengthily as

$$[F] = \begin{bmatrix} F_x^1 \\ F_y^1 \\ F_z^1 \\ M_y^{O(1)} \\ M_y^{E(1)} \\ M_z^{O(1)} \\ M_z^{E(1)} \\ F_x^2 \\ \ldots \end{bmatrix}. \tag{14}$$

Previous expressions of the resultants and moments still involve the unknown displacements $\boldsymbol{u}^n$ and rotations $\boldsymbol{\phi}^n$, which are determined for all nodes by solving the equilibrium of efforts and moments Eqs. (1) and (2). As a consequence, these expressions are functions of the macroscopic strain tensor $[\boldsymbol{E}]$; one can then construct a matrix $[Q_e]$ relating the force vector on the unit cell to the homogenized strain, so that

$$[F] = [Q_e][\boldsymbol{E}]. \tag{15}$$

Additionally, we can relate the effective homogenized strain and stress tensors by the following relation $[\boldsymbol{E}] = [S][\boldsymbol{\Sigma}]$, leading in turn to a relation between the stresses at the micro (forces and moments) and macro levels,

$$[F] = [Q_e][S][\boldsymbol{\Sigma}]. \tag{16}$$

The previous expression of the microscopic stress is next involved in defining the criteria of onset of plastic yielding and fracture surface for 3D lattices under triaxial and shear loadings.

2.3.2 Plastic Collapse and Yield Surface

Yielding in the struts of such cellular materials type is concentrated in small zones. When a section becomes completely plastic, it is usually referred to as formation of plastic hinges. The initial yield strength is based on the first cell strut(s) to reach the fully plastic limit moment of the cell struts, representing complete loss to carry additional load.

We here consider a uniform strut thickness of the unit cell; hence, the plastic hinge occurs at a cross-section of maximum bending moment. Based on the plastic hinge concept, lattice collapses plastically when the bending moment exerted by the loads on the individual struts reaches the fully plastic moment, creating plastic hinges at the corners. The fully plastic moment for a perfectly plastic beam under combined bending moment and extensional stress receives the expression (Stronge and Yu 1993)

$$|M^b| = \sigma_{ys} Z_{\lambda b}\left(1 - \left(\frac{\sigma_a^b}{\sigma_{ys}}\right)^2\right) \tag{17}$$

with M^b the bending moment at the beam extremity nodes with respect to y or z axis, σ_{ys} the initial yield strength of the individual struts (bulk material), $Z_{\lambda b}$ the plastic section modulus at the plastic hinge section of the struts, which is adopted as $Z_{\lambda b} = D_b^3/6$ assuming a circular cross sectional area of the struts with the diameter D_b. We here consider a 3D geometry of the circular struts, hence the extensional stress acting on them is elaborated as $\sigma_a^b = F_x^b/A^b$. Therefore, the criterion that defines the failure surface for plastic yield is described from (17) as

$$\frac{6|M^b|}{\sigma_{ys} D_b^3} + \left(\frac{F_x^b}{A^b \sigma_{ys}}\right)^2 = 1 \tag{18}$$

with M^b the bending moments and F_x^b the axial forces within struts, arising from the external applied normal (Σ_x, Σ_y, Σ_z) and shear stresses (Σ_{xy}, Σ_{yz}, Σ_{xz}) acting on the unit cell.

Introducing the homogenized stress components of $[\boldsymbol{\Sigma}]$ after normalization by the initial yield strength of the bulk material σ_{ys} into Eq. (16) leads to the following relation

$$\left[\frac{F}{\sigma_{ys}}\right] = [F^N] = \begin{bmatrix} F_x^1 \\ F_y^1 \\ F_z^1 \\ M_y^{O(1)} \\ M_y^{E(1)} \\ M_z^{O(1)} \\ M_z^{E(1)} \\ F_x^2 \\ \cdots \end{bmatrix} = [Q_e][S]\left[\frac{\boldsymbol{\Sigma}}{\sigma_{ys}}\right] = [Q_e][S]\begin{bmatrix} \Sigma_x/\sigma_{ys} \\ \Sigma_y/\sigma_{ys} \\ \Sigma_z/\sigma_{ys} \\ \Sigma_{xy}/\sigma_{ys} \\ \Sigma_{yz}/\sigma_{ys} \\ \Sigma_{xz}/\sigma_{ys} \end{bmatrix}. \tag{19}$$

As a final step, the system of previous equations in (18) then rewrites for the set of struts b that belong to the reference unit cell by which the whole structure is generated as

$$\frac{6|M^{bN}|}{D_b^3} + \left(\frac{F_x^{bN}}{A^b}\right)^2 = 1 \tag{20}$$

The plastic collapse strength domain of the structure is then the surface delimited by previous equality in the normalized stress space. As a consequence, the criteria of plastic collapse are derived in three dimensional stress states, thereby addressing the criteria for in-plane and out-of-plane yielding: in-plane stresses Σ_x and Σ_y combined with out-of-plane normal stress Σ_z in addition to in-plane Σ_{xy} and out-of-plane shears Σ_{yz}, Σ_{xz} are considered in the analysis. Note that the normalized axial forces and moments, quantities F_x^{bN} and M^{bN}, are extracted from $[F^N]$ in Eq. (19).

2.3.3 Brittle Failure and Fracture Surfaces

Brittle failure occurs when the maximum stress in a cell strut of lattice subjected to both bending and axial loads exceeds the modulus of rupture σ_{fs} of the cell materials. This criterion has been used by Gibson and Ashby (1999). The moment that causes the maximum stress writes

$$\sigma_{\max} = M^b / S_E^b \tag{21}$$

with S_E^b the elastic section modulus adopted as $S_E^b = \pi D_b^3/32$ for circular struts. Therefore, failure occurs when the previous stress cumulated to the axial stress exceeds σ_{fs}, given by

$$\sigma_{fs} = \sigma_a^b + M^b / S_E^b, \tag{22}$$

where σ_a^b is the axial stress in the cell strut adopted as $\sigma_a^b = F_x^b / A^b$, M^b the bending moment at the strut ends either with respect to y or z axis. The brittle failure condition now becomes

$$\frac{F_x^b}{\sigma_{fs} A^b} + \frac{M^b}{\sigma_{fs} S_E^b} = 1. \tag{23}$$

Inserting the macroscopic stress components $[\boldsymbol{\Sigma}]$ normalized by the modulus of rupture of cell strut into Eq. (16) yields the following relation

$$\left[\frac{F}{\sigma_{fs}}\right] = [F^N] = \begin{bmatrix} F_x^1 \\ F_y^1 \\ F_z^1 \\ M_y^{O(1)} \\ M_y^{E(1)} \\ M_z^{O(1)} \\ M_z^{E(1)} \\ F_x^2 \\ \cdots \end{bmatrix} = [Q_e][S]\left[\frac{\boldsymbol{\Sigma}}{\sigma_{fs}}\right] = [Q_e][S]\begin{bmatrix} \Sigma_x/\sigma_{fs} \\ \Sigma_y/\sigma_{fs} \\ \Sigma_z/\sigma_{fs} \\ \Sigma_{xy}/\sigma_{fs} \\ \Sigma_{yz}/\sigma_{fs} \\ \Sigma_{xz}/\sigma_{fs} \end{bmatrix}. \tag{24}$$

It is obvious that the combination of applied normal and shear stresses required to cause tensile failure in the cell strut is found from the applied moments and the axial stresses acting on the cell struts.

Finally, based on the system of previous Eq. (23), the failure surface of lattices under multiaxial stress conditions accounting for anisotropy is described by a brittle failure criterion condition rewritten for the set of struts b that belong to the reference unit cell as

$$\frac{F_x^{bN}}{A^b} + \frac{32M^{bN}}{\pi D_b^3} = 1. \tag{25}$$

A dedicated code has been constructed from the previous developments for the calculation of the plastic yield and brittle fracture surfaces of 3D porous open-celled structure in the homogenized stress space. The code uses an input file including the topology and mechanical properties within a selected unit cell, and delivers the plastic collapse and brittle fracture stresses.

Although the developed models have the ability to describe the plastic and brittle failure within the microstructure of cellular materials in a broad sense subjected to multiaxial loadings, we are focusing in this contribution on vertebral trabecular bone.

2.4 3D Plastic Collapse and Brittle Fracture Surface of Trabecular Bone

Trabeculae form a 3D porous lattice in cancelous bone, whose micro-architecture determines its mechanical performance including the macroscopic strength. Therefore, a 3D geometric model of a hexagonal structure is considered as an idealized representation of vertebral trabecular bone, for which the yield domain and fracture surface at the macroscopic level are identified from a micromechanical analysis. The whole structure is generated from the repetition of this unit cell using the three periodicity vectors defined in a Cartesian basis as shown in Fig. 2a. This work examines the plastic collapse and brittle failure of a hexagonal model of rod-like columnar structure considered as a prototype topology of vertebral trabecular bone. The model is comprised of struts of length L_v with diameter D_h for the horizontal ones and

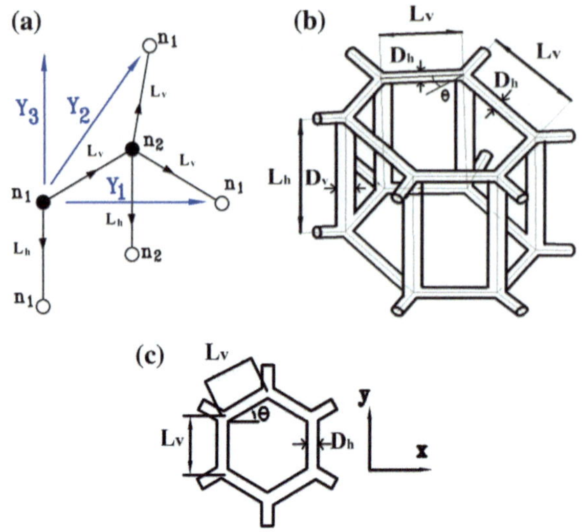

Fig. 2 Representative model of vertebral trabecular bone and its parameters. **a** Reference unit cell and its periodicity vectors Y_1, Y_2 and Y_3, **b** 3D hexagonal unit cell with thick vertical trabeculae and thinner horizontal ones rendering of human vertebral bone, and **c** 2D hexagonal lattice

Table 1 Micro-structural parameters for vertebral bone based on morphology-age relationships by Mosekilde (1988), age is measured in years

Parameter	Relation
Horizontal trabeculae thickness (μm)	$D_h = -1.03\times$ Age $+ 189$
Vertical trabeculae thickness (μm)	$D_v = 0.14\times$ Age $+ 208$
Distance between horizontal trabeculae (μm)	$L_h = 13.74\times$ Age $+ 288$
Distance between vertical trabeculae (μm)	$L_v = -6.75\times$ Age $+ 456$

length L_h with diameter D_v for the vertical struts (see Table 1). The architecture of the model is based on studies of samples taken from the central part of vertebral bodies from normal Individuals aged from 30 to 90 years (Mosekilde 1988, 1989).

Two different types of analysis have to be performed in the so-called ductile and brittle failure modes in order to assess effects of the tissue level material behavior on the apparent trabecular bone behavior. In the case of ductile response, tissue failure occurs by yielding, while in the brittle case, tissue failure occurs via brittle fracture.

The overall collapse surface in the macroscopic stress space consists of intersecting collapse surfaces which are associated with particular collapse modes. The initial yield strength of periodic trabecular structure is based on the first cell wall(s) reaching the fully plastic limit moment of the cell walls in either tension or compression. The plastic yield surface is the inner envelope of the intersecting surfaces for the plastic collapse mechanism. We define the onset of plastic yielding under a multiaxial state of stress for vertebral trabecular bones relying on the homogenization model established in Sect. 2.3 with the plasticity criterion defined by Eq. (20). The

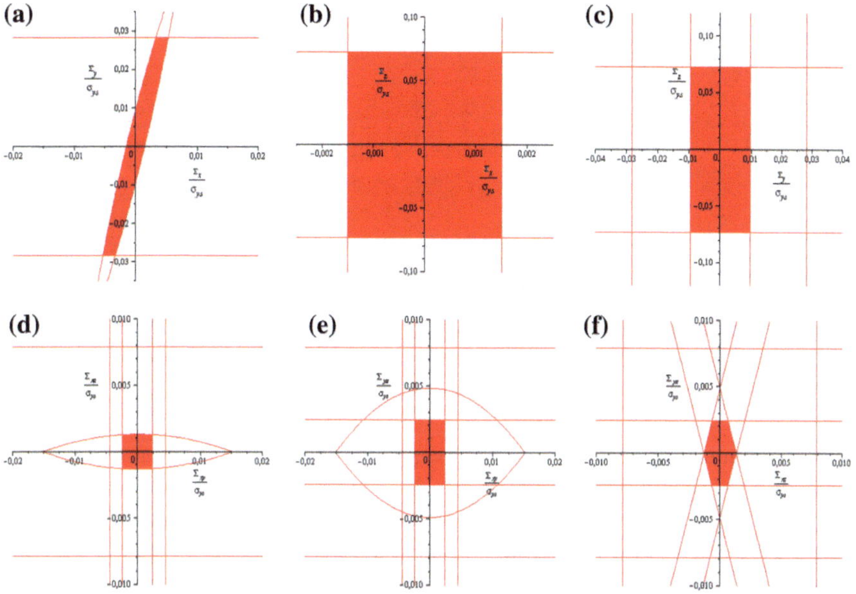

Fig. 3 Plastic yield surfaces in the stress spaces of **a** $\Sigma_x - \Sigma_y$, **b** $\Sigma_x - \Sigma_z$, **c** $\Sigma_y - \Sigma_z$, **d** $\Sigma_{xy} - \Sigma_{xz}$, **e** $\Sigma_{xy} - \Sigma_{yz}$, and **f** $\Sigma_{xz} - \Sigma_{yz}$, all the stresses are normalized with respect to the yield strength of the bulk material. Age is 40 years with relative density $\rho^*/\rho_s = 0.12$

micro-architecture parameters of the trabeculae are selected for a person at the age of 40, with an inclination angle of the oblique trabeculae $\theta = 60^0$, corresponding to a relative density of about 12 %. The combination of the applied normal and shear stresses which causes plastic collapse, plotted on the stress axes $\Sigma_{ij} (i = x, y, z)$, is a closed surface called the yield surface for vertebral trabecular bone, as illustrated on Fig. 3. The yield surfaces are plotted according to Eq. (20) in various stress subspaces, normalized by the trabecula yield strength σ_{ys} (the yield strength is 136 MPa, based on the value reported in Gibson (2005)).

From the results illustrated in Fig. 3, some key features can be noticed: the plastic yield domains are closed, anisotropic, and consist of several convex curved surfaces, with some planar facets. All the plastic failure envelopes are symmetrical with respect to tensile and compressive loadings. This may derive from the fact that the initial yield is based on the first cell strut to collapse in bending or reaching yield by extension. If we would consider the post-buckling behavior past initial yield, the behavior would exhibit a tension-compression asymmetry (in this study, we assume that the first compressive buckling load is greater than the initial yield limit of any strut within the lattice, so that no buckling is likely to arise).

At very low bone volume fraction (high slenderness ratios of trabeculae), trabecular bone may fail in a brittle mode. Additionally, at low slenderness ratios dry individual trabeculae can also fracture in a brittle way (fracture with no tissue ductility); the stress reaches the ultimate point very quickly (only a small amount of

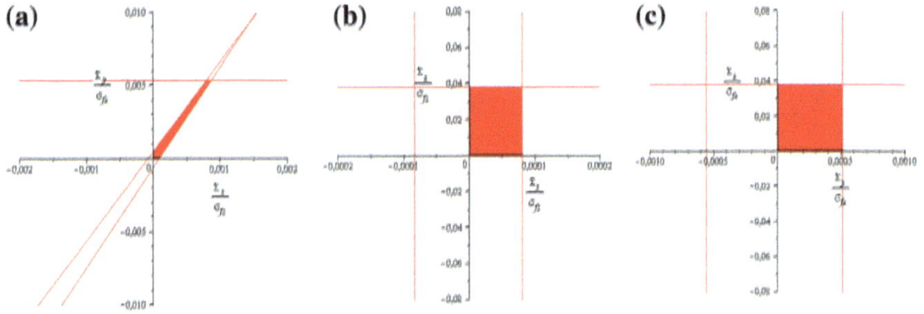

Fig. 4 Brittle fracture surfaces in the stress spaces of **a** $\Sigma_x - \Sigma_y$, **b** $\Sigma_x - \Sigma_z$, **c** $\Sigma_y - \Sigma_z$, for a person aged 85 years with bone relative density $\rho^*/\rho_s = 0.05$. The tension–tension quadrant is colored

plastic strain can be sustained, and consequently only a very small amount of load is enough to break the whole specimen). We consider trabecular architectures corresponding to a typical lattice for a person aged 85 years. Such an (old) trabecular lattice is selected since it has a low relative density of about 5 % and at very low bone relative density, trabecular bone is likely to fail in a brittle manner. Relying on the brittle failure criterion Eq. (25), we evaluate in Fig. 4 the brittle failure response of this lattice under a multiaxial loading by determining the brittle fracture surfaces (the complete failure envelope).

From the evaluated brittle fracture surfaces (Fig. 4), some key points should be noticed: from the in-plane stress state (Fig. 4a), it can be seen that the fracture surface is elongated in the direction of tension and it is truncated in the biaxial tension zone because of the brittle fracture of in-plane vertical trabeculae under the axial stress Σ_y. Moreover, the fracture surfaces in stress spaces $\Sigma_x - \Sigma_z$, and $\Sigma_y - \Sigma_z$ are composed of several planar facets; they exhibit anisotropic properties and are closed in the tension–tension zone.

3 Evolution of the Yield Surface with Ongoing Hardening for 2D Extensional Lattices

As one can infer from the expression of the yield domain (Eq. (18)), the yield criterion accounts for the extensional and flexural energies stored within the lattice. We are further entitled to define a simpler although less complete criterion for 2D lattices subjected to pure tensile/compressive loadings, as

$$\frac{F_x^b}{t} < \sigma_{ys}. \tag{26}$$

Based on this criterion we elaborate stress-strain relation when strain hardening takes place. We assume that the first compressive buckling load is greater than the initial yield limit of any beam within the lattice, so that no buckling will occur.

3.1 Homogenized Macroscopic Cauchy Stress and Micro-Stress Relationship

Assuming a negligible tangential force, a realistic assumption for tension dominated lattices, the expression of the resultant reduces to

$$\boldsymbol{F}^b = \frac{E_s t}{L^b}\left(\boldsymbol{e}_x \cdot \left(\boldsymbol{u}^{E(b)} - \boldsymbol{u}^{O(b)} + \frac{\partial \boldsymbol{u}}{\partial \lambda^i}\right)\right)\boldsymbol{e}_x = \sigma_b t \boldsymbol{e}_x. \tag{27}$$

Thus, stress vector writes

$$\boldsymbol{S}^i = \sum \boldsymbol{F}^b \delta^{ib} = \sum_{b=1}^{n} \boldsymbol{e}_x \sigma^b t \delta^{ib} \tag{28}$$

with n the number of beams within the unit cell. Continuing further, the homogenized stress tensor writes

$$[\boldsymbol{\Sigma}] = \frac{1}{g}\boldsymbol{S}^i \otimes \boldsymbol{e}_i^{\lambda} = \frac{1}{g}\left(\sum_{b=1}^{n} \boldsymbol{e}_x \sigma^b t \delta^{ib}\right) \otimes \boldsymbol{e}_i^{\lambda} \equiv [Q_p][\sigma]. \tag{29}$$

We have thereby obtained a matrix $[Q_p]$ linking the microscopic stress vector of the beams $[\boldsymbol{\sigma}]$ to the macroscopic homogenized $[\Sigma]$ stress tensor.

3.2 Macroscopic Strain Related to Microscopic Stress

The one-dimensional equation for the beam strain in the case of elastoplastic lattices submitted to pure traction writes

$$\sigma^b = E_s(e_e^b - e_p^b) = \frac{E_s}{L^b}\left(\boldsymbol{e}_x \cdot \left(\boldsymbol{u}^{E(b)} - \boldsymbol{u}^{O(b)} + \frac{\partial \boldsymbol{u}}{\partial \lambda^i}\delta^{ib}\right) - L^b e_p^b\right), \tag{30}$$

where e_e^b is the elastic strain and e_p^b is the plastic slip associated with any beam b.

After determining the unknown kinematic variables from the equilibrium Eqs (1) and (2), previous Eq. (30) delivers

$$[\sigma] = [K_e][\boldsymbol{E}] - [K_p][e_p] \tag{31}$$

with $[K_e]$ and $[K_p]$ the elastic and plastic rigidity matrices, respectively.

3.3 Constitutive Equations at Microscale

We set up from the constitutive equations of a beam an incremental formulation using an explicit integration scheme: this means the computation at each time step of the tangent stiffness matrix $[K_t]_n$, such that

$$[\mathrm{d}\boldsymbol{\Sigma}]_n = [K_t]_n [\mathrm{d}\boldsymbol{E}]_n. \tag{32}$$

The yield condition for any beam b writes

$$f^b = |\sigma^b| - s^b = 0 \tag{33}$$

denoting therein s^b the deformation resistance, with initial condition: $s^b(t=0) = \sigma_{ys}$; the response is purely elastic if $f^b < 0$, otherwise the lattice may deform plastically. The flow rule, defining the beam flow rate $\dot{\alpha}^b$ in relation to the plastic strain rate for the same beam, variable $\dot{e}_p^b$, writes

$$\dot{e}_p^b = \dot{\alpha}^b \mathrm{sign}(\sigma^b). \tag{34}$$

We assume the following linear hardening law: $\dot{s}^b = H^b \dot{\alpha}^b$, with H^b the hardening modulus of the beam material. The Kuhn-Tucker complementary conditions must be satisfied for plasticity to take place in a given beam:

$$\dot{\alpha}^b \geq 0,\ f^b \leq 0, \dot{\alpha}^b f^b = 0. \tag{35}$$

3.4 Incremental Formulation and Integration Scheme

We use an explicit integration scheme of the constitutive equations, based on the variable γ^b, resulting from the time integration of the plastic flux $\dot{\alpha}^b$ over the time interval $\mathrm{d}t$, as

$$\gamma^b = \dot{\alpha}^b \mathrm{d}t. \tag{36}$$

We summarize the solution algorithm in the global flow chart shown in Fig. 5. The imposed variables in the considered strain driven process are the strain tensor $[\boldsymbol{E}]_n$ and the strain increment $[\mathrm{d}\boldsymbol{E}]_n$ at a generic time step n. The internal variables $[e_p]_n$ and $[s]_n$ result from calculations done at the previous time step $n-1$. The strain driven structure of the algorithm is designed for the purpose to be easily implemented into a FE code as a user material.

The algorithm involves two main steps: firstly, the calculation of the plastic flow rate vector $[\gamma]$, and secondly, the evaluation of the tangent matrix $[K_t]$, which are next detailed.

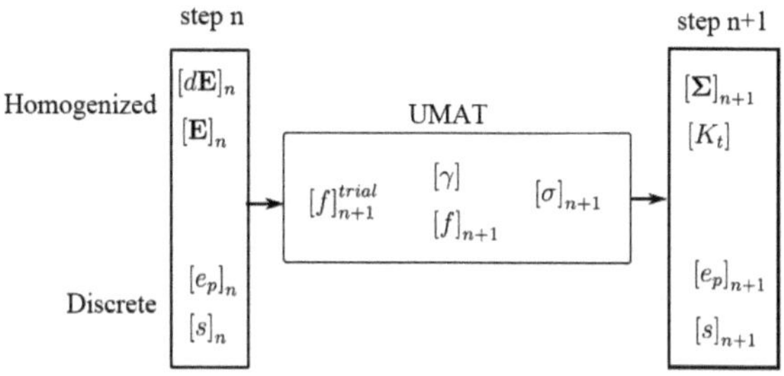

Fig. 5 Strain driven resolution scheme for an elastoplastic homogenized lattice

First step: calculation of the plastic flow rate $[\gamma]_{n+1}$
At time t_n, the lattice is equilibrated, hence Eq. (31) reads

$$[\sigma]_n = [K_e][E]_n - [K_p][e_p]_n. \tag{37}$$

At subsequent time $t_{n+1} = t_n + \mathrm{d}t$, the lattice is also equilibrated, thus it holds

$$\begin{aligned}[\sigma]_{n+1} &= [K_e][E]_{n+1} - [K_p][e_p]_{n+1}\\ &\equiv [K_e]([E]_n + [\mathrm{d}E]_n) - [K_p]([e_p]_n + [\mathrm{sign}(\sigma)]_{n+1}[\gamma]_n)\\ &= [\sigma]^{\mathrm{trial}}_{n+1} - [K_p]([\mathrm{sign}(\sigma)]_{n+1}[\gamma]_n).\end{aligned} \tag{38}$$

We further introduce the trial stress function representing a predicted stress based on a purely elastic behavior

$$[\sigma]^{\mathrm{trial}}_{n+1} = [\sigma]_n + [K_e][\mathrm{d}E]. \tag{39}$$

From Eq. (38), one can write

$$[\mathrm{sign}(\sigma)]_{n+1}([|\sigma|]_{n+1} + [K_p][\gamma]_n) = [\mathrm{sign}\sigma]^{\mathrm{trial}}_{n+1}[|\sigma|]^{\mathrm{trial}}_{n+1}. \tag{40}$$

Assuming that each term in vector $[K_p][\gamma]_n$ is positive, we observe that the components of vector $[|\sigma|]_{n+1} + [K_p][\gamma]_n$ are all positive. Therefore, we require that

$$[\mathrm{sign}(\sigma)]_{n+1} = [\mathrm{sign}(\sigma)]^{\mathrm{trial}}_{n+1} \tag{41}$$

along with the condition

$$[|\sigma|]_{n+1} + [K_p][\gamma]_n = [|\sigma|]^{\mathrm{trial}}_{n+1} \tag{42}$$

and finally

$$[|\sigma|]^{\mathrm{trial}}_{n+1} - [|\sigma|]_{n+1} = [K_p][\gamma]_n. \tag{43}$$

The yield function then writes

$$[f]_{n+1} = [|\sigma|]_{n+1} - [s]_{n+1} \tag{44}$$

with $[s]_{n+1} = [s]_n + [H][\gamma]_n$ and $[H]$ the diagonal matrix of plastic stiffness; this allows rewriting the yield function as

$$[f]_{n+1} = [|\sigma|]_{n+1} - ([s]_n + [H][\gamma]_n). \tag{45}$$

From (43), one can further write the yield function

$$[f]_{n+1} = [|\sigma|]_{n+1}^{\text{trial}} - [K_p][\gamma] - ([s]_n + [H][\gamma]_n) \tag{46}$$

and next

$$[f]_{n+1} = [f]_{n+1}^{\text{trial}} - ([K_p] + [H])[\gamma]_n \tag{47}$$

with the new tangent stiffness matrix

$$[K_\gamma] = [K_p] + [H] \tag{48}$$

and

$$[f]_{n+1}^{\text{trial}} = [|\sigma|]_{n+1}^{\text{trial}} - [s]_n. \tag{49}$$

From the Kuhn-Tucker condition

$$[f]_{n+1}[\gamma]_{n+1} = [0] \tag{50}$$

we may consider the following subsystem of the set of Eq. (47)

$$[K_\gamma]_r[\gamma]_{r,n+1} = [f]_{n+1,r}^{\text{trial}}. \tag{51}$$

with $[K_\gamma]_r$ a square matrix containing the rows and columns for which γ^b is non-nil, or equivalently $f_{n+1}^b > 0$, hence from previous equalities

$$[\gamma]_{r,n+1} = [K_\gamma]_r^{-1}[f]_{n+1,r}^{\text{trial}}. \tag{52}$$

Second step: determination of the tangent matrix
The stress increment $[\mathrm{d}\sigma]$ can be expressed as the difference between two vectors in the space of the micro-stress of the beams

$$[\mathrm{d}\sigma] = [K_e][\mathrm{d}E] - [\text{sign}(\sigma)]_{n+1}^{\text{trial}}([K_p][\gamma]_n) = [\mathrm{d}\sigma_e] - [\mathrm{d}\sigma_\gamma]. \tag{53}$$

We introduce $[\xi]$ the matrix consisting of the diagonal terms $\xi_{ii} = \mathrm{d}\sigma_{\gamma_i}/\mathrm{d}\sigma_{e_i}$, such that

$$[\mathrm{d}\sigma_\gamma] = [\xi][\mathrm{d}\sigma_e]. \quad (54)$$

Introducing the new matrix $[K_r]$ as $[K_r] = [\xi][K_e]$, Eq. (38) can then be reshaped as

$$[\sigma]_{n+1} = [\sigma]_n + ([K_e] - [K_r])[\mathrm{d}E]. \quad (55)$$

Thus we get the new tangent stiffness matrix at $n + 1$

$$[K_t] = [Q_p]([K_e] - [K_r]) \quad (56)$$

All previous calculations are coded in FORTRAN as a user material subroutine in a standard finite element code.

3.5 *Applications Accounting for Ongoing Hardening*

In order to validate the algorithm in the presence of strain hardening, we treat two different lattices presenting a non-uniform deformation, each with a specific loading: the square-star lattice is subjected to an imposed uniaxial displacement, while the asymmetrical lattice is submitted to a biaxial loading.

3.5.1 Square-Star Lattice

The square-star lattice which topology is pictured in Fig. 6 is stretching dominated, and it has internal nodes; it is a relevant application in the presence of strain hardening.

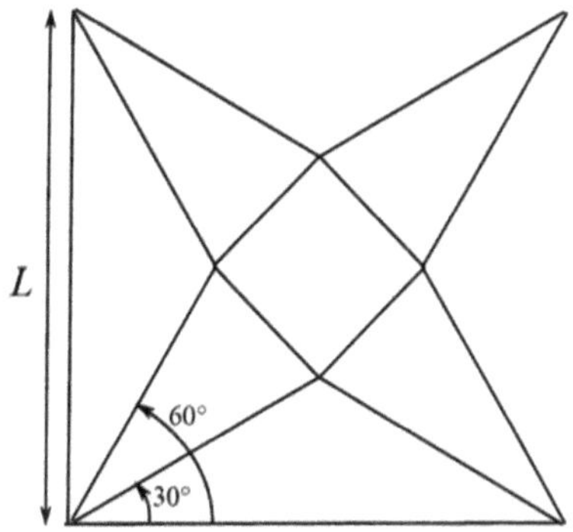

Fig. 6 Topology of the square-star lattice

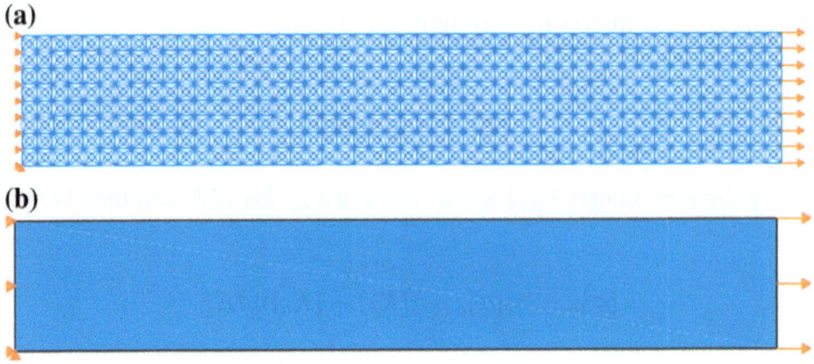

Fig. 7 FE simulation of a beam submitted to displacement on the right edge for **a** a microstructured beam, and **b** a continuous (homogenized) beam

In view of the numerical applications, we choose a relative density of this lattice to be 0.15. Selecting a beam length $L = 1/16\,\mathrm{mm}$, the corresponding width is then calculated as $t = 0.0012\,\mathrm{mm}$.

We perform a numerical tensile test over a macroscale beam of dimension $3\,\mathrm{mm} \times 0.5\,\mathrm{mm}$; the boundary conditions are represented on Fig. 7. We will compare the results of two simulations: (a) A classical FE simulation with a microstructured beam; (b) A continuous homogenized beam with a user material incorporating the homogenized behavior determined from the previous algorithm.

We use the following mechanical data for the constitutive material of the beams: an initial yield strength $\sigma_Y = 20\,\mathrm{MPa}$, an elastic modulus $E_s = 69{,}000\,\mathrm{MPa}$ and a plastic modulus $H = 3000\,\mathrm{MPa}$. We simulate a two steps loading, consisting of an imposed monotonous displacement up to a maximum $\Delta L = 0.003\,\mathrm{mm}$, followed by unloading back to $\Delta L = 0\,\mathrm{mm}$. We use an explicit scheme with an adaptive time step. In order to compute the stress-strain relationship, the strain is calculated from $\varepsilon_{11} = \Delta L/L$ and the uniaxial stress σ_{11} is defined from the applied force. The comparison of the two simulation results is given in Fig. 8, showing that the homogenized response well agrees with the microstructural lattice response.

3.5.2 Asymmetrical Lattice

The asymmetrical lattice is given in Fig. 9. Here we consider a relative density of 0.15, the corresponding width is then calculated as $t = 0.0565L$. We use a lattice of 14×16 cells that constitutes a good approximation of a square (edge effects can be neglected).

The lattice is loaded with an imposed biaxial stress increasing up to a maximum value 1.2 MPa for the continuum model, and an equivalent force for the discrete model is applied to the intersection nodes of each cell at the lattice edge; the nodes of the corners of the edge have been loaded with half the applied force.

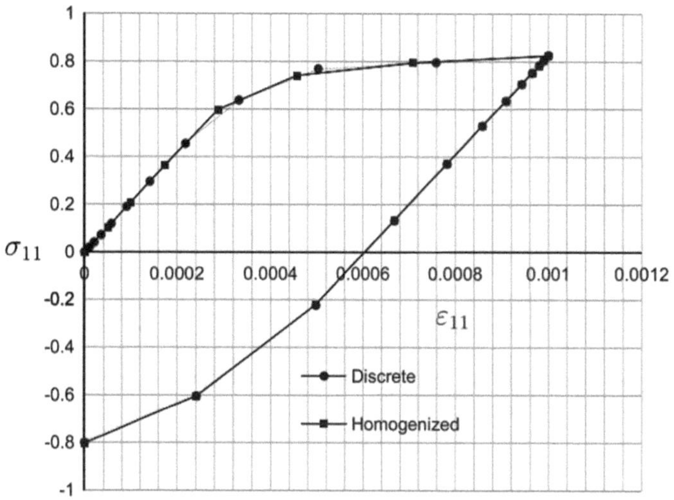

Fig. 8 Simulation of the traction-unloading for a microstructured beam and a homogenized continuous beam

Fig. 9 Topology of the asymmetrical lattice

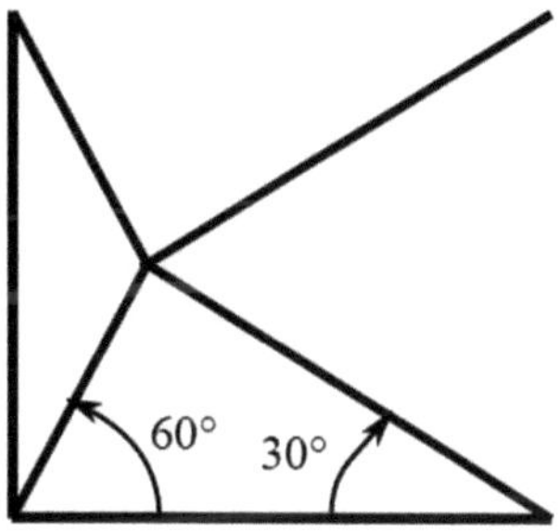

The comparison of the homogenized response with the discrete simulation results in Fig. 10 shows an overall plastic flow triggered for the same stress loading, close to 1.1 MPa in both simulations. We can nevertheless note a small difference between the discrete and continuous responses, presumably due to the fact that the transverse force is neglected in the calculation of the continuous medium.

4 Plastic Yield Surface Based on Cosserat Theory—Size Dependent Plastic Yield Criterion

Several works from the literature show the importance of considering microstructure-related scale effects on the macroscopic properties of bone (Goda et al. 2012, 2014). This motivates to extend previous analyses of collapse mechanisms to micropolar behaviors focusing on the plastic yield behavior in a 2D situation.

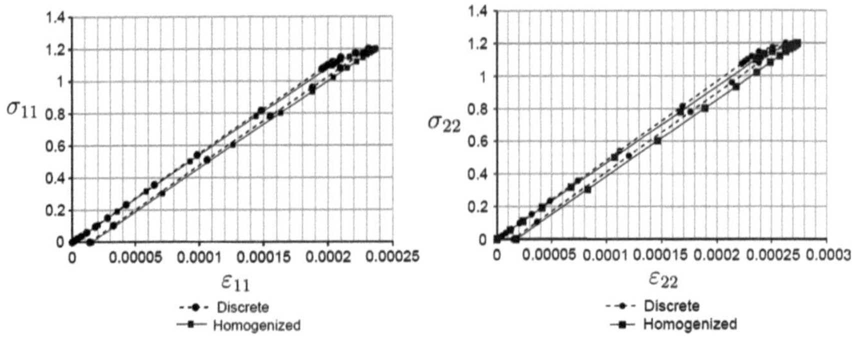

Fig. 10 Stress-strain response due to the biaxial loading-unloading sequence of the asymmetrical lattice

Accounting for the moments would allow the homogenization towards more general effective media, such as micropolar continua, a more complete equivalent continuum incorporating a microrotation in addition to the displacement as kinematic descriptors at the continuum level. This entails that the material can transmit couple stresses in addition to tractions (classical stresses); those couple stresses develop internal work in the variation of microcurvatures, defined as the spatial gradients of the microrotation. As a consequence, the second order development of the displacement is taken into account in this specific theory to capture the micropolar effect and hence the asymptotic expansion of the rotation is limited to the first order. The moments are then expressed at both ends as:

$$
\begin{aligned}
M^{O(b)} &= \frac{6E_s^b I_z^b}{(L^b)^2(1+\Phi)}\left(-\boldsymbol{e}_y\cdot\left(\Delta\boldsymbol{U}_1^b+\Delta\boldsymbol{U}_2^b\right)\right)\\
&\quad+\frac{E_s^b I_z^b}{L^b(1+\Phi)}\Bigg((4+\Phi)\phi_0^{O(b)}+(2-\Phi)\phi_0^{E(b)}\\
&\quad+\left((4+\Phi)\phi_1^{O(b)}+(2-\Phi)\phi_1^{E(b)}+(2-\Phi)\frac{\partial\phi_0}{\partial\lambda^i}\delta^{ib}\right)\Bigg),\qquad(57)\\
M^{E(b)} &= \frac{6E_s^b I_z^b}{(L^b)^2(1+\Phi)}\left(-\boldsymbol{e}_y\cdot\left(\Delta\boldsymbol{U}_1^b+\Delta\boldsymbol{U}_2^b\right)\right)\\
&\quad+\frac{E_s^b I_z^b}{L^b(1+\Phi)}\Bigg((2-\Phi)\phi_0^{O(b)}+(4+\Phi)\phi_0^{E(b)}\\
&\quad+\left((2-\Phi)\phi_1^{O(b)}+(4+\Phi)\phi_1^{E(b)}+(4+\Phi)\frac{\partial\phi_0}{\partial\lambda^i}\delta^{ib}\right)\Bigg)
\end{aligned}
$$

with $\Delta\boldsymbol{U}_2^b$ the second order displacement difference between the ends of each beam, defined as $\Delta\boldsymbol{U}_2^b=\boldsymbol{u}_2^{E(b)}-\boldsymbol{u}_2^{O(b)}$. The unknown displacements u_1^n, u_2^n and rotations ϕ_0^n, ϕ_1^n, are determined for all nodes by solving the equilibrium Eqs. (1) and (2). The solutions lead to u_1^n and ϕ_0^n expressed versus the displacement gradient or deformation tensor, while ϕ_1^n and u_2^n express versus the gradient of micro-rotation or the micro-curvature tensor.

The homogenization of the discrete equilibrium of forces leads to the construction of the stress vectors therein $\boldsymbol{F}^b = \boldsymbol{F}_x^b + \boldsymbol{F}_y^b$ in the 2D case; this in turn leads to the homogenized Cauchy stress tensor. In order to get the couple stress components, the local equilibrium of each individual strut is considered for the purpose of homogenization, written at the center of each strut. After some developments and simplifications, the moment equilibrium equation after homogenization takes the following virtual power form

$$\int_{\Omega} \boldsymbol{\mu}^i \cdot \frac{\partial \boldsymbol{w}}{\partial \lambda^i} \mathrm{d}\lambda = \int_{\Omega} \boldsymbol{\mu}^i \cdot \left(\nabla_x \boldsymbol{w} \cdot \frac{\partial \boldsymbol{R}}{\partial \lambda^i} \right) \mathrm{d}\lambda$$
$$= \int_{\Omega} \left(\boldsymbol{\mu}^i \otimes \frac{\partial \boldsymbol{R}}{\partial \lambda^i} \right) : \frac{1}{g} (\nabla_x \boldsymbol{w}) \mathrm{d}x = \int_{\Omega} (\boldsymbol{m} \cdot \nabla_x) \cdot \boldsymbol{w} \mathrm{d}x = 0 \quad (58)$$

with $\boldsymbol{m}$ the macroscopic (effective) couple stress and $\boldsymbol{w}$ the virtual rotational velocity. The identification of the couple stress vectors $\boldsymbol{\mu}^i$ leads in turn to the expression of the homogenized couple stress according to

$$\boldsymbol{m} = \frac{1}{g} \boldsymbol{\mu}^i \otimes \frac{\partial \boldsymbol{R}}{\partial \lambda^i} \quad (59)$$

with $\boldsymbol{\mu}^i$ therein expressed as the following sum overt all struts forming the lattice

$$\boldsymbol{\mu}^i = \sum_{b \in B_R} \frac{1}{2} (M_2^{E(b)} - M_2^{O(b)}) \delta^{ib}. \quad (60)$$

After solving for all unknowns displacements and rotations, we obtain the effective constitutive equations, written in a 2D context in matrix format

$$\underbrace{\begin{Bmatrix} \sigma_x \\ \sigma_y \\ \sigma_{xy} \\ m_{xz} \\ m_{yz} \end{Bmatrix}}_{[\boldsymbol{\Sigma}]} = [K] \underbrace{\begin{Bmatrix} \varepsilon_x \\ \varepsilon_y \\ \varepsilon_{xy} \\ \chi_{xz} \\ \chi_{yz} \end{Bmatrix}}_{[\boldsymbol{E}]} \quad (61)$$

where $[K]$ is the effective (homogenized) stiffness matrix, σ_{ij} the symmetric part of Cauchy stress tensor and m_{ij} the couple stress tensor.

We consider here the reduced Cosserat theory, where the rotation is related to the displacement gradients $\phi_z = (\partial u_y/\partial x - \partial u_x/\partial y)/2$. As a result, the strain tensor ε_{ij} is symmetrical, with components defined as $\varepsilon_{xy} = \varepsilon_{xy} = (\partial u_y/\partial x + \partial u_x/\partial y)/2$. The couple stress tensor (or moment per unit area) has two components m_{xz}, m_{yz}

and the corresponding micro-curvatures express versus the micro-rotations as $\chi_{xz} = \partial\phi_z/\partial x$, $\chi_{yz} = \partial\phi_z/\partial y$.

We can next construct the matrix relating the microscopic stress tensor to the macroscopic stress tensor; one first elaborates the force vector gathering all resultants and moments for all struts of the reference unit cell

$$[F] = [F_x^1 \ \ F_y^1 \ \ M^{O(1)} \ \ M^{E(1)} \ \ F_x^2 \ \ \ldots]^T. \tag{62}$$

After solving for all kinematic unknowns, we obtain the individual expressions for all forces and moments. Theses expressions involve the macroscopic deformation tensor $[E]$, allowing to build the following relation between the microscopic stress and the macroscopic deformation

$$[F] = [Q_e][E]. \tag{63}$$

This in turn entails the following relation between the microscopic stress and the macroscopic stress

$$[F] = [Q_e][K]^{-1}[\Sigma]. \tag{64}$$

On the basis of the previous calculation of the microscopic stress, we can use the criterion of plasticity initiation in Eq. (17) to define the boundary of the elastic domain of the micropolar continuum medium.

By inserting the macroscopic Cauchy and couple stresses Σ^N normalized by the initial yield strength of strut material into Eq. (64), the following relation is obtained

$$[F^N] = \begin{bmatrix} F_x^1 \\ F_y^1 \\ M^{O(1)} \\ M^{E(1)} \\ F_x^2 \\ \ldots \end{bmatrix}^N = [Q_e][K]^{-1}[\Sigma^N] = [Q_e][K]^{-1} \begin{bmatrix} \sigma_x/\sigma_{ys} \\ \sigma_y/\sigma_{ys} \\ \sigma_{xy}/\sigma_{ys} \\ m_{xz}/\sigma_{ys} \\ m_{yz}/\sigma_{ys} \end{bmatrix}. \tag{65}$$

The plastic yield surface domain of the trabecular lattice is then the surface delimited by the yield criterion; the system of previous equations rewrites for the set of struts that belong to the reference unit cell forming the lattice as

$$\left(\left(F_x^{bN}\right)^2 + 4|M^{bN}|\right)/t^2 = 1. \tag{66}$$

We next plot the plastic yield surfaces of the trabecular bone lattice adopting the following micro-architecture parameters of the 2D trabecular bone cell (see Fig. 2c): the trabecular separation and trabecular thickness are respectively 0.726 and 0.148 mm. To evaluate the overall plastic yield strength relying on micropolar theory, we plot the plastic yield surfaces under the shear stress combined with the couple stress in x–z plane and the couple stress state in both x–z and y–z planes as

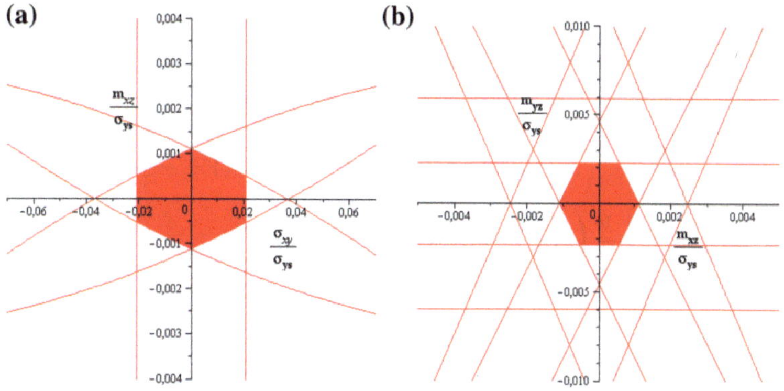

Fig. 11 Plastic yield surfaces under combined shear **a** σ_{xy} and **b** couple stresses states m_{xz}, m_{yz}

well (Fig. 11). The plastic yield surface consists of the intersection of the collapse surfaces of the vertical and inclined struts.

As a matter of fact, the classical yield criterion does not describe the yield size dependency that has been observed in microstructured structures, and thus it may underestimate the yield loads of structures. So, we suggest a specific size-dependent plastic yield criterion relying on micromechanical model on the basis of the micropolar theory. In order to develop the micropolar yield criterion, the deviatoric part of the strain energy density of the microscale structure including both classical and non-classical parts is calculated based on this theory and equated to the deviatoric strain energy of a macro-sample subjected to its yielding tensile load.

The strain energy density consists of one part due to dilation (volume changes) and a second one accounting for distortion (change of shape); the former is called the dilatational strain energy (hydrostatic part) and the latter the distortional energy (deviatoric part). From this decomposition, the hydrostatic and deviatoric energy densities based on micropolar theory take the forms

$$U = \frac{1}{2}(\sigma_{ij}\varepsilon_{ij} + m_{ij}\chi_{ij})\Big|_H + \frac{1}{2}(\sigma_{ij}\varepsilon_{ij} + m_{ij}\chi_{ij})\Big|_D . \tag{67}$$

The hydrostatic stress tensor writes as follows $\sigma_{ij}\big|_H = \frac{1}{3}\delta_{ij}\sigma_{kk}$ and the corresponding hydrostatic strain will be $\varepsilon_{ij}\big|_H = \frac{1}{3}\delta_{ij}\varepsilon_{kk} = \frac{1-2\nu}{3E}\delta_{ij}\sigma_{kk}$. Here, the hydrostatic components of the couple stress and curvature tensors will vanish; however, they will appear later on in the distortional part. Accordingly, the dilatational part of the strain energy can be obtained as

$$U_H = \frac{1-2\nu}{6E}(\sigma_{kk})^2. \tag{68}$$

The deviatoric stress tensor can be obtained by subtracting the hydrostatic stress from the total stress tensor as $\sigma_{ij}\big|_D = \sigma_{ij} - \frac{1}{3}\delta_{ij}\sigma_{kk}$. In the same manner, the deviatoric strain is obtained as

$$\begin{aligned} \varepsilon_{ij}\big|_D &= \varepsilon_{ij} - \frac{1}{3}\delta_{ij}\varepsilon_{kk} = \frac{1}{E}((1+\nu)\sigma_{ij} - \nu\delta_{ij}\sigma_{kk}) - \frac{1-2\nu}{3E}\delta_{ij}\sigma_{kk} \\ &= \frac{1+\nu}{E}\underbrace{\left(\sigma_{ij} - \frac{1}{3}\delta_{ij}\sigma_{kk}\right)}_{\sigma_{ij}\big|_D} \end{aligned}.$$

The non-classical part of the strain energy due to the couple stress and curvature tensors will have a deviatoric nature as will be shown next; therefore, the distortional part of the strain energy can be evaluated by subtracting the dilatational term from the total strain density as follows

$$\begin{aligned} U_D &= \frac{1}{2}(\sigma_{ij}\varepsilon_{ij} + m_{ij}\chi_{ij}) - \frac{1-2\nu}{6E}(\sigma_{kk})^2 \\ &= \frac{1}{2}\left(\frac{\sigma_{ij}}{E}((1+\nu)\sigma_{ij} - \nu\delta_{ij}\sigma_{kk}) + m_{ij}\chi_{ij}\right) - \frac{1-2\nu}{6E}(\sigma_{kk})^2. \end{aligned} \tag{69}$$

According to von Mises's yield theory, a ductile solid will yield when the distortional energy density reaches a critical value for that material; this critical value of the distortional energy can be estimated from a uniaxial tensile test. At the instance of yielding, $\sigma_{xx} = \sigma_Y$ (yield stress), and all other normal and shear stresses will vanish; therefore, the distortion energy density associated with yielding is

$$U_D\big|_Y = \frac{1+\nu}{3E}\sigma_Y^2. \tag{70}$$

We then equate the distortional strain energy density of the micro-level structure which incorporates both the classical and non-classical terms to the distortional strain energy stored in a macro-level sample at the yield point during a tensile test, as follows

$$\left(\sigma_{ij}\sigma_{ij} - \frac{(\sigma_{kk})^2}{3}\right) + \underbrace{\frac{E}{1+\nu}}_{2G} m_{ij}\chi_{ij} = \frac{2}{3}\sigma_Y^2. \tag{71}$$

Consequently, yielding will occur when the stresses and couple stresses produced by loadings satisfy the previous equation. We can also define an equivalent stress σ_{eq} based on the previous relation to express the failure condition under which yield occurs, $\sigma_{eq} \geq \sigma_Y$ as

$$\sigma_{eq} = \sqrt{\frac{1}{2}\left(3\sigma_{ij}\sigma_{ij} - (\sigma_{kk})^2\right) + 3Gm_{ij}\chi_{ij}}. \tag{72}$$

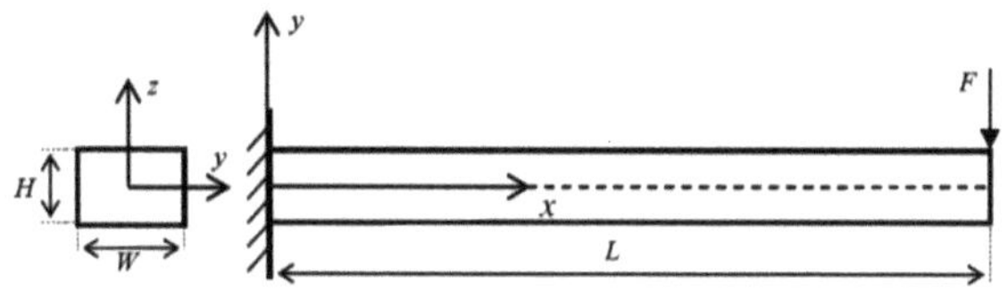

Fig. 12 Schematic diagram of a micro-structured cantilever beam with rectangular cross section exposed to a concentrated force at its free end

It is obvious that when the non-classical term in Eq. (72) vanishes, the classical von Mises yield criterion is recovered.

Consistent with the basic assumptions of Bernoulli-Euler beam and the one-dimensional beam theory (Fig. 12), the nonzero components of the static and kinematic variables write

$$\sigma_{xx} = E_x^* y \frac{\partial^2 v(x,t)}{\partial x^2}, \quad \chi_{xz} = \frac{\partial \phi_z(x,t)}{\partial x} = \frac{\partial^2 v(x,t)}{\partial x^2}, \quad \text{and} \quad m_{xz} = \gamma_{xz} \frac{\partial^2 v(x,t)}{\partial x^2}, \tag{73}$$

where v represents the lateral deflection of the beam in y-direction, x and y refer respectively to the longitudinal and lateral coordinates, E_x^* stands for the effective homogenized elastic Young's modulus of the macroscopic equivalent beam, and γ_{xz} is the homogenized micropolar bending constant. The rotation angle of the centroidal axis of the beam is related to the deflection as:

$$\phi_z(x,t) = \frac{\partial v(x,t)}{\partial x}.$$

Upon substitution of non-zero components of stress, couple stress, and curvature in Eq. (73) into Eq. (72), the equivalent or yield stress of the microstructured beam writes

$$\sigma_Y^* = \sqrt{\left(E_x^* y \frac{\partial^2 v(x,t)}{\partial x^2}\right)^2 + 3G_{xy}\gamma_{xz} \frac{\partial^4 v(x,t)}{\partial x^4}}. \tag{74}$$

The external bending moment acting on a beam based on the micropolar theory can be written as

$$M = \int_A \left(E_x^* y^2 \frac{\partial^2 v(x,t)}{\partial x^2} + \gamma_{xz} \frac{\partial^2 v(x,t)}{\partial x^2}\right) \mathrm{d}A = (E_x^* I_z^* + \gamma_{xz} A) \frac{\partial^2 v(x,t)}{\partial x^2} \tag{75}$$

I_z^* refers to the second moment of cross-sectional area is defined by

$$I_z^* = \int_A y^2 \mathrm{d}A,$$

with A the cross-sectional area of the macroscopic beam and y the perpendicular distance to the neutral axis.

Substituting the second displacement gradient from Eq. (75) into Eq. (74), one can obtain after simplification a relation for the yielding load in the following form

$$(\sigma_Y^*)^2 = \frac{M^2\left((E_x^* y)^2 + 3G_{xy}\gamma_{xz}\right)}{\left(E_x^* I_z^* + \gamma_{xz} A\right)^2}. \tag{76}$$

The size-dependent yielding behavior of the micro-structured beams is analyzed by considering a micro-structured cantilever with uniform rectangular cross section with height H and width W, subjected to a concentrated force F acting on its free end as shown on Fig. 12. For this beam, the maximum bending moment happens at the clamped end based on the relation $M_{\max} = FL$, and the maximum distance on the beam section from the neutral axis is $y_{\max} = H/2$. Hence, the yield moment is determined as $M_{\max} = FL$.

The yielding moment of the beam can be written with the characteristic micropolar bending length l_b using the relation

$$l_b = \sqrt{(\gamma_{xz})/2(2\mu^* + k)},$$

as

$$M_Y = \frac{\sigma_Y^*(E_x^* I_z^* + 4l_b^2 G_{xy} A)}{\sqrt{\left(E_x^* y\right)^2 + 12 G_{xy}^2 l_b^2}}. \tag{77}$$

We here focus on the fully plastic moment of a yielded section of a beam, wherein the plastic moment M_P is always greater than the yielding moment. Rewriting Eq. (77) leads to the fully homogenized plastic moment of a beam with rectangular cross-section as

$$M_P = \frac{3}{2}\frac{\left(\frac{E_x^* H^3}{12} + 4l_b^2 G_{xy} H\right)}{\sqrt{\left(\frac{E_x^* H}{2}\right)^2 + 12 G_{xy}^2 l_b^2}}\sigma_Y^*. \tag{78}$$

We note that when the length scale bending parameter l_b is negligible, the plastic yield load of a classical cantilever beam can be obtained from Eq. (78) as

$$\frac{FL}{H^2} = \frac{M_P}{H^2} = \frac{1}{4}\sigma_Y^*.$$

It should be noted that σ_Y^* refers to the effective homogenized initial yield strength of the beam material; it is determined based on the micromechanical model using the maximum elastic bending moment criteria.

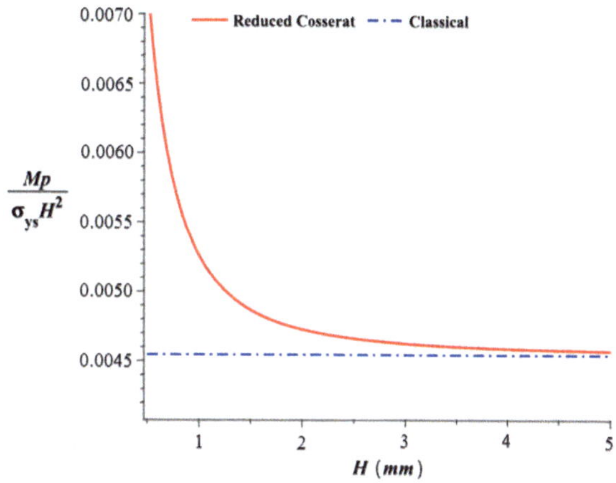

Fig. 13 Effect of bone specimen size on the plastic yield moment of the beam tested in Fig. 12, $l_b = 12\,\text{mm}$

In order to assess the size-dependent plastic yield of micro-structured beams, we consider the microbeams to be made of human trabecular bone; the geometrical parameters of the 2D trabecular lattice are selected as follows: the trabecular separation and trabecular thickness are respectively 0.726 and 0.148 mm, and the inclined angle of the hexagonal trabecular cell is 30^0.

The plastic moment M_P/H^2 normalized by the initial yield stress of bulk material σ_{ys} is evaluated on the basis of the developed size dependent plastic yield criterion and compared to the classical plastic moment (Fig. 13). We observe that contrary to the classical plastic yield criterion, the present non-classical yield criterion is able to capture the size dependency. It is indicated that when the characteristic size of the microscale structure is comparable to the length scale of the structure material, there will be a considerable gap between the results of the classical criterion and the results of the micropolar based criterion. It is also worth noticing that, when the size of the beam samples increases, one recovers the plastic yielding moment predicted by the classical criterion.

5 Conclusions

In the present work, 3D models have been constructed based on micromechanical approaches for describing the multiaxial yield and failure behavior of lattice-like structures. The discrete homogenization technique is developed as a convenient micromechanical approach to construct the plastic yield surfaces of 2D and 3D periodic lattices of articulated beams at a continuum level. The initial lattice is replaced by an effective Cauchy continuous medium at an intermediate scale, endowed with effective properties representative of an identified representative unit cell within the structure. The proposed methodology is quite general, as the representative unit cell includes internal nodes and no assumption of uniform deformation is needed.

The characterization of trabecular bone tissue material and the determination of its failure characteristics and role in the trabecular bone mechanical behavior are of significant clinical importance. In this regard, we have established a complete strength criterion under multiaxial loads for human trabecular bone based on relationships between the morphology and multiaxial failure properties of bone.

The effective elastoplastic response of general 2D beam lattices has been further obtained thanks to an adaptation and extension of the discrete homogenization method. The stress-strain response accounting for ongoing hardening has been constructed. The proposed method has been implemented as a material user subroutine within a standard FE code. We have shown that the method can treat indifferently all stretching dominated lattices that exhibit a non-uniform response in terms of the local deformation pattern; as an illustration, the response of two original lattices, called “square-star” and “asymmetric” has been constructed. The comparison between classical FE simulations and the response of the effective (homogenized) continuum shows a good agreement for a cycle consisting of monotonous tension followed by unloading.

Accounting for the moments allows the homogenization towards the more complete micropolar framework, incorporating a microrotation in addition to the displacement as kinematic descriptors at the continuum level. Such extension and adaptation towards a micropolar effective medium led to construct the plastic yield surfaces for which the material point supports couple stresses in addition to Cauchy-type stresses. As a consequence, the plastic yield surfaces of trabecular bone modeled as open cell structures are determined under the external applied couple stresses. Additionally, a non-classical size-dependent plastic yield criterion has been developed relying on the non-classical continuum theory in order to capture the size dependency observed in the bone structures. It has been observed that when the characteristic size of the bone sample is comparable to the bending length, a considerable difference is shown between the results based on the non-classical and classical theories.

The analysis of the local mesoscopic response of 2D extensional dominated lattices in the plastic regime when strain hardening takes place will be extended to explore the multiaxial response of bending dominated lattices using a more refined yield criterion. The overall modeling of the plastic and brittle collapse behaviors of porous solid materials incorporating size effects shall be considered in future investigations.

References

Alkhader M, Vural M (2009) An energy-based anisotropic yield criterion for cellular solids and validation by biaxial FE simulations. J Mech Phys Solids 57:871–890

Christensen R (2000) Mechanics of cellular and other low-density materials. Int J Solids Struct 37:93–104

Demiray S, Becker W, Hohe J (2007) Numerical determination of initial and subsequent yield surfaces of open-celled model foams. Int J Solids Struct 44:2093–2108

Deshpande V, Fleck N, Ashby M (2001a) Effective properties of the octet-truss lattice material. J Mech Phys Solids 49:1747–1769
Deshpande VS, Ashby MF, Fleck NA (2001b) Foam topology bending versus stretching dominated architectures. Acta mater 49:1035–1040
Doyoyo M, Hu JW (2006) Plastic failure analysis of an auxetic foam or inverted strut lattice under longitudinal and shear loads. Journal of the Mechanics and Physics of Solids 54:1479–1492
Fan HL, Jin FN, Fang DN (2009) Nonlinear mechanical properties of lattice truss materials. Mater Design 30(3):511–517
Florence C, Sab K (2006) A rigorous homogenization method for the determination of the overall ultimate strength of periodic discrete media and an application to general hexagonal lattices of beams. Eur J Mech A/Solids 25:72–97
Gibson LJ (1985) The mechanical behavior of cancellous bone. J Biomech 18:317–328
Gibson LJ (2005) Biomechanics of cellular solids. J Biomech 38:377–399
Gibson LJ, Ashby MF (1999) Cellular solids—structures and properties, 2nd edn. Cambridge University Press, Cambridge
Goda I, Assidi M, Belouettarb S, Ganghoffer JF (2012) A micropolar anisotropic constitutive model of cancellous bone from discrete homogenization. J Mech Behav Biomed Mater 16:87–108
Goda I, Assidi M, Ganghoffer JF (2014) A 3d elastic micropolar model of vertebral trabecular bone from lattice homogenization of the bone microstructure. Biomech Model Mechanobiol 13(1):53–83
Keaveny TM, Wachtel EF, Ford CM, Hayes WC (1994) Differences between the tensile and compressive strengths of bovine tibial trabecular bone depend on modulus. J Biomech 27(9):1137–1146
Kim HS, Al-Hassani STS (2002) The effect of doubly tapered strut morphology on the plastic yield surface of cellular materials. Int J Mech Sci 44(8):1559–1581
Lakes RS (1986) Experimental microelasticity of two porous solids. Int J Solids Struct 22:55–63
McDonald K, Little J, Pearcy M, Adam C (2010) Development of a multi-scale finite element model of the osteoporotic lumbar vertebral body for the investigation of apparent level vertebra mechanics and micro-level trabecular mechanics. Med Eng Phys 32:653–661
Mohr D (2005) Mechanism-based multi-surface plasticity model for ideal truss lattice materials. Int J Solids Struct 42(11–12):3235–3260
Mosekilde L (1988) Age-related changes in vertebral trabecular bone architecture-assessed by a new method. Bone 9:247–250
Mosekilde L (1989) Sex differences in age-related loss of vertebral trabecular bone mass and structure-biomechanical consequences. Bone 10:425–432
Onck PR, Andrews EW, Gibson LJ (2001) Size effects in ductile cellular solids. Part I: modeling. Int J Mech Sci 43(3):681–699
Park T, Hwang WS, Hu J (2010) Plastic continuum models for truss lattice materials with cubic symmetry. J Mech Sci Technol 24(3):657–669
Stronge WJ, Yu TX (1993) Dynamic models for structural plasticity. Springer, London
Sullivan RM, Ghosn LJ, Lerch BA (2008) A general tetrakaidecahedron model for open-celled foams. Int J Solids Struct 45(6):1754–1765
Tekoglu C, Onck PR (2008) Size effects in two-dimensional voronoi foams: a comparison between generalized continua and discrete models. J Mech Phys Solids 56:3541–3564
Ushijima K, Cantwell WJ, Chen DH (2013) Prediction of the mechanical properties of micro-lattice structures subjected to multi-axial loading. Int J Mech Sci 68:47–55
Wang AJ, McDowell DL (2005) Yield surfaces of various periodic metal honeycombs at intermediate relative density. Int J Plast 21(2):285–320
Zhang HW, Wu JK, Fu ZD (2010) Extended multiscale finite element method for elasto-plastic analysis of 2d periodic lattice truss materials. Comput Mech 45(6):623–635

An Improved Constitutive Model for Short Fibre Reinforced Cementitious Composites (SFRC) Based on the Orientation Tensor

Heiko Herrmann

Abstract Short fibre composites are becoming increasingly popular in many applications. This is also true for civil engineering, where short fibre cementitious composites are used more often. For the use in load bearing structures, a constitutive mapping is necessary to calculate the design load and to predict cracking behaviour. Here a constitutive mapping based on the use of isotropic tensor functions of the strain tensor and the orientation tensor is proposed. The model solves some issues of other approaches. A comparison with other constitutive mappings based on tensors is provided.

Keywords Constitutive model · Short fibres · Cementitiuos composites · Orientation tensor · Orientation distribution function

1 Introduction

The properties of steel fibre reinforced concrete (SFRC) (Fig. 1) have been investigated for a long time already (Swamy 1975; Bentur and Mindess 1990; Tejchman and Kozicki 2010; Ju et al. 2007), yet the search for a constitutive function which can take into account any fibre orientation distributions continues. The anisotropic nature of SFRC has been emphasized e.g. by Bentur and Mindess (1990), Barragán et al. (2003) and properties of SFRC in connection with fibre orientation have been topic of several Ph.D. theses (Grünewald 2004; Lappa 2007; Laranjeira de Oliveira 2010; Trüb 2011; Baby 2012; Svec 2013; Eik 2014), which came up with novel ideas how to deal with the fibre orientations. Still the problem is unsolved. Consisting dimensioning rules (standards) or even finite element (FEM) software, that can calculate (predict) the strength of a structural element are missing. This situation is common to all kinds of cementitious matrix composites reinforced with short metal fibres,

H. Herrmann (✉)
Centre for Nonlinear Studies, Institute of Cybernetics at Tallinn University of Technology, Akadeemia Tee 21, 12618 Tallinn, Estonia
e-mail: hh@cens.ioc.ee

H. Altenbach and S. Forest (eds.), *Generalized Continua as Models for Classical and Advanced Materials*, Advanced Structured Materials 42,
DOI 10.1007/978-3-319-31721-2_10

Fig. 1 A cut through SFRC, the *white parts* are filled pores, the *silvery-bright spots* are cut fibres

e.g. ordinary concrete (SFRC) and self-compacting concrete (SCSFRC, SFR-SCC) as well. The importance to develop a constitutive function, which can be used for example in FEM software to calculate and predict the strength of a structural element, has been discussed by several authors (Tejchman and Kozicki 2010; Eik and Puttonen 2011; Herrmann and Eik 2011; Herrmann et al. 2014). This constitutive function should take into account the anisotropy and be able to handle any fibre orientation distribution.

Common approaches start from experiments and try to estimate the properties of SFRC. This approach has to cope with fluctuations during the production of the samples, which makes is difficult to differentiate between experimental uncertainties and effects of the theory to be developed. As has been demonstrated in Herrmann et al. (2014) there would need to be about 130,000 experiments made with the same matrix-properties to tackle the problem. This number is the result of assuming that a change of $\Delta S = 0.1$ or $\Delta b_S = 0.1$ in the order parameter or biaxiality (see Sect. 2.1 and Maier and Saupe 1959, 1960; Pardowitz and Hess 1980; Eppenga and Frenke 1984; Herrmann and Eik 2011; Suuronen et al. 2013) or a change of 10° in orientation of the orientation distribution function will produce a notable change in the material properties. Fluctuations in the production process of the samples will increase the number of necessary experiments even more.

In this paper we will start from the theoretical side, placing assumptions about what parameters will influence material properties and how to describe fibre orientations. Starting from the theory-level may initially increase the complexity of the developed theory, but has the advantage that much less experiments will be needed to confirm or reject the approach. If necessary, an attempt to decrease the complexity of the theory for practical application can be made later.

The aim is to formulate a constitutive model for a homogeneous anisotropic material, although concrete is heterogeneous at a small scale, the description can be homogenized at larger scales (dimensions of structural elements). This approach is suitable and typical for a FEM implementation.

Although concrete in general exhibits a non-linear visco-plastic behaviour at large deformations or at long times (creep), in this article an anisotropic linear elastic model will be developed as a first step. There are many applications, where the dimensions of structural elements are determined by other constraints, e.g. acoustics or the ability to fit a tube into the element. This is quite common for floor slabs in office or apartment buildings, which are from a capacity point-of-view over-dimensioned and will not be loaded to the design limit, so cracks will never appear. Nevertheless usually a "minimum re-inforcement" is added. Due to the still brittle nature of SFRC a restriction to small deformations seems also necessary, despite the failure becoming more gradual due to the effect of the fibres. However, using a similar method to the one presented here, non-linear and visco-plastic models can be formulated as future developments. The main concept is to use an additive superposition of an isotropic material with an anisotropic one.

A summary of the used notation and symbols can be found in Table 1.

Table 1 Summary of used notation and symbols

General notation:	
$\mathbf{v}$	Vector (bold small)
$\otimes$	Outer (tensorial) product
$\mathbf{B}$	(second order) tensor (bold capital)
$\mathbf{B}^T$	Transposed second-order tensor
${}^{[4]}\mathbf{C}$	Fourth-order tensor
${}^{[l]}\mathbf{C}$	l-order tensor
$\mathbf{AB}$	Scalar product
$\overline{\mathbf{A}}$	Symmetric irreducible (traceless) tensor
$\underbrace{\overline{\mathbf{n}\otimes\cdots\otimes\mathbf{n}}}_{l-\text{times}}$	l-order Symmetric traceless tensor product
Special symbols:	
$\mathbf{A}$	Orientation tensor
$\overline{\mathbf{A}}$	Second-order alignment tensor The tracefree part of the orientation tensor
$\mathbf{L}$	Structural tensor
$\mathbf{E}$	Lagrange strain tensor (Green–Lagrange strain tensor)
$\mathbf{C}$	Green strain tensor (right Cauchy–Green tensor)
$\mathbf{S}$	Second Piola–Kirchhoff stress tensor
${}^{[4]}\mathbf{C}$	Elasticity tensor
W	Strain-energy function of concrete (W^{C}), fibres (W^{F}) and composite (W^{SFRC})
$\mathbf{n}$, $\mathbf{n}^{(i)}$	Orientation (of a single fibre), unit vector

2 Phenomenological and Theoretical Considerations

In this section some methods and considerations will be discussed, such as how to describe fibre orientation distributions analytically and how to include these into constitutive mappings. Further, assumptions about the applicable range of the constitutive mapping will be made.

2.1 Description of the Fibre Orientation Distribution

As has been discussed in Herrmann and Eik (2011), Herrmann et al. (2014) the orientation distribution of fibres can be described by a spherical orientation distribution function (ODF), analogue to the orientation distribution of liquid crystals (Muschik et al. 1996, 2000). The ODF can be expanded into a series using tensor coefficients $a_{\mu_1 \ldots \mu_l}$, these correspond to spherical harmonical functions. The expansion is given in the following way:

$$f(\mathbf{n}) = \frac{1}{4\pi}\left(1 + \sum_{l=1}^{\infty} \frac{(2l+1)!!}{l!} a_{\mu_1 \ldots \mu_l} \, \overline{n_{\mu_1} \cdots n_{\mu_l}}\right) \tag{1}$$

$$a_{\mu_1 \ldots \mu_l} = \oint_{S^2} f(\mathbf{n}) \, \overline{n_{\mu_1} \cdots n_{\mu_l}} \, \mathrm{d}^2 n. \tag{2}$$

where $\overline{n_{\mu_1} \cdots n_{\mu_l}}$ is the symmetric traceless part of the l-fold tensor product of $\mathbf{n}$ with itself. Vice versa it is possible to approximate the analytical ODF from measurements by calculating several low order tensors of the series expansion from measured data and cutting the series at the desired accuracy. The procedure, criteria and desired accuracy will be discussed elsewhere. Here the accuracy is restricted to the use of the second order tensor. This tensor can be easily calculated from fibre orientation data, which has been obtained e.g. by CT measurements (Suuronen et al. 2013) or DC-conductivity testing (Eik et al. 2013), in the following way:

$$a_{\mu\nu} = \frac{1}{N} \sum_{i=1}^{N} \overline{n_\mu^{(i)} n_\nu^{(i)}} \tag{3}$$

$$= \frac{1}{N} \sum_{i=1}^{N} \left(n_\mu^{(i)} n_\nu^{(i)} - \frac{1}{3} \delta_{\mu\nu} \right). \tag{4}$$

It is important to note, that the orientation distribution of the fibres varies with respect to the position in a structural element. This happens because the fibres tend to orient according to the flow of the fresh concrete mass, the wall effect (formwork) and due to workers moving the concrete mass with shovels etc. Therefore, there is no representative volume element (RVE) whose ODF is valid for the whole structural element, there are only statistical volume elements or meso-volume elements, that are large enough to allow a homogenization, but small enough to have a constant ODF in them. These are representative either only for themselves or for a small area around them.

2.2 *Phenomenological: Hyperelastic Material*

In the present approach no history dependence (aging) should be taken into account, therefore a hyperelastic material model is suitable. The constitutive function (stress–strain relation) will be based on a strain-energy function, whose derivative results in the desired stress–strain relation. The constitutive function of an anisotropic medium can be obtained by help of representation theorems for isotropic tensor functions (Truesdell and Noll 1965; Itskov 2009; Liefeith and Kolling 2007). One should note, that an isotropic tensor function does not lead to an isotropic medium. The anisotropy is taken into account by structural tensors. In the case of short fibre reinforced materials, the anisotropy is caused by the fibre orientation distribution function (ODF) and the structural tensor is related to the second-order alignment-tensor (Herrmann and Eik 2011).

This motivates two approaches

1. the use of the eigenvectors of the alignment tensor to define structural tensors and to formulate relations analog to an orthotropic material
2. the use of the alignment-tensor directly

The first approach is discussed in Herrmann et al. (2014), the second approach will be discussed in detail in the following. Both approaches will be compared in Sect. 4.

It should be noted, that the second-order alignment tensor is not idempotent, which is a quality that is usually desired for a structural tensor.

2.3 *Material Symmetry, Structural Tensors and Alignment Tensor*

Material symmetry is usually described by the group of transformations under which the material properties are invariant. In this context structural tensors, describing the main symmetry axes (Fig. 2), are introduced. A structural tensor is defined by

$$\mathbf{L}^{(i)} = \mathbf{l}^{(i)} \otimes \mathbf{l}^{(i)}, \tag{5}$$

where $\mathbf{l}^{(i)}$ is a unit vector pointing in the direction of the ith symmetry axis. Usually structural tensors are idempotent ($\mathbf{L}^2 = \mathbf{L}$) and form a partition of unity

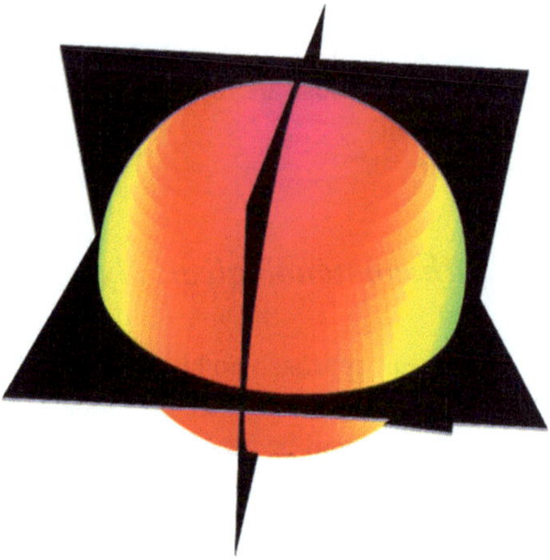

Fig. 2 ODF and symmetry planes defined by the eigenvectors of the second order alignment tensor

($\sum_i \mathbf{L}^{(i)} = \mathbf{I}$). In the case of an orthotropic material the structural tensors are also orthogonal to each other. Here the structure and symmetry of the material (not of the structural element) are described by the second order orientation tensor **A**. Note that the orientation tensors are build in a similar way as the structural tensors (sum of outer products of a unit vector with itself). The second order orientation tensor, however, is not idempotent.

3 Constitutive Relations for SFRC in the Elastic Range

Now the decision needs to be made if the elastic constitutive mapping should be linear or non-linear. If *linear* is chosen, then the elasticity tensor C_{ijkl} does *not* depend on **E**. This means, that the ansatz for the strain-energy function W needs to be quadratic in **E**, higher order terms are not allowed and lower order terms will disappear due to the second derivative with respect to **E**.

Several factors influence the properties of SFRC, one possibility to characterize them is as follows:

- matrix strength (influenced by concrete recipe, porosity),
- fibre-matrix bond (influenced by porosity, concrete recipe, fibre surface),
- local fibre amount (influenced by rheology/flow/vibration),
- local fibre orientations (influenced by rheology/flow/vibration) and
- fibre strength (fibre material and shape).

All factors together determine the strength of the composite and can be independent of each other. It is clear that the fibre orientation distribution is only one of several factors, however, it is the main cause for anisotropy and needs a tensorial treatment. Here, the focus is on how to include the fibre orientations into the constitutive function, the other factors are taken into account by scalar coefficients.

In the following an ansatz for the strain-energy function as an isotropic tensor function will be given and the stress–strain relation will be derived from this function. An isotropic tensor function can be represented as a function of the invariants of its tensor arguments (Truesdell and Noll 1965) (citing theorem by Spencer and Rivlin 1959 and proof by Smith 1960 and Spencer 1961).

In the following an ansatz for a strain-energy function for SFRC is proposed as a superposition of an isotropic strain-energy function for the concrete matrix and an anisotropic strain-energy function taking into account the fibre distribution:

$$W^{\mathrm{SFRC}}(\mathbf{E},\mathbf{A}) = \nu_{\mathrm{m}} W^{\mathrm{C}}(\mathbf{E}) + \nu_{\mathrm{f}} W^{\mathrm{F}}(\mathbf{E},\mathbf{A}), \tag{6}$$

where $W^{\mathrm{C}}(\mathbf{E})$ is the strain-energy function for the matrix, W^{F} is the strain-energy function for the fibres, ν_{m} is the volume fraction of matrix and ν_{f} is the volume fraction of fibres, resp. $\nu_{\mathrm{m}} + \nu_{\mathrm{f}} = 1$.

3.1 The Isotropic Matrix

As stated before, a linear stress–strain relation is desired, this means that a quadratic ansatz for the strain-energy density W is used, this is the standard isotropic St. Venant–Kirchhoff model

$$W^{\mathrm{C}}(\mathbf{E}) = \frac{\lambda}{2}\,\mathrm{tr}^{2}(\mathbf{E}) + \mu\,\mathrm{tr}(\mathbf{E}^{2}). \tag{7}$$

Taking the derivative with respect to $\mathbf{E}$ gives the stress tensor

$$\mathbf{S}^{\mathrm{C}}(\mathbf{E}) = \frac{\partial W^{\mathrm{C}}}{\partial \mathbf{E}} = \lambda\,\mathrm{tr}(\mathbf{E})\mathbf{I} + 2\mu\mathbf{E}. \tag{8}$$

3.2 Anisotropic Fibres: Enrichment by the Orientation Tensor

One of the possibilities mentioned in Sect. 2.2 is the enrichment of the St. Venant–Kirchhoff model by use of the alignment tensors, thus an anisotropic strain-energy function to model the fibres is proposed:

$$W^{\mathrm{F}}(\mathbf{E},\mathbf{A}) = \kappa_{\mathrm{fg}}\left(\frac{\alpha}{2}\,\mathrm{tr}(\mathbf{E}\mathbf{A})^{2}) + \beta\,\mathrm{tr}(\mathbf{E}^{2}\mathbf{A})\right). \tag{9}$$

Inserting this into Eq. (6), the strain-energy function for SFRC results in

$$W^{\mathrm{SFRC}}(\mathbf{E},\mathbf{A}) = \nu_{\mathrm{m}}\left(\frac{\lambda}{2}\,\mathrm{tr}^{\,2}(\mathbf{E}) + \mu\,\mathrm{tr}(\mathbf{E}^2)\right) \\ +\nu_{\mathrm{f}}\kappa_{\mathrm{fg}}\left(\frac{\alpha}{2}\,\mathrm{tr}(\mathbf{EA})^2) + \beta\,\mathrm{tr}(\mathbf{E}^2\mathbf{A})\right), \tag{10}$$

where $\mathbf{A} = \mathbf{a} + \frac{1}{3}\mathbf{I}$ and $\mathbf{a}$ is the second-order alignment-tensor (Hess and Köhler 1980; Muschik et al. 1996; Hess 2015) and ν_{m} is the volume fraction of matrix and ν_{f} is the volume fraction of fibres, resp. $\nu_{\mathrm{m}}+\nu_{\mathrm{f}} = 1$, κ_{fg} is a factor taking into account the fibre geometry and the imperfect bond between fibres and matrix ($\kappa_{\mathrm{fg}} \leq 1$).

Again, the stress-tensor is derived by taking the derivative with respect to $\mathbf{E}$

$$\mathbf{S}^{\mathrm{SFRC}}(\mathbf{E},\mathbf{A}) = \nu_{\mathrm{m}}\left(\lambda\,\mathrm{tr}(\mathbf{E})\mathbf{I} + 2\mu\mathbf{E}\right) \\ +\nu_{\mathrm{f}}\kappa_{\mathrm{fg}}\left(\alpha\,\mathrm{tr}(\mathbf{EA})\mathbf{A}^T + \beta\left((\mathbf{AE})^T + (\mathbf{EA})^T\right)\right), \tag{11}$$

and the elasticity tensor follows as

$$^{[4]}\mathbf{C} = \frac{\partial \mathbf{S}}{\partial \mathbf{E}}, \tag{12}$$

$$C_{ijkl} = \frac{\partial S_{ij}}{\partial E_{kl}}, \tag{13}$$

which results in

$$C_{ijkl}^{\mathrm{SFRC}}(\mathbf{A}) = \nu_{\mathrm{m}}\left(\lambda\delta_{ij}\delta_{kl} + 2\mu\delta_{ik}\delta_{jl} + 2\mu\delta_{il}\delta_{jk}\right) \\ +\nu_{\mathrm{f}}\kappa_{\mathrm{fg}}\left(\alpha A_{ij}A_{kl} + \beta\left(2A_{ik}\delta_{jl} + 2A_{il}\delta_{jk}\right)\right). \tag{14}$$

The constitutive model presented above bears similarity to the model proposed in Altenbach et al. (2003) for short fibre reinforced plastics, a comparison is discussed in the next section.

4 Comparison with Other Models for Short Fibre Reinforced Materials

Previously an orthotropic model where the symmetry planes were derived from the ODF has been proposed in Eik et al. (2015a, b). This approach also starts from a hyperelastic material, proposing a strain-energy density, which contains structural tensors to account for the orthotropy:

$$
\begin{aligned}
W(\mathbf{E}^{(c)}) = {} & \underbrace{V^{(m)}\left(\frac{1}{2}\gamma(\operatorname{tr}(\mathbf{E}^{(c)})\operatorname{tr}(\mathbf{E}^{(c)})) + G\operatorname{tr}((\mathbf{E}^{(c)})^2)\right)}_{\text{concrete, isotropic}} \\
& \underbrace{+V^{(f)}\left(\frac{1}{2}\sum_{i,j=1}^{3}\gamma^{ij}\operatorname{tr}(\mathbf{E}^{(c)}\mathbf{L}^i)\operatorname{tr}(\mathbf{E}^{(c)}\mathbf{L}^j) + \sum_{i,j\neq i}^{3} G^{ij}\operatorname{tr}(\mathbf{E}^{(c)}\mathbf{L}^i\mathbf{E}^{(c)}\mathbf{L}^j)\right)}_{\text{fibres, orthotropic}}.
\end{aligned}
\tag{15}
$$

One obtains the constitutive relation for the second Piola–Kirchhoff pseudo-stress tensor S:

$$
\begin{aligned}
\mathbf{S}^{(c)} & = \frac{\partial}{\partial \mathbf{E}^{(c)}} W(\mathbf{E}^{(c)}) \\
& = \underbrace{V^{(m)}\left(\gamma \mathbf{I}\operatorname{tr}(\mathbf{E}^{(c)}) + 2G\mathbf{E}^{(c)}\right)}_{\text{concrete, isotropic}} \\
& \quad \underbrace{+V^{(f)}\left(\sum_{i,j=1}^{3}\gamma^{ij}\operatorname{tr}(\mathbf{E}^{(c)}\mathbf{L}^j)\mathbf{L}^i + 2\sum_{i,j\neq i}^{3} G^{ij}\mathbf{L}^i\mathbf{E}^{(c)}\mathbf{L}^j\right)}_{\text{concrete, isotropic}}.
\end{aligned}
\tag{16}
$$

Further differentiation of Eq. (16) will give the 4th-order elasticity tensor

$$
\begin{aligned}
{}^{[4]}\mathbf{C}^{(c)} = {} & \underbrace{V^{(m)}\left(\gamma \mathbf{I}\otimes\mathbf{I} + 2G\,{}^{[4]}\mathbf{I}^S\right)}_{\text{concrete, isotropic}} \\
& \underbrace{+V^{(f)}\left(\sum_{i,j}^{3}\gamma^{ij}\mathbf{L}^i\otimes\mathbf{L}^j + \sum_{i,j\neq i}^{3} 2G^{ij}(\mathbf{L}^i\,\tilde{\otimes}\,\mathbf{L}^j)^S\right)}_{\text{concrete, isotropic}}.
\end{aligned}
\tag{17}
$$

The material constants γ^{ij} and G^{ij} can be obtained from the orientation averaged single-fibre elasticity tensor:

$$
C^{(f\mathrm{l})}_{ijkl} = \oint_{S^2} Q_{im}(\mathbf{n})Q_{jn}(\mathbf{n})Q_{ko}(\mathbf{n})Q_{lp}(\mathbf{n})C^{(f\mathrm{id})}_{mnop} f(\mathbf{n})\mathrm{d}^2\mathbf{n}.
$$

In Herrmann et al. (2014), Eik et al. (2015b) only the second order alignment tensor is used to approximate the ODF.

Although developed for a different matrix and fibre material, i.e. short fibre plastics, the model proposed by Altenbach et al. (2003) would be applicable as well, with the correct choice of the coefficients. The motivation of this model uses an orientation averaging of a stress tensor of a transversely isotropic material, i.e. matrix with aligned fibres. This approach can also be used in connection with Eshelby (1957) or Mori and Tanaka (1973) models for the single-fibre-matrix system. By starting with the postulation of the transversely isotropic stress tensor, this model does not follow the hyperelastic strain-energy approach. Although, it could be possible to find a strain-energy density to reproduce the stress tensor.

$$\tilde{\mathbf{S}}(\mathbf{m}) = \lambda \operatorname{tr}(\mathbf{E})\mathbf{I} + 2\mu_T \mathbf{E} + \alpha \mathbf{E} \cdot\cdot \mathbf{N} \otimes \mathbf{I} + \alpha \operatorname{tr}(\mathbf{E})\mathbf{N} + \beta \mathbf{E} \cdot\cdot \mathbf{N} \otimes \mathbf{N} + 2(\mu_L - \mu_T)[\mathbf{N} \cdot \mathbf{E} + \mathbf{E} \cdot \mathbf{N}], \tag{18}$$

$$\mathbf{N} = \mathbf{m} \otimes \mathbf{m}, \tag{19}$$

$$\mathbf{S} = \int_S f(\mathbf{m})\tilde{\mathbf{S}}(\mathbf{m})\mathrm{d}S \tag{20}$$

$$= \lambda_* \operatorname{tr}(\mathbf{E})\mathbf{I} + 2\mu_* \mathbf{E} \cdot\cdot (\mathbf{B} \otimes \mathbf{I} + \mathbf{I} \otimes \mathbf{B}) + 2\xi_* (\mathbf{B} \otimes \mathbf{E} + \mathbf{E} \otimes \mathbf{B}) + \eta_* \mathbf{E} \cdot\cdot \mathbf{B} \otimes \mathbf{B}, \tag{21}$$

$$\lambda_* = \lambda + \frac{2}{3}\alpha + \frac{1}{105}\left(13 - \frac{4}{3}\psi\right), \tag{22}$$

$$\mu_* = \frac{1}{3}(\mu_T + 2\mu_L) + \frac{13}{105}\beta(1 - \psi), \tag{23}$$

$$\beta_* = \alpha + \frac{1}{7}\beta\left(1 + \frac{4}{3}\psi\right), \tag{24}$$

$$\xi_* = \mu_L - \mu_T + \beta\frac{1 - \psi}{7}, \tag{25}$$

$$\eta_* = \psi\beta, \tag{26}$$

$$\psi = 1 - 27(\mathbf{A}), \tag{27}$$

$$\mathbf{B} = \mathbf{A} - \frac{1}{3}\mathbf{I}, \tag{28}$$

$$\operatorname{tr}\mathbf{A} = 1, \tag{29}$$

$$\mathbf{a} \cdot \mathbf{A} = \mathbf{A} \cdot \mathbf{a}. \tag{30}$$

This model offers an advanced precision, but considering the practice on construction sites, the question arises which precision is actually achievable in practical applications.

The new model proposed in Sect. 3.2 maintains the simplicity of the model in Herrmann et al. (2014), Eik et al. (2015b) while solving the problem with negative matrix entries, that have been present in the orthotropic model, due to the approximation of the ODF, as can be seen in the next section.

5 Example Distributions and Resulting Elasticity Tensors

Below the orientation weighted fourth-order elasticity tensors in material symmetry axes of μCT measured cylinder samples are presented. The tensors are given utilizing the Kelvin–Mandel variant of the Voigt notation (Bertram 2005), see Eq. (31), only the fibre-part is presented

$$C^{\mathrm{K-M}} = \begin{pmatrix} C_{1111} & C_{1122} & C_{1133} & \sqrt{2}C_{1123} & \sqrt{2}C_{1131} & \sqrt{2}C_{1112} \\ C_{1122} & C_{2222} & C_{2233} & \sqrt{C}_{2223} & \sqrt{2}C_{2231} & \sqrt{2}C_{2212} \\ C_{1133} & C_{2233} & C_{3333} & \sqrt{C}_{3323} & \sqrt{C}_{3331} & \sqrt{C}_{3312} \\ \sqrt{2}C_{1123} & \sqrt{2}C_{2223} & \sqrt{2}C_{3323} & 2C_{2323} & 2C_{2331} & 2C_{2312} \\ \sqrt{2}C_{1131} & \sqrt{2}C_{2231} & \sqrt{2}C_{3331} & 2C_{2331} & 2C_{3131} & 2C_{3112} \\ \sqrt{2}C_{1112} & \sqrt{2}C_{2212} & \sqrt{2}C_{3312} & 2C_{2312} & 2C_{3112} & 2C_{1212} \end{pmatrix}. \quad (31)$$

The fibre contribution in sample 3a-middle from Suuronen et al. (2013) with the constitutive function from Eik et al. (2015b):

$$C^{(f)\mathrm{K-M}} = \begin{pmatrix} 0.50\,Y & 0.07\,Y & 0.109128\,Y & 0 & 0 & 0 \\ 0.07\,Y & -0.051\,Y & 0.017\,Y & 0 & 0 & 0 \\ 0.11\,Y & 0.017\,Y & 0.16\,Y & 0 & 0 & 0 \\ 0 & 0 & 0 & 0.03\,Y & 0 & 0 \\ 0 & 0 & 0 & 0 & 0.22\,Y & 0 \\ 0 & 0 & 0 & 0 & 0 & 0.15\,Y \end{pmatrix} \quad (32)$$

the same sample with the new constitutive function:

$$C^{(f)\mathrm{K-M}} = \begin{pmatrix} 0.46\,\alpha + 2.72\,\beta & 0.02\,\alpha & 0.19\,\alpha & 0 & 0 & 0 \\ 0.02\,\alpha & 0.001\,\alpha + 0.14\,\beta & 0.01\,\alpha & 0 & 0 & 0 \\ 0.19\,\alpha & 0.01\,\alpha & 0.08\,\alpha + 1.14\,\beta & 0 & 0 & 0 \\ 0 & 0 & 0 & 0.14\,\beta & 0 & 0 \\ 0 & 0 & 0 & 0 & 1.14\,\beta & 0 \\ 0 & 0 & 0 & 0 & 0 & 2.72\,\beta \end{pmatrix}. \quad (33)$$

As one can see, the new constitutive function offers more flexibility, as there are two parameters for the fibres, and most importantly solves the problem with the negative entries in the elasticity matrix.

An artificial example with perfectly aligned fibres in x-direction, with the old constitutive function

$$C^{(f)\mathrm{K-M}} = \begin{pmatrix} 0.77\,Y & 0.114\,Y & 0.114\,Y & 0 & 0 & 0 \\ 0.114\,Y & -0.09\,Y & -0.03\,Y & 0 & 0 & 0 \\ 0.114\,Y & -0.03\,Y & -0.09\,Y & 0 & 0 & 0 \\ 0 & 0 & 0 & -0.06\,Y & 0 & 0 \\ 0 & 0 & 0 & 0 & 0.23\,Y & 0 \\ 0 & 0 & 0 & 0 & 0 & 0.23\,Y \end{pmatrix}. \quad (34)$$

and with the new one

$$C^{(f)_{\mathrm{K-M}}} = \begin{pmatrix} 1\alpha + 4\beta & 0 & 0 & 0 & 0 & 0 \\ 0 & 0 & 0 & 0 & 0 & 0 \\ 0 & 0 & 0 & 0 & 0 & 0 \\ 0 & 0 & 0 & 0 & 0 & 0 \\ 0 & 0 & 0 & 0 & 0 & 0 \\ 0 & 0 & 0 & 0 & 0 & 4\beta \end{pmatrix}. \tag{35}$$

This example demonstrates, that the old constitutive function underestimates the contribution of perfectly aligned fibres. More generally, the old constitutive function has difficulties with fibre orientation distributions that have sharp maxima. To be fully correct, the limitation in the constitutive function published in Eik et al. (2015b) is, that some coefficients are determined by rotation averaging with the ODF, where the ODF is only known approximately, i.e. up to the second order alignment tensor. The same accuracy as the new constitutive function could probably be achieved by using higher order tensors, at the expense of having to calculate these.

6 Outlook: Constitutive Relations for Cracked SFRC

When cracks are developing, in the vicinity of the cracks only the fibres can carry the tensile load and stress (beginning of "range 2" in Fig. 3). A stress redistribution from matrix-plus-fibres to fibres should take place. If there are enough fibres and the anchorage is strong enough, this should work. The chemical bond between fibre and

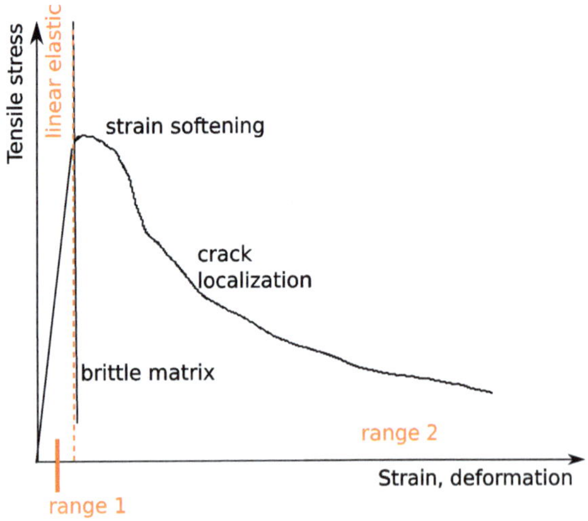

Fig. 3 Sketch of stress–strain measurements and ranges of the constitutive mappings

matrix is lost in this area, hence the fibres will act along their main axis only and their job is to bridge the cracks.

In the parts of a structural element, respectively in meso-volumes, that contain cracks, the constitutive mapping will become strain and strain-history dependent. The first approach to extend the constitutive mapping into the inelastic deformations, can be the following:

$$\mathbf{S}^{\mathrm{SFRC}}(\mathbf{E}, \mathbf{A}, h) = \nu_{\mathrm{f}}\kappa_{\mathrm{fg}}\left(\alpha\,\mathrm{tr}(\mathbf{E}\mathbf{A})\mathbf{A}^T + \beta\left((\mathbf{A}\mathbf{E})^T + (\mathbf{E}\mathbf{A})^T\right)\right). \tag{36}$$

In this case, the matrix contribution has simply been removed from Eq. (11), i.e. by setting $\lambda = \mu = 0$. This can also be done for the constitutive mapping proposed in Herrmann et al. (2014), Eik et al. (2015b) Eq. (16), i.e. setting $\gamma = G = 0$:

$$\mathbf{S}^{(c)} = \underbrace{V^{(f)}\left(\sum_{i,j=1}^{3}\gamma^{ij}\,\mathrm{tr}(\mathbf{E}^{(c)}\mathbf{L}^j)\mathbf{L}^i + 2\sum_{i,j\neq i}^{3} G^{ij}\mathbf{L}^i\mathbf{E}^{(c)}\mathbf{L}^j\right)}_{\text{fibres, orthotropic}}. \tag{37}$$

This is also possible for the constitutive mapping proposed in Altenbach et al. (2003) (Eq. (21)) by setting $\lambda = \mu_\tau = 0$ in Eq. (18), however, this is more difficult to see in Eq. (21), as μ_τ enters μ_* and ξ_*, see Eqs. (23) and (25).

With increasing strain, more and more fibres will be pulled out of the matrix, depending on the anchorage length and their inclination angle. In case of shape-anchorage, as in hooked-end fibres or undulating fibres, the coefficients will more depend on the pull-out force, than on the elastic modulus of the fibres itself. In case of fibres with anchors, like balls, at the end, this may be different. While the mappings in Eqs. (36) and (37) look like linear elastic constitutive mappings, the material will not be going back to smaller strains if the load is reduced. Therefore, the coefficients are strain-history and load-velocity dependent, or at least different for increasing and decreasing load.

7 Conclusion

A constitutive mapping for the elastic state of SFCR was presented in this paper, that uses the second order alignment tensor directly, its accuracy in taking into account the ODF of the fibres is in-between the orthotropic model based on structural tensors and the exact orientation averaging of a transversely isotropic material. The new model based on the second order alignment tensor can be implemented in FEA software, just like the model based on the structural tensors. By using different alignment tensors for the FE cells depending on the position in the structural element, realistic fibre orientation distributions including the variation in the element can be taken into account in simulations.

Further, an outlook into models for the cracked state of SFRC has been given. Also these models can be used in FE software, by switching from the elastic constitutive mapping to the mapping for the cracked state, depending on the maximum strain in that element.

Acknowledgments This research was supported by the European Union through the European Regional Development Fund, in particular through funding for the "Centre for Nonlinear Studies" as an Estonian national centre of excellence. The research was also supported by the Estonian Research Council grant PUT1146. I'd like to thank Marika Eik, Emiliano Pastorelli and Jari Puttonen for discussions and joint work on fibre orientation measurements and for the cooperation in developing the constitutive model based on the structural tensors (Eik et al. 2015b).

References

Altenbach H, Naumenko K, L'vov G, Pilipenko SN (2003) Numerical estimation of the elastic properties of thin-walled structures manufactured from short-fiber-reinforced thermoplastics. Mech Compos Mater 39(3):221–234

Baby F (2012) Contribution à l'identification et la prise en compte du comportement en traction des bfup à l'échelle de la structure (Contribution to identification of uhpfrc tensile constitutive behaviour and accounting for structural design). Ph.D. thesis, Université Paris-Est, in French

Barragán BE, Gettu R, Martin MA, Zerbino RL (2003) Uniaxial tension test for steel fibre reinforced concrete - a parametric study. Cem Concr Compos 25(7):767–777

Bentur A, Mindess S (1990) Fibre reinforced cementitious composites. Spon, Routledge

Bertram A (2005) Elasticity and plasticity of large deformations—an introduction. Springer, Heidelberg

Eik M (2014) Orientation of short steel fibres in concrete: measuring and modelling. Ph.D. thesis, Institute of Cybernetics at Tallinn University of Technology, Faculty of Civil Engineering and Aalto University School of Engineering. http://digi.lib.ttu.ee/i/file.php?DLID=965

Eik M, Puttonen J (2011) Challenges of steel fibre reinforced concrete in load bearing structures. Rakenteiden Mekaniikka 44:44–64

Eik M, Lõhmus K, Tigasson M, Listak M, Puttonen J, Herrmann H (2013) DC-conductivity testing combined with photometry for measuring fibre orientations in SFRC. J Mater Sci 48(10):3745–3759

Eik M, Herrmann H, Puttonen J (2015a) Orthotropic constitutive model for steel fibre reinforced concrete: linear-elastic state and bases for the failure. In: Kouhia R, Mäkinen J, Pajunen S, Saksala T (eds) Proceedings of the XII Finnish mechanics days, 4–5 June 2015, Tampere, Finland, pp 255–260

Eik M, Puttonen J, Herrmann H (2015b) An orthotropic material model for steel fibre reinforced concrete based on the orientation distribution of fibres. Compos Struct 121:324–336. doi:10.1016/j.compstruct.2014.11.018

Eppenga R, Frenke D (1984) Monte Carlo study of the isotropic and nematic phases of infinitely thin hard platelets. Mol Phys 52:1303–1334

Eshelby JD (1957) The determination of the elastic field of an ellipsoidal inclusion, and related problems. Proc R Soc Lond Ser A: Math, Phys Eng Sci 241(1226):376–396

Grünewald S (2004) Performance-based design of self-compacting fibre reinforced concrete. Ph.D. thesis, Technische Universiteit Delft

Herrmann H, Eik M (2011) Some comments on the theory of short fibre reinforced material. Proc Est Acad Sci 60(3):179–183

Herrmann H, Eik M, Berg V, Puttonen J (2014) Phenomenological and numerical modelling of short fibre reinforced cementitious composites. Meccanica 49(8):1985–2000

Hess S (2015) Tensors for physics. Springer International Publishing, Heidelberg
Hess S, Köhler W (1980) Formeln zur Tensor-Rechnung. Palm and Enke, Erlangen
Itskov M (2009) Tensor algebra and tensor analysis for engineers, 2nd edn. Springer, Heidelberg
Ju Y, Jia Y, Liu H, Chen J (2007) Mesomechanism of steel fiber reinforcement and toughening of reactive powder concrete. Sci China Ser E: Technol Sci 50(6):815–832
Lappa ES (2007) High strength fibre reinforced concrete: static and fatigue behaviour in bending. Ph.D. thesis, Technische Universiteit Delft. http://repository.tudelft.nl/view/ir/uuid:0f7ea161-1bbe-4d6b-b2bd-9adfae98323c/
Laranjeira de Oliveira F (2010) Design-oriented constitutive model for steel fiber reinforced concrete. Ph.D. thesis, Universitat Politecnica de Catalunya. http://www.tdx.cat/TDX-0602110-115910
Liefeith D, Kolling S (2007) An anisotropic material model for finite rubber viscoelasticity. 6. LS-DYNA Anwenderforum, Frankenthal 2007. http://www.dynamore.de/de/fortbildung/konferenzen/vergangene/forum07/material02/an-anisotropic-material-model-for-finite-rubber/view
Maier W, Saupe A (1959) Eine einfache molekular-statistische Theorie der nematischen kristallinflüssigen Phase, Teil I. Zeitschrift für Naturforschung A - J Phys Sci 14a:882–889
Maier W, Saupe A (1960) Eine einfache molekular-statistische Theorie der nematischen kristallinflüssigen Phase, Teil II. Zeitschrift für Naturforschung A - J Phys Sci 15a:287–292
Mori T, Tanaka K (1973) Average stress in matrix and average elastic energy of materials with misfitting inclusions. Acta Metall 21(5):571–574
Muschik W, Papenfuss C, Ehrentraut H (1996) Concepts of continuum thermodynamics. Kielce University of Technology, Technische Universität, Berlin
Muschik W, Ehrentraut H, Papenfuss C (2000) Concepts of mesoscopic continuum physics with application to biaxial liquid crystals. J Non-Equilib Thermodyn 25:179–197
Pardowitz I, Hess S (1980) On the theory of irreversible processes in molecular liquids and liquid crystals, nonequilibrium phenomena associated with the second and fourth rank alignment tensors. Phys A: Stat Mech Appl 100(3):540–562
Smith G (1960) On the minimality of integrity bases for symmetric 3x3 matrices. Arch Ration Mech Anal 5:382–389
Spencer A (1961) The invariants of six symmetric 3x3 matrices. Arch Ration Mech Anal 7:64–77
Spencer A, Rivlin R (1959) Further results in the theory of matrix polynomials. Arch Ration Mech Anal 4:214–230
Suuronen JP, Kallonen A, Eik M, Puttonen J, Serimaa R, Herrmann H (2013) Analysis of short fibres orientation in steel fibre reinforced concrete (SFRC) using x-ray tomography. J Mater Sci 48(3):1358–1367
Svec O (2013) Flow modelling of steel fibre reinforced self-compacting concrete—simulating fibre orientation and mechanical properties. Ph.D. thesis, DTU, Department of Civil Engineering. http://www.byg.dtu.dk/english/~/media/Institutter/Byg/publikationer/PhD/byg-r289.ashx
Swamy R (1975) Fibre reinforcement of cement and concrete. Mater Struct 8(3):235–254
Tejchman J, Kozicki J (2010) Experimental and theoretical investigations of steel-fibrous concrete. Springer series in geomechanics and geoengineering, 1st edn. Springer, Heidelberg
Trüb MC (2011) Numerical modeling of high performance fiber reinforced cementitious composites. Ph.D. thesis, ETH Zurich. doi:10.3929/ethz-a-006620566
Truesdell C, Noll W (1965) The non-linear field theories of mechanics (die nicht-linearen feldtheorien der mechanik). In: Flügge S (ed) Encyclopedia of physics (Handbuch der Physik), vol III/3. Springer, Heidelberg

Isogeometric Static Analysis of Gradient-Elastic Plane Strain/Stress Problems

Sergei Khakalo, Viacheslav Balobanov and Jarkko Niiranen

Abstract In the present contribution, isogeometric methods are used to analyze the statics of the plane strain and plane stress problems based on the theory of strain gradient elasticity. The adopted strain gradient elasticity models, in particular, include only one length scale parameter enriching the classical strain energy expression and resulting in fourth order partial differential equations instead of the corresponding second order ones based on the classical elasticity. The problems are discretized by an isogeometric NURBS based C^1 continuous Galerkin method which is implemented as a user subroutine into a commercial software Abaqus. Computational results for benchmark problems, a square plate in tension and a Lamé problem, demonstrate the applicability of the method and verify the implementation.

Keywords Strain gradient elasticity · Plane stress/strain problem · Lamé problem · Isogeometric method

1 Equation of Motion of the Generalized Continua

In this contribution, the static gradient-elastic plane strain/stress problems (Fischer et al. 2011) are studied by applying isogeometric analysis (Hughes et al. 2005; Niiranen et al. 2015) which, in particular, provides a C^1 continuous discretization in a natural way. Namely, the adopted strain gradient elasticity models result in fourth order partial differential equations with corresponding variational formulations of H^2 regularity which, in turn, require C^1 continuity for ensuring the conformity of the discrete method. A brief description of isogeometric analysis and the weak formulations of gradient-elastic problems (omitted in this contribution) can be found in

S. Khakalo (✉) · V. Balobanov · J. Niiranen
Department of Civil and Structural Engineering, Aalto University,
P. O. Box 12100, 00076 Aalto, Espoo, Finland
e-mail: sergei.khakalo@aalto.fi

V. Balobanov
e-mail: viacheslav.balobanov@aalto.fi

J. Niiranen
e-mail: jarkko.niiranen@aalto.fi

H. Altenbach and S. Forest (eds.), *Generalized Continua as Models for Classical and Advanced Materials*, Advanced Structured Materials 42,
DOI 10.1007/978-3-319-31721-2_11

another contribution "Isogeometric analysis of gradient-elastic 1D and 2D problems" of the current collection.

Within the first strain gradient elasticity theory the stress-equation of motion of the generalized continua can be written as follows (Mindlin 1964)

$$\nabla \cdot \boldsymbol{\tau} - l_s^2 \nabla \cdot \Delta \boldsymbol{\tau} + \boldsymbol{F} = \rho(\ddot{\boldsymbol{u}} - l_d^2 \Delta \ddot{\boldsymbol{u}}) \text{ in } \Omega, \tag{1}$$

where $\boldsymbol{\tau}$ is the Cauchy stress tensor, $\boldsymbol{u}$ stands for the displacement vector, $\boldsymbol{F}$ denotes the density of the body forces, ρ is the material volume density, l_s and l_d denote the micro-structural and the micro-inertia parameters, respectively. For the linear elastic isotropic material Cauchy stress tensor $\boldsymbol{\tau}$ can be written as

$$\boldsymbol{\tau} = 2\mu\boldsymbol{\varepsilon} + \lambda tr(\boldsymbol{\varepsilon})\boldsymbol{I}, \tag{2}$$

where the strain tensor $\boldsymbol{\varepsilon}$ is taken as $\boldsymbol{\varepsilon} = \frac{1}{2}(\nabla \boldsymbol{u} + (\nabla \boldsymbol{u})^T)$, λ and μ are the Lamé parameters. The traction forces (Mindlin 1964; Polizzotto 2012) on smooth body surface can be written in the following form

$$\boldsymbol{P} = \boldsymbol{n} \cdot \boldsymbol{\sigma} - \nabla_s \cdot (\boldsymbol{n} \cdot \boldsymbol{\mu}) + (\nabla_s \cdot \boldsymbol{n})\boldsymbol{nn} : \boldsymbol{\mu} + \rho l_d^2 \; \boldsymbol{n} \cdot \nabla \ddot{\boldsymbol{u}} \text{ on } \partial\Omega, \tag{3}$$

$$\boldsymbol{R} = \boldsymbol{nn} : \boldsymbol{\mu} \text{ on } \partial\Omega, \tag{4}$$

where $\nabla_s = \nabla - \boldsymbol{n}\frac{\partial}{\partial n}$ denotes the surface gradient and $\boldsymbol{n}$ is the unit vector normal to the boundary. Here we neglect the wedge loads (Mindlin 1964; Polizzotto 2012) and possible double body forces (Bleustein 1967). Double stress tensor $\boldsymbol{\mu}$ and total stress tensor (Aifantis 1992) $\boldsymbol{\sigma}$ are introduced as follows

$$\boldsymbol{\mu} = l_s^2 \nabla \boldsymbol{\tau}, \tag{5}$$

$$\boldsymbol{\sigma} = \boldsymbol{\tau} - l_s^2 \Delta \boldsymbol{\tau}. \tag{6}$$

The displacement-equation of motion takes the following form (Mindlin 1964; Papargyri-Beskou et al. 2009)

$$(1 - l_s^2 \Delta)[\mu \Delta \boldsymbol{u} + (\lambda + \mu)\nabla\nabla \cdot \boldsymbol{u}] + \boldsymbol{F} = \rho(\ddot{\boldsymbol{u}} - l_d^2 \Delta \ddot{\boldsymbol{u}}) \text{ in } \Omega. \tag{7}$$

2 Numerical Results

In this section, a numerical analysis of two problems within the theory of strain gradient elasticity are considered. The first problem is a square domain in the field of body forces. Convergence results in the $H^2(\Omega)$-norm of displacements are presented demonstrating the reliability of the isogeometric method. The second problem is a hollow thick cylinder under external tension or the Lamé problem. The total and Cauchy stresses are compared and analyzed for different choices of the microstructural parameter l_s. Numerical results are supported by corresponding analytical solutions.

Numerical analysis has been accomplished in Abaqus by using Users Subroutines (Dassault Systemes Simulia 2015). This tool allows us to modify the finite element implementation for the isogeometric method and use the Abaqus as a solver and a post-processor.

2.1 Square Domain in the Field of Body Forces

Let us consider a square domain $\Omega = (0, L) \times (0, L) \subset \mathbb{R}^2$ in the field of the body forces $\boldsymbol{F} = F_x \boldsymbol{e}_x + F_y \boldsymbol{e}_y$, where

$$F_x(x, y) = 4\frac{\pi^2}{L^2} \sin\left(2\pi\frac{x}{L}\right)\left[(2\mu+\lambda)\left(1+4\pi^2\frac{l_s^2}{L^2}\right) - 2\mu\cos\left(2\pi\frac{y}{L}\right)\left(1+8\pi^2\frac{l_s^2}{L^2}\right)\right], \tag{8}$$

$$F_y(x, y) = 4\frac{\pi^2}{L^2} \sin\left(2\pi\frac{y}{L}\right)\left[2\mu\cos\left(2\pi\frac{x}{L}\right)\left(1+8\pi^2\frac{l_s^2}{L^2}\right) - (2\mu+\lambda)\left(1+4\pi^2\frac{l_s^2}{L^2}\right)\right]. \tag{9}$$

The boundary conditions are taken as follows

$$\boldsymbol{u} = 0, \ \boldsymbol{nn} : \nabla\boldsymbol{\tau} = 0 \quad \text{on } \partial\Omega. \tag{10}$$

The analytical solution of the static gradient-elastic displacement-equation of equilibrium with boundary conditions above is the vector function $\boldsymbol{u} = u_x \boldsymbol{e}_x + u_y \boldsymbol{e}_y$, where

$$u_x(x, y) = \sin\left(2\pi\frac{x}{L}\right)\left(1 - \cos\left(2\pi\frac{y}{L}\right)\right), \tag{11}$$

$$u_y(x, y) = \sin\left(2\pi\frac{y}{L}\right)\left(-1 + \cos\left(2\pi\frac{x}{L}\right)\right). \tag{12}$$

For numerical analysis, the square domain is divided into $N = 4, 16, 64, 256, 1024$, 4096 and 16384 elements corresponding to mesh sizes $h = 5$, 2.5, 1.25, 0.625, 0.3125, 0.15625 and 0.078125 mm, respectively. NURBS basis functions of 2nd, 3rd, 4th and 5th order are taken with C^{p-1} continuity across the element boundaries, where p is the NURBS order. Distributions of the u_x and u_y displacement fields for $N = 64$ are shown in Fig. 1.

Next, we consider the convergence results in the $H^2(\Omega)$-norm of displacements shown in Fig. 2. As can be seen, the convergence rates of displacements for quadratic, cubic, quartic and quintic NURBS are approximately 1, 2, 3 and 4 in the H^2-norm with respect to the mesh size h.

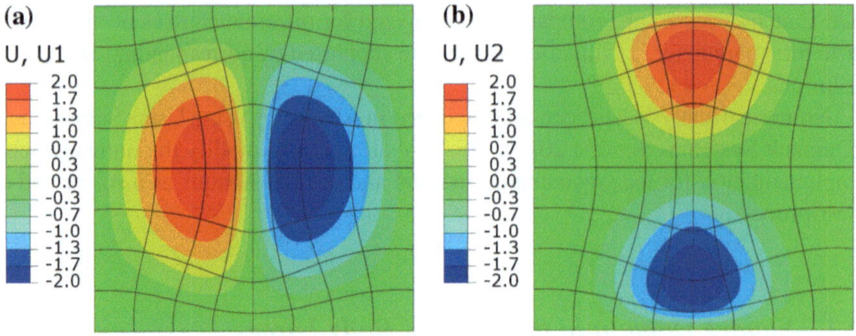

Fig. 1 Distribution of the displacement fields. **a** Distribution of the u_x displacement field. **b** Distribution of the u_y displacement field

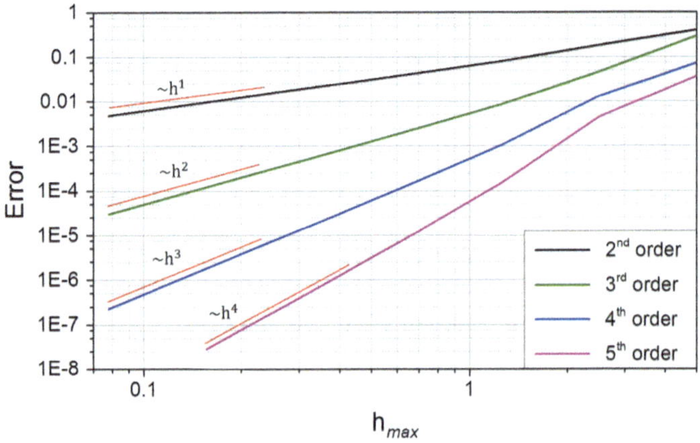

Fig. 2 Error measured in the H^2-norm of displacements

2.2 Lamé Problem

As the next example, the classical Lamé problem of an annulus is considered. Due to the symmetry of the problem, one quarter is modelled. Geometry and boundary conditions are shown in Fig. 3a. The symmetry conditions are utilized on the left and bottom edges and written as

$$\boldsymbol{u} \cdot \boldsymbol{n} = 0 \quad \text{and} \quad \frac{\partial(\boldsymbol{u} \cdot \boldsymbol{s})}{\partial n} = 0, \tag{13}$$

where $\boldsymbol{n}$ and $\boldsymbol{s}$, respectively, denote the unit vectors normal and tangential to the boundary. The ratio of the inner radius a to the outer radius b is set to $a/b = 0.1$, where $a = 1$ mm and $b = 10$ mm. The inner hole surface is free of loading. Tension $p = 100$ MPa is applied to the outer surface. The classical elastic material parameters

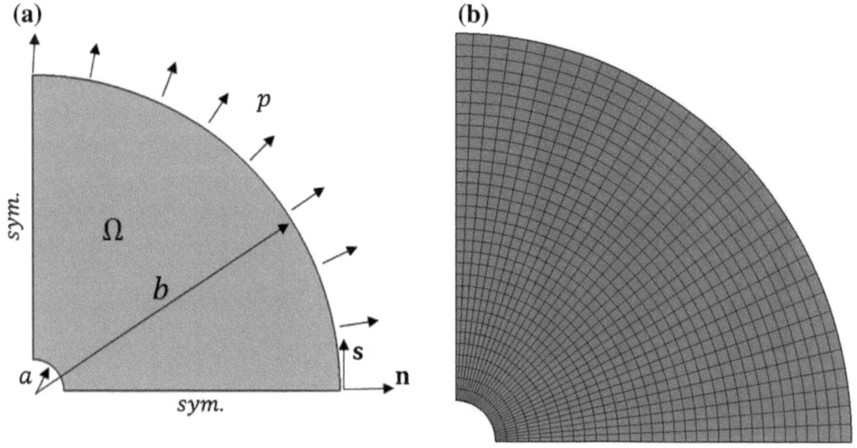

Fig. 3 Lamé problem. **a** Geometry and boundary conditions. **b** Isogeometric finite element mesh

are taken to be $E = 210\,\text{GPa}$ and $\nu = 0.3$. The ratio of the micro structural parameter l_s to the inner radius a is varied from 0 to 1 ($0 \leq l_s/a \leq 1$). Within the gradient theory, the partial differential equation written in polar coordinates has the following form

$$u'' + \frac{u'}{r} - \frac{u}{r^2} - l_s^2 \left(u'''' + 2\frac{u'''}{r} - 3\frac{u''}{r^2} + 3\frac{u'}{r^3} - 3\frac{u}{r^4} \right) = 0, \tag{14}$$

whereas the boundary conditions are given as follows

$$r = a: \quad \begin{cases} \boldsymbol{n} \cdot (\boldsymbol{\tau} - l_s^2 \Delta \boldsymbol{\tau}) - l_s^2 \boldsymbol{s} \cdot \frac{\partial}{\partial s} (\boldsymbol{n} \cdot \nabla \boldsymbol{\tau}) + l_s^2 \left(\boldsymbol{s} \cdot \frac{\partial \boldsymbol{n}}{\partial s} \right) \boldsymbol{nn} : \nabla \boldsymbol{\tau} = 0, \\ l_s^2 \boldsymbol{nn} : \nabla \boldsymbol{\tau} = 0, \end{cases} \tag{15}$$

$$r = b: \quad \begin{cases} \boldsymbol{n} \cdot (\boldsymbol{\tau} - l_s^2 \Delta \boldsymbol{\tau}) - l_s^2 \boldsymbol{s} \cdot \frac{\partial}{\partial s} (\boldsymbol{n} \cdot \nabla \boldsymbol{\tau}) + l_s^2 \left(\boldsymbol{s} \cdot \frac{\partial \boldsymbol{n}}{\partial s} \right) \boldsymbol{nn} : \nabla \boldsymbol{\tau} = p\boldsymbol{n}, \\ l_s^2 \boldsymbol{nn} : \nabla \boldsymbol{\tau} = 0. \end{cases} \tag{16}$$

The analytical solution of the problem can be found in the following form (Gao and Park 2007; Zervos et al. 2009)

$$u(r) = \frac{A}{r} + Br + CI_1 \left(\frac{r}{l_s} \right) + DK_1 \left(\frac{r}{l_s} \right), \tag{17}$$

where I_1 and K_1 denote the first order modified Bessel functions of the first and second kind, respectively.

For numerical calculations, the domain is divided into 1024 elements presented in Fig. 3b. NURBS of 5th order with C^4-continuity across the element boundaries are taken as basis functions. The distributions of the hoop Cauchy and total stress fields

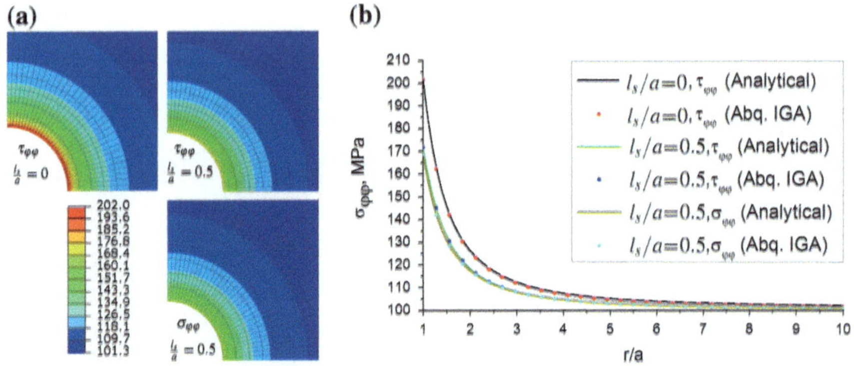

Fig. 4 Results for Lamé problem. **a** Hoop stress distribution. **b** Hoop stresses along the radial coordinate line

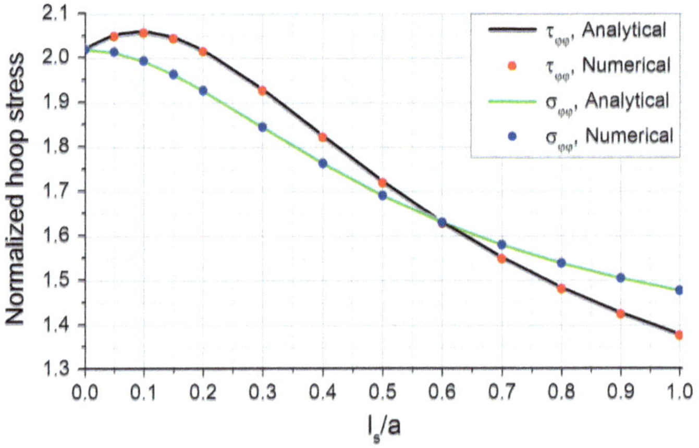

Fig. 5 Normalized hoop stresses at $r = a$

for the cases $l_s/a = 0$ and $l_s/a = 0.5$ obtained by using Abaqus are shown in Fig. 4a. The plots of hoop stresses $\tau_{\varphi\varphi}$ and $\sigma_{\varphi\varphi}$ along the radial coordinate line for the cases $l_s/a = 0$ and $l_s/a = 0.5$ are shown in Fig. 4b. For the classical case $l_s/a = 0$, the normalized stress maximum is 2.02 while for the gradient case $l_s/a = 0.5$ the total and Cauchy stress maximum is 1.69 and 1.72, respectively. For particular choice of $l_s/a = 0.5$, the total and Cauchy stresses are indistinguishable from each other. The normalization is performed with respect to the absolute value of the applied loading p.

The variation of the normalized total and Cauchy stresses at $r = a$ with respect to l_s/a is presented in Fig. 5. It can be seen, that the numerical and analytical results coincide exactly. For $0 < l_s/a < 0.2$, the maximal level of the Cauchy stresses (black line and red dots) exceeds the classical one, while the total stresses (green line and blue dots) constantly monotonically decrease. For $0 < l_s/a < 0.6$, the total stress level is less than the Cauchy stress level and for $l_s/a > 0.6$ this relation changes. Also, it should be mentioned that the total and Cauchy stress maximum is always achieved at the periphery of the hole (results are not included in this contribution).

3 Conclusion

The applicability of isogeometric Galerkin methods for solving plane stress/strain problems within strain gradient elasticity has been studied by two model problems. With the first problem, the convergence properties of the method, in an energy equivalent norm, are shown to be optimal with respect to the NURBS order of the method, as predicted by theoretical results. For the second problem, the analytical solution is captured by the numerical one with a reasonable mesh size. These results serve as a confirmation for the applicability, reliability and efficiency of the numerical approach and verify the implementation accomplished as a user subroutine into a commercial finite element software.

References

Aifantis EC (1992) On the role of gradients in the localization of deformation and fracture. Int J Eng Sci 30:1279–1299

Bleustein JL (1967) A note on the boundary conditions of Toupin's strain-gradient theory. Int J Solids Struct 3:1053–1057

Dassault Systemes Simulia (2015) ABAQUS Analysis User's Guide, 32.15 User-defined elements. http://50.16.225.63/v6.14/books/usb/default.htm

Fischer P, Klassen M, Mergheim J, Steinmann P, Müller R (2011) Isogeometric analysis of 2D gradient elasticity. Comput Mech 47:325–334

Gao XL, Park S (2007) Variational formulation of a simplified strain gradient elasticity theory and its application to a pressurized thick-walled cylinder problem. Int J Solids Struct 44:7486–7499

Hughes TJR, Cottrell JA, Bazilevs Y (2005) Isogeometric analysis: CAD, finite elements, nurbs, exact geometry and mesh refinement. Comput Methods Appl Mech Eng 194:4135–4195

Mindlin RD (1964) Micro-structure in linear elasticity. Arch Ration Mech Anal 16:51–78

Niiranen J, Khakalo S, Balobanov V, Niemi AH (2015) Variational formulation and isogeometric analysis for fourth-order boundary value problems of gradient-elastic bar and plane strain/stress problems. Comput Methods Appl Mech Eng

Papargyri-Beskou S, Polyzos D, Beskos DE (2009) Wave dispersion in gradient elastic solids and structures: a unified treatment. Int J Solids Struct 46:3751–3759

Polizzotto C (2012) A gradient elasticity theory for second-grade materials and higher order inertia. Int J Solids Struct 49:2121–2137

Zervos A, Papanicolopulos SA, Vardoulakis I (2009) Two finite-element discretizations for gradient elasticity. J Eng Mech 135:203–213

Applications of Higher-Order Continua to Size Effects in Bending: Theory and Recent Experimental Results

Christian Liebold and Wolfgang H. Müller

Abstract In the context of static elasticity theory for isotropic materials under small deformations, different approaches of higher-order continuum mechanics are described. Considering generalized continua, the strain gradient-, micropolar- and surface elasticity theory are explained. Analytical solutions, such as the bending line, are derived for each extended theory, using the Euler–Bernoulli beam model. Atomic Force Microscopy investigations of the materials epoxy and the polymer SU-8, as well as flexural vibration analysis of aluminum foams were performed, to determine several additional material parameters. As a result, positive as well as negative size effects in dependency of the thickness and length are observed for micro-cantilevers.

Keywords Strain gradient elasticity · Second gradient continuum · Size effect · Micropolar theory

1 Introduction

With the ever growing applications of simulation technologies, driven by miniaturization and saving of materials, the whole process from design phase up to the valuation of reliability grows in importance. For a more accurate evaluation of engineering materials, a qualitative but also quantitative understanding of size effects needs to be included, either in a physically reasonable manner, or as an alternative technique of homogenization regarding to materials with intrinsic micro- or macro-structure. Size effects are, for example, reflected in a stiffer or softer elastic response to external forces when the size of a body is reduced. Such relations have been recognized in several experiments on metals and polymers, as there are, for

C. Liebold (✉) · W.H. Müller
Berlin University of Technology, Institute of Mechanics, Einsteinufer 5, 10587 Berlin, Germany
e-mail: christian.liebold@tu-berlin.de

W.H. Müller
e-mail: wolfgang.h.mueller@tu-berlin.de

H. Altenbach and S. Forest (eds.), *Generalized Continua as Models for Classical and Advanced Materials*, Advanced Structured Materials 42,
DOI 10.1007/978-3-319-31721-2_12

example, copper (Fleck et al. 1994), silver (Ma and Clarke 1995), zinc oxide (Stan et al. 2007), lead (Cuenot et al. 2004), carbon nanotubes (Salvetat et al. 1999a, b), epoxy (Chong 2002) and polypropylene (McFarland and Colton 2005), cf. Table 1. Actually, physical reasoning for size-dependent material behavior can be manifold, such as non-negligible interactions of molecular chains in polymers (Nikolov et al. 2007), rearrangement of atoms or molecules near the surface, influence of grain-size in polycrystals (Smyshlyaev and Fleck 1996) or even long-ranging force distributions from dislocations, voids or some inner micro- or nanostructure, and it is poorly understood especially in context of their complex combinations. For the reason that conventional continuum theory is unable to predict size effects, different non-conventional continua are proposed in literature, like *non-local theories* (Peddieson et al. 1996; Eringen 2010), *strain-gradient theories* (Toupin 1962; Mindlin and Eshel 1968), *micropolar theories* (Eringen 1966; Nowacki 1970), *theories of material surfaces* (Gurtin and Murdoch 1975) or *fractional continuum mechanics* (Carpinteri 1994; Klimek 2001). In this work, attention will be paid to just a few higher-order continuum theories in the scope of elastic material behavior.

In Sect. 2 *micropolar theory*, *strain gradient theory*, *couple stress theory*, as well as *theories of material surfaces* will be presented in order to analyze the problem of simple beam bending. Current experimental investigations of size effects of the materials *SU-8*, *epoxy*, *aluminum foams* and *aluminum with artificial heterogeneities* are carried out in Sect. 3 using an Atomic Force Microscope (AFM) to record force as well as deflection data of micro-beams, and a vibration analysis system to measure flexural eigenfrequencies of specimens of macroscopic size.

2 Selected Higher-Order Continuum Mechanical Theories

Generalized continua are driven by the question:

> Is it possible to construct continuum theories that can predict physical phenomena on the atomic, molecular, or nano scales?

(quote from A.C. Eringen 2009, in the preface to Maugin and Metrikine 2010). Many publications about the origin of generalized continua cite the work of Cosserat and Cosserat (1909), describing for the first time a polar character of material points. Half a century later, this idea has been recapitulated e.g. in the works of Günther (1958), Schaefer (1967). Among others, the linearized Cosserat theory was essentially influenced by Eringen (1966), Eringen (1976), Eringen (1999), Eringen (2010), Toupin (1962), Koiter (1964), Mindlin and Tiersten (1962), Mindlin (1964), Mindlin and Eshel (1968), Nowacki (1970). A more detailed history of generalized continua is included in Altenbach et al. (2010), Maugin and Metrikine (2010). Basically, today's concept allows to separate between micromorphic and strain gradient theories as an extension to the conventional Cauchy continuum. With the help of specific restrictions, the micromorphic continuum can be simplified to the micropolar continuum and to the Couple Stress model (CS), cf. Fig. 1. The same applies to the Strain

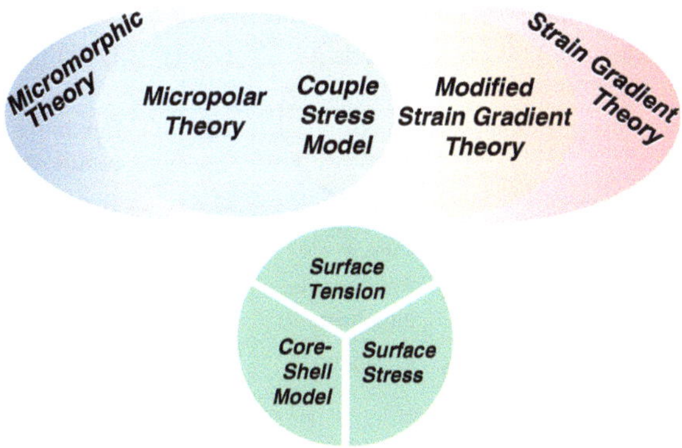

Fig. 1 Overview of selected continuum mechanical theories of higher-order including models of material surfaces

Gradient (SG) theory, which can be Modified (MSG) and transferred into the Couple Stress (CS) theory as well.

Additionally, some theories of material surfaces can be taken into account as good candidates to describe size dependent material behavior (cf. Fig. 1). These are based on the idea that the elastic behavior of the surface of a solid could be separated from its volume. Therefore, a core-surface model with respect to the theory of *Surface Elasticity* (SE) and a core-shell model of an elastic *Surface Layer* (SL) will be used in Sect. 2.3. Table 1 summarizes some of the additional material constants that are already given in literature, like surface elastic moduli or material length scale parameters for certain materials.

2.1 Strain Gradient Theories

First works on the development of so-called couple stress theories by, e.g., Toupin (1962), Mindlin and Tiersten (1962), Koiter (1964), Mindlin and Eshel (1968) contain second order derivatives of displacements to describe quantities of curvature or rotation. The introduction of second order derivatives in terms of constitutive relations and energy considerations was generalized by Eringen (2010), who claimed

> nonsimple materials of gradient type

and derived respective higher-order material dependencies. In the following, the Einstein summation convention on repeated indices is used and spatial partial derivatives in the Cartesian coordinate system are denoted by comma-separated indices:

$$i, j, k, l, m, n, A \in \{1, 2, 3\}, \quad (\cdot)_{,j} = \frac{\partial (\cdot)}{\partial x_j}, \quad (\cdot)_{,A} = \frac{\partial (\cdot)}{\partial X_A}, \tag{1}$$

where x_i belongs to the actual configuration of the body and X_i to the reference configuration. Considering a cascade of principles of analytical mechanics and taking

Table 1 Some literature values for the size effect in elasticity

Additional parameters	Material	Theory	Testing method	Reference
$\ell = 9.4\,\mu\text{m}$	Epoxy	CS	Bending of cantilevers	Chong (2002)[a], Lam et al. (2003)[a]
$\ell = 3.0\,\mu\text{m}$	Copper	MSG	Torsion of wires	Yang et al. (2002)
$\ell = 7.0\,\text{nm}$	Zinc oxide	MSG	Torsion	Stan et al. (2007)
$\ell = 57.0\,\text{nm}$	Polypyrrole nanotubes	MSG	Bending	Cuenot et al. (2000)
$E_\text{B} = 169\,\text{GPa}$ $E_\text{S} = -12.14\,\text{N/m}$	sc-silicon[b]	SE	Bending	Miller and Shenoy (2000), Sadeghian et al. (2009)[a]
$E_\text{S} = 5.8\,\text{N/m}$	Silver	SE	Bending	Jing et al. (2006)[a]
$\ell = 3.2\,\text{nm}$	sw-CNT[b]	SG	Bending	Jing et al. (2006), Sadeghian et al. (2009)
$\ell = 5.7\,\text{nm}$	mw-CNT[b]	SG	Bending	Jing et al. (2006), Sadeghian et al. (2009)
$\ell = 9.3\,\text{nm}$	mw-CNT[b]	SG	Bending	Jing et al. (2006), Poncharal et al. (1999)
$E = 15.0\,\text{GPa}$ $\ell = 17.7\,\text{nm}$	Lead	CS	Bending	Cuenot et al. (2004)[a]

[a]The raw experimental data that are included in the given reference have been used to extract the respective additional parameter.
[b]sc = single-crystalline, sw = single-walled, mw = multi-walled, CNT = Carbon Nanotube

nonsimple materials of gradient type into account (cf. Liebold and Müller 2015), the internal energy density u of a body reads:

$$u = \hat{u}\left(F_{ij}, F_{ij,A}\right), \tag{2}$$

where the fundamental variables are identified as the deformation gradient F_{ij} (the material gradient of the motion χ_i) and its gradient:

$$F_{ij} = \frac{\partial \chi_i}{\partial X_j} \quad \text{and} \quad F_{ij,A} = \frac{\partial \chi_i}{\partial X_j \partial X_A}. \tag{3}$$

By utilizing objective strain measures, like Green's strain tensor E_{ij}^{G} and a higher-order strain tensor K_{ijk} (cf. Bertram 2015):

$$E_{ij}^{\text{G}} = \frac{1}{2}\left(F_{ik}F_{kj} - \delta_{ij}\right) \quad \text{and} \quad K_{ijk} = F_{il}^{-1}\frac{\partial F_{lj}}{\partial X_k}, \tag{4}$$

where δ_{ij} represents the components of the identity tensor, so-called reduced forms are developed:

$$u = \tilde{u}\left(E_{ij}^{\mathrm{G}}, K_{ijk}\right). \tag{5}$$

With the assumption of the existence of a specific elastic energy (cf. Bertram 2015) of the form:

$$p := \frac{1}{\rho_0}\left[\frac{1}{2}T_{ij}^{\mathrm{2PK}}\dot{E}_{ij}^{\mathrm{G}} + S_{ijk}\dot{K}_{ijk}\right], \tag{6}$$

the respective work-conjugated stress measures T_{ij}^{2PK} (second Piola–Kirchhoff stress tensor) and S_{ijk} (higher-order material stress tensor) can be defined as:

$$T_{ij}^{\mathrm{2PK}} = \frac{\partial \tilde{u}}{\partial E_{ij}^{\mathrm{G}}} \quad \text{and} \quad S_{ijk} = \frac{\partial \tilde{u}}{\partial K_{ijk}}. \tag{7}$$

2.1.1 Mindlin's Forms of Strain Energy Density

Following the restriction to small deformations by means of:

$$\frac{\partial(\cdot)}{\partial X_i} \approx \frac{\partial(\cdot)}{\partial x_i}, \tag{8}$$

the reduced forms of the higher-order strain energy density in Eq. (5) yield in at least three forms, postulated by Mindlin and Tiersten (1962):

$$u^{\mathrm{SG}} = u^{\mathrm{1ST}}\left(\varepsilon_{ij}, \eta_{ijk}\right) = u^{\mathrm{2ND}}\left(\varepsilon_{ij}, \tilde{\eta}_{ijk}\right) = u^{\mathrm{3RD}}\left(\varepsilon_{ij}, \bar{\eta}_{ij}, \bar{\bar{\eta}}_{ijk}\right). \tag{9}$$

Proposing linear elastic material behavior, the respective work-conjugated stress measures of the first form are obtained as:

$$\sigma_{ij} = \frac{\partial u^{\mathrm{1ST}}}{\partial \varepsilon_{ij}} \quad \text{and} \quad \mu_{ijk} = \frac{\partial u^{\mathrm{1ST}}}{\partial \eta_{ijk}}, \tag{10}$$

where σ_{ij} and μ_{ijk} are the Cauchy stress tensor and the hyper- (or double-) stress tensor, respectively. For isotropic nonsimple materials of gradient type a linear strain energy density results in (Mindlin and Eshel 1968):

$$\begin{aligned} 2u^{\mathrm{1ST}} &= \alpha_1\varepsilon_{ij}\varepsilon_{ij} + \alpha_2\varepsilon_{kk}\varepsilon_{mm} + \beta_1\eta_{ijk}\eta_{ijk} + \beta_2\eta_{iik}\eta_{jjk} \\ &\quad + \beta_3\eta_{iik}\eta_{kjj} + \beta_4\eta_{ijj}\eta_{ikk} + \beta_5\eta_{ijk}\eta_{kji}, \end{aligned} \tag{11}$$

Table 2 Kinematic variables of small deformations in linear elastic higher-gradient continuum, acc. Mindlin and Eshel (1968)

Symbol	Meaning
$\varepsilon_{ij} = u_{(i,j)} = \frac{1}{2}\left(u_{i,j} + u_{j,i}\right)$	Small strain tensor (symmetric part of the gradient of displacement)
$\eta_{ijk} = u_{k,ij}$	Second gradient of displacement
$\bar{\bar{\eta}}_{ijk} = \frac{1}{3}\left(u_{k,ij} + u_{i,jk} + u_{j,ki}\right)$	Symmetric part of the second gradient of displacement
$\tilde{\eta}_{ijk} = \frac{1}{2}\left(u_{k,ij} + u_{j,ki}\right) = \varepsilon_{kj,i}$	Gradient of strain
$\varphi_i = \frac{1}{2}\epsilon_{ijk}u_{k,j}$	Macroscopic rotation vector
$\bar{\eta}_{ij} = \frac{1}{2}\epsilon_{ilk}u_{k,lj}$	Gradient of rotation
ϵ_{ijk}	Alternating tensor (Levi–Civita symbol)

where α_1 and α_2 can be identified as Lamé's constants, whereas $\beta_{1,\dots,5}$ are five unknown additional material parameters. The number of additional material parameters is going to be reduced in the next section.

2.1.2 Modified Strain Gradient Theory

A decomposition of the second gradient of displacement in combination with utilizing the macroscopic rotation vector yields in a reduction of independent additional material parameters from five to three. Independent metrics of η_{ijk} are first introduced by Fleck and Hutchinson (1997), who decomposed the second order deformation gradient into its symmetric part $\bar{\bar{\eta}}_{ijk}$ and anti-symmetric part η^{A}_{ijk} (see Fig. 2, cf. Table 3). The symmetric part of η_{ijk} is further decomposed into its spherical and deviatoric part $\eta^{(0)}_{ijk}$ and $\eta^{(1)}_{ijk}$. Now, caused by the definition of the macroscopic rotation vector φ_i and its gradient $\bar{\eta}_{ij}$, two important relations will appear: (i) The spherical part of $\bar{\bar{\eta}}_{ijk}$ can be decomposed into $\varepsilon_{mm,i}$ and χ^{A}_{ij} by using the relations:

$$\eta^{(0)}_{ijk} = \frac{1}{5}\left(\delta_{ij}\bar{\bar{\eta}}_{mmk} + \delta_{jk}\bar{\bar{\eta}}_{mmi} + \delta_{ki}\bar{\bar{\eta}}_{mmj}\right) \text{ and } \bar{\bar{\eta}}_{mmi} = \varepsilon_{mm,i} + \tfrac{2}{3}\epsilon_{iln}\chi^{A}_{ln}, \qquad (12)$$

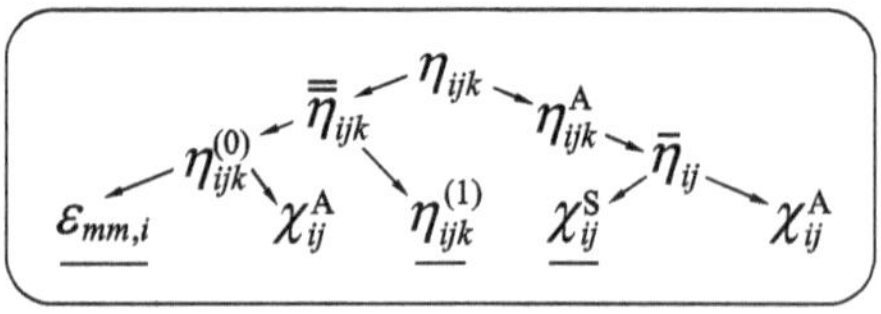

Fig. 2 Scheme of splitting the second gradient of displacement in the MSG theory

Table 3 Modified kinematic variables of small deformations in linear elastic strain-gradient continuum, acc. Lam et al. (2003)

Symbol	Meaning
$\chi_{ij}^{\mathrm{S}} = \frac{1}{4}\left(\epsilon_{ilk}u_{k,lj} + \epsilon_{jlk}u_{k,li}\right)$	Symmetric part of the gradient of rotation
$\chi_{ij}^{\mathrm{A}} = \frac{1}{4}\left(\epsilon_{ilk}u_{k,lj} - \epsilon_{jlk}u_{k,li}\right)$	Anti-symmetric part of the gradient of rotation
$\bar{\bar{\eta}}_{ijk} = \frac{1}{3}\left(u_{k,ij} + u_{i,jk} + u_{j,ki}\right)$	Symmetric part of the second gradient of displacement
$\eta_{ijk}^{\mathrm{A}} = \frac{2}{3}\left(\epsilon_{ikl}\bar{\eta}_{lj} + \epsilon_{jkl}\bar{\eta}_{li}\right)$	Anti-symmetric part of the second gradient of displacement
$\eta_{ijk}^{(0)} = \frac{1}{5}\left(\delta_{ij}\bar{\bar{\eta}}_{mmk} + \delta_{jk}\bar{\bar{\eta}}_{mmi} + \delta_{ki}\bar{\bar{\eta}}_{mmj}\right)$	Spherical part of $\bar{\bar{\eta}}_{ijk}$
$\eta_{ijk}^{(1)} = \bar{\bar{\eta}}_{ijk} - \eta_{ijk}^{(0)}$	Deviatoric part of $\bar{\bar{\eta}}_{ijk}$
$\varepsilon_{mm,i} = u_{m,mi}$	Dilatation gradient

and (ii) it can be shown that the anti-symmetric part of η_{ijk} will completely reduce to the gradient of rotation $\bar{\eta}_{ij}$, cf. Table 3$_4$. The latter is decomposed itself into the symmetric and anti-symmetric part χ_{ij}^{S} and χ_{ij}^{A}. The anti-symmetric part of the gradient of rotation is crossed out, by assuming the symmetry of the couple stress tensor μ_{ij} (the work-conjugate of the gradient of rotation) as proposed by Yang et al. (2002). Taking all remaining kinematic variables of the modified strain-gradient model into account (see underlined entries in Fig. 2), a linear strain energy density for nonsimple materials of modified gradient type reads:

$$
\begin{aligned}
u^{\mathrm{MSG}} &= \hat{u}\left(\varepsilon_{ij}, \varepsilon_{mm,i}, \eta_{ijk}^{(1)}, \chi_{ij}^{\mathrm{S}}\right) \\
&= \tfrac{1}{2}\sigma_{ij}\varepsilon_{ij} + \tfrac{1}{2}p_i\varepsilon_{mm,i} + \tfrac{1}{2}\mu_{ijk}^{(1)}\eta_{ijk}^{(1)} + \tfrac{1}{2}\mu_{ij}\chi_{ij}^{\mathrm{S}} \\
&= \tfrac{1}{2}\lambda\varepsilon_{ii}\varepsilon_{kk} + \mu\varepsilon_{ij}\varepsilon_{ij} + \mu\ell_0^2\varepsilon_{mm,i}\varepsilon_{nn,i} + \mu\ell_1^2\eta_{ijk}^{(1)}\eta_{ijk}^{(1)} + \mu\ell_2^2\chi_{ij}^{\mathrm{S}}\chi_{ij}^{\mathrm{S}},
\end{aligned}
\tag{13}
$$

where the respective work-conjugated stress measures are:

$$
\begin{aligned}
\sigma_{ij} &= \frac{\partial u^{\mathrm{MSG}}}{\partial \varepsilon_{ij}} = \lambda\varepsilon_{kk}\delta_{ij} + 2\mu\varepsilon_{ij}, \quad p_i = \frac{\partial u^{\mathrm{MSG}}}{\partial \varepsilon_{nn,i}} = 2\mu\ell_0^2\varepsilon_{mm,i}, \\
\mu_{ijk}^{(1)} &= \frac{\partial u^{\mathrm{MSG}}}{\partial \eta_{ijk}^{(1)}} = 2\mu\ell_1^2\eta_{ijk}^{(1)}, \qquad \mu_{ij} = \frac{\partial u^{\mathrm{MSG}}}{\partial \chi_{ij}^{\mathrm{S}}} = 2\mu\ell_2^2\chi_{ij}^{\mathrm{S}}.
\end{aligned}
\tag{14}
$$

Application of the Euler–Bernoulli Beam Model

To derive an analytical formula for simple beam bending, the displacement field:

$$
u_x = -z\frac{\mathrm{d}w(x)}{\mathrm{d}x}, \quad u_y = 0, \quad u_z = w(x), \tag{15}
$$

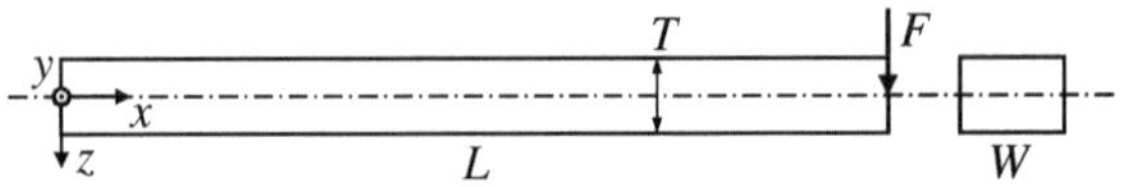

Fig. 3 Orientation of the coordinate system for the beam bending problem. L, W, T are the length, width and thickness, respectively

is used according to the Euler–Bernoulli beam assumptions, following the relations described in Tables 2 and 3. $w(x)$ represents the bending line in the coordinate system given in Fig. 3.

The principle of virtual displacements is applied by using the variation of the function $w(x)$ and integration by parts to derive the differential equation of the problem in the following manner. The variation of the strain energy of the body is given by:

$$
\begin{aligned}
U^{\mathrm{MSG}} &= \iiint\limits_{x\,y\,z} u^{\mathrm{MSG}} \mathrm{d}z\mathrm{d}y\mathrm{d}x, \\
\delta U^{\mathrm{MSG}} &= \int\limits_{x=0}^{x=L} \left(S\frac{\mathrm{d}^4 w}{\mathrm{d}x^4} - K\frac{\mathrm{d}^6 w}{\mathrm{d}x^6} \right) \delta w \mathrm{d}x + K\frac{\mathrm{d}^5 w}{\mathrm{d}x^5}\delta w\Big|_0^L - S\frac{\mathrm{d}^3 w}{\mathrm{d}x^3}\delta w\Big|_0^L \\
&\quad + S\frac{\mathrm{d}^2 w}{\mathrm{d}x^2}\frac{\mathrm{d}(\delta w)}{\mathrm{d}x}\Big|_0^L - K\frac{\mathrm{d}^4 w}{\mathrm{d}x^4}\frac{\mathrm{d}(\delta w)}{\mathrm{d}x}\Big|_0^L + K\frac{\mathrm{d}^3 w}{\mathrm{d}x^3}\frac{\mathrm{d}^2(\delta w)}{\mathrm{d}x^2}\Big|_0^L,
\end{aligned}
\tag{16}
$$

where the abbreviations $S = EI\left(1 + \frac{4308}{225}\frac{\ell^2}{T^2}\right)$ and $K = \frac{7}{5}EI\ell^2$ are given, if a rectangular cross-section and $\ell_0 = \ell_1 = \ell_2 = \ell$ is assumed without further reasoning. E denotes Young's modulus and I the second moment of inertia. By comparing Eq. (16) to the variation of the work W^{MSG} done by the external forces $q(x)$ and V, moments M and higher moments M^{h} (cf. Kong et al. 2009):

$$
\delta W^{\mathrm{MSG}} = \int\limits_{x=0}^{x=L} q(x)\delta w \mathrm{d}x + V\delta w\Big|_0^L + M\frac{\mathrm{d}(\delta w)}{\mathrm{d}x}\Big|_0^L + M^{\mathrm{h}}\frac{\mathrm{d}^2(\delta w)}{\mathrm{d}x^2}\Big|_0^L, \tag{17}
$$

the differential equation of the problem is identified as:

$$
S\frac{\mathrm{d}^4 w}{\mathrm{d}x^4} - K\frac{\mathrm{d}^6 w}{\mathrm{d}x^6} = q(x), \ \forall\, x \in [0, L] \ . \tag{18}
$$

For a solution of the homogeneous differential equation Eq. (18) of the form:

$$
w(x) = C_1 x^3 + C_2 x^2 + C_3 x + C_4 + C_5 \mathrm{e}^{\sqrt{S/K}x} + C_6 \mathrm{e}^{-\sqrt{S/K}x}, \tag{19}
$$

two sets of boundary conditions can be applied to determine the constants $C_{1,\dots,6}$. The five boundary conditions $V(L) = F$, $M(L) = 0$, $M^{\mathrm{h}}(L) = 0$, $w(0) = 0$ and $\frac{\mathrm{d}w}{\mathrm{d}x}|_{x=0} = 0$ may be extended by either:

$$\begin{aligned} \text{BC1:}\ K\left.\frac{\mathrm{d}^3 w}{\mathrm{d}x^3}\right|_{x=0} &= M^{\mathrm{h}}(0) = 0, \text{ or} \\ \text{BC2:}\ EI\left.\frac{\mathrm{d}^2 w}{\mathrm{d}x^2}\right|_{x=0} &= M^{\mathrm{class}}(0) = 0 \end{aligned} \tag{20}$$

where M^{class} represents the description of a moment from classical beam theory. The difference in evaluation of deflections for all values of $0 < \ell < L$ is below 4 %. With adjusted constants $C_{1,\dots,6}$, the deflection line of an Euler–Bernoulli beam in the modified strain gradient model (Eq. 19) is evaluated numerically. To fit the material length scale parameter ℓ to experiments, the least square method is used to find the minimum error between experimental data (Sect. 3) and the present model.

2.2 *Micromorphic Continuum*

To quote Eringen (cf. pp.33 Eringen 1976),

> a micromorphic continuum may be thought of as a classical continuum to each point of which is associated another continuum.

The sub-continuum is restricted to homogeneous deformations[1] only and may be described as a continuous distribution of deformable point particles. The description of the intrinsic deformation of the point particles succeeds via so-called directors, who are attached to each material point and map the orientation and the particle's deformation.

2.2.1 Micropolar Theory

The micropolar continuum theory can be classified into the micromorphic one, whereas in the micropolar case the directors are proper orthogonal and do not change their length with respect to each other. Thus, only rotational degrees of freedom are assigned to the particles, making them "rigid". Considering a cascade of principles of analytical mechanics taking simple materials and additional degrees of freedom into account (cf. Liebold and Müller 2015), the internal energy density u of a body reads:

$$u = \breve{u}\left(F_{ij}, Q_{ij}, Q_{ij,A}\right), \tag{21}$$

[1]The displacement gradient is constant over the body.

where the fundamental variables are identified as the deformation gradient F_{ij}, the orientation tensor of the particles Q_{ij} (a proper orthogonal tensor) and its gradient $Q_{ij,A}$. Only by utilizing special strain measures, like (cf. Eremeyev et al. 2013):

$$\begin{aligned} E^Q_{ij} &= Q_{ik}F_{kj} - \delta_{ij}, \\ \Gamma_{ij} &= -\tfrac{1}{2}\epsilon_{irm}(Q_{mA}Q_{rj,A}), \end{aligned} \tag{22}$$

the relative stretch tensor E^Q_{ij} and the relative wryness tensor Γ_{ij}, reduced forms can be developed for the non-symmetric material stress tensor $\overline{T}^{2PK}_{ij}$ (similar to the second Piola–Kirchhoff stress tensor) and the material couple stress tensor M_{ij}:

$$\begin{aligned} \overline{T}^{2PK}_{ij}(x_n, t) &= k(E^Q_{ij}, \Gamma_{ij}), \\ M_{ij}(x_n, t) &= K(E^Q_{ij}, \Gamma_{ij}). \end{aligned} \tag{23}$$

With the assumption of small deformations, the constitutive equations for the non-symmetric Cauchy stress tensor $\overline{\sigma}_{ij} \neq \overline{\sigma}_{ji}$ and the couple stress tensor μ_{ij} read (cf. Lakes 1995):

$$\begin{aligned} \overline{\sigma}_{ij} &= \lambda\varepsilon_{kk}\delta_{ij} + (2\mu + \kappa)\varepsilon_{ij} + \kappa\epsilon_{ijk}(\varphi_k - \phi_k), \\ \mu_{ij} &= \gamma_1\phi_{r,r}\delta_{ij} + \gamma_2\phi_{i,j} + \gamma_3\phi_{j,i}. \end{aligned} \tag{24}$$

γ_1, γ_2, γ_3, and κ denote the additional material coefficients, called Cosserat elastic constants. The micro-rotation vector ϕ_k is kinematically distinct from the macro-rotation φ_k here.

2.2.2 Couple Stress Theory

The next level of simplification is to equalize the microscopic and macroscopic rotation vector in the geometrically linearized micropolar theory. The intrinsic degree of freedom of rotation is completely assigned to the macroscopic deformation and this is called pseudo- or indeterminate Cosserat theory. Considering the total kinetic and internal energy of a micropolar body, claiming that $\phi_k = \varphi_k$, it can be shown, that the stress tensor turns out to be symmetric (cf. Liebold and Müller 2015) and the constitutive equations (24) reduce to:

$$\begin{aligned} \sigma_{ij} &= \lambda\varepsilon_{kk}\delta_{ij} + 2\mu\varepsilon_{ij}, \\ \mu_{ij} &= 2\mu\ell^2\chi^S_{ij}. \end{aligned} \tag{25}$$

Application of the Euler–Bernoulli Beam Model

To derive an analytical formula for simple beam bending, the displacement field according to the Euler–Bernoulli beam assumptions is used, following Eq. (15). To identify the differential equation of the problem by a comparison of the virtual work done by external forces and the couple stress strain energy:

$$U^{CS} = \int_V (\sigma_{ij}\varepsilon_{ij} + \mu_{ij}\chi_{ij}^{S})\mathrm{d}V, \tag{26}$$

the principle of virtual displacements is applied by using the variation of the function $w(x)$ and integration by parts. In analogy to the derivation of Eq. (18), the Euler–Bernoulli differential equation in the case of couple stress theory yields in:

$$(EI + \mu A\ell^2)\frac{\mathrm{d}^4 w}{\mathrm{d}x^4} = q(x), \quad \forall x \in [0, L], \tag{27}$$

where A denotes the cross-sectional area of the beam. The solution of the homogeneous differential equation Eq. (27) is given to be a polynomial of fourth-order, whose four constants are obtained by using the classical boundary conditions for a cantilever beam, as described in Sect. 2.1.2. Consequently, the deflection line results in:

$$w^{CS}(x) = \frac{F}{(EI + \mu A\ell^2)}\left[\frac{Lx^2}{2} - \frac{x^3}{6}\right], \tag{28}$$

where a comparison of the rigidity to bending from present to classical theory yields in a formulation of a general elastic modulus E^{CS} of the form:

$$E^{CS}I = (EI + \mu A\ell^2), \quad E^{CS} = E\left(1 + 6\frac{\ell^2}{T^2}\right), \tag{29}$$

where $\mu = E/2$ is used, regarding to the assumptions stated at the beginning. Equation $(29)_2$ serves as a target function in a least square method, to find E and ℓ from experimental data from force and deflection measurements.

2.3 *Theories of Material Surfaces*

The surface of a solid typically exhibits characteristics, that may vary from the properties of its volume. These differences are caused for example by surface oxidation, aging, coating, atomic and molecular rearrangement (see Fig. 4) or even surface roughness. To capture these issues, different continuum-mechanical theories have been developed. From a first point of view they may be divided into (i) a core-surface model, called Surface Elasticity (SE), in which a surface layer is introduced with a thickness of zero, and (ii) a core-shell model with a non-zero thickness of a Surface Layer (SL).

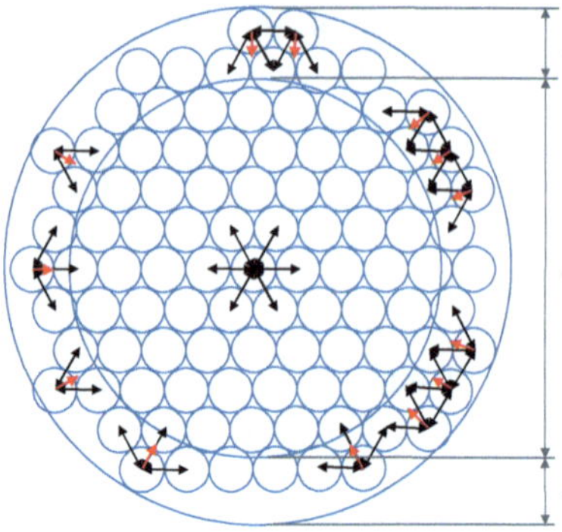

Fig. 4 An example of the origin of deviating surface characteristics: Molecular rearrangements, caused by the completion of the molecular bonds close to the surface, and resultant intermolecular forces in a relatively *thin* boundary layer

2.3.1 Core-Surface Model (Surface Elasticity)

Most works on surface elasticity, starting from the works of Shuttleworth (1950), Orowan (1970), have been worked out following the Gibbsian thermodynamics. The applied continuum mechanical description of a surface stress tensor stated here is based on Gurtin and Murdoch (1975), and has been picked up in various recent works, such as Miller and Shenoy (2000), Javili et al. (2013), Wang et al. (2010), Ru (2010). Starting from the most general case of an arbitrarily curved and smooth surface, that is parameterized with the surface coordinates z^1 and z^2, the concept of co- and contravariant components of tensor representation will be used. In general, the kinematics and tensorial arithmetic operations used here, are known from differential geometry. Important surface coefficients are indicated by "S," whereas tensorial quantities of the surface are indicated by $\alpha, \beta, \gamma, \delta \in \{1, 2\}$, due to the two-dimensionality. Corresponding unnormalized tangential vectors and the surface identity $\delta^{\alpha\cdot}_{\cdot\beta}$ read:

$$\underline{s}_\alpha = \frac{\partial x_i}{\partial z^\alpha}\underline{e}_i, \quad \underline{s}^\alpha = \frac{\partial z^\alpha}{\partial x_i}\underline{e}_i \quad \text{and} \quad \delta^{\alpha\cdot}_{\cdot\beta} = \underline{s}^\alpha \cdot \underline{s}_\beta. \tag{30}$$

Indices that are separated by a semicolon denote spatial derivatives of tensorial symbols, that incorporate additional derivatives of transformations between surface ($\underline{s}_\alpha$)- and Cartesian ($\underline{e}_i$) coordinate system (known as Christoffel symbols Γ^i_{jk}). A Cauchy-like surface stress tensor $\tau^{\alpha\cdot}_{\cdot\beta}$ can be described, using the concept of the Gibbs free surface energy (cf. Vermaak et al. 1968):

$$\tau^{\alpha\cdot}_{\cdot\beta} = \sigma\delta^{\alpha\cdot}_{\cdot\beta} + \frac{\partial\sigma(\varepsilon^{\delta\cdot}_{\cdot\gamma})}{\partial\varepsilon^{\cdot\beta}_{\alpha\cdot}}, \tag{31}$$

where σ is the surface energy density and $\varepsilon_{\alpha\beta}$ the small strain tensor of the surface (cf. Flügge 1972):

$$\varepsilon^{\alpha\cdot}_{\cdot\beta} = \frac{1}{2}\left(u^{\alpha}_{;\beta} + u^{\beta}_{;\alpha}\right) . \tag{32}$$

If elastically isotropic material behavior is assumed, the surface stress tensor is represented as:

$$\tau^{\alpha\cdot}_{\cdot\beta} = \gamma_0 \delta^{\alpha\cdot}_{\cdot\beta} + \lambda^{S} \varepsilon^{\gamma\cdot}_{\cdot\gamma} \delta^{\alpha\cdot}_{\cdot\beta} + 2\mu^{S} \varepsilon^{\alpha\cdot}_{\cdot\beta}, \tag{33}$$

where λ^S and μ^S are the Lamé's constants of the surface, and γ_0 the component of a possible spherical residual stress. A *surface strain energy density* u^{SE} for linear elastic material surfaces is then given by:

$$u^{SE} = \hat{u}(\gamma_0, \varepsilon_{\alpha\beta}) = \frac{1}{2}\tau_{\alpha\beta}\varepsilon^{\alpha\beta} = \frac{1}{2}\left(\gamma_0 + \lambda^{S}\varepsilon^{\gamma\cdot}_{\cdot\gamma}\right)\varepsilon^{\delta\cdot}_{\cdot\delta} + \mu^{S}\varepsilon_{\alpha\beta}\varepsilon^{\alpha\beta}. \tag{34}$$

Application of the Euler–Bernoulli Beam Model

In the following, a beam with rectangular cross-section, flat surfaces, no surface Poisson's ratio and without residual stresses is assumed, in order to derive an advanced formula for beam bending. As a result of these assumptions, the tangential vectors of the surface are always orthogonal, co- and contravariant notation correspond to each other and the surface strain energy density of the body reduces to:

$$u^{SE} = \mu^{S}\varepsilon_{\alpha\beta}\varepsilon_{\alpha\beta}. \tag{35}$$

The non-zero components of the surface strain tensor per surface area are:

$$\begin{aligned} \varepsilon_{11}(x, y, -\frac{T}{2}) &= \frac{T}{2}\frac{\mathrm{d}^2 w}{\mathrm{d}x^2}, \quad \varepsilon_{11}(x, y, \frac{T}{2}) = -\frac{T}{2}\frac{\mathrm{d}^2 w}{\mathrm{d}x^2}, \\ \varepsilon_{11}(x, \frac{W}{2}, z) &= -z\frac{\mathrm{d}^2 w}{\mathrm{d}x^2}, \quad \varepsilon_{11}(x, -\frac{W}{2}, z) = -z\frac{\mathrm{d}^2 w}{\mathrm{d}x^2}. \end{aligned} \tag{36}$$

The total strain energy U of this beam model consists of the superposition of U^S and U^{Vol} (the classical strain energy of simple beam theory):

$$U = U^{S} + U^{Vol} = \oint\limits_{\partial V} \frac{1}{2}E^{S}\varepsilon_{11}\varepsilon_{11}\mathrm{d}A + \int\limits_{V} \frac{1}{2}E^{B}\varepsilon_{11}\varepsilon_{11}\mathrm{d}V, \tag{37}$$

where $E^S = 2\mu^S$ and E^B denote the surface and bulk Young's moduli, respectively. The integration and variation of Eq. (37) with respect to the function $w(x)$ yields in:

$$\delta U = \left[E^{S}\left(\frac{WT^2}{2} + \frac{T^3}{6}\right) + E^{B}\frac{WT^3}{12}\right]\int\limits_{x=0}^{x=L} \delta\left(\frac{\mathrm{d}^2 w}{\mathrm{d}x^2}\right)^2 \mathrm{d}x. \tag{38}$$

In analogy to the derivation of Eq. (18), the Euler–Bernoulli differential equation in the case of surface elasticity yields in:

$$\left[E^{\mathrm{S}}\left(\frac{WT^2}{2}+\frac{T^3}{6}\right)+E^{\mathrm{B}}\frac{WT^3}{12}\right]\frac{\mathrm{d}^4 w}{\mathrm{d}x^4}=q(x),\ \forall\, x\in[0,L]\,, \tag{39}$$

where a comparison of the rigidity to bending from present to classical theory yields in a formulation of a general elastic modulus E^{SE}:

$$\begin{aligned} E^{\mathrm{SE}}I &= \left[E^{\mathrm{S}}\left(\frac{WT^2}{2}+\frac{T^3}{6}\right)+E^{\mathrm{B}}\frac{WT^3}{12}\right], \\ E^{\mathrm{SE}} &= E^{\mathrm{B}}+E^{\mathrm{S}}\left(\frac{6}{T}+\frac{2}{W}\right). \end{aligned} \tag{40}$$

Equation $(40)_2$ is identical to the result of Miller and Shenoy (2000) and serves as a target function in a least square method, to find E^{B} and $E^{\mathrm{S}}>0$ from experimental data from force and deflection measurements for the materials SU-8 and epoxy.

2.3.2 Core-Shell (Surface Layer) Model

The core-shell model implies a Surface Layer (SL) with non-zero thickness. In classical core-shell models, this layer is treated like a laminate like structure, incorporating a jump of mechanical properties at the interface to the bulk (cf. Yao et al. 2012). For that reason, the model works well, if a softening of the overall material behavior is concerned. The most simple case of the SL model is, if the elastic modulus of the surface layer of the constant thickness r (see Fig. 5) is assumed to be equal to zero, $E^*=0$.

Concerning rectangular cantilever beams, the general elastic modulus E^{SL}:

$$E^{\mathrm{SL}}=E\frac{(L-2r)^3}{(W-2r)(T-2r)^3}\frac{WT^3}{L^3}, \tag{41}$$

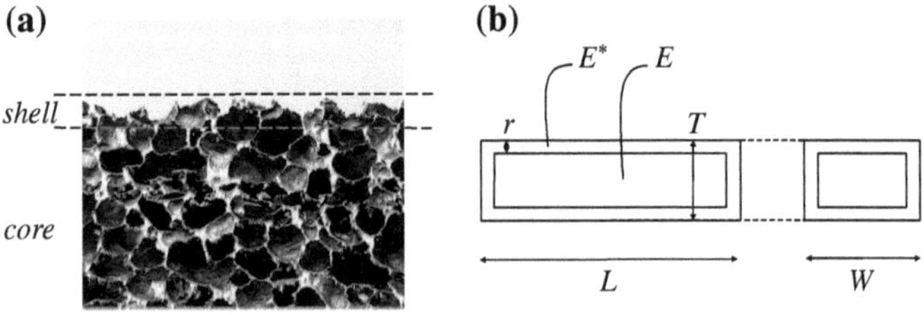

Fig. 5 **a** Application of the core-shell model to aluminum foams with open surface. **b** Illustration of the symbols for a cantilever beam

is derived, based on geometric considerations and serves as a target function in a least square method, to find r and E from experimental data for aluminum foams. In contrast to the higher-order theories stated above, the SL-model as defined here, is able to characterize negative size effects in dependency of the thickness, width and length of the beams.

3 Experimental Analysis

In the linear theory of elasticity of isotropic materials, the elastic modulus is a central property of the material behavior that can be quantified by standardized measurement methods such as the uniaxial tensile test, simple bending or flexural vibration analysis. Higher-order continuum theories are beneficial, if in measurements of the elastic modulus based on a conventional continuum theory, results are dependent on the external dimensions of the body. For this reason, the following experimental analysis focus on the determination of Young's modulus using conventional approaches. External dimensions as the thickness of a beam are successively reduced to obtain results of different sizes. This procedure has been described by Lakes (1995) as an experimental *method of size effect*, which makes it possible to determine some of the material parameters of extended theories.

3.1 *Atomic Force Microscopy*

To measure the elastic modulus of structures with the outer dimensions in the order of a few micrometers, simple static bending tests are performed on freestanding beam structures. The load is applied using the off-axis laser-reflective Atomic Force Microscope (AFM) MultiView-1000 from Nanonics Imaging Ltd.[2] The system consists of a flat scanner, including a fine thread that is driven by piezo-elements using high-voltage power supply. The detection device works with four Photo-Sensitive Diods (PSD) which are interconnected as a Wheatstone bridge to monitor deflections of the laser beam path. The laser reflects in an obtuse angle from a fixed AFM-cantilever, such that the system directly monitors its deflections w_c, when it is deformed by the piezo uplift (referred to as separation z). The raw PSD-signal obtained in [mV] was able to be converted into forces $F = k_c w_c$ in [μN] by the knowledge of the well calibrated spring constant of the AFM-cantilever of $k_c = 31.4\ \mathrm{N\ m^{-1}}$. The calibration process is detailed in Varenberg et al. (2005) and was performed, using a precise silicon normal that was provided by the PTB.[3] In micro-beam bending tests it can be assumed, that the raw AFM-data consist of a combined signal of the deflection of the AFM-cantilever and the micro-beam's deflection w in the following manner:

[2]www.nanonics.co.il, Jerusalem, Israel.

[3]Physikalisch-Technische Bundesanstalt – Braunschweig, Germany.

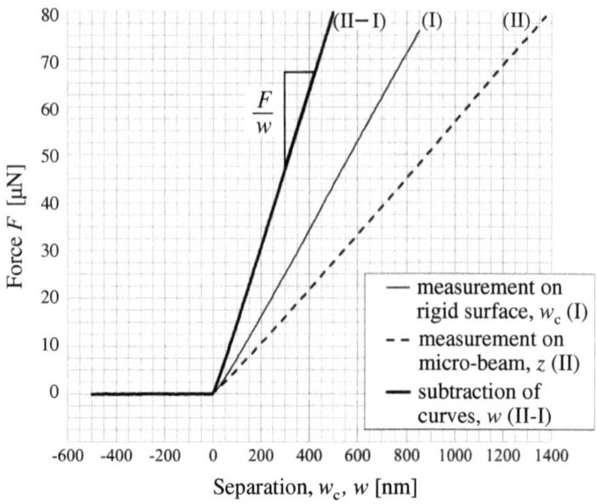

Fig. 6 Identification of the deflection data w of a micro-beam: *Deflection curves*, shifted to the zero point and converted into micro-Newton are substracted in a comparison approach

$w = z - w_c$ (see Fig. 6). Presuming rectangular cross-sections of the specimens, the following classical relation between the ratio F/w and the elastic modulus E is used:

$$E = \frac{4L^3}{WT^3}\frac{F}{w}. \tag{42}$$

Length and width of the samples have been measured in an optical microscope with a magnification of 500 times, whereas the value for the thickness consists of the mean value from two different optical determination systems together with the value of a Scanning Electron Microscope (SEM).

3.1.1 AFM Experiments on Epoxy

The commercially available resin HT 2 from Poxy Systems, consisting of the basic ingredients bisphenol-F-diglycidylether and 4,4'methylendiphenyldiglycidylether was mixed with the appropriate hardener in the ratio 100:48 and put between two preparation glasses within a 45-minute processing time. Different spacers between the preparation glasses assured an adjustable film thickness of 17–170 μm. By the help of a parallel cutting tool, which allows to realize widths of 100–400 μm, the cured epoxy film was cut into stripes (see Fig. 7). The single stripes were glued over the edges of cover glasses, such that about two-thirds of a strip overlap. The effective bending length was determined between the edge of the glass support and the force application point and varied between 180–4400 μm. The influence of the slightly viscous material behavior of epoxy was tested by applying different loading rates between 0.1–$20\,\mu\mathrm{m\,s^{-1}}$, and only a one percent scatter of the measured values was verified. The mean-values of the measured elastic moduli and the corresponding standard deviations are given in Fig. 8.

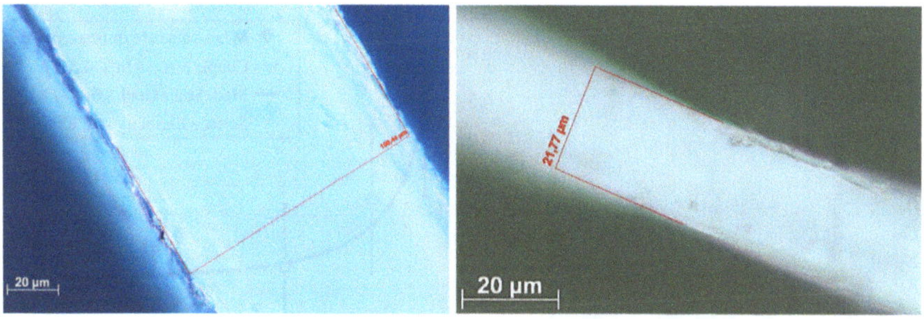

Fig. 7 *Left* Measurement of the width of an exemplary micro-beam of epoxy by optical microscopy. *Right* Exemplary measurement of the thickness of the epoxy film

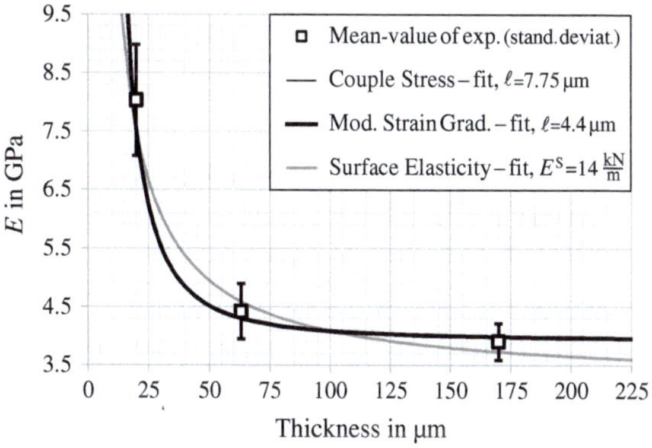

Fig. 8 Averaged results of the measurements of the elastic modulus of epoxy and evaluated fit functions of the CS, MSG and SE theory

3.1.2 AFM Experiments on SU-8 Polymer

SU-8 is known as the photo-resist Nano^{TM}-SU-8, used in the micro-system technology (cf. Lorenz et al. 1997) and produced by the company MicroChem. Structuring of the samples was carried out in the following steps:

- A 4-inch silicon wafer was coated with a thin metal film as a barrier layer.
- A conventional spin-coating machine was applied to disperse the fluid resist on the metal-coated silicon wafer. The rotational speed of the coating machine determined the thickness of the homogeneous layer (between 8–40 µm).
- The solvent evaporated in the rotary process in large part and in a subsequent drying process the remaining part evaporated at temperatures of 60–95 °C, whereby the material finally received its rigidity.
- Structuring has been achieved by using a Laser Direct Imaging (LDI). After exposure to light, an additional heat treatment was carried out at 60–95 °C to assist the chemical reaction of the illumination process.
- By using a proper developer, the regions of exposed SU-8 were dissolved from unexposed regions.

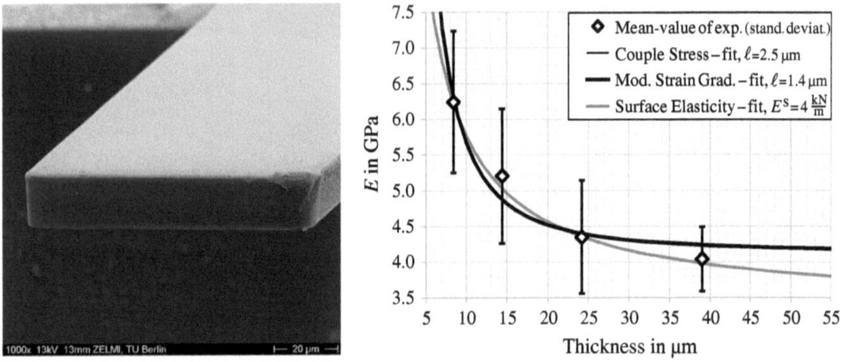

Fig. 9 *Left* Exemplary measurement of the outer dimensions by scanning electron microscopy (SEM) of a micro-beam made of the polymer SU-8. *Right* Averaged results of the measurements of the elastic modulus of the SU-8 polymer and evaluated fit functions of the CS, MSG and SE theory

- By a chemical etch process, that does not attack the SU-8 material, the thin metal film on the silicon wafer was dissolved and the micro-beam structures were finally peeled off and glued on a glass support.

Widths and lengths of the samples, realized in this way, ranged from 80–124 and 82–920 μm, respectively. The mean-values of the measured elastic moduli and the corresponding standard deviations are given in Fig. 9, on the right hand side.

3.2 *Flexural Vibration Analysis*

A simple method for evaluating the elastic modulus of a certain material is given by a measurement of the flexural resonance frequency of a structure. The frequency detecting system used here is a GrindoSonic-device from Lemmens, N.V.. The macroscopic samples that consist of rectangular bars, lie on two soft supports in a three point bending mode and are excited manually by an impact tool. The acoustic signal of the flexural modes are analyzed by a Fast-Fourier-Transformation (FFT) in order to detect the first eigenfrequency f_1. Assuming small amplitudes, linear material behavior and slender beam structures, the solution of the dynamic Euler–Bernoulli differential equation in the case of a free-free boundary condition gives:

$$E = 0.946 \frac{\rho L^4}{T^2} f_1^2, \tag{43}$$

where ρ is the mass density of the material.

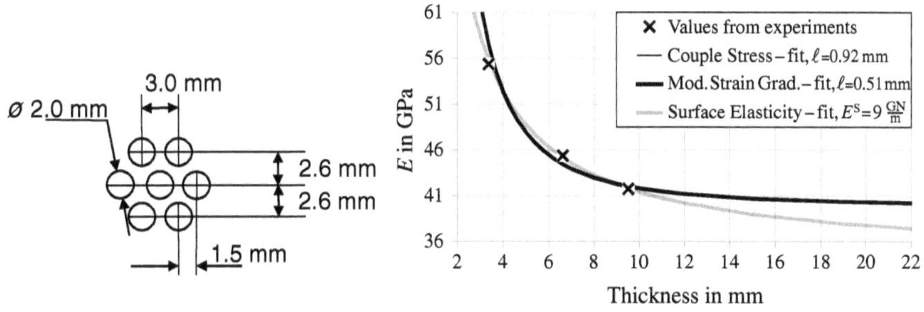

Fig. 10 *Left* The constant arrangement of the artificial heterogeneities of the aluminum samples. *Right* Results of the frequency measurements of the elastic modulus and the evaluation of the fit functions of the CS, MSG and SE theory

3.2.1 Vibration Analysis of Aluminum with Artificial Heterogeneities

Out of the total of three bars of different sizes, that were provided with drilled holes as artificial heterogeneities in a constant arrangement (see Fig. 10, left), the flexural eigenfrequencies were measured. The thicknesses and the widths of the rectangular samples were about 3.5, 6.6 and 9.6 mm, and the ratio of lengths to thicknesses were constant at $L/T \approx 13$.

3.2.2 Vibration Analysis of Aluminum Foams

The aluminum foams used here are ultralight materials with very high porosity. The mean size of the closed-pores is about three millimeters. Two different foams, having two different densities due to different production methods, have been investigated. Distinction is made between powder metallurgical production (foam-I) and gas injection method (foam-II). Samples have been cut in different thicknesses of 6.6, 15 and 40 mm (foam-I), and 7, 14 and 44 mm (foam-II), whereas the ratio of the length to thickness was $L/T \approx 20$. The results of the frequency analysis reveal a negative size effect and show an additional influence of the width (see Fig. 11). A fit of the results for the aluminum foam samples succeeded only with the model of a surface layer as described in Sect. 2.3.2 (core-shell model). The arrows point to two results, which show that the measurement results and the results of the model respond to differences in single widths in good accordance. A least square fit using the SL-model reveals an elastic modulus of $E = 3.98$ GPa and $r = 1.0$ mm for foam-I and $E = 1.06$ GPa and $r = 0.41$ mm for foam-II.

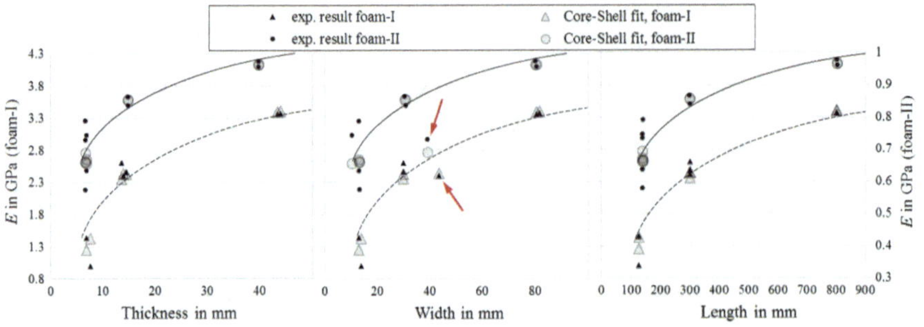

Fig. 11 Results of the flexural vibration analysis of aluminum foam-I and -II: The axis of ordinates on the *left hand side* belongs to the values of foam-I, whereas the axis of ordinates on the *right hand side* belongs to foam-II. The *arrows* indicate the results of those samples that differ from the ratio of the length to width of $L/W \approx 10$. The *solid* and *dotted lines* represent inversely proportional functions that were built with the assumption of constant ratios

4 Results

Bending rigidities F/w from static bending tests, and flexural vibration frequencies f_1 have been measured for four different types of materials. As results of the experimental work, the additional parameters of higher-order continuum theories were determined with a least square method, using the derived target functions of the CS, MSG, SE and SL theories.

A multi-scale graph of normalized Young's moduli for all different thicknesses of all materials used in this work, is presented in Fig. 12. The normalization that is

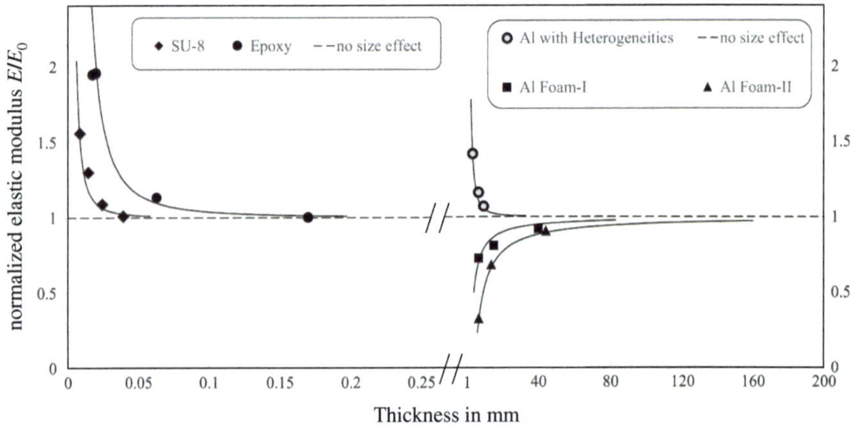

Fig. 12 A multi-scale graph of normalized Young's moduli for all different thicknesses of all materials used in this work

Table 4 Results of the evaluation of the measured data

Material	Theory	Parameter I	Parameter II
Epoxy	Couple stress theory	$E = 3.93\,\text{GPa}$	$\ell = 7.75\,\mu\text{m}$
	Mod. strain gradient theory	$E = 3.93\,\text{GPa}$	$\ell = 4.35\,\mu\text{m}$
	Surface elasticity theory	$E^{\text{B}} = 3.22\,\text{GPa}$	$E^{\text{S}} = 14.1\,\text{kN m}^{-1}$
SU-8	Couple stress theory	$E = 4.13\,\text{GPa}$	$\ell = 2.5\,\mu\text{m}$
	Mod. strain gradient theory	$E = 4.14\,\text{GPa}$	$\ell = 1.39\,\mu\text{m}$
	Surface elasticity theory	$E^{\text{B}} = 3.36\,\text{GPa}$	$E^{\text{S}} = 3.95\,\text{kN m}^{-1}$
Drilled aluminum	Couple stress theory	$E = 39.8\,\text{GPa}$	$\ell = 0.92\,\text{mm}$
	Mod. strain gradient theory	$E = 39.8\,\text{GPa}$	$\ell = 0.51\,\text{mm}$
	Surface elasticity theory	$E^{\text{B}} = 34.1\,\text{GPa}$	$E^{\text{S}} = 9.15\,\text{GN m}^{-1}$
Al foam-I	Surface layer model	$E = 3.98\,\text{GPa}$	$r = 1.0\,\text{mm}$
Al foam-II	Surface layer model	$E = 1.06\,\text{GPa}$	$r = 0.41\,\text{mm}$

used in the multi-scale graph refers to the specific elastic modulus (parameter I from Table 4), called E_0.

A more detailed analysis of the data of the materials epoxy and SU-8 shows an additional and clear relationship of measured elastic moduli with the length of the beams, see Fig. 13. The solid and dotted lines in Fig. 13 are inversely proportional functions, that are optimized separately to assure a minimum error to the experimental data.

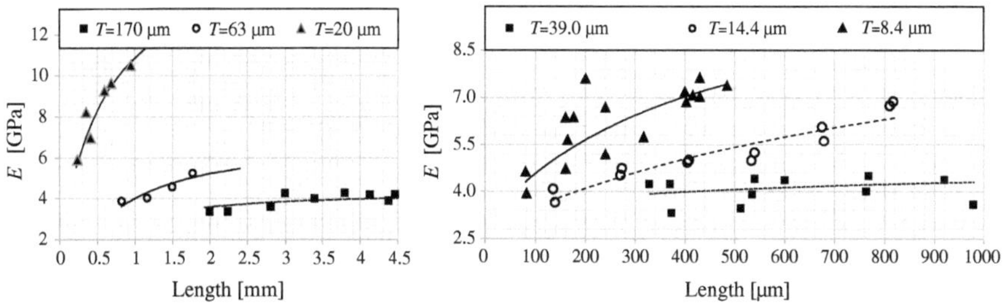

Fig. 13 Detailed graphs of the measurement values versus the lengths of the micro-beams. *Left* epoxy. *Right* SU-8

5 Conclusions

While the sizes of structures that were made of epoxy, SU-8, drilled aluminum and aluminum foams were reduced, positive as well as negative size effects have occurred. A positive size effect reflects in a stiffer elastic response to external forces, whereas a negative effect is reflected in a softer elastic response when outer dimensions of a body are reduced. Selected higher-order continuum mechanical theories have been explained, as for example strain gradient theories, micropolar theory and theories of material surfaces. For any of these theories, an analytical solution of the Euler–Bernoulli beam model has been derived and presented. The assumption of a variable elastic modulus allowed an adjustment of the higher-order coefficients ℓ, r and E^{S}. With the example of the macroscopic samples of the drilled aluminum it could be shown, in comparison to the micro-beams of epoxy, for example, that the value of the parameter ℓ is in the order of the physical dimension of the internal structure (e.g. the diameter and distance of the drillings, cf. Fig. 10, left). The value of the material length scale parameter for epoxy that has been measured in the present work, can be compared with the literature value from Table 1, where there is a deviation of 17 %, that may be attributed to different manufacturing processes or other base materials for the epoxy. As reported for epoxy, there is no size effect in tensile testing with specimens of the same thicknesses that were tested here, cf. Lam et al. (2003). None of the present approaches alone is able to include the pronounced dependence of the elastic modulus from both, lengths and thicknesses of a beam made of epoxy or SU-8 (see Fig. 13).

Acknowledgments The present work is supported by (Deutsche Forschungsgemeinschaft) DFG under Grant MU 1752/33-1. The author like to thank the Fraunhofer Institute for Reliability and Microintegration Berlin for sample preparation and PTB-Braunschweig for AFM calibration.

References

Altenbach J, Altenbach H, Eremeyev VA (2010) On generalized Cosserat-type theories of plates and shells: a short review and bibliography. Arch Appl Mech 80:73–92

Bertram A (2015) Finite gradient elasticity and plasticity: a constitutive mechanical framework. Contin Mech Thermodyn 27(6):1039–1058

Carpinteri A (1994) Fractal nature of material microstructure and size effects on apparent mechanical properties. Mech Mater 18(2):89–101, (Special issue on microstructure and strain localization in geomaterials)

Chong CM (2002) Experimental investigation and modeling of size effect in elasticity. Phd thesis, Hong Kong University of Science and Technology

Cosserat E, Cosserat F (1909) Theorie des Corps Deformables. Hermann et Fils, Paris

Cuenot S, Demoustier-Champagne S, Nysten B (2000) Elastic modulus of polypyrrole nanotubes. Phys Rev Lett 85(8):1690–1693

Cuenot S, Fretigny C, Demoustier-Champagne S, Nysten B (2004) Surface tension effect on the mechanical properties of nanomaterials measured by atomic force microscopy. Phys Rev B 69:01–05

Eremeyev VA, Lebedev LP, Altenbach H (2013) Foundations of micropolar mechanics. Springer, Heidelberg
Eringen AC (1966) A unified theory of thermomechanical materials. Int J Eng Sci 4:179–202
Eringen AC (1976) Continuum Physics, vol IV—Polar and nonlocal field theories. Academic Press, New York
Eringen AC (1999) Microcontinuum field theories, vol I—Foundations and solids. Springer, New York
Eringen AC (2010) Nonlocal continuum field theories. Springer, New York
Fleck NA, Hutchinson JW (1997) Strain gradient plasticity. In: Hutchinson JW, Wu TY (eds) Advances in applied mechanics, vol 33. Academic Press, New York, pp 295–361
Fleck NA, Muller GM, Ashby MF, Hutchinson JW (1994) Strain gradient plasticity: theory and experiment. Acta Metall Mater 42(2):475–487
Flügge W (1972) Tensor analysis and continuum mechanics. Springer, Berlin
Günther W (1958) Zur Statik und Kinematik des Cosseratschen Kontinuums. Abh Braunschweig Wiss Ges 10:195–213
Gurtin ME, Murdoch AI (1975) A continuum theory of elastic material surfaces. Arch Ration Mech Anal 57(4):291–323
Javili A, McBride A, Steinmann P (2013) Thermomechanics of solids with lower-dimensional energetics: on the importance of surface, interface, and curve structures at the nanoscale. A unifying review. Appl Mech Rev 65(1):1–31
Jing GY, Duan HL, Sun XM, Zhang ZS, Xu J, Li YD, Wang JX, Yu DP (2006) Surface effects on elastic properties of silver nanowires: contact atomic-force microscopy. Phys Rev B 73(235):409
Klimek M (2001) Fractional sequential mechanics—models with symmetric fractional derivative. Czechoslovak J Phys 51:1348–1354
Koiter WT (1964) Couple-stresses in the theory of elasticity. Pt. I-II. Proc Koninkl Nederland Akad Wetensh 67:17–44
Kong S, Zhou S, Nie Z, Wang K (2009) Static and dynamic analysis of micro beams based on strain gradient elasticity theory. Int J Eng Sci 47(4):487–498
Lakes R (1995) Experimental methods for study of cosserat elastic solids and other generalized elastic continua. In: Mühlhaus H (ed) Continuum models for materials with micro-structure. Wiley, New York, pp 1–22
Lam DCC, Yang F, Chong CM, Wang J, Tong P (2003) Experiments and theory in strain gradient elasticity. J Mech Phys Solids 51(8):1477–1508
Liebold C, Müller WH (2015) Are microcontinuum field theories of elasticity amenable to experiments?—A review of some recent results (in press). In: Knops RJ, Chen GQ, Grinfeld M (eds) Springer proceedings in mathematics and statistics (PROMS), differential geometry and continuum mechanics, Springer, Heidelberg
Lorenz H, Despont M, Fahrni N, LaBianca N, Renaud P, Vettiger P (1997) SU-8: a low-cost negative resist for MEMS. J Micromech Microeng 7(3):121–124
Ma Q, Clarke DR (1995) Size dependent hardness of silver single crystals. J Mater Res 10:853–863
Maugin GA, Metrikine AV (eds) (2010) Mechanics of generalized continua—one hundred years after the cosserats, advances in mechanics and mathematics, vol 21. Springer, Heidelberg
McFarland AW, Colton JS (2005) Role of material microstructure in plate stiffness with relevance to microcantilever sensors. J Micromech Microeng 15(5):1060–1067
Miller RE, Shenoy VB (2000) Size-dependent elastic properties of nanosized structural elements. Nanotechnology 11:139–147
Mindlin RD (1964) Micro-structure in linear elasticity. Arch Ration Mech Anal 16:51–78
Mindlin RD, Eshel NN (1968) On first strain-gradient theories in linear elasticity. Int J Solids Struct 4:109–124
Mindlin RD, Tiersten HF (1962) Effects of couple-stresses in linear elasticity. Arch Ration Mech Anal 11(1):415–448
Nikolov S, Han CS, Raabe D (2007) On the origin of size effects in small-strain elasticity of solid polymers. Int J Solids Struct 44(5):1582–1592

Nowacki W (1970) Theory of micropolar elasticity, CISM Courses, vol 25. Springer, Vienna
Orowan E (1970) Surface energy and surface tension in solids and liquids. Proc R Soc A 316(1527):473–491
Peddieson J, Buchanan GR, McNitt RP (1996) Application of nonlocal continuum models to nanotechnology. Int J Eng Sci 41:305–312
Poncharal P, Wang ZL, Ugarte D, De Heer WA (1999) Electrostatic deflections and electromechanical resonances of carbon nanotubes. Science 283:1513–1516
Ru CQ (2010) Simple geometrical explanation of Gurtin–Murdoch model of surface elasticity with clarification of its related versions. Sci China Phys Mech Astron 53(3):536–544
Sadeghian H, Yang CK, Goosen JFL, van der Drift E, Bossche A, French PJ, van Keulen F (2009) Characterizing size-dependent effective elastic modulus of silicon nanocantilevers using electrostatic pull-in instability. Appl Phys Lett 94(22):01–03
Salvetat JP, Andrew G, Briggs D, Bonard JM, Bacsa RR, Kulik AJ, Stöckli T, Burnham NA, Forró L (1999a) Elastic and shear moduli of single-walled carbon nanotube ropes. Phys Rev Lett 82:944–947
Salvetat JP, Kulik AJ, Bonard JM, Briggs D, Stöckli T, Metenier K, Bonnamy S, Beguin F, Burnham NA, Forró L (1999b) Elastic modulus of ordered and disordered multiwalled carbon nanotubes. Phys Rev Lett 11:161–165
Schaefer H (1967) Das Cosserat Kontinuum. ZAMM - Journal of Applied Mathematics and Mechanics / Zeitschriftü Angewandte Mathematik und Mechanik 47(8):485–498
Shuttleworth R (1950) The surface tension of solids. Proc Phys Soc A 63(5):444–457
Smyshlyaev VP, Fleck NA (1996) The role of strain gradients in the grain size effect for polycrystals. J Mech Phys Solids 44(4):465–495
Stan G, Ciobanu CV, Parthangal PM, Cook RF (2007) Diameter-dependent radial and tangential elastic moduli of ZnO nanowires. Nano Lett 7(12):3691–3697
Toupin RA (1962) Elastic materials with couple-stresses. Arch Ration Mech Anal 1:365–414
Varenberg M, Etsion I, Halperin G (2005) Nanoscale fretting wear study by scanning probe microscopy. Tribol Lett 18(4):493–498
Vermaak JS, Mays CW, Kuhlmann-Wilsdorf D (1968) On surface stress and surface tension: I. theoretical considerations. Surf Sci 12(2):128–133
Wang ZQ, Zhao YP, Huang ZP (2010) The effects of surface tension on the elastic properties of nano structures. Int J Eng Sci 48(2):140–150
Yang F, Chong CM, Lam DCC, Tong P (2002) Couple stress based strain gradient theory for elasticity. Int J Solids Struct 39(10):2731–2743
Yao H, Yun G, Bai N, Li J (2012) Surface elasticity effect on the size-dependent elastic property of nanowires. J Appl Phys 111(8):01–06

Classification of Gradient Adhesion Theories Across Length Scale

Sergey Lurie, Petr Belov and Holm Altenbach

Abstract The sequence of continuum theories of adhesion is discussed. We give a brief analysis of the existing theories of adhesion and present a continuum theory of adhesion as a natural generalization of appropriate options for the theory of elasticity and gradient theories of elasticity. We offer a sequence of variational formulations of theories of adhesion and constitutive equations. In addition, the analysis of the structures of the tensors of adhesive elastic modules is presented. As a result, we propose a classification of theories of adhesion and gradient theories of elasticity in terms accounting for scale effects. The classification is based on the qualitative analysis of scale effects of different orders depending on the physical properties of the continuum.

1 Introduction

Currently, there is a large number of publications devoted to the history of the theory of surface and interface interactions, starting with the pioneering contributions of Young, Laplace, Poisson, and Gauß discussing the effects of surface tension, of surface properties on the basis of perceptions of intermolecular interactions and of surface stresses in solids (which was initiated by Gibbs). Recent investigations of adhesive properties of surfaces and interfaces in deformable solids, in the mechan-

S. Lurie (✉)
Institute for Problem of Mechanics of RAS and Institute of Applied Mechanics of RAS, Moscow, Russia
e-mail: salurie@mail.ru

P. Belov
Institute of Research, Development and Technology Transfer, Moscow, Russia
e-mail: belovpa@yandex.ru

H. Altenbach
Lehrstuhl für Technische Mechanik, Institut für Mechanik,
Fakultät für Maschinenbau, Otto-von-Guericke-Universtät Magdeburg,
Universitätsplatz 2, 39106 Magdeburg, Germany
e-mail: holm.altenbach@ovgu.de

H. Altenbach and S. Forest (eds.), *Generalized Continua as Models for Classical and Advanced Materials*, Advanced Structured Materials 42,
DOI 10.1007/978-3-319-31721-2_13

ics of heterogeneous structures and in the mechanics of composites are developed in various publications and analyzed in detail (see, for example Steigmann and Ogden 1997a, 1999a; Ostoja-Starzewski 2002; Duan et al. 2005, 2008; Zhu et al. 2009; Altenbach et al. 2011; Povstenko 2008; Huang and Wang 2013; Eremeyev 2016). The first adhesion continuum theories were developed in the framework of the classical theory of elasticity (Gurtin and Murdoch 1975a, b; Steigmann and Ogden 1997a, 1999a; Eremeyev et al. 2009; Javili and Steinmann 2010; Kim et al. 2011; Javili et al. 2012, among others). A generalization of the adhesion model of Murdoch-Gurtin was suggested in Belov and Lurie (2007). In addition to the adhesive analogues of the Lamé's coefficients, two additional physical parameters were introduced in this generalized model. They were based on the analysis of the structure of the adhesive elastic moduli. The first parameter can be identified by the Laplace constant for capillary pressure. The second one is related to the possible non-symmetry of the adhesive stresses in the adhesive interaction of surfaces. Usually, the second one is set to zero for reasons of material indifference of model.

With the development of models for continua with fields of defects and nonlocal theories that allow to take into account the scale effects, the models of mechanics of continua which take into account the scale effects in the volume, concentrated in the vicinity of the boundary of interfaces, and at the same time the adhesive effects became interesting (Mindlin 1965; Lurie and Belov 2008; Lurie and Tuchkova 2009; Lurie et al. 2009, 2010; Belov and Lurie 2009, 2014). In Belov and Lurie (2009, 2014), a model was formulated in which the adhesive properties were attributed to the newly formed surface connected with a field of defects. A variational model that takes into account the adhesive interactions of perfect (not damaged) surfaces, surfaces damaged by defects, and their interaction was presented. In Lurie et al. (2010), it was shown that the inclusion of scale effects due to the adhesive properties of the surface of the body in gradient theories can be of great importance (scale effects of the first order) in comparison with the scale effects in nonlocal theory (second order effects). A gradient theory of second order, which can be considered as a generalization of the theory of Steigmann and Ogden (1997a, 1999a), is described in Belov and Lurie (2014).

The purpose of this paper is the sequential analysis of variational formulations of the theories of adhesive interactions, the classification of adhesion models by the degree of accuracy of accounted scale effects. We present the theorems that allow to lift the restrictions associated with the requirements of the material frame indifference.

2 Theory of Perfect Adhesion of Surface in Classical Elasticity

Let us consider a linear elastic body which occupies the volume V. We assume that the body is limited by an adhesively active surface F. We write the Lagrangian, which allows to simulate the adhesive properties of an ideal (defect-free) classic media surfaces as follows:

$$L = A_P - \iiint\limits_V U_V \,\mathrm{d}V - \iint\limits_F U_F \,\mathrm{d}F, \tag{1}$$

where

$$A_P = \iiint\limits_V P_i^V R_i \,\mathrm{d}V + \iint\limits_F P_i^F R_i \,\mathrm{d}F$$

is the work of external body forces P_i^V, distributed in the volume of the elastic body V, and of surface forces P_i^F, defined on the surface of the body F on the displacements R_i, U_V is the density of the potential energy, U_F is the surface density of the potential energy of adhesively active surface F.

Note that in the case of contacting bodies, interfacial interactions are determined by the difference of the potential energy of the surface of contacting bodies at each contact point. Indeed, consider the interfacial interaction between two bodies in contact with the volumes V_i, $i = 1, 2$ and bounded by surfaces F_i, $(i = 1, 2)$, correspondingly, the contact between the bodies occurs through the common contact surface C: $F_i = \bar{F}_i \cup C$ $(i = 1, 2)$, $\bar{F}_i$ are the body surfaces that are free of contact. The Lagrangian for the composed body is obviously determined by the following equation:

$$\begin{aligned}
L = L^{(1)} + L^{(2)} = A_P^{(1)} - \iiint\limits_{V_1} U_V^{(1)} \,\mathrm{d} - \iint\limits_{F_1} U_F^{(1)} \,\mathrm{d}F \\
+ A_P^{(2)} - \iiint\limits_{V_2} U_V^{(2)} \,\mathrm{d}V - \iint\limits_{F_2} U_F^{(2)} \,\mathrm{d}F \\
= A_P^{(1)} + A_P^{(2)} - \iiint\limits_{V_2} U_V^{(2)} \,\mathrm{d}V - \iiint\limits_{V_2} U_V^{(2)} \,\mathrm{d}V - \iint\limits_{\bar{F}_1} U_F^{(1)} \,\mathrm{d}F - \iint\limits_{\bar{F}_2} U_F^{(2)} \,\mathrm{d}F \\
- \iint\limits_C (U_F^{(1)} - U_F^{(2)}) \,\mathrm{d}F.
\end{aligned} \tag{2}$$

Here

$$A_P^k = \iiint\limits_{V_k} P_i^{(V_k)} R_i^{(1)} \,\mathrm{d}V + \iint\limits_{\bar{F}_k} P_i^{(\bar{F}_k)} R_i^{(1)} \,\mathrm{d}F; \quad U_V^{(k)}, U_F^{(k)} \quad (k = 1, 2)$$

is the density of the potential strain energy in the volume, and $U_F^{(k)}$ $(k = 1, 2)$ are the energy densities of the adhesively active surface of each of the contacting bodies.

The last expression in Eq. (2) proves given assertion. It determines the interfacial interactions at the contact boundary. Consequently, the classification of adhesion theories can be realized by relying on the expression of the surface density of potential energy, inherent in insulated body with adhesively active surface.

Let us further assume that the density of the potential energy of the surface U_F as well as the density of strain energy in the volume depend on the first derivatives of the displacement vectors. We will not involve any hypothesis regarding the structure of the surface density of the potential energy U_F. Only the quadratic form of surface density of potential energy is postulated, which corresponds to the physical linearity of the theory:

$$2U_F = A_{ijmn} R_{i,j} R_{m,n}.$$

In this case, the adhesion stresses are determined by Green's formulas

$$a_{ij} = \frac{\partial U_F}{\partial R_{i,j}} = A_{ijmn} R_{m,n}. \tag{3}$$

Note that considering Eq. (2), we can write the following expression for the strain energy of the contact surface of the two contacting bodies

$$\iint\limits_C (U_F^{(1)} - U_F^{(2)})\,\mathrm{d}F = \iint\limits_C (A_{ijnm}^{(1)} - A_{ijnm}^{(2)})\,d_{ij}\,d_{nm}\,\mathrm{d}F,\, d_{ij} = \partial R_i/\partial x_j,\, A_{ijnm}^{(1)},\, A_{ijnm}^{(2)}.$$

$A_{ijnm}^{(l)}$ are the elastic moduli of the interfacial interactions of contact surface for each of the bodies.

We have the following statement regarding the general structure of the adhesion elastic moduli for the classical linearly elastic body (Belov and Lurie 2007; Lurie and Belov 2008; Lurie and Tuchkova 2009; Lurie et al. 2009).

Theorem 13.1 *Let us consider a linearly elastic body with adhesively active surface. Then*

1. *The elastic moduli of the interfacial interactions are as follows:*

$$A_{ijmn} = \lambda^F \delta_{ij}^* \delta_{mn}^* + \mu^F (\delta_{im}^* \delta_{jn}^* + \delta_{in}^* \delta_{jm}^*) + \chi^F (\delta_{im}^* \delta_{jn}^* - \delta_{in}^* \delta_{jm}^*) + \delta^F n_i n_m \delta_{jn}^*, \tag{4}$$

where $\delta_{ij}^* = (\delta_{ij} - n_i n_j)$, δ_{ij} is Kronecker's delta λ^F, μ^F are the elastic constants similar to Lamé's coefficients (bulk properties) in Hooke's law that determine the interfacial interactions in the model of Gurtin and Murdoch (1975a, b), χ^F is the adhesive elastic modulus characterizing the rigidity of the surface at the asymmetric deformation of the surface, δ^F is the modulus of elasticity that defines the surface resistance in bending.

2. *In the case of a surface having isotropic properties with a symmetric tensor of adhesion stresses $a_{ij} e_{ijk} n_k = 0$ (e_{ijk} is the permutation symbol), the adhesive*

elastic moduli have the structure

$$A_{ijmn} = \lambda^F \delta^*_{ij}\delta^*_{mn} + \mu^F(\delta^*_{im}\delta^*_{jn} + \delta^*_{in}\delta^*_{jm}) + \delta^F n_i n_m \delta^*_{jn}.$$

The proof of (4) can be found in Belov and Lurie (2007), Lurie and Tuchkova (2009). It is based on the representation of the tensors of adhesive moduli A_{ijmn} as an expansion in the system of basic fourth-rank tensors. These tensors are built as products of planar Kronecker tensors $\delta^*_{ij} = (\delta_{ij} - n_i n_j)$ and tensors of the type $(n_i n_j)$ with all possible permutations of the indices.

In general, they are supplemented by the tensors of the form $n_i n_j n_m n_n$ and are conditionally divided into three groups. The first group includes only those tensors that contain the planar Kronecker tensors as factors, the second one includes those which contain planar Kronecker tensor and tensor $(n_i n_j)$ as factors, the third one includes those that contain two tensors of $(n_i n_j)$ type. As a result, it is established a common form of the tensor A_{ijmn}. The most common structure of the fourth-rank transversely isotropic tensor contains eight adhesive moduli and has the form:

$$\begin{aligned} A_{ijmn} &= \lambda^F \delta^*_{ij}\delta^*_{mn} + (\mu^F + \chi^F)\delta^*_{im}\delta^*_{jn} + (\mu^F - \chi^F)\delta^*_{in}\delta^*_{jm} \\ &+ \delta^F n_i n_m \delta^*_{jn} + (\alpha^F + \eta^F) n_i n_n \delta^*_{jm} + (\alpha^F - \eta^F) n_m n_j \delta^*_{in} \\ &+ (\beta^F + \xi^F) n_i n_j \delta^*_{mn} + (\beta^F - \xi^F) n_m n_n \delta^*_{ij} + B^F \delta^*_{im} n_j n_n \\ &+ A^F n_i n_j n_m n_n. \end{aligned} \tag{5}$$

The condition of the existence of the potential energy of adhesion $A_{ijmn} = A_{mnij}$ leads to the requirement of vanishing coefficients $\eta^F = 0$ and $\xi^F = 0$ in Eq. (5). Finally, we must fulfill the conditions for which the boundary value problems for a classic elastic body contain in each regular point of the surface three boundary conditions. To satisfy this requirement of the consistency of the adhesion model with the general formulation of boundary value problems of classical theory of elasticity, the following should be inserted in Eq. (5)

$$\alpha^F = \beta^F = B^F = A^F = 0.$$

Indeed, these conditions correspond to the conditions $a_{ij} n_j = A_{ijmn} n_j R_{m,n} \equiv 0$. Obviously, in this case the adhesive elastic moduli of the classical elastic body satisfy the following conditions $A_{ijmn} n_j \equiv 0$:

$$A_{ijmn} n_j = \alpha^F n_m \delta^*_{in} + \beta^F n_i \delta^*_{mn} + B^F \delta^*_{im} n_n + A^F n_i n_m n_n = 0.$$

In other words, the density of the potential energy of surface in the case of the classical elastic body has the form:

$$2U_F = A_{ijmn}(R_{i,k}\delta^*_{kj})(R_{m,p}\delta^*_{pj}).$$

A special case of Eqs. (2)–(4) for $\lambda^F \neq 0$, $\mu^F \neq 0$, $\chi^F = \delta^F = 0$ is the "theory of elasticity of the surface" of Gurtin and Murdoch (1975a, b). The question of the material frame indifference of the Murdoch-Gurtin adhesion theory defined by (4) with condition $\delta^F = 0$ can be raised here. We will show in Sect. 6 that for the generalized theory of adhesion the material frame indifference is always satisfied.

We give an analysis of the potential energy of elastic body with the adhesive properties of the free surface in terms of scale effects. Let us pay attention to the fact that the adhesive moduli A_{ijmn} differ from the components of the tensor of the classical moduli C_{ijmn} on the scale of length. We can define the scale parameter l, associated with the interfacial interactions, as the ratio of the norms of the tensor components A_{ijmn} and C_{ijmn}, for example, $A = \sqrt{A_{ijmn}A_{ijmn}}$ and $C = \sqrt{C_{ijmn}C_{ijmn}}$. Then, $l = A/C$, $\bar{A}_{ijmn} = A_{ijmn}C/A$, and the tensor can be represented as $A_{ijmn} = \bar{A}_{ijmn}l$, where tensor $\bar{A}_{ijmn}$ has the same dimension as the tensor of classical moduli. The Lagrangian (1) can be written now as a linear expansion of the parameter l:

$$L = \left(A_P - \frac{1}{2}\iiint C_{ijmn}R_{i,j}R_{m,n}\,\mathrm{d}V\right) + \left(-\frac{1}{2}\bar{A}_{ijmn}R_{i,j}R_{m,n}\,\mathrm{d}F\right)l = L_0 l^0 + L_1 l^1 \tag{6}$$

with

$$L_0 = A_P - \frac{1}{2}\iiint C_{ijmn}R_{i,j}R_{m,n}\,\mathrm{d}V$$

and

$$L_1 = -\frac{1}{2}\bar{A}_{ijmn}R_{i,j}R_{m,n}\,dF.$$

Thus, any kind of continuum theory of adhesion of a type (2)–(4) determines the scale effects of first order.

3 Theory of Adhesion Interactions in Gradient Elasticity

Let us consider nonlocal Mindlin and Eshel (1968) theory. We assume that there are no defects in a deformed elastic body and on its surface. In this case, the Lagrangian that allows us to formulate a gradient continuum theory with adhesive interactions, i.e. to introduce the constitutive equations and to provide a complete mathematical formulation of the boundary value problem, can be represented as follows:

$$
\begin{aligned}
L &= A_P - U_V(R_{i,j}, R_{i,jk}) - U_F(R_{i,j}), \\
U_V(R_{i,j}, R_{i,jk}) &= \frac{1}{2} \iiint\limits_V (C_{ijmn} R_{i,j} R_{m,n} + C_{ijkmnl} R_{i,jk} R_{m,nl})\, \mathrm{d}V, \\
U_F(R_{i,j}) &= \frac{1}{2} \iint\limits_F A_{ijmn} R_{i,j} R_{m,n}\, \mathrm{d}F, \\
A_P &= \iiint\limits_V P_i^V R_i \mathrm{d}V + \iint\limits_F [P_i^F R_i + q_i R_{i,j} n_j]\, \mathrm{d}F,
\end{aligned}
\tag{7}
$$

where A_P is the work done by external body force P_i^V and Cauchy traction P_i^F, q_i, $R_{i,j}$ and $R_{i,jk}$ are the components of the double stress traction vector and the distortion tensor, gradient of the distortion (or second gradient of displacement) tensor, respectively, the coefficients C_{ijmn} are the components of the forth rank tensor of elastic parameters, which we assume to be homogeneous throughout the body, the coefficients C_{ijkmnl} are the components of the sixth rank tensor of distortion gradient elastic parameters.

The following theorem is valid.

Theorem 13.2 *1. For a linearly elastic body bounded by a smooth surface, the Lagrange principle*

$$
\delta L = \delta A_P - \delta U_V(R_{i,j}, R_{i,jk}) - \delta U_F(R_{i,j}) = 0 \tag{8}
$$

with the functional (7) completely determines the mathematical formulation of gradient theory of elasticity of a body with the volume V, limited with adhesively active surface F.

2. *Generalized equations of Hooke's law are defined by Green's formulas. In the volume of the body, the constitutive equations have the structure:*

$$
\sigma_{ij} = \frac{\partial U_V(R_{i,j}, R_{i,jk})}{\partial R_{i,j}} = C_{ijmn} R_{m,n}, \qquad \mu_{ijk} = \frac{\partial U_V(R_{i,j}, R_{i,jk})}{\partial R_{i,jk}} = C_{ijkmnl} R_{m,nl}, \tag{9}
$$

where σ_{ij} are the components of the Cauchy stress tensor, μ_{ijk} are the components of the double stress tensor.

3. *On the adhesively active surface of the body F for the adhesion stresses, Green's formulas give the following defining relations for the tensor of adhesive stresses a_{ij}:*

$$
a_{ij} = \frac{\partial U_F(R_{i,j})}{\partial R_{i,j}} = A_{ijmn} R_{m,n}, \tag{10}
$$

where A_{ijmn} are the components of the forth rank tensor of adhesion elastic parameters. Tensor A_{ijmn} has the most common structure and includes eight physical parameters:

$$
\begin{aligned}
A_{ijmn} &= \lambda^F \delta^*_{ij}\delta^*_{mn} + \mu^F(\delta^*_{im}\delta^*_{jn} + \delta^*_{in}\delta^*_{jm}) + \chi^F(\delta^*_{im}\delta^*_{jn} - \delta^*_{in}\delta^*_{jm}) + \delta^F n_i n_m \delta^*_{jn} \\
&+ \alpha^F(n_i n_n \delta^*_{jm} + n_m n_j \delta^*_{in}) + \beta^F(n_i n_j \delta^*_{mn} + n_m n_n \delta^*_{ij}) \\
&+ B^F \delta^*_{im} n_j n_n + A^F n_i n_j n_m n_n. \qquad (11)
\end{aligned}
$$

4. *The variational equation (8) with use of the procedure of integration by parts, yields the following equilibrium equations in V*

$$\sigma_{ij,j} - \mu_{ijk,kj} + P^V_i = 0 \qquad (12)$$

and both natural, von Neumann-like, and essential, Dirichlet-like, boundary conditions on F;

$$\sigma_{ij}n_j - \mu_{ijk,k}n_j - (\mu_{ijk}n_k)_{,j} + (\mu_{ijk}n_k n_l)_{,l}n_j = P^F_i, \qquad \mu_{ijk}n_j n_k = q_i$$

and

$$R_i = \bar{R}_i, \qquad R_{i,j}n_j = \overline{\partial R_i/\partial n},$$

where the overbar stands for prescribed functions.

The proof of the theorem is obvious. It is based on the consistent use of the procedure of integration by parts and on the analysis of the structure of tensor of adhesive elastic moduli (see Sect. 2) and Lurie and Belov (2008), Lurie and Tuchkova (2009), Lurie et al. (2009). The above theorem completely defines a generalized gradient theory of elasticity and the theory of adhesion of gradient defect-free media. The question of the material frame indifference of the presented generalized gradient theory of elasticity and the theory of adhesion is completely solved on the basis of theorem given in the Sect. 6.

In Eq. (11), the moduli λ^F, μ^F, χ^F, δ^F can be related to the adhesive properties of the surface of the classical body, and moduli A^F, B^F can be attributed to the properties of the gradient medium surface. Adhesion moduli α^F, β^F can be interpreted as the moduli of adhesive interactions between the surfaces of classical and gradient medium.

We give the analysis of the Lagrange functional of an elastic body with adhesion properties of the free surface (6) in terms of scale effects. To this end, in a similar way as has been done previously, we normalize both the tensor of the adhesion elastic moduli $\bar{A}_{ijmn} = A_{ijmn}C/A$ and the tensor of gradient moduli of the elasticity of the sixth grade $\bar{C}_{ijkmnl} = C_{ijkmnl}(C/A)^2, l = A/C$. As a result, the Lagrangian of adhesive generalization of Toupin's theory takes the form of the square decomposition in the scale parameter l:

$$L = \left(A_P - \frac{1}{2}\iiint C_{ijmn} R_{i,j} R_{m,n}\, dV\right) + \left(-\frac{1}{2}\oint \bar{A}_{ijmn} R_{i,j} R_{m,n}\, dF\right) l + \left(-\frac{1}{2}\iiint \bar{C}_{ijkmnl} R_{i,jk} R_{m,nl}\, dV\right) l^2$$
$$= L_0 l^0 + L_1 l^1 + L_2 l^2. \tag{13}$$

Thus, any continuum theory of elasticity, modified by accounting adhesive interactions on the surface of gradient media of the form (13), determines the scale effects of both first and second order. The gradient theories are modified by introducing the potential energy in the Lagrangian additionally

$$U_F(R_{i,j}) = \frac{1}{2}\iint_F A_{ijmn} R_{i,j} R_{m,n}\, \mathrm{d}F,$$

defined at the adhesively active part of the surface (see Eqs. (6), (8)–(10)), and may relate to any of the gradient theory as a general theory of Mindlin and Eshel (1968) or to applied gradient theories (Lurie et al. 2009, 2010; Gao and Park 2007; Altan and Aifantis 1992; Gusev and Lurie 2015).

4 Theory of Gradient Adhesion in Gradient Elasticity

Note that the Lagrangian (7) is formally asymmetric with respect to the structures of the volume and surface potential energy densities. Indeed, both the density of energy in the volume and the density of energy on the surface contain quadratic forms of the first derivatives of the displacements $C_{ijmn} R_{i,j} R_{m,n}$ and $A_{ijmn} R_{i,j} R_{m,n}$. However, only the density of strain energy in the volume includes a quadratic form of the second derivatives of the displacements $C_{ijkmnl} R_{i,jk} R_{m,nl}$. Let us extend formally the list of arguments for the density of the potential energy on the surface so that the lists of arguments of potential energy densities in the volume and on the surface match. This generalization is provided by addition of a quadratic form of second derivatives of the displacements $A_{ijkmnl} R_{i,jk} R_{m,nl}$ to the expression of the potential energy density on the surface. As a result, the Lagrangian of the gradient theory of adhesion has the form (Belov and Lurie 2014):

$$L = A_P - \frac{1}{2}\iiint (C_{ijmn} R_{i,j} R_{m,n} + C_{ijkmnl} R_{i,jk} R_{m,nl})\, \mathrm{d}V - \frac{1}{2}\iint (A_{ijmn} R_{i,j} R_{m,n} + A_{ijkmnl} R_{i,jk} R_{m,nl})\, \mathrm{d}F. \tag{14}$$

The coefficients A_{ijkmnl} are the components of the sixth rank tensor of moduli of the gradient elastic adhesion theory.

The formal approach used above allows us to consider the theory of adhesion determined by the use of the Lagrangian (14) as a natural generalization of the theory of adhesion formulated for the classical theory of elasticity (10)–(12). The proposed gradient adhesion theory is completely determined by the variational equation that results from the principle of Lagrange with the Lagrangian (14). The following statements are valid:

1. The equilibrium equations in the volume represent Eq. (12).
2. The generalized equations of Hooke's law in the volume of the body are determined by Eq. (9).
3. The generalized constitutive equations on the surface can be found with the help of Green's formulas and have the form:

$$a_{ij} = \frac{\partial U_F(R_{i,j}, R_{i,jk})}{\partial R_{i,j}} = A_{ijmn} R_{m,n}, \qquad b_{ijk} = \frac{\partial U_V(R_{i,j}, R_{i,jk})}{\partial R_{i,jk}} = A_{ijkmnl} R_{m,nl}, \tag{15}$$

   where a_{ij} are the components of the adhesion stress tensor, b_{ijk} are the components of the adhesion double stress tensor.
4. The tensor of the fourth-rank adhesive moduli A_{ijmn} depending on eight elastic constants has the form of (11), and the sixth rank tensor A_{ijkmnl} has the following structure:

$$\begin{aligned} A_{ijkmnl} = A_1 \Big(& \delta^*_{ij}\delta^*_{km}\delta^*_{nl} + \delta^*_{mn}\delta^*_{li}\delta^*_{jk} + \delta^*_{ij}\delta^*_{kn}\delta^*_{ml} + \delta^*_{mn}\delta^*_{lj}\delta^*_{ik} + \delta^*_{ij}\delta^*_{kl}\delta^*_{mn} \\ & + \delta^*_{ik}\delta^*_{jm}\delta^*_{nl} + \delta^*_{ml}\delta^*_{ni}\delta^*_{jk} + \delta^*_{in}\delta^*_{km}\delta^*_{jl} + \delta^*_{mj}\delta^*_{li}\delta^*_{nk} + \delta^*_{in}\delta^*_{lk}\delta^*_{jm} \\ & + \delta^*_{ik}\delta^*_{jn}\delta^*_{ml} + \delta^*_{il}\delta^*_{km}\delta^*_{nj} \Big) \\ & + A_2(\delta^*_{im}\delta^*_{kj}\delta^*_{nl} + \delta^*_{im}\delta^*_{lj}\delta^*_{nk} + \delta^*_{im}\delta^*_{nj}\delta^*_{kl}) \\ & + A_3(n_i n_j \delta^*_{km}\delta^*_{nl} + n_m n_n \delta^*_{li}\delta^*_{jk} + n_i n_j \delta^*_{kn}\delta^*_{ml} + n_m n_n \delta^*_{lj}\delta^*_{ik} + n_i n_j \delta^*_{kl}\delta^*_{mn} \\ & + n_m n_n \delta^*_{lk}\delta^*_{ij}) \\ & + A_4(n_i n_n \delta^*_{km}\delta^*_{jl} + n_m n_j \delta^*_{li}\delta^*_{nk} + n_i n_n \delta^*_{ml}\delta^*_{jk} + n_m n_j \delta^*_{ik}\delta^*_{nl} + n_i n_n \delta^*_{lk}\delta^*_{jm} \\ & + n_m n_j \delta^*_{kl}\delta^*_{ni}) \\ & + A_5(n_i n_m \delta^*_{kj}\delta^*_{nl} + n_i n_m \delta^*_{lj}\delta^*_{nk} + n_i n_m \delta^*_{nj}\delta^*_{kl}) \\ & + A_6(n_j n_n \delta^*_{ik}\delta^*_{ml} + n_j n_n \delta^*_{il}\delta^*_{km}) \\ & + A_7 n_j n_n \delta^*_{im}\delta^*_{kl} \\ & + A_8 \delta^*_{kl} n_i n_j n_m n_n. \end{aligned} \tag{16}$$

5. Both the natural, von Neumann-like, and the essential, Dirichlet-like, boundary conditions on F are written as

$$P_i^F - (\sigma_{ij} - \mu_{ijk,k})n_j + (\mu_{ijk}n_k + a_{ij} - b_{ijk,k})_{,p}\delta^*_{pj} = 0, \quad \delta^*_{pj} = \delta_{pj} - n_p n_j, \tag{17}$$

$$(\mu_{ijk}n_k + a_{ij} - b_{ijk,k})n_i = q_j$$

   and

$$R_i = \bar{R}_i, \qquad R_{i,j}n_j = \overline{\partial R_i / \partial n}.$$

The dimension of the tensor of the sixth rank adhesive moduli is different from the dimension of the tensor of the fourth-rank adhesive moduli on the square of length dimension. Therefore, after applying the normalization procedure $\bar{A}_{ijkmnl} = A_{ijkmnl}/(A/C)^3$, the Lagrangian (14) can be represented as a cubic polynomial in the parameter l (characteristic length of scale effects):

$$
\begin{aligned}
L &= \left(A_P - \frac{1}{2}\iiint C_{ijmn} R_{i,j} R_{m,n}\,\mathrm{d}V\right) + \left(-\frac{1}{2}\iint \bar{A}_{ijmn} R_{i,j} R_{m,n}\,\mathrm{d}F\right) l \\
&\quad + \left(-\frac{1}{2}\iiint \bar{C}_{ijkmnl} R_{i,jk} R_{m,nl}\,\mathrm{d}V\right) l^2 + \left(-\frac{1}{2}\iint \bar{A}_{ijkmnl} R_{i,jk} R_{m,nl}\,\mathrm{d}F\right) l^3 \\
&= L_0 l^0 + L_1 l^1 + L_2 l^2 + L_3 l^3. \qquad (18)
\end{aligned}
$$

Thus in general, the gradient theories of adhesion of gradient media allow to take into account the scale effects up to the third order, see Eq. (18).

5 On the Unified Nature of Cohesive–Adhesive Interactions

Let us consider the density of potential energy of the Mindlin–Toupin gradient elasticity (Mindlin and Eshel 1968) with the tensor of the sixth order elastic moduli C_{ijkmnl} that are written with five physical constants for isotropic media (Gusev and Lurie 2015). It can be proved (Belov and Lurie 2014) that the density of potential energy of the gradient elasticity can be represented as a sum of positive definite quadratic forms and a divergence term.

$$
\begin{aligned}
C_{ijkmnl} R_{i,jk} R_{m,nl} &= (2\mu + \lambda) l_\theta^2 R_{i,ik} R_{j,jk} + \mu l_\omega^2 (\Delta R_k - R_{i,ik})(\Delta R_k - R_{j,jk}) \\
&\quad + \{2C_1 R_{i,i}(\Delta R_k - R_{j,jk}) \\
&\quad + 2C_2 [R_{i,j}(R_{k,ji} + R_{j,ki}) - R_{k,j}(R_{i,ij} + \Delta R_j)] \\
&\quad + 2C_3 (R_{i,j} R_{i,jk} - R_{i,k} \Delta R_i)\}_{,k}. \qquad (19)
\end{aligned}
$$

Here l_θ^2, l_ω^2, C_1, C_2, C_3 are five independent strain gradient coefficients which are linear combinations of five components of the Mindlin–Toupin tensor for isotropic media.

The divergence term is determined by the divergence of the vector (19). This term can be interpreted on one hand as a part of the volume density of potential energy, and on the other hand as part of the density of the surface potential energy (potential energy of adhesion). For this, it is enough to use the Ostrogradskii–Gauß theorem.

It is important to note that the divergence term is of second order in the decomposition of the scale parameter. However, in Eqs. (6), (13), (17) given above, there are no terms corresponding to the decomposition of the second order with respect to the scale parameter and associated with the density of potential energy of adhesion. Indeed, the density of the potential energy of perfect adhesion gives the first order in the decomposition in the scale parameter, and the density of potential energy of gradient adhesion gives, respectively, the third order. Therefore, it is natural in gradient theories to take into account the adhesion potential energy defined by the bilinear form of the first and second derivatives of the displacements $A_{ijmnl} R_{i,j} R_{m,nl}$, which also gives the second order in the decomposition in the scale parameter. The Lagrangian of such generalized theory has the following structure:

$$
\begin{aligned}
L = A_P - \frac{1}{2} \iiint (C_{ijmn} R_{i,j} R_{m,n} + C_{ijkmnl} R_{i,jk} R_{m,nl})\, \mathrm{d}V \\
- \frac{1}{2} \iint (A_{ijmn} R_{i,j} R_{m,n} + 2A_{ijmnl} R_{i,j} R_{m,nl} + A_{ijkmnl} R_{i,jk} R_{m,nl})\, \mathrm{d}F. \quad (20)
\end{aligned}
$$

Through the proper procedure of normalization $\bar{A}_{ijmnl} = A_{ijmnl}/(A/C)^2$, the Lagrangian of the theory (20) can also be represented as a finite decomposition in the characteristic length of scale effects l:

$$
\begin{aligned}
L &= (A_P - \frac{1}{2} \iiint C_{ijmn} R_{i,j} R_{m,n} \mathrm{d}V) + (-\frac{1}{2} \iint \bar{A}_{ijmn} R_{i,j} R_{m,n}\, \mathrm{d}F) l \\
&+ \left(-\frac{1}{2} \iiint \bar{C}_{ijkmnl} R_{i,jk} R_{m,nl} \mathrm{d}V - \iint \bar{A}_{ijmnl} R_{i,j} R_{m,nl}\, \mathrm{d}F\right) l^2 \\
&+ \left(-\frac{1}{2} \iint \bar{A}_{ijkmnl} R_{i,jk} R_{m,nl}\, \mathrm{d}F\right) l^3 \\
&= L_0 l^0 + L_1 l^1 + L_2 l^2 + L_3 l^3. \quad (21)
\end{aligned}
$$

In this version of the theory (21), the Lagrangian is a cubic polynomial of the characteristic length of scale effects, the term with L_2 contains not only the quadratic form of the second derivatives in the volume

$$
\iiint \bar{C}_{ijkmnl} R_{i,jk} R_{m,nl}\, \mathrm{d}V,
$$

but also a bilinear form of first and second derivatives on the surface

$$
\iint \bar{A}_{ijmnl} R_{i,j} R_{m,nl} \mathrm{d}F.
$$

Obviously, these terms are quite equivalent and the neglect of the adhesive effects of the second one can not be considered as reasonable. On the other hand, this analysis is the argumentation of generalization in the theory of adhesion which involves the consideration of the deformation energy

$$
\iint \bar{A}_{ijmnl} R_{i,j} R_{m,nl}\, \mathrm{d}F
$$

defined on the adhesive-active surface. We assume that if the potential energy of gradient theory contains the term

$$l^2 \iiint \bar{C}_{ijkmnl} R_{i,jk} R_{m,nl} \, \mathrm{d}V,$$

then there should be taken into account the equitable terms

$$l \iint \bar{A}_{ijmn} R_{i,j} R_{m,n} \, \mathrm{d}F$$

and

$$l^2 \iint \bar{A}_{ijmnl} R_{i,j} R_{m,nl} \, \mathrm{d}F.$$

It can be shown that the general structure of the tensor of the fifth rank adhesive modules is as follows:

$$\begin{aligned} A_{ijmnl} = {} & G_1(n_i \delta^*_{jm} \delta^*_{nl} - n_m \delta^*_{jl} \delta^*_{in} + n_i \delta^*_{jn} \delta^*_{lm} - n_m \delta^*_{jn} \delta^*_{li} + n_i \delta^*_{jl} \delta^*_{mn} - n_m \delta^*_{ji} \delta^*_{nl}) \\ & + G_2(n_j \delta^*_{in} \delta^*_{lm} - n_n \delta^*_{jm} \delta^*_{il} + n_j \delta^*_{il} \delta^*_{mn} - n_n \delta^*_{ji} \delta^*_{lm}) \\ & + G_3(n_j \delta^*_{im} \delta^*_{nl} - n_n \delta^*_{jl} \delta^*_{mi}) + \\ & + G_4(\delta^*_{il} n_j n_m n_n - \delta^*_{ml} n_i n_j n_n) \\ & + G_5(\delta^*_{jl} n_i n_m n_n - \delta^*_{nl} n_i n_j n_m). \end{aligned}$$

Note that for this generalized theory of adhesion, both the equilibrium equation and the static boundary conditions retain the form (12), (17). However, the constitutive equations change on the surface (Hooke's law). They have a more general form:

$$a_{ij} = A_{ijmn} R_{m,n} + A_{ijmnl} R_{m,nl}, \qquad b_{ijk} = A_{mnijk} R_{m,n} + A_{ijkmnl} R_{m,nl}. \tag{22}$$

Therefore, taking into account the relations (12), (17), (22), the whole statement of boundary value problems in displacements will be different.

Note that in this theory the Lagrangian is a cubic polynomial of the characteristic length of scale effects. However, the term L_2 contains not only the quadratic form of the second derivatives in volume, but also a bilinear form of first and second derivatives on the surface. Note also that formally we cannot neglect the adhesive effects of third order. It can be easily seen that the significant adhesive corrections of the second order can exist only in the presence of quadratic terms due to the requirement of positive definiteness of the potential energy. Consequently, the adhesive effects of third order should be taken into account when we consider gradient problems so that it would be possible to describe the adhesion second order effects.

6 Discussion and Conclusions

With formulations of non-classical theories we consider that the question of their physical objectivity arises (Liu 2009; Murdoch 2000, 2003, 2005; Gurtin et al. 2010). The problem of the material frame indifference of the generalized Murdoch-Gurtin theory of adhesion (see (4) with condition $\delta^F = 0$) was mentioned earlier in Sect. 2. The following theorem is valid:

Theorem 13.3 *For generalized theories of adhesion, the material frame indifference is always satisfied.*

To prove this statement, note that the problem reduces to the analysis of invariance of the Lagrangian with respect to translation and rotation of the body as a rigid body.

The invariance of any physically linear theory reduces to the invariance of the quadratic distortion of the Lagrangian:

$$L = A_P - \frac{1}{2}\iiint C_{ijmn} R_{i,j} R_{m,n}\,\mathrm{d}V - \frac{1}{2}\iint A_{ijmn} R_{i,j} R_{m,n}\,\mathrm{d}F. \tag{23}$$

All the terms of the Lagrangian that contain the second derivatives of the displacements, are automatically invariant under translations and rotations of the body as a rigid body.

Let us define a random field of body displacements in the form of the decomposition:

$$\begin{gathered} R_i = R_i^0 - \omega_p^0 (x_q - x_q^0) e_{pqi} + u_i, \\ R_{i,j} = -\omega_p^0 e_{pij} + u_{i,j}, \qquad R_{m,n} = -\omega_q^0 e_{qmn} + u_{m,n}. \end{gathered} \tag{24}$$

Substituting (24) into (23) we obtain:

$$L(R_i) = P_i^0 R_i^0 + m_k^0 \omega_k^0 - \frac{1}{2} D_{pq} \omega_p^0 \omega_q^0 + L(u_i), \tag{25}$$

here:

$$\begin{gathered} \iiint P_i^V\,\mathrm{d}V + \iint P_i^F\,\mathrm{d}F = P_i^0, \\ \iiint P_i^V (x_q - x_q^0) e_{pqi}\,\mathrm{d}V + \iint [P_i^F (x_q - x_q^0) e_{pqi}]\,\mathrm{d}F = M_p^0, \\ \iiint P_i^V u_i\,\mathrm{d}V + \iint P_i^F u_i\,\mathrm{d}F - \frac{1}{2}\iiint C_{ijmn} u_{i,j} u_{m,n}\,\mathrm{d}V \\ - \frac{1}{2}\iint [A_{ijmn} u_{i,j} u_{m,n}]\,\mathrm{d}F = L(u_i), \\ M_k^0 + \iiint [C_{ijmn} e_{kmn} u_{i,j}]\,\mathrm{d}V + \iint [A_{ijmn} e_{kmn} u_{i,j}]\,\mathrm{d}F = m_k^0, \\ C_{ijmn} e_{pij} e_{qmn} V + \iint [A_{ijmn} e_{pij} e_{qmn}]\mathrm{d}F = D_{pq}. \end{gathered} \tag{26}$$

The use of the Lagrange principle $\delta L(R_i) = 0$ and Eq. (25) leads to the equations:

$$P_i^0 = 0, \qquad m_p^0 - D_{pq}\omega_q^0 = 0, \qquad \delta L(u_i) = 0. \tag{27}$$

Taking into account Eq. (27) we rewrite the Lagrangian (25) in the form of:

$$L(R_i) = \frac{1}{2} D_{ij}^{-1} m_i^0 m_j^0 + L(u_i), \qquad D_{ik}^{-1} D_{kj} = \delta_{ij}. \tag{28}$$

Therefore, according to equation (28) the necessary and sufficient conditions of invariance of the Lagrangian are to perform isoperimetric equations:

$$m_k^0 = M_k^0 + \iiint [C_{ijmn} e_{kmn} u_{i,j}] \, \mathrm{d}V + \iint [A_{ijmn} e_{kmn} u_{i,j}] \, \mathrm{d}F = 0. \tag{29}$$

The isoperimetric conditions (29) have a physical meaning of generalized global equations of the equilibrium of moments. They may be satisfied as well before the solutions constructing as after it. Before the solution constructing, they are met by introducing them as isoperimetric conditions on the pseudo-vector of undefined Lagrange multipliers according to the known procedure. It is important to note that these conditions of orthogonality (isoperimetric relations) do not violate the algorithm for the construction of the solutions of boundary value problems. Indeed, these relationships can always be satisfied after the construction of the boundary value problem solution R_i and substitution of the following relation to the isoperimetric expressions:

$$\bar{R}_i = R_i - R_i^0 + \omega_m^0 (x_n - x_n^0) e_{mni}.$$

As a result, after integration, we obtain a linear system of algebraic equations to determine the values R_i^0 and ω_m^0 that provide the orthogonality of the space of kinematic states of an elastic body $\bar{R}_i$ to the spaces of translations and rotations of the body as a rigid body.

The article describes a sequence of the theories of adhesion of ideal (defect-free) surfaces of bodies of varying degree of generality. We indicated the relation between the models of adhesion and gradient models of media and offered a new classification of theories of adhesion. It was proved that the adhesive and cohesive (gradient) effects should be considered together. The proposed classification shows that the presence of the moduli of different dimensions determines the scale effects of different orders, depending on what physical properties this medium is endowed with. The study of the behavior of the body under gradient models should be carried out not only considering the adhesion effects of first order, but also the adhesion effects of second and even third order. Otherwise, taking into account the gradient effects can produce a smaller contribution to the solution compared with the contribution of unrecorded adhesion effects.

Acknowledgments This work was supported by the Russian Foundation for Basic Research project No. 15-01-03649-a.

References

Altan BS, Aifantis EC (1992) On the structure of the mode iii crack-tip in gradient elasticity. Scripta Met 26:319–324

Altenbach H, Eremeyev VA, Lebedev LP (2011) On the spectrum and stiffness of an elastic body with surface stresses. ZAMM 91(9):699–710

Belov PA, Lurie SA (2007) Theory of ideal adhesion interactions. J Compos Mech Des 14:545–561

Belov PA, Lurie SA (2009) Continual theory of adhesion interactions of damaged media. J Compos Mech Des 15(4):610–629

Belov PA, Lurie SA (2014) Mathematical theory of damaged media. Gradient theory of elasticity. Formulations hierarchy comparative analysis. Palmarium Academic Publishing, Germany

Duan HL, Wang J, Huang ZP, Karihaloo BL (2005) Size-dependent effective elastic constants of solids containing nanoinhomogeneities with interface stress. J Mech Phys Solids 53:1574–1596

Duan HL, Wang J, Karihaloo BL (2008) Theory of elasticity at the nanoscale. In: Aref H, van der Giessen E (eds) Advances in applied mechanics, vol 42. Elsevier, Amsterdam, pp 1–68

Eremeyev VA (2016) On effective properties of materials at the nano- and microscales considering surface effects. Acta Mech 227(1):29–42

Eremeyev VA, Altenbach H, Morozov NF (2009) The influence of surface tension on the effective stiffness of nanosize plates. Dokl Phys 54(2):98–100

Gao XL, Park SK (2007) Variational formulation of a simplified strain gradient elasticity theory and its application to a pressurized thick-walled cylinder problem. Int J Solids Struct 44:7486–7499

Gurtin ME, Murdoch AI (1975a) Addenda to our paper a continuum theory of elastic material surfaces. Arch Ration Mech Anal 59(4):389–390

Gurtin ME, Murdoch AI (1975b) A continuum theory of elastic material surfaces. Arch Ration Mech Anal 57(4):291–323

Gurtin ME, Fried E, Anand L (2010) The mechanics and thermodynamics of Continua. Cambridge University Press, New York

Gusev AA, Lurie SA (2015) Symmetry conditions in strain gradient elasticity. Math Mech Solids Math pp 1–9

Huang Z, Wang J (2013) Micromechanics of nanocomposites with interface energy effect. In: Li S, Gao XL (eds) Handbook on micromechanics and nanomechanics, vol 42. Pan Stanford Publishing, Stanford, pp 303–348

Javili A, Steinmann P (2010) On thermomechanical solids with boundary structures. Int J Solids Struct 47(24):3245–3253

Javili A, McBride A, Steinmann P (2012) Thermomechanics of solids with lower-dimensional energetics: on the importance of surface, interface, and curve structures at the nanoscale. A unifying review. Appl Mech Rev 65(10):802–1–3

Kim CI, Schiavone P, Ru CQ (2011) Effect of surface elasticity on an interface crack in plane deformations. Proc Roy Soc A 467(2136):3530–3549

Liu IS (2009) Continuum mechanics. UNESCO Publications, Oxdford

Lurie S, Tuchkova N (2009) A continuous adhesion model for deformed solid bodies and media with nanostructures. Kompoz Nanostruct 2(2):25–43

Lurie S, Belov PA, Tuchkova NP (2010) Gradient theory of media with conserved dislocations: application to microstruc-tured materials. In: Maugin GA, Metrikine AV (eds) One hundred years after the Cosserats, advances in mechanics and mathematics, vol 21. Springer, Heidelberg, pp 223–234

Lurie SA, Belov PA (2008) Cohesion field: Barenblatt's hypothesis as formal corollary of theory of contonuous media with conserved dislocations. Int J Fract 50(1–2):181–194

Lurie SA, Volkov-Bogorodsky DB, Zubov VI, Tuchkova NP (2009) Advanced theoretical and numerical multiscale modeling of cohesion/adhesion interactions in continuum mechanics and its applications for filled nanocomposites. Int J Compos Mater Sci 45(3):709–741

Mindlin RD (1965) Second gradient of strain and surface-tension in linear elasticity. Int J Solids Struct 1(4):417–439

Mindlin RD, Eshel NN (1968) On first strain-gradient theories in linear elasticity. Int J Solids Struct 4:109–112

Murdoch AI (2000) On objectivity and material symmetry for simple elastic solids. J Elast. 60:233–242

Murdoch AI (2003) Objectivity in classical continuum physics: a rationale for discarding the principle of invariance under superposed rigid body motions in favor of purely objective considerations. Continuum Mech Thermodyn 15:309–320

Murdoch AI (2005) On criticism of the nature of objectivity in classical continuum physics. Continuum Mech Thermodyn 17:135–148

Ostoja-Starzewski M (2002) Lattice models in micromechanics. Appl Mech Rev 55(1):35–59

Povstenko Y (2008) Mathematical modeling of phenomena caused by surface stresses in solids. In: Altenbach H, Morozov NF (eds) Surface effects in solid mechanics, vol 42. Springer, Heidelberg, pp 135–153

Steigmann DJ, Ogden RW (1997a) Plane deformations of elastic solids with intrinsic boundary elasticity. Proc Roy Soc A 453(1959):853–877

Steigmann DJ, Ogden RW (1997b) Plane deformations of elastic solids with intrinsic boundary elasticity. Proc Roy Soc A 453(1959):853–877

Steigmann DJ, Ogden RW (1999a) Elastic surface-substrate interactions. Proc Roy Soc A 455(1999):437–474

Steigmann DJ, Ogden RW (1999b) Elastic surface-substrate interactions. Proc Roy Soc A 455(1982):437–474

Zhu HX, Wang JX, Karihaloo BL (2009) Effects of surface and initial stresses on the bending stiffness of trilayer plates and nanofilms. J Mech Mater Struct 4(3):589–604

Eigenvalue Problems of a Tensor and a Tensor-Block Matrix (TMB) of Any Even Rank with Some Applications in Mechanics

Mikhail U. Nikabadze

Abstract In this work, the eigenvalue problems of the symmetric tensor-block matrix of any even rank and sizes $m \times m, m \geq 1$ (sizes 2×2) is formulated (studied). Some definitions and theorems are formulated concerning the tensor-block matrix. Formulas expressing the classical invariants of the tensor-block matrix of any even rank and sizes 2×2 through the first invariants of the powers of this tensor-block matrix are given. We also obtain formulas which are inverse to the latter. A complete orthonormal system of eigentensor columns for the tensor-block matrix of any even rank and sizes 2×2 is constructed. We formulate the generalized eigenvalue problems of the tensor-block matrix. As a special case, we consider the tensor-block matrix of the elastic modulus tensors. The canonical representation of the tensor-block matrix is given. Using this representation, we get the canonical forms of the elastic strain energy and the constitutive relations. Besides, a classification of the micropolar linear elastic anisotropic bodies that do not have a center of symmetry is given.

Keywords Block matrix · Tensor-block matrix · Tensor column · Tensor row · Eigentensor · Eigenvalue problem · Generalized eigenvalue problem · Orthonormal system · Complete orthonormal system · Eigentensor column · Eigentensor row · Symbol of anisotropy · Symbol of structure

1 Introduction

The eigenmoduli (eigenvalues) and eigenstates (eigentensors) for isotropic materials are known since Stokes (see in Love 2013). Under other names, these terms were introduced by Kelvin (Rychlewski 1984; Todhunter and Pearson 1960) for anisotropic materials in the mid-nineteenth century. However, these terms were forgotten for a long time; only about 30 years ago, researchers turned to this prob-

M.U. Nikabadze (✉)
Lomonosov Moscow State University, Moscow, Russia
e-mail: munikabadze@yandex.ru

H. Altenbach and S. Forest (eds.), *Generalized Continua as Models for Classical and Advanced Materials*, Advanced Structured Materials 42,
DOI 10.1007/978-3-319-31721-2_14

lem again (Rychlewski 1983, 1984; Minkevich 1973; Tolokonnikov and Matchenko 1974; Alexandrov 1967; Lur'e 1979; Chanyshev 1984a, b; Revuzhenko et al. 1985; Ostrosablin 1984, 1986b, a, 2000; Annin and Ostrosablin 2008), see also Chen (1984), Mehrabadi and Cowin (1990, 1991), Theocaris (1989) Theocaris and Philippidis (1989, 1990, 1991), Sutcliffe (1992). In Ostrosablin (1986a, 2000), a classification of anisotropic linear elastic classical materials is proposed. The eigenvalue and eigentensor problem is considered in Nikabadze (2008, 2009a, b, 2015), Vekua (1978) for a temperature of even rank. The concept of eigenstates is used to develop the theory of plasticity (Pobedrya 1984, 1990) and the flow theory (Pobedrya 1986). However, publications on this problem for micropolar materials, except Nikabadze (2014, 2015), are not known to the author of this paper.

2 On Tensors of Module $\mathbb{R}_{2p}(\Omega)$

Let us consider some questions related to the tensors of the module $\mathbb{R}_{2p}(\Omega)$. They will help us in the future presentation of the material. Here, $\mathbb{R}_{2p}(\Omega)$ is a set of real $2p$th rank tensors, p is some non-negative integer, Ω is some domain of the n-dimensional Riemannian space $\mathbb{R}^n$. Therefore, $\mathbb{R}_{2p}(\Omega)$ and $\mathbb{R}_p(\Omega)$ (a set of real pth rank tensors) are the modules over the ring of scalars $\mathbb{R}_0(\Omega)$ (a set of real zero-rank tensors), i.e. $\mathbb{R}_{2p}(\Omega)$ and $\mathbb{R}_p(\Omega)$ are $\mathbb{R}_0(\Omega)$-module (Nikabadze 2009a, 2015; Vekua 1978). Tensors of the module $\mathbb{R}_{2p}(\Omega)$ can be represented in different multibases. For example, if ${}^{2p}\mathbf{A}$ is a tensor of the module $\mathbb{R}_{2p}(\Omega)$, then it may have different representations:

$$
{}^{2p}\mathbf{A} = A^{i_1 i_2 \cdots i_p}{}_{j_1 j_2 \cdots j_p} \mathbf{R}_{i_1 i_2 \cdots i_p} \mathbf{R}^{j_1 j_2 \cdots j_p} = A^{i\cdot}_{\cdot j} \mathbf{R}_i \mathbf{R}^j, \tag{1}
$$

$$
\begin{aligned}
&\mathbf{R}_{i_1 \cdots i_p} = \mathbf{r}_{i_1} \ldots \mathbf{r}_{i_p}, \quad \mathbf{R}^{j_1 \cdots j_p} = \mathbf{r}^{j_1} \cdots \mathbf{r}^{j_p}, \quad i_1, \ldots, i_p, j_1, \ldots, j_p = \overline{1, n}, \quad \mathbf{r}_s \cdot \mathbf{r}^t = \delta_s^t, \\
&s, t = \overline{1, n}, \quad i = \mathrm{N}^{\circ}\{i_1, \ldots, i_p\}, \ \ldots, j = \mathrm{N}^{\circ}\{j_1, \ldots, j_p\}, \\
&\mathbf{R}_i \overset{p}{\otimes} \mathbf{R}^j = \delta_i^j, \quad i, j = \overline{1, N}, \quad N = n^p.
\end{aligned}
$$

Here δ_i^j is Kronecker delta; $\overset{p}{\otimes}$ is the inner p-product (Nikabadze 2009a, 2015; Vekua 1978). Therefore, $\mathbf{r}_s$ and $\mathbf{r}^t$, $s, t = \overline{1, n}$, are the biorthonormal systems of bases with respect to the scalar product or inner 1-product $\overset{1}{\otimes}$; $\mathbf{R}_i$, $\mathbf{R}^j$, $i, j = \overline{1, N}$, are the biorthonormal systems of multibases or bases of module $\mathbb{R}_p(\Omega)$ with respect to the operation of the inner p-product (p-times inner product), $\overline{1, m} = 1, 2, \ldots, m$.

Note that for the tensors of module $\mathbb{R}_p(\Omega)$, we apply the p-index or one-index representations. So if ${}^p\mathbf{U} \in \mathbb{R}_p(\Omega)$, then

$$
\begin{aligned}
&{}^p\mathbf{U} = U_{i_1 i_2 \ldots i_p} \mathbf{R}^{i_1 i_2 \cdots i_p} = U^{i_1 i_2 \ldots i_p} \mathbf{R}_{i_1 i_2 \cdots i_p} = U_i \mathbf{R}^i = U^i \mathbf{R}_i, \\
&i_1, i_2, \ldots, i_p = \overline{1, n}, \quad i = \overline{1, N}.
\end{aligned}
$$

Note that Eq. (1) give ($2p$)-index and two-index representations of a tensor ${}^{2p}\mathbf{A}$ of the module $\mathbb{R}_{2p}(\Omega)$. The rank of the tensor does not change under the two-index representation. We also note that the numbering of the elements of the set of numerical sequences $\{i_1, \ldots, i_p\}, i_1, \ldots, i_p = \overline{1, n}$, can be produced, for example, as follows: if $i = \text{N}^{\text{o}}\{i_1, \ldots, i_p\}$ then $i = i_1 + \sum_{k=2}^{p} n^{k-1}(i_k - 1)$ or $i = i_p + \sum_{k=1}^{p-1} n^k(i_{p-k} - 1)$, $i_1, \ldots, i_p = \overline{1, n}$.

Let us introduce some definitions.

Definition 14.1 The tensor of the module $\mathbb{R}_{2p}(\Omega)$ denoted by ${}^{2p}\mathbf{A}^T$ and defined by formula

$$\begin{aligned} {}^{2p}\mathbf{A}^T &= (A^{i}_{\cdot j}\mathbf{R}_i\mathbf{R}^j)^T \\ &= A^{i}_{\cdot j}\mathbf{R}^j\mathbf{R}_i = A_{j}^{\cdot i}\mathbf{R}_i\mathbf{R}^j = A^{i_1 i_2 \cdots i_p}_{ j_1 j_2 \cdots j_p}\mathbf{R}^{j_1 j_2 \cdots j_p}\mathbf{R}_{i_1 i_2 \cdots i_p},\ i, j = \overline{1, N}, \end{aligned} \tag{2}$$

$$N = n^p,\ i = \text{N}^{\text{o}}\{i_1, i_2, \ldots, i_p\},\ j = \text{N}^{\text{o}}\{j_1, j_2, \ldots, j_p\}\ \ i_1, i_2, \ldots, i_p, j_1, j_2, \ldots, j_p = \overline{1, n},$$

is called transposed tensor to ${}^{2p}\mathbf{A} \in \mathbb{R}_{2p}(\Omega)$.

Definition 14.2 The tensor of the module $\mathbb{R}_{2p}(\Omega)$, which is equal to its transposed tensor is called symmetric.

Definition 14.3 The tensor of the module $\mathbb{R}_{2p}(\Omega)$, which commutes with its transposed tensor is called normal tensor.

Based on Definition of 14.3, we conclude that a symmetric tensor of module $\mathbb{R}_{2p}(\Omega)$ is normal.

In the module $\mathbb{R}_{2p}(\Omega)$ besides operations of the addition of tensors and multiplication of a tensor by a scalar, the operations of the inner p-product and the inner ($2p$)-product (scalar product) of two tensors can be introduced. Let us define these operations. Let ${}^{2p}\mathbf{A}$ and ${}^{2p}\mathbf{B}$ be two tensors of the module $\mathbb{R}_{2p}(\Omega)$.

Definition 14.4 The inner p-product of tensors ${}^{2p}\mathbf{A}$ and ${}^{2p}\mathbf{B}$ of the module $\mathbb{R}_{2p}(\Omega)$ is said to be a tensor denoted by ${}^{2p}\mathbf{A} \overset{p}{\otimes} {}^{2p}\mathbf{B}$, whose components are defined as follows:

$$({}^{2p}\mathbf{A} \overset{p}{\otimes} {}^{2p}\mathbf{B})^{i_1 i_2 \cdots i_p}_{ j_1 j_2 \cdots j_p} = A^{i_1 i_2 \cdots i_p}_{ k_1 k_2 \cdots k_p} B^{k_1 k_2 \cdots k_p}_{ j_1 j_2 \cdots j_p}. \tag{3}$$

With (3), we have

$${}^{2p}\mathbf{A} \overset{p}{\otimes} {}^{2p}\mathbf{B} = A^{i_1 i_2 \cdots i_p}_{ k_1 k_2 \cdots k_p} B^{k_1 k_2 \cdots k_p}_{ j_1 j_2 \cdots j_p}\mathbf{R}_{i_1 i_2 \cdots i_p}\mathbf{R}^{j_1 j_2 \cdots j_p}.$$

From (3), it is seen that only tensors of rank not less than p can take part in the inner p-product operation. The number p is called the multiplicity of the inner p-product. If $p = 0$, i.e. if there is no reduction of the indices under the product of the tensors, then the product is called the direct product. In this case, instead of the ${}^{2p}\mathbf{A} \overset{0}{\otimes} {}^{2p}\mathbf{B}$, we write ${}^{2p}\mathbf{A} \otimes {}^{2p}\mathbf{B}$. If ${}^p\mathbf{A}$ and ${}^p\mathbf{B}$ are p-rank tensors of the module $\mathbb{R}_p(\Omega)$, then under inner p-product of these tensors we often omit a sign $\overset{p}{\otimes}$ and simply write ${}^p\mathbf{A}{}^p\mathbf{B}$ or $({}^p\mathbf{A}, {}^p\mathbf{B})$. In this case, the inner p-product is called the inner or scalar product. It is, of course, expressed by the formula

$$
{}^p\mathbf{A}{}^p\mathbf{B} = ({}^p\mathbf{A}, {}^p\mathbf{B}) = A_{i_1 i_2 \cdots i_p} B^{i_1 i_2 \cdots i_p}. \tag{4}
$$

Therefore, as in (4) we have

$$
{}^{2p}\mathbf{A}{}^{2p}\mathbf{B} = ({}^{2p}\mathbf{A}, {}^{2p}\mathbf{B}) = {}^{2p}\mathbf{A} \overset{2p}{\otimes} {}^{2p}\mathbf{B} = A^{i_1 i_2 \cdots i_p}{}_{j_1 j_2 \cdots j_p} B_{i_1 i_2 \cdots i_p}{}^{j_1 j_2 \cdots j_p}.
$$

It should be noted that if the inner product ${}^m\mathbf{A}{}^m\mathbf{B}$ becomes zero for any tensor ${}^m\mathbf{B}$, then ${}^m\mathbf{A} = \mathbf{0}$. Here, ${}^m\mathbf{A}$ and ${}^m\mathbf{B}$ are tensors of the module $\mathbb{R}_m(\Omega)$, where m is an arbitrary non-negative integer.

Definition 14.5 Two tensors ${}^m\mathbf{A}$ and ${}^m\mathbf{B}$ of the module $\mathbb{R}_m(\Omega)$, where m is an arbitrary non-negative integer, are called orthogonal if their inner product is zero, i.e. $({}^m\mathbf{A}, {}^m\mathbf{B}) = 0$.

Note that there is the unit tensor in the module $\mathbb{R}_{2p}(\Omega)$ with respect to the inner p-product introduced above. Indeed, in the module $\mathbb{R}_{2p}(\Omega)$ the $(2p)$-rank unit tensor is the following tensor

$$
\begin{aligned}
{}^{2p}\mathbf{E} &= \mathbf{R}_{i_1 i_2 \cdots i_p} \mathbf{R}^{i_1 i_2 \cdots i_p} = \mathbf{r}_{i_1} \mathbf{r}_{i_2} \cdots \mathbf{r}_{i_p} \mathbf{r}^{i_1} \mathbf{r}^{i_2} \cdots \mathbf{r}^{i_p} = \mathbf{R}_i \mathbf{R}^i = g_i^j \mathbf{R}^i \mathbf{R}_j \\
&= g_{i_1}^{j_1} g_{i_2}^{j_2} \cdots g_{i_p}^{j_p} \mathbf{r}^{i_1} \mathbf{r}^{i_2} \cdots \mathbf{r}^{i_p} \mathbf{r}_{j_1} \mathbf{r}_{j_2} \cdots \mathbf{r}_{j_p}, \\
& i_1, i_2, \ldots, i_p, j_1, j_2, \ldots, j_p = \overline{1, n}, \quad i, j = \overline{1, N}.
\end{aligned} \tag{5}
$$

It is easy to prove that for any tensor ${}^{2p}\mathbf{A} \in \mathbb{R}_{2p}(\Omega)$ we have

$$
{}^{2p}\mathbf{A} \overset{p}{\otimes} {}^{2p}\mathbf{E} = {}^{2p}\mathbf{E} \overset{p}{\otimes} {}^{2p}\mathbf{A} = {}^{2p}\mathbf{A}. \tag{6}
$$

Using ${}^{2p}\mathbf{E}$, the first invariant of the tensor ${}^{2p}\mathbf{A} \in \mathbb{R}_{2p}(\Omega)$ is expressed by

$$
I_1({}^{2p}\mathbf{A}) = {}^{2p}\mathbf{E} \overset{2p}{\otimes} {}^{2p}\mathbf{A} = {}^{2p}\mathbf{A} \overset{2p}{\otimes} {}^{2p}\mathbf{E} = ({}^{2p}\mathbf{E}, {}^{2p}\mathbf{A}) = A^{i_1 i_2 \cdots i_p}{}_{i_1 i_2 \cdots i_p}. \tag{7}
$$

The determinant of a tensor is the determinant of the mixed components of this tensor, i.e.

$$
\det({}^{2p}\mathbf{A}) = \det(A^{i_1 \cdots i_p}{}_{j_1 \cdots j_p}) = \det(A^{j}_{\cdot\, i}), \quad i_1, \ldots, i_p, j_1, \ldots, j_p = \overline{1, n}, \quad i, j = \overline{1, N}, \quad N = n^p. \tag{8}
$$

The determinant of a tensor is an invariant quantity. The expressions for the other classical invariants of a tensor of the module $\mathbb{R}_{2p}(\Omega)$ can be easily obtained from the formulas for the invariants of a tensor-block matrix given below (Nikabadze 2015). Therefore, for brevity, we will not dwell on this.

3 Eigenvalue Problem and Construction of a Complete System of Eigentensor Columns of Symmetric Tensor-Block Matrix

The eigenvalue problem is considered and a complete system of eigentensor columns of symmetric tensor-block matrix (TBM) consisting of four tensors of the module $\mathbb{R}_{2p}(\Omega)$ is constructed. We introduce a definition of the TBM.

Definition 14.6 A block matrix, whose blocks are composed of the various rank tensors, is called the TBM.

TBM of sizes $q \times m$ can be written as

$$\mathbb{M} = \begin{pmatrix} \mathbf{A}_{11} & \mathbf{A}_{12} & \cdots & \mathbf{A}_{1m} \\ \cdots & \cdots & \cdots & \cdots \\ \mathbf{A}_{q1} & \mathbf{A}_{q2} & \cdots & \mathbf{A}_{qm} \end{pmatrix}, \tag{9}$$

where m and q are some natural numbers; $\mathbf{A}_{kl}$, $k = \overline{1, q}$, $l = \overline{1, m}$ are the arbitrary tensors, also called the subtensors of the matrix (9).

Definition 14.7 The matrix

$$\mathbb{M}^T = \begin{pmatrix} \mathbf{A}^T_{11} & \mathbf{A}^T_{21} & \cdots & \mathbf{A}^T_{q1} \\ \cdots & \cdots & \cdots & \cdots \\ \mathbf{A}^T_{1m} & \mathbf{A}_{2m} & \cdots & \mathbf{A}^T_{qm} \end{pmatrix} \tag{10}$$

is called transposed matrix with the TBM (9).

Definition 14.8 A TBM, which coincides with its transposed matrix, is called symmetric.

Below it is to our interest to study the internal structure of a TBM (9) for which $q = m$, and subtensors are tensors of the same even rank, say, of the module $\mathbb{R}_{2p}(\Omega)$. Such matrices are often used in the application. One of them can be written as

$$\mathbb{M} = \mathbb{M}_{i_1 i_2 \cdots i_p}^{\;\;j_1 j_2 \cdots j_p} \mathbf{R}^{i_1 i_2 \cdots i_p} \mathbf{R}_{j_1 j_2 \cdots j_p} = \mathbb{M}_i^j \mathbf{R}^i \mathbf{R}_j = \begin{pmatrix} \mathbf{A}_{11} & \mathbf{A}_{12} & \cdots & \mathbf{A}_{1m} \\ \cdots & \cdots & \cdots & \cdots \\ \mathbf{A}_{m1} & \mathbf{A}_{m2} & \cdots & \mathbf{A}_{mm} \end{pmatrix}, \tag{11}$$

$$\mathbf{A}_{kl} = A_{kl,\, j_1 j_2 \cdots j_p}^{\;\;\;i_1 i_2 \cdots i_p} \mathbf{R}_{i_1 i_2 \cdots i_p} \mathbf{R}^{j_1 j_2 \cdots j_p} = A_{kl,\;\cdot j}^{\;\;\;i\,\cdot} \mathbf{R}_i \mathbf{R}^j, \;\; i_1, \ldots i_p, j_1, \ldots j_p = \overline{1, n}, \;\; i, j = \overline{1, N}.$$

Note that the sequence of the subtensors $\mathbf{A}_{11}, \mathbf{A}_{22}, \ldots, \mathbf{A}_{mm}$ is called the main diagonal of the TBM (11). Each tensor, which stands on the main diagonal, is called the diagonal subtensor. Note also that the matrix (11) consists of square tensors, therefore the matrix of its components is square.

Definition 14.9 The TBM, in which all subtensors besides the diagonal are zero tensors, is called the tensor-block diagonal matrix.

By Definition 14.9, it follows that the tensor-block diagonal matrix (TBDM) has the form

$$\mathbb{M} = \begin{pmatrix} \mathbf{A}_{11} & \mathbf{0} & \cdots & \mathbf{0} \\ \cdots & \cdots & \cdots & \cdots \\ \mathbf{0} & \mathbf{0} & \cdots & \mathbf{A}_{mm} \end{pmatrix}. \tag{12}$$

Definition 14.10 The TBM is called the upper or right (lower or left) tensor-block-triangular matrix if all tensors lying below (above) the main diagonal are zero.

The right (left) tensor-block-triangular matrix has the form

$$\mathbb{M} = \begin{pmatrix} \mathbf{A}_{11} & \mathbf{A}_{12} & \cdots & \mathbf{A}_{1m} \\ \cdots & \cdots & \cdots & \cdots \\ \mathbf{0} & \mathbf{0} & \cdots & \mathbf{A}_{mm} \end{pmatrix} \quad \left(\mathbb{M} = \begin{pmatrix} \mathbf{A}_{11} & \mathbf{0} & \cdots & \mathbf{0} \\ \cdots & \cdots & \cdots & \cdots \\ \mathbf{A}_{m1} & \mathbf{A}_{m2} & \cdots & \mathbf{A}_{mm} \end{pmatrix} \right). \tag{13}$$

In (12) and (13), the symbol $\mathbf{0}$ means the zero tensor of the module $\mathbb{R}_{2p}(\Omega)$.

It is easy to see that the components of the matrices (11)–(13) will change the same way as the components of a tensor of the module $\mathbb{R}_{2p}(\Omega)$ under a change of coordinates. In addition, they are composed of $m^2 = m \times m$ tensors. Therefore, we shall call such matrix the $(2p)$-rank and $m \times m$ sizes TBM. It is easy to see that the set of such tensor-block matrices forms a module over the ring of scalars. We denote this module by $\mathbb{R}_{2p}^{m\times m}(\Omega)$. Then we can write $\mathbb{M} \in \mathbb{R}_{2p}^{m\times m}(\Omega)$. Of course, we introduce designation $\mathbb{R}_{2p}^{q\times m}(\Omega)$ for a module consisting of the $(2p)$-rank and $q \times m$ sizes tensor-block matrices. Note that the order of matrix of components of the TBM (11) equal to mN, where $N = n^p$. Below we consider the TBM of the form (11) and we will assume that it is symmetric, i.e. the equality $\mathbb{M}^T = \mathbb{M}$ is true, which obviously holds if and only if $\mathbf{A}_{ij} = \mathbf{A}_{ji}^T$.

We also introduce the concept of a tensor column (tensor row).

Definition 14.11 Column matrix (row matrix) whose elements are the tensors of various rank, is called the tensor column (tensor row).

The tensor column $\mathbb{U}$, whose elements are the p-rank tensors, is represented in the form

$$\mathbb{U} = (\mathbf{U}_1, \ldots, \mathbf{U}_m)^T = \left(\mathrm{U}_{1,\, i_1 i_2 \cdots i_p}, \ldots, \mathrm{U}_{m,\, i_1 i_2 \cdots i_p}\right)^T \mathbf{R}^{i_1 i_2 \cdots i_p}. \tag{14}$$

It is easy to see that the tensor columns of the form (14) generate the module, denoted by $\mathbb{R}_p^{m\times 1}(\Omega)$. Then we can write that $\mathbb{U} \in \mathbb{R}_p^{m\times 1}(\Omega)$. Of course, the tensor rows of the form $\mathbb{V} = (\mathbf{V}_1, \mathbf{V}_2, \ldots, \mathbf{V}_m)$ also form the module, which is denoted by $\mathbb{R}_p^{1\times m}(\Omega)$.

The tensor row is the transposed tensor column, and the tensor column is the transposed tensor row. Therefore below we will speak about the tensor column, having in mind that everything, said about its, is equally true for the tensor row. We also introduce the definitions, similar to the tensor of the module $\mathbb{R}_{2p}(\Omega)$ (see also Vekua 1978; Nikabadze 2008, 2009a, b, 2014, 2015).

Definition 14.12 Let $\mathbb{U}$ and $\mathbb{V}$ be two tensor columns of the module $\mathbb{R}_p^{m\times 1}(\Omega)$, see (14). The value determined by the formula

$$(\mathbb{U}, \mathbb{V}) = \mathbb{U}^T \overset{p}{\otimes} \mathbb{V} = \sum_{k=1}^{m} \mathbf{U}_k \overset{p}{\otimes} \mathbf{V}_k = \sum_{k=1}^{m} U_{k,\, i_1 i_2 \cdots i_p} V_{k,}^{\; i_1 i_2 \cdots i_p}, \tag{15}$$

where $\overset{p}{\otimes}$ is the inner p-product, is called the scalar product of the tensor columns $\mathbb{U}$ and $\mathbb{V}$.

Definition 14.13 Two tensor columns $\mathbb{U}$ and $\mathbb{V}$ of the module $\mathbb{R}_p^{m\times 1}(\Omega)$ are called orthogonal if their scalar product is zero, i.e. $(\mathbb{U}, \mathbb{V}) = \mathbb{U}^T \overset{p}{\otimes} \mathbb{V} = 0$.

Definition 14.14 The norm of the tensor column $\mathbb{U}$ of the module $\mathbb{R}_p^{m\times 1}(\Omega)$ is defined to be

$$||\mathbb{U}|| = \sqrt{(\mathbb{U}, \mathbb{U})} = \sqrt{\mathbb{U}^T \overset{p}{\otimes} \mathbb{U}} = \sqrt{\sum_{k=1}^{m} \mathbf{U}_k \overset{p}{\otimes} \mathbf{U}_k} = \sqrt{\sum_{k=1}^{m} U_{k,\, i_1 i_2 \cdots i_p} U_{k,}^{\; i_1 i_2 \cdots i_p}}. \tag{16}$$

Now the eigenvalue problem of the TBM (11) can be formulated as follows. *Eigenvalue problem.* It is required to find all tensor columns

$$\mathbb{U} = (\mathbf{U}_1, \ldots, \mathbf{U}_m)^T \quad \big(\mathbb{U}^T = (\mathbf{U}_1, \ldots, \mathbf{U}_m)\big), \tag{17}$$

satisfying the equation

$$\mathbb{M} \overset{p}{\otimes} \mathbb{U} = \lambda \mathbb{U}, \tag{18}$$

where λ is a scalar.

The relations (15), (16), and (18) contains the following two operations: the matrix multiplication and the inner 2-product. Note that Eq. (18) always has the trivial solution $\mathbb{U} = \mathbb{O}$, where $\mathbb{O}$ is the zero tensor column, consisting of the m zero tensors of pth rank. Moreover, it is clear that eigentensor column is determined up to a scalar factor, so it can always be normalized. Therefore, we shall assume that the solution of Eq. (18) satisfies the conditions

$$\mathbb{U} \neq \mathbb{O}, \quad ||\mathbb{U}|| = \sqrt{\mathbb{U}^T \overset{p}{\otimes} \mathbb{U}} = \sqrt{\sum_{k=1}^{m} \mathbf{U}_k \overset{p}{\otimes} \mathbf{U}_k} = \sqrt{\sum_{k=1}^{m} U_{k,\, i_1 i_2 \cdots i_p} U_{k,}^{\; i_1 i_2 \cdots i_p}} = 1.$$

If the Eq. (18) has a non-trivial solution $\mathbb{U}$ for a scalar λ, then λ is called an eigenvalue of the TBM $\mathbb{M}$, and $\mathbb{U}$ is called the eigentensor column corresponding to λ. Since $\mathbb{M}^T = \mathbb{M}$, it is easy to prove that $\mathbb{U}^T \overset{p}{\otimes} \mathbb{M} = \lambda \mathbb{U}^T$, i.e. $\mathbb{U}^T$ is an eigentensor row for $\mathbb{M} \in \mathbb{R}_{2p}^{m\times m}(\Omega)$ corresponding to λ.

Definition 14.15 A symmetric TBM $\mathbb{M}$ is said to be positive-definite if the quadratic form $\mathbb{U}^T \overset{p}{\otimes} \mathbb{M} \overset{p}{\otimes} \mathbb{U}$ is positive for any non-zero tensor column $\mathbb{U}$.

Based on Definition 14.15, we conclude that the matrix of components of the constituting tensors of positive-definite TBM is positive-definite, and hence, we have the following theorem.

Theorem 14.1 *The diagonal subtensors of positive-definite TBM are positive-definite.*

Note that if we will study the internal structure of the TBM consisting of four tensors of the module $\mathbb{R}_{2p}(\Omega)$, then, the results of research can be easily generalized to the case of the matrix (11). Therefore, below we consider the TBM of the form

$$\mathbb{M} = \begin{pmatrix} \mathbf{A} & \mathbf{B} \\ \mathbf{C} & \mathbf{D} \end{pmatrix}, \tag{19}$$

where **A**, **B**, **C**, and **D** are the tensors of the module $\mathbb{R}_{2p}(\Omega)$.

From the above it follows that (19) is a TBM of the module $\mathbb{R}_{2p}^{2\times 2}(\Omega)$. Further if the opposite is not specified, we shall mainly deal with the TBM of the module $\mathbb{R}_{2p}^{2\times 2}(\Omega)$, and $\mathbb{M} \in \mathbb{R}_{2p}^{2\times 2}(\Omega)$ means that $\mathbb{M}$ is the TBM consisting of four tensors of the module $\mathbb{R}_{2p}(\Omega)$ or the TBM of the module $\mathbb{R}_{2p}^{2\times 2}(\Omega)$.

Of course, the transposed TBM to (19) has the form (see also Eq. (10))

$$\mathbb{M}^T = \begin{pmatrix} \mathbf{A}^T & \mathbf{C}^T \\ \mathbf{B}^T & \mathbf{D}^T \end{pmatrix}. \tag{20}$$

Here are given some definitions and theorems from Nikabadze (2014, 2015) for a TBM of the module $\mathbb{R}_{2p}^{2\times 2}(\Omega)$ (in Nikabadze 2014, 2015 this module is denoted also by $\mathbb{R}_{2p}^{4}(\Omega)$).

Definition 14.16 The TBM is called the left or lower (right or upper) triangular [unitriangular] TBM if the corresponding block-matrix of matrix of components of the constituting tensors is the left or lower (right or upper) triangular [unitriangular] matrix.

If **A**, **0**, **C**, and **D** are tensors of the module $\mathbb{R}_{2p}(\Omega)$, then according to Definition 14.16

$$\mathbb{M} = \begin{pmatrix} \mathbf{A} & \mathbf{0} \\ \mathbf{C} & \mathbf{D} \end{pmatrix}, \tag{21}$$

is the left triangular (unitriangular) matrix of the module $\mathbb{R}_{2p}^{2\times 2}(\Omega)$, if the matrices of components of tensors **A** and **D** are the left triangular (unitriangular) matrices.

Similarly, if **A**, **B**, **0**, and **D** are tensors of the module $\mathbb{R}_{2p}(\Omega)$, then

$$\mathbb{M} = \begin{pmatrix} \mathbf{A} & \mathbf{B} \\ \mathbf{0} & \mathbf{D} \end{pmatrix}, \tag{22}$$

is the right triangular (unitriangular) matrix of the module $\mathbb{R}_{2p}^{2\times 2}(\Omega)$, if the matrices of components of tensors **A** and **D** are right triangular (unitriangular) matrices.

Definition 14.17 The matrix of the form

$$\mathbb{M} = \begin{pmatrix} \mathbf{A} & \mathbf{0} \\ \mathbf{0} & \mathbf{D} \end{pmatrix}, \tag{23}$$

is called the TBDM of the module $\mathbb{R}_{2p}^{2\times 2}(\Omega)$, where **A**, **0**, and **D** are tensors of the module $\mathbb{R}_{2p}(\Omega)$.

Definition 14.18 The determinant of the TBM is the determinant of a block matrix of mixed components of the constituting tensors.

Definition 14.19 The TBM is called non-singular if its determinant is non-zero.

Based on Definition 14.16, we conclude that if in (21) **A** and **D** are left triangular (unitriangular) tensors, and **C** is square tensor, then (21) is a left triangular (unitriangular) TBM. Similarly, if in (22) **A** and **D** are right triangular (unitriangular) tensors, and **B** is square tensor, then (22) is a right triangular (unitriangular) TBM.

In the future, it is also interesting to consider a tensor column, which consists of the same rank tensors. For example, if the TBM has the form (19), then the tensor column (tensor row) can be written as

$$\mathbb{W} = \begin{pmatrix} \mathbf{U} \\ \mathbf{V} \end{pmatrix} \quad (\mathbb{W}^T = (\mathbf{U}, \mathbf{V})), \tag{24}$$

where **U** and **V** are tensors of the module $\mathbb{R}_p(\Omega)$, i.e.

$$\mathbf{U} = U^{i_1 i_2 \cdots i_p} \mathbf{R}_{i_1 i_2 \cdots i_p} = U^i \mathbf{R}_i, \quad \mathbf{V} = V^{i_1 i_2 \cdots i_p} \mathbf{R}_{i_1 i_2 \cdots i_p} = V^i \mathbf{R}_i,$$
$$i_1, \ldots i_p = \overline{1, n}, \quad i, j = \overline{1, N}.$$

Analogous to the TBM (19), we can say that the tensor column (tensor row) (24) has pth rank and 2×1 (1×2) sizes. Moreover, it is easy to see that the set of tensor columns such as (24) over the ring of scalars forms a module denoted by $\mathbb{R}_p^{2\times 1}(\Omega)$ $(\mathbb{R}_p^{1\times 2}(\Omega))$. Then $\mathbb{W} \in \mathbb{R}_p^{2\times 1}(\Omega)$ means that $\mathbb{W}$ is the tensor column composed of two tensors of the module $\mathbb{R}_p(\Omega)$ or the tensor column of the module $\mathbb{R}_p^{2\times 1}(\Omega)$. The same thing is true for the tensor rows. In particular, the pth rank and 1×2 sizes tensor row is an element of the module $\mathbb{R}_p^{1\times 2}(\Omega)$.

Definition 14.20 The tensor column is called normalized if its norm is equal to one.

Definition 14.21 A system of tensor columns $\mathbb{W}_1, \mathbb{W}_2, \ldots, \mathbb{W}_m$ is said to be orthonormal if

$$(\mathbb{W}_k, \mathbb{W}_l) = \mathbb{W}_k^T \overset{p}{\otimes} \mathbb{W}_l = \delta_{kl}, \quad k, l = \overline{1, m}.$$

Definition 14.22 Two systems of tensor columns $\mathbb{W}_1, \mathbb{W}_2, \ldots, \mathbb{W}_m$ and $\mathbb{W}^1$, $\mathbb{W}^2, \ldots, \mathbb{W}^m$ $(1 \le m \le 2N)$ of the module $\mathbb{R}_p^{2\times 1}(\Omega)$ are said to be biorthonormal if

$$(\mathbb{W}_k, \mathbb{W}^l) = \mathbb{W}_k^T \overset{p}{\otimes} \mathbb{W}^l = \delta_k^l, \quad k, l = \overline{1, m}, \quad 1 \le m \le 2N.$$

Note that we can build the biorthonormal system for the linearly independent system of tensor columns in the same way as for a linearly independent system of tensors (Vekua 1978; Nikabadze 2009a, b, 2014, 2015). In particular, the following theorem is valid.

Theorem 14.2 *For any linearly independent system of tensor columns of the module* $\mathbb{R}_p^{2\times 1}(\Omega)$*, there exists the biorthonormal system.*

This theorem is proved like the theorem for linearly independent system of tensors (Vekua 1978; Nikabadze 2009a, 2015).

3.1 Eigenvalue Problem for $\mathbb{M} \in \mathbb{R}_{2p}^{2\times 2}(\Omega)$

It is required for any $\mathbb{M} \in \mathbb{R}_{2p}^{2\times 2}(\Omega)$ to find the tensor columns $\mathbb{W} \in \mathbb{R}_p^{2\times 1}(\Omega)$ satisfying the following equation:

$$\mathbb{M} \overset{p}{\otimes} \mathbb{W} = \lambda \mathbb{W}, \tag{25}$$

where λ is a scalar.

Equation (25) always has the trivial solution $\mathbb{W} = \mathbb{O}$ where $\mathbb{O} \in \mathbb{R}_p^{2\times 1}(\Omega)$ is the zero tensor column ($\mathbb{O} = (\mathbf{0}, \mathbf{0})^T$). Later, speaking about the solution of Eq. (25), we have in mind only the non-trivial solution $\mathbb{W} \ne \mathbb{O}$. Moreover, since an eigentensor column is defined up to a scalar factor, it can always be normalized. Therefore, talking about the solution of Eq. (25) (see also (18)) we will have in mind only non-trivial normalized solutions in mind. If Eq. (25) has a solution $\mathbb{W}$ for some λ, then λ is called an eigenvalue of TBM $\mathbb{M} \in \mathbb{R}_{2p}^{2\times 2}(\Omega)$, and $\mathbb{W} \in \mathbb{R}_p^{2\times 1}(\Omega)$ is an eigentensor column corresponding to λ.

Naturally, we may consider the following problem: for any $\mathbb{M} \in \mathbb{R}_{2p}^{2\times 2}(\Omega)$ it is required to find the tensor rows $\mathbb{W}' = (\mathbf{U}', \mathbf{V}')$ of the module $\mathbb{R}_p^{1\times 2}(\Omega)$ which satisfy the following equation:

$$\mathbb{W}' \overset{p}{\otimes} \mathbb{M} = \mu \mathbb{W}', \tag{26}$$

where μ is a scalar.

If for some μ Eq. (26) has a non-trivial solution $\mathbb{W}'$, then μ is called an eigenvalue of the TBM $\mathbb{M}$ (19), and $\mathbb{W}'$ is an eigentensor row corresponding to μ. Note that in the general case $\lambda = \mu$ and $\mathbb{W}' \neq \mathbb{W}^T$. But if $\mathbb{M}^T = \mathbb{M}$, then $\mathbb{W}' = \mathbb{W}^T$.

Let us formulate also some statements and theorems concerning eigentensor columns of a TBM (Nikabadze 2014, 2015).

Statement 14.1 If λ and λ' are two different eigenvalues of a TBM $\mathbb{M}$, then the corresponding two eigentensor columns $\mathbb{W}$ and $\mathbb{W}'$ are linearly independent.

Theorem 14.3 *Eigentensor columns of TBM corresponding to the pairwise different eigenvalues are linearly independent.*

Theorem 14.4 *If λ and λ' are two different eigenvalues of the symmetric TBM $\mathbb{M}$, then the any two corresponding eigentensor columns $\mathbb{W}$ and $\mathbb{W}'$ are orthogonal.*

Theorem 14.5 *Eigentensor columns of symmetric TBM corresponding to the pairwise different eigenvalues are pairwise orthogonal.*

Theorem 14.6 *The eigenvalues of positive-definite symmetric TBM are positive.*

It is not difficult to show that (25) can be rewritten as

$$\begin{pmatrix} \mathbf{A} - \lambda\mathbf{E} & \mathbf{B} \\ \mathbf{C} & \mathbf{D} - \lambda\mathbf{E} \end{pmatrix} \overset{p}{\otimes} \begin{pmatrix} \mathbf{U} \\ \mathbf{V} \end{pmatrix} = \mathbb{O} \quad \left((\mathbb{M} - \lambda\mathbb{E}) \overset{p}{\otimes} \mathbb{W} = \mathbb{O} \right), \tag{27}$$

where $\mathbf{E}$ is the $2p$th rank unit tensor (5), $\mathbb{E}$ is the following unit TBM:

$$\mathbb{E} = \begin{pmatrix} \mathbf{E} & \mathbf{0} \\ \mathbf{0} & \mathbf{E} \end{pmatrix} = \begin{pmatrix} g_i^j \mathbf{R}^i \mathbf{R}_j & \mathbf{0} \\ \mathbf{0} & g_j^i \mathbf{R}_i \mathbf{R}^j \end{pmatrix} = \begin{pmatrix} g_i^j & 0 \\ 0 & g_i^j \end{pmatrix} \mathbf{R}^i \mathbf{R}_j, \quad i, j = \overline{1, N}, \tag{28}$$

and $\mathbf{0}$ is the $2p$th rank zero tensor.

The homogeneous system (27) consists of $2N$ ($N = n^p$) equations with $2N$ unknowns (the components of two asymmetric temperatures $\mathbf{U}$ and $\mathbf{V}$). This system should have a non-trivial solution. Hence, the determinant of its matrix should be equal to zero:

$$\det \begin{pmatrix} \mathbf{A} - \lambda\mathbf{E} & \mathbf{B} \\ \mathbf{C} & \mathbf{D} - \lambda\mathbf{E} \end{pmatrix} = 0 \quad \left(\det(\mathbb{M} - \lambda\mathbb{E}) = 0 \right). \tag{29}$$

The equality expressed by (29) is called the characteristic η of the TBM $\mathbb{M} \in \mathbb{R}_{2p}^{2\times 2}(\Omega)$. The power of this η is equal to $2N$ with respect to λ. If the symmetric TBM $\mathbb{M}$ is positive-definite, then by Theorem 14.6 the characteristic η (29) has $2N$ positive roots (eigenvalues). Here we take the possible multiplicity of these roots into account.

Below we formulate a number of definitions and theorems to discuss the properties of TBM (Nikabadze 2015) (similar questions for matrices can be seen in Gantmacher 1959).

Definition 14.23 A real TBM is said to be normal if this matrix is commutative with its transposed matrix.

Based on this definition, we conclude, that the real symmetric TBM, the real skew-symmetric and orthogonal tensor-block matrices, the unit TBM are normal.

Definition 14.24 The orthonormal system of eigentensor columns of the TBM $\mathbb{M} \in \mathbb{R}_{2p}^{2\times 2}(\Omega)$ consisting of $2N$ eigentensor columns is said to be a complete orthonormal system of eigentensor columns.

Definition 14.25 $\mathbb{M} \in \mathbb{R}_{2p}^{2\times 2}(\Omega)$ is a TBM of simple structure if it has $2N$ linearly independent eigentensor columns.

Statement 14.2 TBM has a simple structure if all roots of its characteristic equation are distinct.

Note that the converse is not true. There are tensor-block matrices of simple structure, characteristic equations of which have the multiple roots. For example, the unit TBM.

Theorem 14.7 *A real symmetric TBM always has a complete orthonormal system of eigentensor columns.*

Note that Theorem 14.7 is true for any real symmetric TBM and normal complex TBM. By Theorem 14.7, a positive-definite symmetric TBM $\mathbb{M} \in \mathbb{R}_{2p}^{2\times 2}(\Omega)$ always has a complete orthonormal system of eigentensor columns and by Theorem 14.6 its eigenvalues are positive.

We denote this system by $\mathbb{W}_k = (\mathbf{U}_k, \mathbf{V}_k)^T$, $k = \overline{1, 2N}$. The kth eigentensor column of this system corresponds to the eigenvalue λ_k. Then, the matrix $\mathbb{M} \in \mathbb{R}_{2p}^{2\times 2}(\Omega)$ can be represented in the canonical form

$$\mathbb{M} = \sum_{k=1}^{2N} \lambda_k \mathbb{W}_k \mathbb{W}_k^T = \sum_{k=1}^{2N} \lambda_k \begin{pmatrix} \mathbf{U}_k \\ \mathbf{V}_k \end{pmatrix} \otimes (\mathbf{U}_k, \mathbf{V}_k) = \sum_{k=1}^{2N} \lambda_k \begin{pmatrix} \mathbf{U}_k\mathbf{U}_k & \mathbf{U}_k\mathbf{V}_k \\ \mathbf{V}_k\mathbf{U}_k & \mathbf{V}_k\mathbf{V}_k \end{pmatrix}, \quad (30)$$

where the following orthonormality condition is valid:

$$(\mathbb{W}_k, \mathbb{W}_l) = \mathbb{W}_k^T \overset{p}{\otimes} \mathbb{W}_l = \mathbf{U}_k \overset{p}{\otimes} \mathbf{U}_l + \mathbf{V}_k \overset{p}{\otimes} \mathbf{V}_l = \delta_{kl}, \quad k, l = \overline{1, 2N}. \quad (31)$$

It should be noted that the dimensions of the modules $\mathbb{R}_p^{2\times 1}(\Omega)$ and $\mathbb{R}_{2p}^{2\times 2}(\Omega)$ introduced above equal to $2N$ and $4N^2$, respectively, while the dimensions of the modules $\mathbb{R}_p(\Omega)$, $\mathbb{R}_{2p}(\Omega)$ are N and N^2, respectively. Consequently, there are $2N$ linearly independent tensor columns in the module $\mathbb{R}_p^{2\times 1}(\Omega)$. They form the basis of this module. There are $4N^2$ linearly independent tensor-block matrices in the module $\mathbb{R}_{2p}^{2\times 2}(\Omega)$ which constitute the basis of this module. Solving the system of Eq. (27), say, for TBM of simple structure $\mathbb{M} \in \mathbb{R}_{2p}^{2\times 2}(\Omega)$ we get $2N$ linearly independent eigentensor columns. They can be taken as a basis of the module $\mathbb{R}_p^{2\times 1}(\Omega)$. If $\mathbb{M} \in$

$\mathbb{R}_{2p}^{2\times 2}(\Omega)$ is a symmetric TBM, then its complete system of orthonormal eigentensor columns will be an orthonormal basis of the module $\mathbb{R}_p^{2\times 1}(\Omega)$. Hence, any tensor column of the module $\mathbb{R}_p^{2\times 1}(\Omega)$ can be expanded in the above-mentioned bases. Knowing the basis of the module $\mathbb{R}_p^{2\times 1}(\Omega)$ (module of smaller dimension), we can construct a basis of the module $\mathbb{R}_{2p}^{2\times 2}(\Omega)$ (module of greater dimension) in the same way as the basis of the module $\mathbb{R}_{2p}(\Omega)$ using the basis of the module $\mathbb{R}_p(\Omega)$ (Vekua 1978; Nikabadze 2009a, 2015). In particular, if we consider a tensor product of a tensor column of the module $\mathbb{R}_p^{2\times 1}(\Omega)$ by a tensor row of the module $\mathbb{R}_p^{1\times 2}(\Omega)$, we obtain a TBM of the module $\mathbb{R}_{2p}^{2\times 2}(\Omega)$. Therefore, for multiplying tensorially each tensor column of some basis of the module $\mathbb{R}_p^{2\times 1}(\Omega)$ by each of the tensor rows, received by transposition of all tensor columns of considered basis, we get $4N^2$ tensor-block matrices. These matrices form a basis of the module $\mathbb{R}_{2p}^{2\times 2}(\Omega)$. So, if tensor columns $\mathbb{W}_k = (\mathbf{U}_k, \mathbf{V}_k)^T$, $k = \overline{1, 2N}$, form, for example, an orthonormal basis of the module $\mathbb{R}_p^{2\times 1}(\Omega)$, where $\mathbb{W}_k = (\mathbf{U}_k, \mathbf{V}_k)^T$, $k = \overline{1, 2N}$, can be considered as a complete system of orthonormal eigentensor columns of a symmetric TBM $\mathbb{M} \in \mathbb{R}_{2p}^{2\times 2}(\Omega)$, then tensor-block matrices $\mathbb{W}_k\mathbb{W}_l^T$, $k, l = \overline{1, 2N}$, constitute the basis of the module $\mathbb{R}_{2p}^{2\times 2}(\Omega)$ and it is easy to verify that the unit TBM $\mathbb{E} \in \mathbb{R}_{2p}^{2\times 2}(\Omega)$ can be represented in the form

$$\mathbb{E} = \sum_{k=1}^{2N} \mathbb{W}_k\mathbb{W}_k^T = \sum_{k=1}^{2N} \begin{pmatrix} \mathbf{U}_k \\ \mathbf{V}_k \end{pmatrix} \otimes (\mathbf{U}_k, \mathbf{V}_k) = \sum_{k=1}^{2N} \begin{pmatrix} \mathbf{U}_k\mathbf{U}_k & \mathbf{U}_k\mathbf{V}_k \\ \mathbf{V}_k\mathbf{U}_k & \mathbf{V}_k\mathbf{V}_k \end{pmatrix}. \tag{32}$$

If tensor columns $\mathbb{Z}_k = (\mathbf{X}_k, \mathbf{Y}_k)^T$, $k = \overline{1, 2N}$, form a basis of the module $\mathbb{R}_p^{2\times 1}(\Omega)$, and the tensor columns $\mathbb{Z}^l = (\mathbf{X}^l, \mathbf{Y}^l)^T$, $l = \overline{1, 2N}$, constitute a corresponding biorthonormal basis, then tensor-block matrices $\mathbb{Z}_k\mathbb{Z}^{l^T}$, $k, l = \overline{1, 2N}$, will be the basis of the module $\mathbb{R}_{2p}^{2\times 2}(\Omega)$. Then the unit TBM $\mathbb{E} \in \mathbb{R}_{2p}^{2\times 2}(\Omega)$ can be written as

$$\mathbb{E} = \sum_{k=1}^{2N} \mathbb{Z}_k\mathbb{Z}^{k^T} = \sum_{k=1}^{2N} \begin{pmatrix} \mathbf{X}_k \\ \mathbf{Y}_k \end{pmatrix} \otimes (\mathbf{X}^k, \mathbf{Y}^k) = \sum_{k=1}^{2N} \begin{pmatrix} \mathbf{X}_k\mathbf{X}^k & \mathbf{X}_k\mathbf{Y}^k \\ \mathbf{Y}_k\mathbf{X}^k & \mathbf{Y}_k\mathbf{Y}^k \end{pmatrix}. \tag{33}$$

Note that in the last relation we can raise or lower the corresponding indices.

Knowing the representation, see (28), (32), and (33), of the unit TBM $\mathbb{E} \in \mathbb{R}_{2p}^{2\times 2}(\Omega)$ it is not difficult to give a canonical representation of the TBM $\mathbb{M} \in \mathbb{R}_{2p}^{2\times 2}(\Omega)$ of a simple structure, as well as its inverse matrix. Note that if $\mathbb{M} \in \mathbb{R}_{2p}^{2\times 2}(\Omega)$ is a non-singular TBM, then there is only one inverse TBM $\mathbb{M}^{-1}$ to this matrix such that we have the relation

$$\mathbb{M} \overset{p}{\otimes} \mathbb{M}^{-1} = \mathbb{M}^{-1} \overset{p}{\otimes} \mathbb{M} = \mathbb{E} \tag{34}$$

and the theorem is valid.

Theorem 14.8 *If* $\mathbb{W}$ *(*$\mathbb{W}'$*) is the eigentensor column (eigentensor row) of the module* $\mathbb{R}_p^{2\times 1}(\Omega)$ *(*$\mathbb{R}_p^{1\times 2}(\Omega)$*) of a non-singular TBM* $\mathbb{M} \in \mathbb{R}_{2p}^{2\times 2}(\Omega)$ *corresponding to* λ*, then it will also be an eigentensor column (eigentensor row) of TBM* $\mathbb{M}^{-1}$*, which is inverse to* $\mathbb{M}$*, corresponding to* λ^{-1}*.*

Based on Theorem 14.8, we can prove the following theorem.

Theorem 14.9 *The eigenvalues of the inverse TBM* $\mathbb{M}^{-1}$ *equal to the inverse values of the eigenvalues of the initial non-singular TBM* $\mathbb{M}$ *and the eigentensor columns and rows of the tensor-block matrices* $\mathbb{M}^{-1}$ *and* $\mathbb{M}$ *are the same.*

By Theorem 14.9, the TBM $\mathbb{M}^{-1}$, inverse to (30), will take the form

$$\mathbb{M}^{-1} = \sum_{k=1}^{2N} \lambda_k^{-1} \mathbb{W}_k \mathbb{W}_k^T = \sum_{k=1}^{2N} \lambda_k^{-1} \begin{pmatrix} \mathbf{U}_k \\ \mathbf{V}_k \end{pmatrix} \otimes (\mathbf{U}_k, \mathbf{V}_k) = \sum_{k=1}^{2N} \lambda_k^{-1} \begin{pmatrix} \mathbf{U}_k\mathbf{U}_k & \mathbf{U}_k\mathbf{V}_k \\ \mathbf{V}_k\mathbf{U}_k & \mathbf{V}_k\mathbf{V}_k \end{pmatrix}. \tag{35}$$

Now we can prove the following theorem for $\mathbb{M} \in \mathbb{R}_{2p}^{2\times 2}(\Omega)$.

Theorem 14.10 *If* $\mathbb{M} \in \mathbb{R}_{2p}^{2\times 2}(\Omega)$ *is the TBM of a simple structure, it has a complete system of eigentensor columns and eigentensor rows. In addition, if the complete system of eigentensor columns (rows) is found and then a corresponding biorthonormal system of tensor columns (rows) is constructed, then by transposing each tensor column (tensor row) in the constructed biorthonormal system of tensor columns (rows), we obtain a complete system of eigentensor rows (columns).*

In particular, the following theorem is valid.

Theorem 14.11 *Let* $\mathbb{M} \in \mathbb{R}_{2p}^{2\times 2}(\Omega)$ *be the simple structure TBM,* $\mathbb{Z}_k = (\mathbf{X}_k, \mathbf{Y}_k)^T$*,* $k = \overline{1, 2N}$*, is a complete system of eigentensor columns corresponding to the system of eigenvalues* λ_k*,* $k = \overline{1, 2N}$*, respectively, and* $\mathbb{Z}^l = (\mathbf{X}^l, \mathbf{Y}^l)^T$*,* $l = \overline{1, 2N}$*, is the biorthonormal for* $\mathbb{Z}_k$*,* $k = \overline{1, 2N}$*, system of tensor columns* $\left((\mathbb{Z}_k, \mathbb{Z}^l) = \delta_k^l\right)$*. Then the system* ${\mathbb{Z}^l}^T = (\mathbf{X}^l, \mathbf{Y}^l)$*,* $l = \overline{1, 2N}$*, will be a complete system of eigentensor rows, and the TBM* $\mathbb{M} \in \mathbb{R}_{2p}^{2\times 2}(\Omega)$ *has the representation*

$$\mathbb{M} = \sum_{k=1}^{2N} \lambda_k \mathbb{Z}_k {\mathbb{Z}^k}^T = \sum_{k=1}^{2N} \lambda_k \begin{pmatrix} \mathbf{X}_k \\ \mathbf{Y}_k \end{pmatrix} \otimes (\mathbf{X}^k, \mathbf{Y}^k) = \sum_{k=1}^{2N} \lambda_k \begin{pmatrix} \mathbf{X}_k\mathbf{X}^k & \mathbf{X}_k\mathbf{Y}^k \\ \mathbf{Y}_k\mathbf{X}^k & \mathbf{Y}_k\mathbf{Y}^k \end{pmatrix}. \tag{36}$$

Having the representation (36) for $\mathbb{M} \in \mathbb{R}_{2p}^{2\times 2}(\Omega)$, it is easy to prove that its power for any integer α has the form

$$\mathbb{M}^\alpha = \overbrace{\mathbb{M} \overset{p}{\otimes} \mathbb{M} \overset{p}{\otimes} \cdots \overset{p}{\otimes} \mathbb{M}}^{\alpha} = \sum_{k=1}^{2N} \lambda_k^\alpha \mathbb{Z}_k {\mathbb{Z}^k}^T. \tag{37}$$

In particular, when $\alpha = -1$ and $\alpha = 0$, we have

$$\mathbb{M}^{-1} = \sum_{k=1}^{2N} \lambda_k^{-1} \mathbb{Z}_k \mathbb{Z}^{k^T}, \quad \mathbb{E} = \mathbb{M}^0 = \sum_{k=1}^{2N} \mathbb{Z}_k \mathbb{Z}^{k^T}. \tag{38}$$

Now we write the characteristic equation (29) in expanded form

$$\lambda^{2N} - I_1(\mathbb{M})\lambda^{2N-1} + \cdots + (-1)^s I_s(\mathbb{M})\lambda^{2N-s} + \cdots + (-1)^{2N} I_{2N}(\mathbb{M}) = 0$$
$$(I_{2N}(\mathbb{M}) = \det\mathbb{M}). \tag{39}$$

It is not difficult to prove the Hamilton–Cayley theorem.

Theorem 14.12 *Any TBM satisfies its characteristic equation.*

Thus, we have the relation

$$\mathbb{M}^{2N} - I_1(\mathbb{M})\mathbb{M}^{2N-1} + \cdots + (-1)^s I_s(\mathbb{M})\mathbb{M}^{2N-s} + \cdots + (-1)^{2N} I_{2N}(\mathbb{M})\mathbb{E} = 0.$$

This theorem is proved similarly to theorem given in Nikabadze (2015) (see also Nikabadze 2009a).

Next it should be noted that the invariants of the TBM $\mathbb{M} \in \mathbb{R}_{2p}^{2\times 2}(\Omega)$ in (39) as the invariants of a matrix (Gantmacher 1959; Faddeev and Sominckii 1999; Korn and Korn 2000) or tensor $\mathbf{A} \in \mathbb{R}_{2p}(\Omega)$ (Nikabadze 2015) are computed as follows (Nikabadze 2014, 2015)

$$S_k = I_k(\mathbb{M}) = \frac{1}{k!}\begin{vmatrix} s_1 & 1 & 0 & \cdots & 0 & 0 \\ s_2 & s_1 & 2 & \cdots & 0 & 0 \\ \cdots & \cdots & \cdots & \cdots & \cdots & \cdots \\ s_{k-1} & s_{k-2} & s_{k-3} & \cdots & s_1 & k-1 \\ s_k & s_{k-1} & s_{k-2} & \cdots & s_2 & s_1 \end{vmatrix}, \quad k = \overline{1, 2N}. \tag{40}$$

Here $s_k = I_1(\mathbb{M}^k)$, $k = \overline{1, 2N}$, $\mathbb{M}^k = \overbrace{\mathbb{M}\overset{p}{\otimes}\mathbb{M}\overset{p}{\otimes}\cdots\overset{p}{\otimes}\mathbb{M}}^{k}$, and the first invariant for $\mathbb{M} \in \mathbb{R}_{2p}^{2\times 2}(\Omega)$ is defined as follows:

$$I_1(\mathbb{M}) = \mathbb{E} \overset{2p}{\otimes} \mathbb{M} = I_1\begin{pmatrix} \mathbf{A} & \mathbf{B} \\ \mathbf{C} & \mathbf{D} \end{pmatrix} = I_1(\mathbf{A} + \mathbf{D}) = I_1(\mathbf{A}) + I_1(\mathbf{D}). \tag{41}$$

The inverse relations to (40) are represented in the form (Nikabadze 2014, 2015)

$$s_k = I_1(\mathbb{M}^k) = \begin{vmatrix} S_1 & 1 & 0 & \cdots & 0 & 0 \\ 2S_2 & S_1 & 1 & \cdots & 0 & 0 \\ \cdots & \cdots & \cdots & \cdots & \cdots & \cdots \\ (k-1)S_{k-1} & S_{k-2} & S_{k-3} & \cdots & S_1 & 1 \\ kS_k & S_{k-1} & S_{k-2} & \cdots & S_2 & S_1 \end{vmatrix}, \quad k = \overline{1, 2N}.$$

From (36) (see also (30)), it is easy to see that to give the TBM $\mathbb{M} \in \mathbb{R}_{2p}^{2\times 2}(\Omega)$ we must give the $2N$ eigentensor columns $\mathbb{Z}_k$, $k = \overline{1, 2N}$, and $2N$ eigenvalues λ_k, $k = \overline{1, 2N}$. It should be noted that the complete system of eigentensor columns $\mathbb{Z}_k \in \mathbb{R}_p^{2\times 1}(\Omega)$, $k = \overline{1, 2N}$, and biorthonormal system $\mathbb{Z}^l$, $l = \overline{1, 2N}$, in the representation $\mathbb{M} \in \mathbb{R}_{2p}^{2\times 2}(\Omega)$ (see (36)) satisfies the conditions of biorthonormality (see also (31))

$$(\mathbb{Z}_k, \mathbb{Z}^l) = \mathbb{Z}_k^T \overset{p}{\otimes} \mathbb{Z}^l = \mathbf{X}_k \overset{p}{\otimes} \mathbf{X}^l + \mathbf{Y}_k \overset{p}{\otimes} \mathbf{Y}^l = \delta_k^l, \quad k, l = \overline{1, 2N}. \tag{42}$$

Note also that if the complete system of eigentensor columns $\mathbb{Z}_k \in \mathbb{R}_p^{2\times 1}(\Omega)$, $k = \overline{1, 2N}$, of TBM $\mathbb{M} \in \mathbb{R}_{2p}^{2\times 2}(\Omega)$ of the simple structure is constructed, then by Theorem 14.2 it is always possible to construct the biorthonormal system $\mathbb{Z}^l$, $l = \overline{1, 2N}$ (tensors $\mathbb{Z}^l, l = \overline{1, 2N}$, are defined by using tensors $\mathbb{Z}_k, k = \overline{1, 2N}$), and by Theorem 14.11, the matrix $\mathbb{M} \in \mathbb{R}_{2p}^{2\times 2}(\Omega)$ can be represented as (36). Therefore, we may say that in the relations (42) the eigentensor columns $\mathbb{Z}_k \in \mathbb{R}_p^{2\times 1}(\Omega), k = \overline{1, 2N}$ are unknown. Since $\mathbb{Z}_k = (\mathbf{X}_k, \mathbf{Y}_k)^T, k = \overline{1, 2N}$, then the complete system of eigentensor columns $\mathbb{Z}_k$, $k = \overline{1, 2N}$, is given by $4N^2$ components of the temperatures $\mathbf{X}_k$ and $\mathbf{Y}_k$, $k = \overline{1, 2N}$, of the module $\mathbb{R}_p(\Omega)$. It is easy to see that the number of relations in (42) equals to $4N^2$. However, only $N(2N + 1)$ relations from them are independent.

Now we consider a symmetric $\mathbb{M} \in \mathbb{R}_{2p}^{2\times 2}(\Omega)$, which by Theorem 14.7 always has a complete orthonormal system of eigentensor columns and is represented in the form (30), where not all λ_k, $k = \overline{1, 2N}$, are positive (by Theorem 14.6 they are positive for positive-definite TBM). Moreover, the conditions of orthonormality (31) are valid. Our goal is to construct a complete orthonormal system of eigentensor columns in explicit form for the symmetric $\mathbb{M} \in \mathbb{R}_{2p}^{2\times 2}(\Omega)$.

3.2 Construction of the Eigentensor Columns of a TBM

From (30), it can be seen that to give the symmetric (not necessarily the positive-definite) TBM $\mathbb{M} \in \mathbb{R}_{2p}^{2\times 2}(\Omega)$ we must have $2N$ eigenvalues λ_k, $k = \overline{1, 2N}$, and $2N$ eigentensor columns $\mathbb{W}_k$, $k = \overline{1, 2N}$, which form a complete orthonormal system and, therefore, satisfy the conditions of orthonormality (31). It is easy to see that the number of independent relations in (31) equal to $(2N + 1)N$. These relations correlate $4N^2$ components of the eigentensor columns $\mathbb{W}_k = (\mathbf{U}_k, \mathbf{V}_k)$, $k = \overline{1, 2N}$. Hence, $(2N - 1)N$ components remain independent and can be used to construct the complete orthonormal system of tensor columns of the TBM $\mathbb{M} \in \mathbb{R}_{2p}^{2\times 2}(\Omega)$. In this connection, we consider the following tensors of the module $\mathbb{R}_{2p}(\Omega)$:

$$\begin{aligned} &\mathbf{Q}_{11} = \mathfrak{e}_s \mathbf{U}_s = U_{st} \mathfrak{e}_s \mathfrak{e}_t, && \mathbf{Q}_{12} = \mathfrak{e}_s \mathbf{V}_s = V_{st} \mathfrak{e}_s \mathfrak{e}_t, \\ &\mathbf{Q}_{21} = \mathfrak{e}_s \mathbf{U}_{N+s} = U_{N+st} \mathfrak{e}_s \mathfrak{e}_t, && \mathbf{Q}_{22} = \mathfrak{e}_s \mathbf{V}_{N+s} = V_{N+st} \mathfrak{e}_s \mathfrak{e}_t, \quad s, t = \overline{1, N}, \end{aligned} \tag{43}$$

where $\mathbb{W}_k = (\mathbf{U}_k, \mathbf{V}_k)^T, k = \overline{1, 2N}; \mathbf{e}_s, s = \overline{1, N}$, is an orthonormal basis of the module $\mathbb{R}_p(\Omega)$, i.e. $\mathbf{e}_s \overset{p}{\otimes} \mathbf{e}_t = \delta_{st}, s, t = \overline{1, N}$.

Further, by (43) we construct the following TBM of the module $\mathbb{R}_{2p}^{2\times 2}(\Omega)$:

$$\mathbb{Q} = \begin{pmatrix} \mathbf{Q}_{11} & \mathbf{Q}_{12} \\ \mathbf{Q}_{21} & \mathbf{Q}_{22} \end{pmatrix}. \tag{44}$$

If the tensor columns $\mathbb{W}_k = (\mathbf{U}_k, \mathbf{V}_k)^T, k = \overline{1, 2N}$, satisfy the orthonormality conditions (31), then taking into account (32), (43) and (44), it is easy to prove that the TBM $\mathbb{Q} \in \mathbb{R}_{2p}^{2\times 2}(\Omega)$ (44) satisfies the conditions

$$\mathbb{Q} \overset{p}{\otimes} \mathbb{Q}^T = \mathbb{Q}^T \overset{p}{\otimes} \mathbb{Q} = \mathbb{E}. \tag{45}$$

By (45) we conclude that $\mathbb{Q} \in \mathbb{R}_{2p}^{2\times 2}(\Omega)$ is an orthogonal TBM.

It is easy to check that relations (31) are equivalent to the relation

$$\mathbb{Q} \overset{p}{\otimes} \mathbb{Q}^T = \mathbb{E}. \tag{46}$$

Thus, the orthogonal TBM (44) of the module $\mathbb{R}_{2p}^{2\times 2}(\Omega)$ consists of $4N^2$ components related by equality (46), which is equivalent to $(2N + 1)N$ relations (31). Among the $4N^2$ components , we have $(2N - 1)N$ independent components. These components can be used to construct the complete orthonormal system of eigentensor columns of the symmetric TBM $\mathbb{M} \in \mathbb{R}_{2p}^{2\times 2}(\Omega)$. To accomplish this, we apply the theorem formulated in Nikabadze (2015) for the triangular decomposition of a square nonsingular tensor of even rank, which can be formulated for a square nonsingular TBM as follows (Nikabadze 2014, 2015).

Theorem 14.13 *In order for a square non-singular TBM* $\mathbb{M} \in \mathbb{R}_{2p}^{2\times 2}(\Omega)$ *to be decomposed into the product of a left triangular (unitriangular) TBM and a right unitriangular (triangular) TBM, it is necessary and sufficient that the determinants of all leading principal subtemperatures of this TBM (submatrices of component matrix) are not equal to zero.*

Note that there is a more general theorem obtained from Theorem 14.13 if in its formulation $\mathbb{R}_{2p}^{2\times 2}(\Omega)$ is replaced to $\mathbb{R}_{2p}^{m\times m}(\Omega)$, where m is an arbitrary natural number.

By Theorem 14.13, the TBM (44) can be represented in the form

$$\mathbb{Q} = \mathbb{L} \overset{p}{\otimes} \mathbb{R}, \tag{47}$$

where the left triangular TBM $\mathbb{L}$ and the right unitriangular TBM $\mathbb{R}$ have the form

$$\mathbb{L} = \begin{pmatrix} \mathbf{L}_{11} & 0 \\ \mathbf{L}_{21} & \mathbf{L}_{22} \end{pmatrix}, \quad \mathbb{R} = \begin{pmatrix} \mathbf{R}_{11} & \mathbf{R}_{12} \\ 0 & \mathbf{R}_{22} \end{pmatrix}. \tag{48}$$

Let the matrices L, L_{11}, L_{21}, L_{22} and R, R_{11}, R_{12}, R_{22} correspond to the tensor objects $\mathbb{L}$, $\mathbf{L}_{11}$, $\mathbf{L}_{21}$, $\mathbf{L}_{22}$ and $\mathbb{R}$, $\mathbf{R}_{11}$, $\mathbf{L}_{12}$, $\mathbf{L}_{22}$. Then, they can be represented as:

$$L = \begin{pmatrix} L_{11} & 0 \\ L_{21} & L_{22} \end{pmatrix}; \; R = \begin{pmatrix} R_{11} & R_{12} \\ 0 & R_{22} \end{pmatrix}; \; L_{11} = \text{matr}(l_{ij}), \; l_{ij} = 0, \quad i < j, \; i,j = \overline{1,N};$$
$$L_{21} = \text{matr}(l_{N+ij}), \; i,j = \overline{1,N}; \; L_{22} = \text{matr}(l_{N+iN+j}), \; l_{N+iN+j} = 0, \; i < j, \; i,j = \overline{1,N};$$
$$R_{11} = \text{matr}(r_{ij}), \; r_{ii} = 1, \; r_{ij} = 0, \; i > j, \; i,j = \overline{1,N}; \; R_{12} = \text{matr}(r_{iN+j}), \; i,j = \overline{1,N};$$
$$R_{22} = \text{matr}(r_{N+iN+j}), \quad r_{N+iN+i} = 1, \quad r_{N+iN+j} = 0, \quad i > j, \quad i,j = \overline{1,N}.$$

The left triangular tensor $\mathbf{L}_{11}$, the square tensor $\mathbf{L}_{21}$, the left triangular tensor $\mathbf{L}_{22}$, the right unitriangular tensor $\mathbf{R}_{11}$, the square tensor $\mathbf{R}_{12}$, and the right unitriangular tensor $\mathbf{R}_{22}$ can be represented as

$$\begin{aligned} &\mathbf{L}_{11} = \underset{\sim}{\mathbf{e}}_s \mathbf{l}_{s\cdot}, \quad \mathbf{L}_{21} = \underset{\sim}{\mathbf{e}}_s \mathbf{m}_{s\cdot}, \quad \mathbf{L}_{22} = \underset{\sim}{\mathbf{e}}_s \mathbf{l}_{N+s\cdot}, \\ &\mathbf{R}_{11} = \underset{\sim}{\mathbf{e}}_s \mathbf{r}_{s\cdot}, \quad \mathbf{R}_{12} = \underset{\sim}{\mathbf{e}}_s \mathbf{n}_{s\cdot}, \quad \mathbf{R}_{22} = \underset{\sim}{\mathbf{e}}_s \mathbf{r}_{N+s\cdot}, \quad s = \overline{1,N}, \end{aligned} \tag{49}$$

$$\begin{aligned} &\mathbf{l}_{s\cdot} = \sum_{t=1}^{s} l_{st} \underset{\sim}{\mathbf{e}}_t, \quad \mathbf{m}_{s\cdot} = \sum_{t=1}^{N} l_{N+st} \underset{\sim}{\mathbf{e}}_t, \quad \mathbf{l}_{N+s\cdot} = \sum_{t=1}^{s} l_{N+sN+t} \underset{\sim}{\mathbf{e}}_t, \quad s = \overline{1,N}, \\ &\mathbf{r}_{s\cdot} = \underset{\sim}{\mathbf{e}}_s + \sum_{t=s+1}^{N} r_{st} \underset{\sim}{\mathbf{e}}_t, \quad s = \overline{1,N-1}, \quad \mathbf{r}_{N\cdot} = \underset{\sim}{\mathbf{e}}_N; \quad \mathbf{n}_{s\cdot} = \sum_{t=1}^{N} r_{sN+t} \underset{\sim}{\mathbf{e}}_t, \quad s = \overline{1,N}, \\ &\mathbf{r}_{N+s\cdot} = \underset{\sim}{\mathbf{e}}_s + \sum_{t=s+1}^{N} r_{N+sN+t} \underset{\sim}{\mathbf{e}}_t, \quad s = \overline{1,N-1}, \quad \mathbf{r}_{2N\cdot} = \underset{\sim}{\mathbf{e}}_N. \end{aligned} \tag{50}$$

Taking into account (44) and (48) from (47), we get

$$\begin{pmatrix} \mathbf{Q}_{11} & \mathbf{Q}_{12} \\ \mathbf{Q}_{21} & \mathbf{Q}_{22} \end{pmatrix} = \begin{pmatrix} \mathbf{L}_{11} & \mathbf{0} \\ \mathbf{L}_{21} & \mathbf{L}_{22} \end{pmatrix} \overset{p}{\otimes} \begin{pmatrix} \mathbf{R}_{11} & \mathbf{R}_{12} \\ \mathbf{0} & \mathbf{R}_{22} \end{pmatrix} = \begin{pmatrix} \mathbf{L}_{11} \overset{p}{\otimes} \mathbf{R}_{11} & \mathbf{L}_{11} \overset{p}{\otimes} \mathbf{R}_{12} \\ \mathbf{L}_{21} \overset{p}{\otimes} \mathbf{R}_{11} & \mathbf{L}_{21} \overset{p}{\otimes} \mathbf{R}_{12} + \mathbf{L}_{22} \overset{p}{\otimes} \mathbf{R}_{22} \end{pmatrix}.$$

Hence, we find

$$\mathbf{Q}_{11} = \mathbf{L}_{11} \overset{p}{\otimes} \mathbf{R}_{11}, \;\; \mathbf{Q}_{12} = \mathbf{L}_{11} \overset{p}{\otimes} \mathbf{R}_{12}, \;\; \mathbf{Q}_{21} = \mathbf{L}_{21} \overset{p}{\otimes} \mathbf{R}_{11}, \;\; \mathbf{Q}_{22} = \mathbf{L}_{21} \overset{p}{\otimes} \mathbf{R}_{12} + \mathbf{L}_{22} \overset{p}{\otimes} \mathbf{R}_{22}. \tag{51}$$

By (43), (49), and (50) from (51), we have

$$\begin{aligned} &\mathbf{U}_s = \sum_{t=1}^{s} l_{st} \mathbf{r}_{t\cdot}, \quad \mathbf{U}_{N+s} = \sum_{t=1}^{N} l_{N+st} \mathbf{r}_{t\cdot}, \quad \mathbf{V}_s = \sum_{t=1}^{s} l_{st} \mathbf{n}_{t\cdot}, \\ &\mathbf{V}_{N+s} = \sum_{t=1}^{N} l_{N+st} \mathbf{n}_{t\cdot} + \sum_{t=1}^{s} l_{N+sN+t} \mathbf{r}_{N+t\cdot}, \quad s = \overline{1,N}. \end{aligned} \tag{52}$$

It should be noted that the systems of tensors

$$\mathbf{r}_{1\cdot}, \mathbf{r}_{2\cdot}, \ldots, \mathbf{r}_{N\cdot}; \quad \mathbf{r}_{N+1\cdot}, \mathbf{r}_{N+2\cdot}, \ldots, \mathbf{r}_{2N\cdot}; \quad \mathbf{U}_{1\cdot}, \mathbf{U}_{2\cdot}, \ldots, \mathbf{U}_{N\cdot}$$

in (52) are linearly independent. Let us introduce the tensor columns

$$\mathbb{T}_s = (\mathbf{r}_{s\cdot}, \mathbf{n}_{s\cdot})^T, \quad s = \overline{1, N}, \quad \mathbb{T}_t = (\mathbf{0}, \mathbf{r}_{t\cdot})^T, \quad t = \overline{N+1, 2N}, \tag{53}$$

where $\mathbf{0}$ is the pth rank zero tensor; $\mathbf{r}_{s\cdot}$, $\mathbf{n}_{s\cdot}$ and $\mathbf{r}_{N+s\cdot}$ are given by (50). It is not difficult to prove that the system of the tensor columns $\mathbb{T}_s$, $s = \overline{1, 2N}$, is linearly independent. Hence, the subsystems $\mathbb{T}_s, s = \overline{1, N}$, and $\mathbb{T}_t, t = \overline{N+1, 2N}$, are linearly independent.

It is easy to see that we have the following relations

$$\mathbb{W}_s = (\mathbf{U}_s, \mathbf{V}_s)^T = \sum_{t=1}^{s} l_{st}\mathbb{T}_t, \quad s = \overline{1, 2N}. \tag{54}$$

From (54) we conclude that the orthonormal system of the tensor columns $\mathbb{W}_s = (\mathbf{U}_s, \mathbf{V}_s)^T$, $s = \overline{1, 2N}$, can be obtained by applying the Gram–Schmidt orthogonalization procedure (Gantmacher 1959; Vekua 1978; Nikabadze 2009a) to the linearly independent system of tensor columns $\mathbb{T}_p, p = \overline{1, 18}$ (see (53)). Applying this procedure to the tensor columns $\mathbb{T}_s$, $s = \overline{1, 2N}$, we obtain the following expressions (Nikabadze 2014, 2015) for the coefficients l_{st}, $s = \overline{1, 2N}$, $t = \overline{1, s}$:

$$l_{11} = \frac{d_0}{\pm\sqrt{d_0 d_1}}, \quad l_{mt} = \frac{S^{(m)}_{mt}}{\pm\sqrt{d_{m-1} d_m}}, \quad S^{(m)}_{mt} = (-1)^{m+t} S\begin{pmatrix} 1\,2 \cdots t-1\;t\;t+1 \cdots m-1 \\ 1\,2 \cdots t-1\;t+1 \;\; \cdots \;\; m \end{pmatrix},$$
$$m = \overline{2, N}, \quad t = \overline{1, m}, \tag{55}$$

$$d_0 = 1, \quad d_m = \det S_m = \begin{vmatrix} s_{11} & s_{12} & \cdots & s_{1m} \\ \cdots & \cdots & \cdots & \cdots \\ s_{m1} & s_{m2} & \cdots & s_{mm} \end{vmatrix}, \quad s_{mt} = (\mathbb{T}_m, \mathbb{T}_t) = \mathbb{T}^T_m \overset{p}{\otimes} \mathbb{T}_t, \quad m, t = \overline{1, 2N},$$

$$S^{(m)}_{mt} = \begin{vmatrix} s_{11} & s_{12} & \cdots & s_{1t-1} & 0 & s_{1t+1} & \cdots & s_{1m} \\ \cdots & \cdots & \cdots & \cdots & \cdots & \cdots & \cdots & \cdots \\ s_{m-11} & s_{m-12} & \cdots & s_{m-1t-1} & 0 & s_{m-1t+1} & \cdots & s_{m-1m} \\ 0 & 0 & \cdots & 0 & 1 & 0 & \cdots & 0 \end{vmatrix}, \quad \langle m = \overline{1, 2N} \rangle, \quad t = \overline{1, m},$$

$$S\begin{pmatrix} 1\,2 \cdots t-1\;t\;t+1 \cdots m-1 \\ 1\,2 \cdots t-1\;t+1 \;\; \cdots \;\; m \end{pmatrix} = \begin{vmatrix} s_{11} & s_{12} & \cdots & s_{1t-1} & s_{1t+1} & \cdots & s_{1m} \\ \cdots & \cdots & \cdots & \cdots & \cdots & \cdots & \cdots \\ s_{m-11} & s_{m-12} & \cdots & s_{m-1t-1} & s_{m-1t+1} & \cdots & s_{m-1m} \end{vmatrix},$$

$S^{(m)}_{mt}$ is the cofactor of the element s_{mt} of the submatrix $S_m = \mathrm{matr}(s_{kl})$, $k, l = \overline{1, m}$;

$$S\begin{pmatrix} 1\,2 \cdots t-1\;t\;t+1 \cdots m-1 \\ 1\,2 \cdots t-1\;t+1 \;\; \cdots \;\; m \end{pmatrix}$$

is the $(m-1)$th order minor obtained by deleting the mth row and the tth column of the determinant $\det S_m$.

Taking into account (55), we can obtain from (54)

$$\mathbb{W}_1 = \frac{1}{\pm\sqrt{d_0 d_1}}\mathbb{T}_1, \quad \mathbb{W}_m = \frac{1}{\pm\sqrt{d_{m-1}d_m}}\begin{vmatrix} s_{11} & s_{12} & \cdots & s_{1m-1} & \mathbb{T}_1 \\ s_{21} & s_{22} & \cdots & s_{2m-1} & \mathbb{T}_2 \\ \cdots & \cdots & \cdots & \cdots & \cdots \\ s_{m1} & s_{m2} & \cdots & s_{mm-1} & \mathbb{T}_m \end{vmatrix}, \quad m = \overline{2, 2N}. \tag{56}$$

Using (56), the complete system of eigentensor columns $\mathbb{W}_m$, $m = \overline{1, 2N}$, of the symmetric TBM $\mathbb{M} \in \mathbb{R}_{2p}^{2\times 2}(\Omega)$ can be determined with the aid of the $(2N-1)N$ independent parameters being the elements of the unitriangular TBM $\mathbb{R}$ with respect to an arbitrary coordinate system (an arbitrary orthonormal basis of n-dimensional space). By choosing an appropriate coordinate system, we can decrease the number of independent parameters. Below, this is discussed in more detail in the particular case. Here, we shall not dwell on this.

In order to describe the inner structure of a symmetric TBM $\mathbb{M} \in \mathbb{R}_{2p}^{2\times 2}(\Omega)$, thus, it is sufficient to give the invariant characteristics of the TBM in some coordinate system rather than the components of the tensors $\mathbf{A}$, $\mathbf{B} = \mathbf{C}^T$ and $\mathbf{D}$. In other words, it is sufficient to give the eigenvalues λ_k, $k = \overline{1, 2N}$, and the corresponding eigentensor columns $\mathbb{W}_m$, $m = \overline{1, 2N}$, specified by $(2N-1)N$ independent parameters with respect to an arbitrary coordinate system. The number of independent parameters can be decreased in the case of a special coordinate system. These invariant characteristics can be used to compare and classify symmetric TBM of the module $\mathbb{R}_{2p}^{2\times 2}(\Omega)$. Here it is advisable to introduce a definition.

Definition 14.26 The symbol $\{\alpha_1, \alpha_2, \ldots, \alpha_k\}$, where k is the number of different eigenvalues of the TBM and α_i is the multiplicity of the eigenvalue λ_i $(i = 1, 2, \ldots, k)$, is called the symbol of the anisotropy (structure) of the TBM.

Below, we give a classification of the tensor-block matrices of the module $\mathbb{R}_4^{2\times 2}(\Omega)$, from which it is easy to see how to carry out a similar classification of the tensor-block matrices of the module $\mathbb{R}_{2p}^{2\times 2}(\Omega)$, as well as of the module $\mathbb{R}_{2p}^{m\times m}(\Omega)$, $m \geq 1$.

Note that such research for the symmetric TBDM of the module $\mathbb{R}_{2p}^{2\times 2}(\Omega)$, as a special case of symmetric TBM of the module $\mathbb{R}_{2p}^{2\times 2}(\Omega)$, does not require much labor.

3.3 *Eigenvalue Problem of a Tensor-Block Diagonal Matrix*

Above, it was determined that the TBDM $\mathbb{M} \in \mathbb{R}_{2p}^{m\times m}(\Omega)$, $m \geq 1$, has the form (12), and $\mathbb{M} \in \mathbb{R}_{2p}^{2\times 2}(\Omega)$, as a special case, is represented in the form (23). Thus, if we study the internal structure of the TBDM of the module $\mathbb{M} \in \mathbb{R}_{2p}^{2\times 2}(\Omega)$, then the results of its study can easily be extended to a similar matrix of the module $\mathbb{M} \in \mathbb{R}_{2p}^{m\times m}(\Omega)$. Therefore, we consider the matrix (23). It is easy to see that the characteristic equation of the TBDM (23) of the module $\mathbb{M} \in \mathbb{R}_{2p}^{2\times 2}(\Omega)$ has the form

$$\det(\mathbb{M} - \lambda\mathbb{E}) = \det(\mathbf{A} - \lambda\mathbf{E})\det(\mathbf{D} - \lambda\mathbf{E}) = 0. \tag{57}$$

By (57), we conclude that λ is an eigenvalue of the TBDM (23), if and only if it is an eigenvalue of tensor $\mathbf{A}$, or of tensor $\mathbf{D}$. If (23) is a positive-definite TBDM, then by Theorem 14.1 its subtensors $\mathbf{A}$ and $\mathbf{D}$ are also positive-definite, i.e. the eigenvalues of the tensors $\mathbf{A}$ and $\mathbf{D}$ are positive. In general, we have the following theorem.

Theorem 14.14 *A tensor-block-diagonal (tensor-block-triangular) matrix of the module $\mathbb{M} \in \mathbb{R}_{2p}^{m\times m}(\Omega)$ is a positive-definite matrix if and only if its subtensors (diagonal subtensors) are positive-definite.*

Here m is an arbitrary natural number. In this case $m = 2$.

Finding solutions of the characteristic equation (57) is equivalent to finding the solutions of the equations

$$\det(\mathbf{A} - \lambda\mathbf{E}) = 0, \quad \det(\mathbf{D} - \lambda\mathbf{E}) = 0. \tag{58}$$

It can be seen that each of Eq. (58) is the equation of Nth power with respect to λ. In the case of symmetric (positive-definite) TBDM (23), each from the equations has N real (positive) roots; here we take into account the possible multiplicity of these roots.

We note that the theorems for a triangular matrix given in Watkins (2002) in this case can be formulated as follows.

Theorem 14.15 *If λ_k, $1 \le k \le N$ is an eigenvalue of the subtensor $\mathbf{A}$, but it is not an eigenvalue of $\mathbf{D}$, then the eigentensor column $\mathbb{W}_k$ of the TBDM $\mathbb{M}$ (23) corresponding to λ_k has the form $\mathbb{W}_k = (\mathbf{U}_k, \mathbf{0})^T$, where $\mathbf{U}_k$ is the eigentensor of subtensor $\mathbf{A}$ corresponding to λ_k, and $\mathbf{0}$ is the zero pth rank tensor.*

Theorem 14.16 *If λ_m, $N + 1 \le m \le 2N$ is an eigenvalue of subtensor $\mathbf{D}$, but it is not an eigenvalue of $\mathbf{A}$, then the eigentensor column $\mathbb{W}_m$ of the TBDM $\mathbb{M}$ (23) corresponding to λ_m has the form $\mathbb{W}_m = (\mathbf{0}, \mathbf{V}_m)^T$, where $\mathbf{V}_m$ is the eigentensor of subtensor $\mathbf{D}$ corresponding to λ_m, and $\mathbf{0}$ is the zero pth rank tensor.*

Theorem 14.17 *If λ_k is a common eigenvalue of subtensors $\mathbf{A}$ and $\mathbf{D}$, then the eigentensor column $\mathbb{W}_k$ of the TBDM $\mathbb{M}$ (23) corresponding to λ_k has the form $\mathbb{W}_k = (\mathbf{U}_k, \mathbf{V}_k)^T$, where $\mathbf{U}_k$ and $\mathbf{V}_k$ are the eigentensors of subtensors $\mathbf{A}$ and $\mathbf{D}$ respectively, corresponding to the eigenvalue λ_k.*

If $\lambda_1, \lambda_2, \ldots, \lambda_N$ and $\mathbf{U}_1, \mathbf{U}_2, \ldots, \mathbf{U}_N$ are the eigenvalues and the corresponding eigentensors of the subtensor $\mathbf{A}$, and $\mu_1, \mu_2, \ldots, \mu_N$ and $\mathbf{V}_1, \mathbf{V}_2, \ldots, \mathbf{V}_N$ are the eigenvalues and the corresponding eigentensors of the subtensor $\mathbf{D}$, and $\{\lambda_1, \lambda_2, \ldots, \lambda_N\} \cap \{\mu_1, \mu_2, \ldots, \mu_N\} = \emptyset$, where $\emptyset$ is the empty set, then by Theorems 14.15 and 14.16 the TBDM $\mathbb{M}$ can be written as

$$\mathbb{M} = \sum_{k=1}^{N} \lambda_k \begin{pmatrix} \mathbf{U}_k \\ \mathbf{0} \end{pmatrix} (\mathbf{U}_k, \mathbf{0}) + \sum_{l=1}^{N} \mu_l \begin{pmatrix} \mathbf{0} \\ \mathbf{V}_l \end{pmatrix} (\mathbf{0}, \mathbf{V}_l) = \begin{pmatrix} \sum\limits_{k=1}^{N} \lambda_k \mathbf{U}_k \mathbf{U}_k & \mathbf{0} \\ \mathbf{0} & \sum\limits_{l=1}^{N} \mu_l \mathbf{V}_l \mathbf{V}_l \end{pmatrix}. \tag{59}$$

From (59) we have

$$\mathbf{A} = \sum_{k=1}^{N} \lambda_k \mathbf{U}_k \mathbf{U}_k, \quad \mathbf{D} = \sum_{l=1}^{N} \mu_l \mathbf{V}_l \mathbf{V}_l, \tag{60}$$

as it is required. If $\lambda_1 = \cdots = \lambda_N \equiv \lambda$, $\mu_1 = \cdots = \mu_N \equiv \mu$, then from (60) we have

$$\mathbf{A} = \lambda \sum_{k=1}^{N} \mathbf{U}_k \mathbf{U}_k = \lambda \mathbf{E}, \quad \mathbf{D} = \mu \sum_{l=1}^{N} \mathbf{V}_l \mathbf{V}_l = \mu \mathbf{E}. \tag{61}$$

Taking into account (61), in this case, the TBDM can be written as

$$\mathbb{M} = \begin{pmatrix} \lambda \mathbf{E} & \mathbf{0} \\ \mathbf{0} & \mu \mathbf{E} \end{pmatrix}. \tag{62}$$

If $\lambda = \mu$, then from (62) we get

$$\mathbb{M} = \lambda \begin{pmatrix} \mathbf{E} & \mathbf{0} \\ \mathbf{0} & \mathbf{E} \end{pmatrix} = \lambda \mathbb{E}. \tag{63}$$

Based on (62) and (63), is advisable to introduce a definition.

Definition 14.27 A TBDM, which has the form (62) and (63), is called isotropic (ideal isotropic).

Now assume that among the eigenvalues $\lambda_1, \ldots, \lambda_N$ and $\mu_1, \ldots, \mu_N$ of the tensors $\mathbf{A}$ and $\mathbf{D}$, the first m eigenvalues are equal, i.e. $\lambda_k = \mu_k$, $k = \overline{1, m}$. The eigentensors of these tensors are denoted as above, by $\mathbf{U}_k$ and $\mathbf{V}_k$, $k = \overline{1, N}$ respectively. Then by Theorem 14.17, the first m eigentensor columns of the TBDM $\mathbb{M} \in \mathbb{R}_{2p}^{2\times 2}(\Omega)$ corresponding to the eigenvalues $\lambda_k = \mu_k$, $k = \overline{1, m}$, get the form $\mathbb{W}_k = (\mathbf{U}_k, \mathbf{V}_k)^T$, $k = \overline{1, m}$, the tensors $\mathbf{A}$ and $\mathbf{D}$ are represented in the form

$$\mathbf{A} = \sum_{k=1}^{N} \lambda_k \mathbf{U}_k \mathbf{U}_k = \sum_{k=1}^{m} \lambda_k \mathbf{U}_k \mathbf{U}_k + \sum_{k=m+1}^{N} \lambda_k \mathbf{U}_k \mathbf{U}_k, \ \mathbf{D} = \sum_{l=1}^{m} \lambda_l \mathbf{V}_l \mathbf{V}_l + \sum_{l=m+1}^{N} \mu_l \mathbf{V}_l \mathbf{V}_l,$$

but the TBDM $\mathbb{M} \in \mathbb{R}_{2p}^{2\times 2}(\Omega)$ can be written as

$$\mathbb{M} = \sum_{k=1}^{m} \lambda_k \begin{pmatrix} \mathbf{U}_k \\ \mathbf{V}_k \end{pmatrix} (\mathbf{U}_k, \mathbf{V}_k) + \sum_{k=m+1}^{N} \left[\lambda_k \begin{pmatrix} \mathbf{U}_k \\ \mathbf{0} \end{pmatrix} (\mathbf{U}_k, \mathbf{0}) + \mu_k \begin{pmatrix} \mathbf{0} \\ \mathbf{V}_k \end{pmatrix} (\mathbf{0}, \mathbf{V}_k) \right].$$

Thus, from the above it follows that to study the internal structure of a TBDM $\mathbb{M} \in \mathbb{R}_{2p}^{2\times 2}(\Omega)$, it is sufficient to study the eigenvalue problem of a tensor of the module $\mathbb{R}_{2p}(\Omega)$, which has been studied in some detail in Nikabadze (2015) (see also Nikabadze 2009a). Therefore, we will not stop at this. Although, from the above, the similar research for any even rank symmetric tensor can be obtained as a special case. In fact, taking into account that if $\mathbf{A} \in \mathbb{R}_{2p}(\Omega)$ and $\mathbf{W} \in \mathbb{R}_p(\Omega)$, then $\mathbf{A} \in \mathbb{R}_{2p}^{1\times 1}(\Omega) = \mathbb{R}_{2p}(\Omega)$ and $\mathbf{W} \in \mathbb{R}_p^{1\times 1}(\Omega) = \mathbb{R}_p(\Omega)$, it is sufficient to replace the TBM of the module $\mathbb{R}_{2p}^{2\times 2}(\Omega)$ and tensor column of the module $\mathbb{R}_p^{2\times 1}(\Omega)$ with the tensors of the modules $\mathbb{R}_{2p}(\Omega) = \mathbb{R}_{2p}^{1\times 1}(\Omega)$ and $\mathbb{R}_p(\Omega) = \mathbb{R}_p^{1\times 1}(\Omega)$ in the above relations, respectively. Obviously, in this case, the tensor row is replaced with the tensor of the module $\mathbb{R}_p(\Omega)$, i.e. there is no need to introduce the concept of the tensor row.

The above material can be easily generalized to the eigenvalue problem for a TBM of the module $\mathbb{R}_{2p}^{m\times m}(\Omega)$ (see (11)). In fact, it is sufficient to replace the TBM of the module $\mathbb{R}_{2p}^{2\times 2}(\Omega)$ and tensor column (tensor row) of the module $\mathbb{R}_p^{2\times 1}(\Omega)$ ($\mathbb{R}_p^{1\times 2}(\Omega)$) with the TBM of the module $\mathbb{R}_{2p}^{m\times m}(\Omega)$ and tensor column (tensor row) of the module $\mathbb{R}_p^{m\times 1}(\Omega)$ ($\mathbb{R}_p^{1\times m}(\Omega)$) in the above relations, respectively. In addition, the values of the indices have to be changed accordingly.

Note also that, if necessary, it is not difficult to carry out similar studies for a generalized eigenvalue problem that can be formulated as follows. For some $\mathbb{M} \in \mathbb{R}_{2p}^{m\times m}(\Omega)$ and $\mathbb{N} \in \mathbb{R}_{2p}^{m\times m}(\Omega)$ find all tensor columns $\mathbb{W} \in \mathbb{R}_p^{m\times 1}(\Omega)$, which satisfy the equation

$$\mathbb{M} \overset{p}{\otimes} \mathbb{W} = \lambda \mathbb{N} \overset{p}{\otimes} \mathbb{W},$$

where λ is a scalar; m is an arbitrary fixed natural number.

4 Some Applications to Mechanics

4.1 Representations of the Specific Strain Energy and Constitutive Relations in the Linear Micropolar Theory of Elasticity

In the linear micropolar theory of elasticity, the specific strain energy and constitutive relations for the case of an anisotropic material with no symmetry center in the sense of elastic properties (Eringen 1999; Kupradze et al. 1976) can be written in the following form (isothermic processes are considered)[1]:

$$\Phi(\underset{\sim}{\gamma}, \underset{\sim}{\varkappa}) = \frac{1}{2}\left(A^{ijkl}\gamma_{ij}\gamma_{kl} + 2B^{ijkl}\gamma_{ij}\varkappa_{kl} + D^{ijkl}\varkappa_{ij}\varkappa_{kl}\right)$$

$$= \frac{1}{2}\left(\underset{\sim}{\gamma}\overset{2}{\otimes}\underset{\approx}{\mathbf{A}}\overset{2}{\otimes}\underset{\sim}{\gamma} + 2\underset{\sim}{\gamma}\overset{2}{\otimes}\underset{\approx}{\mathbf{B}}\overset{2}{\otimes}\underset{\sim}{\varkappa} + \underset{\sim}{\varkappa}\overset{2}{\otimes}\underset{\approx}{\mathbf{D}}\overset{2}{\otimes}\underset{\sim}{\varkappa}\right), \tag{64}$$

$$\underset{\sim}{\mathbf{P}} = \frac{\partial\Phi}{\partial\underset{\sim}{\gamma}} = \underset{\approx}{\mathbf{A}}\overset{2}{\otimes}\underset{\sim}{\gamma} + \underset{\approx}{\mathbf{B}}\overset{2}{\otimes}\underset{\sim}{\varkappa}, \quad \underset{\sim}{\mu} = \frac{\partial\Phi}{\partial\underset{\sim}{\varkappa}} = \underset{\approx}{\mathbf{C}}\overset{2}{\otimes}\underset{\sim}{\gamma} + \underset{\approx}{\mathbf{D}}\overset{2}{\otimes}\underset{\sim}{\varkappa}, \tag{65}$$

where $\underset{\sim}{\gamma} = \nabla\mathbf{u} - \underset{\simeq}{\mathbf{C}}\cdot\boldsymbol{\varphi}$ is the strain tensor; $\underset{\sim}{\varkappa} = \nabla\boldsymbol{\varphi}$ is the bending-torsion tensor; $\mathbf{u}$ is the displacement vector; $\boldsymbol{\varphi}$ is the internal rotation vector; $\underset{\simeq}{\mathbf{C}}$ is the third-rank discriminant tensor; $\underset{\approx}{\mathbf{A}}^T = \underset{\approx}{\mathbf{A}}$, $\underset{\approx}{\mathbf{D}}^T = \underset{\approx}{\mathbf{D}}$, $\underset{\approx}{\mathbf{C}}^T = \underset{\approx}{\mathbf{B}}$ are the fourth-rank material temperatures known as the elastic motion tensors; $\underset{\sim}{\mathbf{P}}$ is the stress tensor; $\underset{\sim}{\mu}$ is the couple-stress tensor; $\overset{2}{\otimes}$ is the inner 2-product.

Now we introduce the temperature column of the strain tensor and the bending-torsion tensor, the temperature column of the stress tensor and the couple-stress tensor, as well as the tensor-block matrix of the elastic motion tensors:

$$\underset{\sim}{\mathbb{X}} = \begin{pmatrix}\underset{\sim}{\gamma}\\ \underset{\sim}{\varkappa}\end{pmatrix} \left(\underset{\sim}{\mathbb{X}}^T = (\underset{\sim}{\gamma}, \underset{\sim}{\varkappa})\right), \quad \underset{\sim}{\mathbb{Y}} = \begin{pmatrix}\underset{\sim}{\mathbf{P}}\\ \underset{\sim}{\mu}\end{pmatrix} \left(\underset{\sim}{\mathbb{Y}}^T = (\underset{\sim}{\mathbf{P}}, \underset{\sim}{\mu})\right), \tag{66}$$

$$\underset{\approx}{\mathbb{M}} = \begin{pmatrix}\underset{\approx}{\mathbf{A}} & \underset{\approx}{\mathbf{B}}\\ \underset{\approx}{\mathbf{C}} & \underset{\approx}{\mathbf{D}}\end{pmatrix} \quad (\underset{\approx}{\mathbb{M}}^T = \underset{\approx}{\mathbb{M}}). \tag{67}$$

Hence, the specific strain energy and the constitutive relations can be rewritten as

$$2\Phi(\underset{\sim}{\gamma}, \underset{\sim}{\varkappa}) = \underset{\sim}{\mathbb{X}}^T\overset{2}{\otimes}\underset{\approx}{\mathbb{M}}\overset{2}{\otimes}\underset{\sim}{\mathbb{X}}, \quad \underset{\sim}{\mathbb{Y}} = \underset{\approx}{\mathbb{M}}\overset{2}{\otimes}\underset{\sim}{\mathbb{X}}. \tag{68}$$

If the material has the symmetry center in the sense of elastic properties, then $\underset{\approx}{\mathbf{B}} = \underset{\approx}{\mathbf{0}}$, where $\underset{\approx}{\mathbf{0}}$ is the fourth-rank zero tensor. As a result, the matrix expressed by (67) takes the block-diagonal form (see (23)).

It should be noted that the tensor columns $\underset{\sim}{\mathbb{X}}$ and $\underset{\sim}{\mathbb{Y}}$ are elements of the module $\mathbb{R}_2^{2\times 1}(\Omega)$, and the TBM $\underset{\approx}{\mathbb{M}}$ is an element of the module $\mathbb{R}_4^{2\times 2}(\Omega)$. Therefore, all of the above with respect to $\mathbf{A} \in \mathbb{R}_{2p}(\Omega)$ and $\mathbb{M} \in \mathbb{R}_{2p}^{2\times 2}(\Omega)$ is equally true for $\underset{\approx}{\mathbf{A}} \in \mathbb{R}_4(\Omega)$ and $\underset{\approx}{\mathbb{M}} \in \mathbb{R}_4^{2\times 2}(\Omega)$, respectively (here Ω is any domain of n-dimensional space in $\mathbb{R}_{2p}(\Omega)$ and $\mathbb{R}_{2p}^{2\times 2}(\Omega)$, and Ω is any domain of three-dimensional space in $\mathbb{R}_4(\Omega)$ and $\underset{\approx}{\mathbb{M}} \in \mathbb{R}_4^{2\times 2}(\Omega)$). In this regard, we shall not dwell on the consideration of some questions relating to $\underset{\approx}{\mathbf{A}} \in \mathbb{R}_4(\Omega)$ and $\underset{\approx}{\mathbb{M}} \in \mathbb{R}_4^{2\times 2}(\Omega)$. We will write the

[1] Further to indicate the second-rank, third-rank and fourth-rank tensors, we will use a wave, a wave and a line and the two waves from below, respectively.

necessary relations based on similar relations obtained above while we will refer to the definitions and theorems formulated above.

Hereafter, the two-index and one-index representations are used for asymmetric second-rank tensors, whereas the four-index and two-index representations are used for fourth-rank tensors. For example, let $\underset{\sim}{\mathbf{P}}$ be a second-rank tensor and $\underset{\approx}{\mathbf{A}}$ be a fourth-rank tensor. Then,

$$\underset{\sim}{\mathbf{P}} = P_{ij}\mathbf{e}_i\mathbf{e}_j - \sum_{m=1}^{9} P_m \underset{\sim}{\mathbf{e}}_m = P_m \underset{\sim}{\mathbf{e}}_m, \quad \mathbf{c}_i \cdot \mathbf{e}_j = \delta_{ij},$$

$$\underset{\approx}{\mathbf{A}} = A_{ijkl}\mathbf{e}_i\mathbf{e}_j\mathbf{e}_k\mathbf{e}_l = \sum_{m=1}^{9}\sum_{n=1}^{9} A_{mn}\underset{\sim}{\mathbf{e}}_m\underset{\sim}{\mathbf{e}}_n = A_{mn}\underset{\sim}{\mathbf{e}}_m\underset{\sim}{\mathbf{e}}_n, \quad i,j,k,l = 1,2,3, \quad m,n = \overline{1,9},$$

where $\underset{\sim}{\mathbf{e}}_m, m = \overline{1,9}$, are the temperatures of the orthonormalized basis for the second-rank temperature with respect to the inner 2-product:

$$\underset{\sim}{\mathbf{e}}_1 = \mathbf{e}_1\mathbf{e}_1, \quad \underset{\sim}{\mathbf{e}}_2 = \mathbf{e}_2\mathbf{e}_2, \quad \underset{\sim}{\mathbf{e}}_3 = \mathbf{e}_3\mathbf{e}_3, \quad \underset{\sim}{\mathbf{e}}_4 = \frac{1}{\sqrt{2}}(\mathbf{e}_1\mathbf{e}_2 + \mathbf{e}_1\mathbf{e}_2),$$

$$\underset{\sim}{\mathbf{e}}_5 = \frac{1}{\sqrt{2}}(\mathbf{e}_2\mathbf{e}_3 + \mathbf{e}_3\mathbf{e}_2), \quad \underset{\sim}{\mathbf{e}}_6 = \frac{1}{\sqrt{2}}(\mathbf{e}_3\mathbf{e}_1 + \mathbf{e}_1\mathbf{e}_3), \quad \underset{\sim}{\mathbf{e}}_7 = \frac{1}{\sqrt{2}}(\mathbf{e}_1\mathbf{e}_2 - \mathbf{e}_1\mathbf{e}_2),$$

$$\underset{\sim}{\mathbf{e}}_8 = \frac{1}{\sqrt{2}}(\mathbf{e}_2\mathbf{e}_3 - \mathbf{e}_3\mathbf{e}_2), \quad \underset{\sim}{\mathbf{e}}_9 = \frac{1}{\sqrt{2}}(\mathbf{e}_3\mathbf{e}_1 - \mathbf{e}_1\mathbf{e}_3), \quad \underset{\sim}{\mathbf{e}}_m \overset{2}{\otimes} \underset{\sim}{\mathbf{e}}_m = \delta_{mn}, m,n = \overline{1,9}.$$

Here, δ_{pq} is the Kronecker symbol.

Further, we note that due to the positive definiteness of the specific strain energy (Eringen 1999; Kupradze et al. 1976) and Definition 14.15, the TBM (67) of the elastic modulus tensors is positive-definite, and based on Theorem 14.1, the elastic modulus tensors (subtensors) $\underset{\approx}{\mathbf{A}}$, and $\underset{\approx}{\mathbf{D}}$ are positive-definite. We also note that the TBM (67) differs from (19). In this case $\underset{\approx}{\mathbf{A}}$, $\underset{\approx}{\mathbf{B}}$, $\underset{\approx}{\mathbf{C}}$ and $\underset{\approx}{\mathbf{D}}$ are the tensors of the module $\mathbb{R}_4(\Omega)$, where Ω is a domain of the three-dimensional space, i.e. $\mathbb{M} \in \mathbb{R}_4^{2\times2}(\Omega)$, $N = 9$. Hence, the characteristic equation (39) for the positive-definite TBM (67) will have 18th power and 18 positive roots (eigenvalues). Here, we take into account the possible multiplicity of these roots. Let $\lambda_1, \ldots, \lambda_{18}$ are the roots of the equation. Numbering them in descending order, we have $\lambda_1 \geq \cdots \geq \lambda_{18} > 0$. In this case, a complete orthonormal system of eigentensor columns of the TBM (67) consists of 18 tensor columns. Let $\mathbb{W}_p = (\underset{\sim}{\mathbf{u}}_p, \underset{\sim}{\mathbf{v}}_p)^T, p = \overline{1,18}$, be the complete orthonormal system of eigentensor columns of the TBM (67), corresponding to λ_p, $p = \overline{1,18}$, respectively. Then, the matrix (67) can be represented as follows:

$$\underset{\approx}{\mathbb{M}} = \sum_{p=1}^{18} \lambda_p \mathbb{W}_p \otimes \mathbb{W}_p^T = \sum_{p=1}^{18} \lambda_p \begin{pmatrix} \underset{\sim}{\mathbf{u}}_p \\ \underset{\sim}{\mathbf{v}}_p \end{pmatrix} \otimes \begin{pmatrix} \underset{\sim}{\mathbf{u}}_p & \underset{\sim}{\mathbf{v}}_p \end{pmatrix} = \sum_{p=1}^{18} \lambda_p \begin{pmatrix} \underset{\sim}{\mathbf{u}}_p \otimes \underset{\sim}{\mathbf{u}}_p & \underset{\sim}{\mathbf{u}}_p \otimes \underset{\sim}{\mathbf{v}}_p \\ \underset{\sim}{\mathbf{v}}_p \otimes \underset{\sim}{\mathbf{u}}_p & \underset{\sim}{\mathbf{v}}_p \otimes \underset{\sim}{\mathbf{v}}_p \end{pmatrix}, \tag{69}$$

where the orthonormality conditions are valid

$$(\mathbb{W}_p, \mathbb{W}_q) = \mathbb{W}_p^T \overset{2}{\otimes} \mathbb{W}_q = \underset{\sim}{\mathbf{u}}_p \overset{2}{\otimes} \underset{\sim}{\mathbf{u}}_q + \underset{\sim}{\mathbf{v}}_p \overset{2}{\otimes} \underset{\sim}{\mathbf{v}}_q = \delta_{pq}, \quad p,q = \overline{1,18}. \tag{70}$$

Next, analogous to (37), the relation

$$\underset{\approx}{\mathbb{M}}^{\alpha} = \sum_{p=1}^{18} \lambda_p^{\alpha} \mathbb{W}_p \mathbb{W}_p^T = \sum_{p=1}^{18} \lambda_p^{\alpha} \begin{pmatrix} \mathbb{U}_p \mathbb{U}_p & \mathbb{U}_p \mathbb{V}_p \\ \mathbb{V}_p \mathbb{U}_p & \mathbb{V}_p \mathbb{V}_p \end{pmatrix}.$$

is true for any integer α. In particular, for $\alpha = -1$ and $\alpha = 0$ similar to (38), we have

$$\begin{aligned} \underset{\approx}{\mathbb{M}}^{-1} &= \sum_{p=1}^{18} \lambda_p^{-1} \mathbb{W}_p \mathbb{W}_p^T = \sum_{p=1}^{18} \lambda_p^{-1} \begin{pmatrix} \mathbb{U}_p \mathbb{U}_p & \mathbb{U}_p \mathbb{V}_p \\ \mathbb{V}_p \mathbb{U}_p & \mathbb{V}_p \mathbb{V}_p \end{pmatrix}, \\ \underset{\approx}{\mathbb{E}} = \underset{\approx}{\mathbb{M}}^{0} &= \sum_{p=1}^{18} \mathbb{W}_p \otimes \mathbb{W}_p^T = \sum_{p=1}^{18} \begin{pmatrix} \mathbb{U}_p \\ \mathbb{V}_p \end{pmatrix} \otimes \left(\mathbb{U}_p \ \mathbb{V}_p \right) = \sum_{p=1}^{18} \begin{pmatrix} \mathbb{U}_p \otimes \mathbb{U}_p & \mathbb{U}_p \otimes \mathbb{V}_p \\ \mathbb{V}_p \otimes \mathbb{U}_p & \mathbb{V}_p \otimes \mathbb{V}_p \end{pmatrix}. \end{aligned} \tag{71}$$

Next, we assume that the above relations for the specific strain energy and constitutive relations are written in dimensionless quantities since it can always be done.

4.2 Presentations of the Specific Strain Energy and the Constitutive Relations Using Eigenvalues and Eigentensor Columns

By virtue of (69), from (68) we get

$$2\Phi(\underset{\sim}{\gamma}, \underset{\sim}{\varkappa}) = \sum_{p=1}^{18} \lambda_p \mathbb{X}^T \overset{2}{\otimes} \mathbb{W}_p \mathbb{W}_p^T \overset{2}{\otimes} \mathbb{X}, \quad \mathbb{Y} = \sum_{p=1}^{18} \lambda_p \mathbb{W}_p \mathbb{W}_p^T \overset{2}{\otimes} \mathbb{X}. \tag{72}$$

Multiplying scalarly the second relation of (72) by $\mathbb{W}_\alpha$ and taking into account (70), the constitutive relations can be rewritten as

$$(\mathbb{Y}, \mathbb{W}_\alpha) = \lambda_\alpha (\mathbb{X}, \mathbb{W}_\alpha) \ \left(\mathbb{Y}^T \overset{2}{\otimes} \mathbb{W}_\alpha = \lambda_\alpha \mathbb{X}^T \overset{2}{\otimes} \mathbb{W}_\alpha \right), \ \langle \alpha = \overline{1, 18} \rangle. \tag{73}$$

Note that by the formulas (73) the equivalent records of constitutive relations are given.

Introducing the notations ($\langle \alpha = \overline{1, 18} \rangle$ means that there is no summation over α):

$$\begin{aligned} \mathbb{X}_\alpha &= (\mathbb{X}, \mathbb{W}_\alpha) = \mathbb{X}^T \overset{2}{\otimes} \mathbb{W}_\alpha = \mathbb{W}_\alpha^T \overset{2}{\otimes} \mathbb{X} = \mathbb{U}_\alpha \overset{2}{\otimes} \underset{\sim}{\gamma} + \mathbb{V}_\alpha \overset{2}{\otimes} \underset{\sim}{\varkappa}, \\ \mathbb{Y}_\alpha &= (\mathbb{Y}, \mathbb{W}_\alpha) = \mathbb{Y}^T \overset{2}{\otimes} \mathbb{W}_\alpha = \mathbb{W}_\alpha^T \overset{2}{\otimes} \mathbb{Y} = \mathbb{U}_\alpha \overset{2}{\otimes} \underset{\sim}{\mathbf{P}} + \mathbb{V}_\alpha \overset{2}{\otimes} \underset{\sim}{\boldsymbol{\mu}}, \ \langle \alpha = \overline{1, 18} \rangle, \end{aligned} \tag{74}$$

the specific strain energy (the first relation (72)) and the constitutive relations (73) can be represented in the form

$$2\Phi(\underset{\sim}{\gamma}, \underset{\approx}{\varkappa}) = \sum_{p=1}^{18} \lambda_p \mathbb{X}_p^2, \quad \mathbb{Y}_\alpha = \lambda_\alpha \mathbb{X}_\alpha, \ \langle \alpha = \overline{1, 18} \rangle. \tag{75}$$

Multiplying both sides of the equalities of (74) by $\mathbb{W}_\alpha$ and summing the resulting relations from 1 to 18, by virtue of second relation of (71), we obtain

$$\begin{aligned} \underset{\approx}{\mathbb{X}} &= \sum_{\alpha=1}^{18} \mathbb{X}_\alpha \mathbb{W}_\alpha = \sum_{\alpha=1}^{18} (\underset{\sim}{\mathbf{u}}_\alpha \overset{2}{\otimes} \underset{\sim}{\gamma} + \underset{\sim}{\mathbf{v}}_\alpha \overset{2}{\otimes} \underset{\approx}{\varkappa}) \mathbb{W}_\alpha, \\ \underset{\approx}{\mathbb{Y}} &= \sum_{\alpha=1}^{18} \mathbb{Y}_\alpha \mathbb{W}_\alpha = \sum_{\alpha=1}^{18} (\underset{\sim}{\mathbf{u}}_\alpha \overset{2}{\otimes} \underset{\sim}{\gamma} + \underset{\sim}{\mathbf{v}}_\alpha \overset{2}{\otimes} \underset{\approx}{\varkappa}) \mathbb{W}_\alpha. \end{aligned} \tag{76}$$

The formulas (76) are the decompositions of the temperature columns $\underset{\approx}{\mathbb{X}}$ and $\underset{\approx}{\mathbb{Y}}$ in the orthonormal basis $\mathbb{W}_\alpha, \alpha = \overline{1, 18}$, where $\mathbb{X}_\alpha$ and $\mathbb{Y}_\alpha$ are the projections of $\underset{\approx}{\mathbb{X}}$ and $\underset{\approx}{\mathbb{Y}}$ onto $\mathbb{W}_\alpha$.

Taking into account the first equality of (74), from the second relation of (72) we come to the following representations for the stress temperature and the couple-stress tensor:

$$\begin{aligned} \underset{\sim}{\mathbf{P}} &= \sum_{p=1}^{18} \lambda_p \mathbb{X}_p \underset{\sim}{\mathbf{u}}_p = \sum_{p=1}^{18} \lambda_p (\underset{\sim}{\mathbf{u}}_p \overset{2}{\otimes} \underset{\sim}{\gamma} + \underset{\sim}{\mathbf{v}}_p \overset{2}{\otimes} \underset{\approx}{\varkappa}) \underset{\sim}{\mathbf{u}}_p, \\ \underset{\sim}{\mu} &= \sum_{p=1}^{18} \lambda_p \mathbb{X}_p \underset{\sim}{\mathbf{v}}_p = \sum_{p=1}^{18} \lambda_p (\underset{\sim}{\mathbf{u}}_p \overset{2}{\otimes} \underset{\sim}{\gamma} + \underset{\sim}{\mathbf{v}}_p \overset{2}{\otimes} \underset{\approx}{\varkappa}) \underset{\sim}{\mathbf{v}}_p. \end{aligned} \tag{77}$$

It is not difficult to obtain the inverse constitutive relations. Taking into account the first relation of (71) and the second formula of (75), from the second equality of (68) we get

$$\underset{\approx}{\mathbb{X}} = \underset{\approx}{\mathbb{M}}^{-1} \overset{2}{\otimes} \underset{\approx}{\mathbb{Y}} = \sum_{p=1}^{18} \lambda_p^{-1} \mathbb{W}_p \mathbb{W}_p^T \overset{2}{\otimes} \underset{\approx}{\mathbb{Y}} = \sum_{p=1}^{18} \lambda_p^{-1} \mathbb{Y}_p \mathbb{W}_p, \mathbb{X}_\alpha = \lambda_\alpha^{-1} \mathbb{Y}_\alpha, \ \langle \alpha = \overline{1, 18} \rangle. \tag{78}$$

Taking into account the second equality of (74), from the first relation of (78) we obtain

$$\begin{aligned} \underset{\sim}{\gamma} &= \sum_{p=1}^{18} \lambda_p^{-1} \mathbb{Y}_p \underset{\sim}{\mathbf{u}}_p = \sum_{p=1}^{18} \lambda_p^{-1} (\underset{\sim}{\mathbf{u}}_p \overset{2}{\otimes} \underset{\sim}{\mathbf{P}} + \underset{\sim}{\mathbf{v}}_p \overset{2}{\otimes} \underset{\sim}{\mu}) \underset{\sim}{\mathbf{u}}_p, \\ \underset{\approx}{\varkappa} &= \sum_{p=1}^{18} \lambda_p^{-1} \mathbb{Y}_p \underset{\sim}{\mathbf{v}}_p = \sum_{p=1}^{18} \lambda_p^{-1} (\underset{\sim}{\mathbf{u}}_p \overset{2}{\otimes} \underset{\sim}{\mathbf{P}} + \underset{\sim}{\mathbf{v}}_p \overset{2}{\otimes} \underset{\sim}{\mu}) \underset{\sim}{\mathbf{v}}_p. \end{aligned}$$

4.3 Construction of the Eigentensor Columns of the TBM of Elastic Modulus Tensor

In this case, the expressions for the eigentensor columns can be obtained from the formula (56) for $N = 9$. Therefore, we will not dwell on this. However we note that using (56) for $N = 9$ the eigentensor columns $\mathbb{W}_p \in \mathbb{R}_2^{2\times1}(\Omega), p = \overline{1, 18}$, of the TBM $\underset{\sim}{\mathbb{M}} \in \mathbb{R}_4^{2\times2}(\Omega)$ are determined using 153 independent parameters (components of the unitriangular TBM $\underset{\approx}{\mathbb{R}} \in \mathbb{R}_4^{2\times2}(\Omega)$) with respect to an arbitrary coordinate system (an arbitrary basis $\mathbf{e}_p, p = 1, 2, 3$). Obviously, the number of independent parameters can be reduced by the choice of coordinate system (see, for example, Novozhilov 1961). It depends on the type of roots of the characteristic equation, for example, of the tensor $\mathbf{r}_{1\cdot\cdot}$. If all three roots of the tensor $\mathbf{r}_{1\cdot}$ are a simple real numbers, or among them there are two complex-conjugate roots, or the characteristic equation has the triple root, then the number of independent parameters becomes equal to 147.

Thus, to describe the mechanical properties of a micropolar material that does not have a center of symmetry in the sense of the elastic properties, it is sufficient instead of the components of elastic modulus tensors $\underset{\approx}{\mathbf{A}}$, $\underset{\approx}{\mathbf{B}} = \underset{\approx}{\mathbf{C}}^T$ and $\underset{\approx}{\mathbf{D}}$ to give invariant characteristics of the TBM $\underset{\sim}{\mathbb{M}}$ (67) in some coordinate system, i.e. the eigenvalues $\lambda_k > 0, k = \overline{1, 18}$ and the corresponding eigentensor columns $\underset{\approx}{\mathbb{W}}_p, p = \overline{1, 18}$. They are defined by using 153 independent parameters with respect to an arbitrary coordinate system or 147 independent parameters in the case of a special coordinate system.

Note that these invariant characteristics can be used to compare and classify micropolar linear elastic anisotropic materials without a symmetry center in the sense of elastic properties. Hence, it is not difficult to conduct a similar study for a micropolar material with a symmetry center as a particular case. For the classical linear elastic material, as noted above, similar research was carried out by Ostrosablin (1984, 1986a, b, 2000).

4.4 Micropolar Material with a Center of Symmetry

In this case, $\underset{\approx}{\mathbf{B}} = \underset{\approx}{\mathbf{C}}^T = \underset{\approx}{\mathbf{0}}$ and the constitutive relations (65) and specific strain energy (64) can be written as

$$\underset{\sim}{\mathbf{P}} = \underset{\approx}{\mathbf{A}} \overset{2}{\otimes} \underset{\sim}{\gamma}, \quad \underset{\sim}{\mu} = \underset{\approx}{\mathbf{D}} \overset{2}{\otimes} \underset{\sim}{\varkappa}, \quad 2\Phi(\underset{\sim}{\gamma}, \underset{\sim}{\varkappa}) = \mathbb{X}^T \overset{2}{\otimes} \underset{\sim}{\mathbb{M}} \overset{2}{\otimes} \mathbb{X} = \underset{\sim}{\gamma} \overset{2}{\otimes} \underset{\approx}{\mathbf{A}} \overset{2}{\otimes} \underset{\sim}{\gamma} + \underset{\sim}{\varkappa} \overset{2}{\otimes} \underset{\approx}{\mathbf{D}} \overset{2}{\otimes} \underset{\sim}{\varkappa}.$$

The matrix (66) will receive the form of the TBDM. Then, the characteristic equation is equivalent to the following equations:

$$\det(\underset{\approx}{\mathbf{A}} - \lambda \underset{\approx}{\mathbf{E}}) = 0, \quad \det(\underset{\approx}{\mathbf{D}} - \lambda \underset{\approx}{\mathbf{E}}) = 0. \tag{79}$$

It is seen that each of Eq. (79) is an equation of 9th power with respect to λ and has 9 positive roots. Here we take into account the possible multiplicity of these roots. The set of roots of each Eq. (79) can be arranged in descending order. Let $\lambda_1, \lambda_2, \ldots, \lambda_9$ be the roots of the first Eq. (79) and $\lambda_{10}, \lambda_{11}, \ldots, \lambda_{18}$ be the roots of the second equation. Then, obviously, we have $\lambda_1 \geq \cdots \geq \lambda_9 > 0$, $\lambda_{10} \geq \cdots \geq \lambda_{18} > 0$. In the future, we consider such materials, for which the condition $\{\lambda_1, \ldots, \lambda_9\} \cap \{\lambda_{10}, \ldots, \lambda_{18}\} = \emptyset$ is satisfied, where $\emptyset$ denotes the empty set. In other words, Eq. (79) have no common roots (in general, Eq. (79) may have common roots).

Let $\mathbb{W}_p, p = \overline{1, 18}$ be the eigentensor columns of the TBDM $\underset{\approx}{\mathbb{M}}$ corresponding to $\lambda_p, p = \overline{1, 18}$. Then, we have

$$\underset{\approx}{\mathbb{M}} = \begin{pmatrix} \underset{\approx}{\mathbf{A}} & \underset{\approx}{\mathbf{0}} \\ \underset{\approx}{\mathbf{0}} & \underset{\approx}{\mathbf{D}} \end{pmatrix} = \sum_{p=1}^{18} \lambda_p \mathbb{W}_p \mathbb{W}_p^T.$$

All that was said in Sect. 3.3, remains valid in this case. In particular, from all formulas of Sect. 3.3 we obtain the corresponding formulas if N is replaced by 9. So we will not stop at this. Note only, that for a given anisotropy of material with a center of symmetry, the tensors $\underset{\approx}{\mathbf{A}}$ and $\underset{\approx}{\mathbf{D}}$ have the same structure corresponding to this anisotropy. For example, in the case of an isotropic material, both of these tensors are isotropic, in the case of a transversely isotropic material, these tensors are transversely isotropic, etc. If so, then to give a classification of the micropolar anisotropic elastic materials having a center of symmetry, it is enough to consider, for example, the tensor $\underset{\approx}{\mathbf{A}}$ since for a given material anisotropy the tensor $\underset{\approx}{\mathbf{D}}$ has the same structure as the tensor $\underset{\approx}{\mathbf{A}}$. Although, according to the author, tensors $\underset{\approx}{\mathbf{A}}$ and $\underset{\approx}{\mathbf{D}}$ can have different structures.

Thus, from the above it follows that to study the internal structure of a TBDM $\mathbb{M} \in \mathbb{R}_4^{2\times 2}(\Omega)$ it is sufficient to consider the eigenvalue problem, say, of a tensor $\underset{\approx}{\mathbf{A}} \in \mathbb{R}_4(\Omega)$.

5 Eigenvalue Problem and Construction of the Complete System of Eigentensors for a Symmetric Fourth Rank Tensor

All that has been said above for the tensor $\mathbf{A} \in \mathbb{R}_{2p}(\Omega)$ remains valid for the tensor $\underset{\approx}{\mathbf{A}} \in \mathbb{R}_4(\Omega)$. In particular, the relations for $\underset{\approx}{\mathbf{A}} \in \mathbb{R}_4(\Omega)$ are obtained from the corresponding relations for $\mathbf{A} \in \mathbb{R}_{2p}(\Omega)$ if we assume Ω is a three-dimensional domain and we replace $\mathbf{A}$, n, p, and N with $\underset{\approx}{\mathbf{A}}$, 3, 2, and 9 respectively. Of course, a similar study for the symmetric tensor $\underset{\approx}{\mathbf{A}} \in \mathbb{R}_4(\Omega)$ can be carried out as a special case. In fact, for this it is sufficient to consider that if $\mathbf{A} \in \mathbb{R}_4(\Omega)$ and $\mathbf{W} \in \mathbb{R}_2(\Omega)$, then $\mathbf{A} \in \mathbb{R}_4^{1\times 1}(\Omega) = \mathbb{R}_4(\Omega)$ and $\mathbf{W} \in \mathbb{R}_2^{1\times 1}(\Omega) = \mathbb{R}_2(\Omega)$. And then the TBM of the module $\mathbb{R}_4^{2\times 2}(\Omega)$ and the tensor column of the module $\mathbb{R}_2^{2\times 1}(\Omega)$ is replaced with the tensors of the modules $\mathbb{R}_4(\Omega) = \mathbb{R}_4^{1\times1}(\Omega)$ and $\mathbb{R}_2(\Omega) = \mathbb{R}_2^{1\times1}(\Omega)$ in the above relations, respectively. Obviously, in this case, the tensor row is replaced with a tensor of the module $\mathbb{R}_2(\Omega)$, i.e. there is no need to introduce the concept of the tensor row.

6 Classification of the Micropolar Linearly Elastic Anisotropic Materials Without a Center of Symmetry

We introduce a definition of the symbol of anisotropy (structure) of micropolar linear-elastic anisotropic materials without a center of symmetry.

Definition 14.28 The symbol $\{\alpha_1, \alpha_2, \dots, \alpha_k\}$, where k is the number of distinct eigenvalues of the TBM of the elastic modulus tensors, and α_i is the multiplicity of the eigenvalue λ_i $(i = 1, 2, \dots, k)$, is called the symbol of anisotropy (structure) of micropolar linear-elastic materials without a center of symmetry.

In this case, we have

$$\alpha_1 + \alpha_2 + \ldots + \alpha_k = 18, \quad 1 \le \alpha_i \le 18 - (k-1) = 19 - k, \ i = \overline{1,k}, \ 1 \le k \le 18.$$

There are 18 classes (groups). Each class contains several subclasses (subgroups). All classes are listed below. For each class, a symbol of anisotropy is given. For some materials, the representation of TBM is given. The number of materials in each class is expressed by the corresponding binomial coefficient.

6.1 *Symbol of Anisotropy Consisting of One Element*

$$\{\alpha\}, \quad \alpha = 18, \quad \lambda \equiv \lambda_1 = \lambda_2 = \dots = \lambda_{18}.$$

The total number of such materials is equal to $C_{17}^0 = 1$. The TBM $\underset{\approx}{\mathbb{M}}$ has the form

$$\underset{\approx}{\mathbb{M}} = \sum_{p=1}^{18} \lambda_p \mathbb{W}_p \mathbb{W}_p^T = \lambda \sum_{p=1}^{18} \mathbb{W}_p \mathbb{W}_p^T = \lambda \underset{\approx}{\mathbb{E}}.$$

6.2 *Symbol of Anisotropy Consisting of Two Elements*

$$\{\alpha_1, \alpha_2\}, \quad \alpha_1 + \alpha_2 = 18, \quad 1 \le \alpha_m \le 17, \quad m = 1, 2,$$
$$\{1,17\}, \ \{2,16\}, \ \{3,15\}, \ \{4,14\}, \ \{5,13\}, \ \{6,12\}, \ \{7,11\}, \ \{8,10\},$$
$$\{9,9\}, \ \{10,8\}, \ \{11,7\}, \ \{12,6\}, \ \{13,5\}, \ \{14,4\}, \ \{15,3\}, \ \{16,2\}, \ \{17,1\}.$$

The total number of such materials is equal to $C_{17}^{1} = 17$. The TBM $\underset{\approx}{\mathbb{M}}$ corresponding to, say, the material {1, 17} has the representation

$$\underset{\approx}{\mathbb{M}} = \lambda_1 \mathbb{W}_1 \mathbb{W}_1^T + \lambda_2 \sum_{p=2}^{18} \mathbb{W}_p \mathbb{W}_p^T = (\lambda_1 - \lambda_2) \mathbb{W}_1 \mathbb{W}_1^T + \lambda_2 \underset{\approx}{\mathbb{E}}.$$

6.3 Symbol of Anisotropy Consisting of Three Elements

$$\{\alpha_1, \alpha_2, \alpha_3\}, \quad \alpha_1 + \alpha_2 + \alpha_3 = 18, \quad 1 \leq \alpha_m \leq 16, \quad m = 1, 2, 3,$$
$$\{1, 1, 16\},\ \{1, 2, 15\},\ \{1, 3, 14\},\ \{1, 4, 13\},\ \{1, 5, 12\},\ \{1, 6, 11\},\ \{1, 7, 10\},$$
$$\{1, 8, 9\},\ \{1, 9, 8\},\ \{1, 10, 7\},\ \{1, 11, 6\},\ \{1, 12, 5\},\ \{1, 13, 4\},\ \{1, 14, 3\},$$
$$\{1, 15, 2\},\ \{1, 16, 1\};\ \{2, 1, 15\},\ \{2, 2, 14\}, \{2, 3, 13\},\ \{2, 4, 12\},\ \{2, 5, 11\},$$
$$\{2, 6, 10\},\ \{2, 7, 9\}\ \{2, 8, 8\},\ \{2, 9, 7\},\ \{2, 10, 6\},\ \{2, 11, 5\},\ \{2, 12, 4\},\ \{2, 13, 3\},$$
$$\{2, 14, 2\},\ \{2, 15, 1\};\ \{3, 1, 14\}, \ldots, \{3, 14, 1\}; \ldots,\ \{15, 1, 2\}, \{15, 2, 1\};\ \{16, 1, 1\}.$$

The total number of such materials is equal to $C_{17}^{2} = 136$. The TBM $\underset{\approx}{\mathbb{M}}$, for example, for the material {1, 1, 16} has the form

$$\begin{aligned}\underset{\approx}{\mathbb{M}} &= \lambda_1 \mathbb{W}_1 \mathbb{W}_1^T + \lambda_2 \mathbb{W}_2 \mathbb{W}_2^T + \lambda_3 \sum_{p=3}^{18} \mathbb{W}_p \mathbb{W}_p^T \\ &= (\lambda_1 - \lambda_3) \mathbb{W}_1 \mathbb{W}_1^T + (\lambda_2 - \lambda_3) \mathbb{W}_2 \mathbb{W}_2^T + \lambda_3 \underset{\approx}{\mathbb{E}}.\end{aligned}$$

Other classes of the anisotropy contain a larger number of materials and there is no sense to write them. Therefore, below for the other classes, we indicate the symbol of anisotropy and the corresponding number of materials.

6.4 Symbol of Anisotropy Consisting of Four Elements

$$\{\alpha_1, \alpha_2, \alpha_3, \alpha_4\}, \quad \alpha_1 + \alpha_2 + \alpha_3 + \alpha_4 = 18, \quad 1 \leq \alpha_m \leq 15, \quad m = 1, 2, 3, 4,$$
$$\{1, 1, 1, 15\}, \ldots, \{15, 1, 1, 1\}.$$

The total number of such materials is equal to $C_{17}^{3} = 680$.

6.5 Symbol of Anisotropy Consisting of Five Elements

$$\{\alpha_1, \alpha_2, \ldots, \alpha_5\}, \ \ \alpha_1 + \alpha_2 + \cdots + \alpha_5 = 18, \ \ 1 \leq \alpha_m \leq 14, \ \ m = \overline{1, 5},$$
$$\{1, 1, 1, 1, 14\}, \ldots, \{14, 1, 1, 1, 1\}.$$

The total number of such materials is equal to $C_{17}^{4} = 2380$.

6.6 Symbol of Anisotropy Consisting of Six Elements

$$\{\alpha_1, \alpha_2, \ldots, \alpha_6\}, \ \ \alpha_1 + \alpha_2 + \cdots + \alpha_6 = 18, \ \ 1 \leq \alpha_m \leq 13, \ \ m = \overline{1, 6},$$
$$\{1, 1, 1, 1, 1, 13\}, \ldots, \{13, 1, 1, 1, 1, 1\}.$$

The total number of such materials is equal to $C_{17}^5 = 6188$.

6.7 Symbol of Anisotropy Consisting of Seven Elements

$$\{\alpha_1, \alpha_2, \alpha_3, \ldots, \alpha_7\}, \ \ \alpha_1 + \alpha_2 + \alpha_3 + \cdots + \alpha_7 = 18, \ \ 1 \leq \alpha_m \leq 12, \ \ m = \overline{1, 7},$$
$$\{1, 1, 1, 1, 1, 1, 12\}, \ldots, \{12, 1, 1, 1, 1, 1, 1\}.$$

The total number of such materials is equal to $C_{17}^6 = 12376$.

6.8 Symbol of Anisotropy Consisting of Eight Elements

$$\{\alpha_1, \alpha_2, \alpha_3, \ldots, \alpha_8\}, \ \ \alpha_1 + \alpha_2 + \alpha_3 + \cdots + \alpha_8 = 18, \ \ 1 \leq \alpha_m \leq 11, \ \ m = \overline{1, 8},$$
$$\{1, 1, 1, 1, 1, 1, 1, 11\}, \ldots, \{11, 1, 1, 1, 1, 1, 1, 1\}.$$

The total number of such materials is equal to $C_{17}^7 = 19448$.

6.9 Symbol of Anisotropy Consisting of Nine Elements

$$\{\alpha_1, \alpha_2, \alpha_3, \ldots, \alpha_9\}, \ \ \alpha_1 + \alpha_2 + \alpha_3 + \cdots + \alpha_9 = 18, \ \ 1 \leq \alpha_m \leq 10, \ \ m = \overline{1, 9},$$
$$\{1, 1, 1, 1, 1, 1, 1, 1, 10\}, \ldots, \{10, 1, 1, 1, 1, 1, 1, 1, 1\}.$$

The total number of such materials is equal to $C_{17}^8 = 24310$.

6.10 Symbol of Anisotropy Consisting of Ten Elements

$$\{\alpha_1, \alpha_2, \alpha_3, \ldots, \alpha_{10}\}, \ \ \alpha_1 + \alpha_2 + \alpha_3 + \cdots + \alpha_{10} = 18, \ \ 1 \leq \alpha_m \leq 9, \ \ m = \overline{1, 10},$$
$$\{1, 1, 1, 1, 1, 1, 1, 1, 1, 9\}, \ldots, \{9, 1, 1, 1, 1, 1, 1, 1, 1, 1\}.$$

The total number of such materials is equal to $C_{17}^9 = 24310$.

6.11 Symbol of Anisotropy Consisting of Eleven Elements

$$\{\alpha_1, \alpha_2, \alpha_3, \dots \alpha_{11}\}, \quad \alpha_1 + \alpha_2 + \alpha_3 + \cdots + \alpha_{11} = 18, \quad 1 \le \alpha_m \le 8, \quad m = \overline{1, 11},$$
$$\{1, 1, 1, 1, 1, 1, 1, 1, 1, 1, 8\}, \dots, \{8, 1, 1, 1, 1, 1, 1, 1, 1, 1, 1\}.$$

The total number of such materials is equal to $C_{17}^{10} = 19448$.

6.12 Symbol of Anisotropy Consisting of Twelve Elements

$$\{\alpha_1, \alpha_2, \alpha_3, \dots, \alpha_{12}\}, \quad \alpha_1 + \alpha_2 + \alpha_3 + \cdots + \alpha_{12} = 18, \quad 1 \le \alpha_m \le 7, \quad m = \overline{1, 12},$$
$$\{1, 1, 1, 1, 1, 1, 1, 1, 1, 1, 1, 7\}, \dots, \{7, 1, 1, 1, 1, 1, 1, 1, 1, 1, 1, 1\}.$$

The total number of such materials is equal to $C_{17}^{11} = 12376$.

6.13 Symbol of Anisotropy Consisting of Thirteen Elements

$$\{\alpha_1, \alpha_2, \alpha_3, \dots, \alpha_{13}\}, \quad \alpha_1 + \alpha_2 + \alpha_3 + \cdots + \alpha_{13} = 18, \quad 1 \le \alpha_m \le 6, \quad m = \overline{1, 13},$$
$$\{1, 1, 1, 1, 1, 1, 1, 1, 1, 1, 1, 1, 6\}, \dots, \{6, 1, 1, 1, 1, 1, 1, 1, 1, 1, 1, 1, 1\}.$$

The total number of such materials is equal to $C_{17}^{12} = 6188$.

6.14 Symbol of Anisotropy Consisting of Fourteen Elements

$$\{\alpha_1, \alpha_2, \alpha_3, \dots, \alpha_{14}\}, \quad \alpha_1 + \alpha_2 + \alpha_3 + \cdots + \alpha_{14} = 18, \quad 1 \le \alpha_m \le 5, \quad m = \overline{1, 14},$$
$$\{1, 1, 1, 1, 1, 1, 1, 1, 1, 1, 1, 1, 1, 5\}, \dots, \{5, 1, 1, 1, 1, 1, 1, 1, 1, 1, 1, 1, 1, 1\}.$$

The total number of such materials is equal to $C_{17}^{13} = 2380$.

6.15 Symbol of Anisotropy Consisting of Fifteen Elements

$$\{\alpha_1, \alpha_2, \alpha_3, \dots, \alpha_{15}\}, \quad \alpha_1 + \alpha_2 + \alpha_3 + \cdots + \alpha_{15} = 18, \quad 1 \le \alpha_m \le 4, \quad m = \overline{1, 15},$$
$$\{1, 1, 1, 1, 1, 1, 1, 1, 1, 1, 1, 1, 1, 1, 4\}, \dots, \{4, 1, 1, 1, 1, 1, 1, 1, 1, 1, 1, 1, 1, 1, 1\}.$$

The total number of such materials is equal to $C_{17}^{14} = 680$.

6.16 Symbol of Anisotropy Consisting of Sixteen Elements

$$\{\alpha_1, \alpha_2, \alpha_3, \ldots, \alpha_{16}\}, \ \ \alpha_1 + \alpha_2 + \alpha_3 + \cdots + \alpha_{16} = 18, \ \ 1 \le \alpha_m \le 3, \ \ m = \overline{1, 16},$$
$$\{1, 1, 1, 1, 1, 1, 1, 1, 1, 1, 1, 1, 1, 1, 1, 3\}, \ldots, \{3, 1, 1, 1, 1, 1, 1, 1, 1, 1, 1, 1, 1, 1, 1, 1\}.$$

The total number of such materials is equal to $C_{17}^{15} = 136$.

6.17 Symbol of Anisotropy Consisting of Seventeen Elements

$$\{\alpha_1, \alpha_2, \alpha_3, \ldots, \alpha_{17}\}, \ \ \alpha_1 + \alpha_2 + \alpha_3 + \cdots + \alpha_{17} = 18, \ \ 1 \le \alpha_m \le 2, \ \ m = \overline{1, 17},$$
$$\{1, 1, 1, 1, 1, 1, 1, 1, 1, 1, 1, 1, 1, 1, 1, 1, 2\}, \ldots,$$
$$\{2, 1, 1, 1, 1, 1, 1, 1, 1, 1, 1, 1, 1, 1, 1, 1, 1\}.$$

The total number of such materials is equal to $C_{17}^{16} = 17$.

6.18 Symbol of Anisotropy Consisting of Eighteen Elements

$$\{\alpha_1, \alpha_2, \alpha_3, \ldots, \alpha_{18}\}, \ \ \alpha_1 + \alpha_2 + \alpha_3 + \cdots + \alpha_{18} = 18, \ \ \alpha_m = 1, \ \ m = \overline{1, 18},$$
$$\{1, 1, 1, 1, 1, 1, 1, 1, 1, 1, 1, 1, 1, 1, 1, 1, 1, 1\}.$$

The total number of such materials is equal to $C_{17}^{17} = 1$.

Note that the total number of all micropolar linear elastic anisotropic materials without a center of symmetry, is equal to $\sum_{k=0}^{17} C_{17}^{k} = 2^{17} = 131072$, the total number of all anisotropic materials of Mindlin (1964) is equal to $2^{40} = 1099511627776$, and the number of materials of Toupin (1962) is equal to $2^{25} = 33554432$.

7 Materials with Negative Poisson's Ratio

Now we will calculate the Poisson's ratio for the materials {1, 5, 3} and {5, 1, 3}. To do this, we write the inverse Hooke's law for materials with a center symmetry

$$\underset{\sim}{\gamma} = \underset{\approx}{\mathbf{A}}^{-1} \overset{2}{\otimes} \underset{\sim}{\mathbf{P}}, \quad \underset{\sim}{\varkappa} = \underset{\approx}{\mathbf{D}}^{-1} \overset{2}{\otimes} \underset{\sim}{\boldsymbol{\mu}}. \tag{80}$$

Let us write here the expressions for the tensor $\underset{\approx}{\mathbf{A}}^{-1}$ with structures {1, 5, 3} and {5, 1, 3}

$$\{1,5,3\}:\underset{\approx}{\mathbf{A}}^{-1}=(\lambda_1^{-1}-\lambda_2^{-1})\mathbf{u}_1\mathbf{u}_1+\lambda_2^{-1}\underset{\approx}{\mathbf{C}}_{(2)}+(\lambda_7^{-1}-\lambda_2^{-1})(\mathbf{u}_7\mathbf{u}_7+\mathbf{u}_8\mathbf{u}_8+\mathbf{u}_9\mathbf{u}_9);$$
$$\{5,1,3\}:\underset{\approx}{\mathbf{A}}^{-1}=\lambda_1^{-1}\underset{\approx}{\mathbf{C}}_{(2)}-(\lambda_1^{-1}-\lambda_6^{-1})\mathbf{u}_6\mathbf{u}_6-(\lambda_1^{-1}-\lambda_7^{-1})(\mathbf{u}_7\mathbf{u}_7+\mathbf{u}_8\mathbf{u}_8+\mathbf{u}_9\mathbf{u}_9). \tag{81}$$

Suppose

$$\mathbf{u}_1=\mathbf{u}_6=\pm\frac{\sqrt{3}}{3}\underset{\sim}{\mathbf{E}},\quad \mathbf{u}_m=\underset{\sim}{\mathbf{e}}_m,\quad m=7,8,9. \tag{82}$$

Then with the help of $\sum_{m=7}^{9}\underset{\sim}{\mathbf{e}}_m\underset{\sim}{\mathbf{e}}_m=1/2(\underset{\approx}{\mathbf{C}}_{(2)}-\underset{\approx}{\mathbf{C}}_{(3)})$ and (82) from (81), we get

$$\{1,5,3\}:\underset{\approx}{\mathbf{A}}^{-1}=\frac{1}{3}(\lambda_1^{-1}-\lambda_2^{-1})\underset{\approx}{\mathbf{C}}_{(1)}+\frac{1}{2}(\lambda_2^{-1}+\lambda_7^{-1})\underset{\approx}{\mathbf{C}}_{(2)}+\frac{1}{2}(\lambda_2^{-1}-\lambda_7^{-1})\underset{\approx}{\mathbf{C}}_{(3)};$$
$$\{5,1,3\}:\underset{\approx}{\mathbf{A}}^{-1}=-\frac{1}{3}(\lambda_1^{-1}-\lambda_6^{-1})\underset{\approx}{\mathbf{C}}_{(1)}+\frac{1}{2}(\lambda_1^{-1}+\lambda_7^{-1})\underset{\approx}{\mathbf{C}}_{(2)}+\frac{1}{2}(\lambda_1^{-1}-\lambda_7^{-1})\underset{\approx}{\mathbf{C}}_{(3)}. \tag{83}$$

If we assume $\underset{\sim}{\mathbf{P}}=P_1\underset{\sim}{\mathbf{e}}_1$, then using (83) and the first relation (80), we get

$$\begin{aligned}
&\{1,5,3\}:\underset{\sim}{\boldsymbol{\gamma}}=P_1\underset{\approx}{\mathbf{A}}^{-1}\overset{2}{\otimes}\underset{\sim}{\mathbf{e}}_1=P_1\left[\frac{1}{3}(\lambda_1^{-1}-\lambda_2^{-1})\underset{\sim}{\mathbf{E}}+\lambda_2^{-1}\underset{\sim}{\mathbf{e}}_1\right],\\
&\gamma_1=\underset{\sim}{\boldsymbol{\gamma}}\overset{2}{\otimes}\underset{\sim}{\mathbf{e}}_1=\frac{(\lambda_2+2\lambda_1)P_1}{3\lambda_1\lambda_2},\quad \gamma_2=\gamma_3=\underset{\sim}{\boldsymbol{\gamma}}\overset{2}{\otimes}\underset{\sim}{\mathbf{e}}_2=-\frac{(\lambda_1-\lambda_2)P_1}{3\lambda_1\lambda_2}<0,\\
&\nu=-\frac{\gamma_2}{\gamma_1}=-\frac{\gamma_3}{\gamma_1}=\frac{\lambda_1-\lambda_2}{\lambda_2+2\lambda_1}>0,\quad 0<\nu<\frac{1}{2};\\
&\{5,1,3\}:\underset{\sim}{\boldsymbol{\gamma}}=P_1\underset{\approx}{\mathbf{A}}^{-1}\overset{2}{\otimes}\underset{\sim}{\mathbf{e}}_1=P_1\left[-\frac{1}{3}(\lambda_1^{-1}-\lambda_6^{-1})\underset{\sim}{\mathbf{E}}+\lambda_1^{-1}\underset{\sim}{\mathbf{e}}_1\right],\\
&\gamma_1=\underset{\sim}{\boldsymbol{\gamma}}\overset{2}{\otimes}\underset{\sim}{\mathbf{e}}_1=\frac{(\lambda_1+2\lambda_6)P_1}{3\lambda_1\lambda_6},\quad \gamma_2=\gamma_3=\underset{\sim}{\boldsymbol{\gamma}}\overset{2}{\otimes}\underset{\sim}{\mathbf{e}}_2=\frac{(\lambda_1-\lambda_6)P_1}{3\lambda_1\lambda_6}<0,\\
&\nu=-\frac{\gamma_2}{\gamma_1}=-\frac{\gamma_3}{\gamma_1}=-\frac{\lambda_1-\lambda_6}{\lambda_1+2\lambda_6}<0,\quad -1<\nu<0.
\end{aligned} \tag{84}$$

A similar result is obtained from the second relation (80) for ε

$$\{1,5,3\}:\ 0<\varepsilon<\frac{1}{2},\quad \{5,1,3\}:\ -1<\varepsilon<0\ \left(\varepsilon=-\frac{\varkappa_2}{\varkappa_1}=-\frac{\varkappa_3}{\varkappa_1}\right).$$

8 Orthotropic Micropolar Material with a Center of Symmetry

In this case, the matrix of components of the tensor $\underset{\approx}{\mathbf{A}}$ has the form (the tensor $\underset{\approx}{\mathbf{D}}$ has the similar form)

$$A = \begin{pmatrix} A_{11} & A_{12} & A_{13} & 0 & 0 & 0 & 0 & 0 & 0 \\ A_{12} & A_{22} & A_{23} & 0 & 0 & 0 & 0 & 0 & 0 \\ A_{13} & A_{23} & A_{33} & 0 & 0 & 0 & 0 & 0 & 0 \\ 0 & 0 & 0 & A_{44} & A_{45} & 0 & 0 & 0 & 0 \\ 0 & 0 & 0 & A_{45} & A_{55} & 0 & 0 & 0 & 0 \\ 0 & 0 & 0 & 0 & 0 & A_{66} & A_{67} & 0 & 0 \\ 0 & 0 & 0 & 0 & 0 & A_{67} & A_{77} & 0 & 0 \\ 0 & 0 & 0 & 0 & 0 & 0 & 0 & A_{88} & A_{89} \\ 0 & 0 & 0 & 0 & 0 & 0 & 0 & A_{89} & A_{99} \end{pmatrix}. \tag{85}$$

The number of independent components is 15 (see Nikabadze 2009c; Zheng and Spencer 1993; Eremeyev and Pietraszkiewicz 2012).

The characteristic equation of the tensor $\underset{\approx}{\mathbf{A}}$ using (85) will have the form

$$\begin{aligned} \det(\underset{\approx}{\mathbf{A}} - \underset{\approx}{\mathbf{E}}) = &\begin{vmatrix} A_{11}-\lambda & A_{12} & A_{13} \\ A_{12} & A_{22}-\lambda & A_{23} \\ A_{13} & A_{23} & A_{33}-\lambda \end{vmatrix} \begin{vmatrix} A_{44}-\lambda & A_{45} \\ A_{45} & A_{55}-\lambda \end{vmatrix} \cdot \\ &\cdot \begin{vmatrix} A_{66}-\lambda & A_{67} \\ A_{67} & A_{77}-\lambda \end{vmatrix} \begin{vmatrix} A_{88}-\lambda & A_{89} \\ A_{89} & A_{99}-\lambda \end{vmatrix} = 0. \end{aligned} \tag{86}$$

From (86), it follows that at least one of the following equations is true:

$$\begin{aligned} &\begin{vmatrix} A_{11}-\lambda & A_{12} & A_{13} \\ A_{12} & A_{22}-\lambda & A_{23} \\ A_{13} & A_{23} & A_{33}-\lambda \end{vmatrix} = 0, \quad \begin{vmatrix} A_{44}-\lambda & A_{45} \\ A_{45} & A_{55}-\lambda \end{vmatrix} = 0, \\ &\begin{vmatrix} A_{66}-\lambda & A_{67} \\ A_{67} & A_{77}-\lambda \end{vmatrix} = 0, \quad \begin{vmatrix} A_{88}-\lambda & A_{89} \\ A_{89} & A_{99}-\lambda \end{vmatrix} = 0. \end{aligned} \tag{87}$$

The first equation of (87) is a cubic equation, which has three positive roots. Let us denote these roots by $\lambda_1, \lambda_2, \lambda_3$ and the eigentensors corresponding to these roots by $\underset{\sim}{\mathbf{u}}_1$, $\underset{\sim}{\mathbf{u}}_2$ and $\underset{\sim}{\mathbf{u}}_3$.

The eigenvalues and the eigentensors of the tensor $\underset{\approx}{\mathbf{A}}$ have form

$$\begin{aligned} &\lambda_1, \ \lambda_2, \ \lambda_3, \ \underset{\sim}{\mathbf{u}}_i = u_{i,1}\underset{\sim}{\mathbf{e}}_1 + u_{i,2}\underset{\sim}{\mathbf{e}}_2 + u_{i,3}\underset{\sim}{\mathbf{e}}_3, \ \ i = 1,2,3, \\ &\lambda_{4,5} = \frac{1}{2}(A_{44}+A_{55}) \pm \frac{1}{2}(A_{44}-A_{55})\frac{1}{\cos 2\alpha}, \quad \mathrm{tg}2\alpha = \frac{2A_{45}}{A_{44}-A_{55}}, \\ &\underset{\sim}{\mathbf{u}}_4 = -\sin\alpha\underset{\sim}{\mathbf{e}}_4 + \cos\alpha\underset{\sim}{\mathbf{e}}_5, \quad \underset{\sim}{\mathbf{u}}_5 = \cos\alpha\underset{\sim}{\mathbf{e}}_4 + \sin\alpha\underset{\sim}{\mathbf{e}}_5; \\ &\lambda_{6,7} = \frac{1}{2}(A_{66}+A_{77}) \pm \frac{1}{2}(A_{66}-A_{77})\frac{1}{\cos 2\beta}, \quad \mathrm{tg}2\beta = \frac{2A_{67}}{A_{66}-A_{77}}, \\ &\underset{\sim}{\mathbf{u}}_6 = -\sin\beta\underset{\sim}{\mathbf{e}}_6 + \cos\beta\underset{\sim}{\mathbf{e}}_7, \quad \underset{\sim}{\mathbf{u}}_7 = \cos\beta\underset{\sim}{\mathbf{e}}_6 + \sin\beta\underset{\sim}{\mathbf{e}}_7; \\ &\lambda_{8,9} = \frac{1}{2}(A_{88}+A_{99}) \pm \frac{1}{2}(A_{88}-A_{99})\frac{1}{\cos 2\gamma}, \quad \mathrm{tg}2\gamma = \frac{2A_{89}}{A_{88}-A_{99}}, \\ &\underset{\sim}{\mathbf{u}}_8 = -\sin\gamma\underset{\sim}{\mathbf{e}}_8 + \cos\gamma\underset{\sim}{\mathbf{e}}_9, \quad \underset{\sim}{\mathbf{u}}_9 = \cos\gamma\underset{\sim}{\mathbf{e}}_8 + \sin\gamma\underset{\sim}{\mathbf{e}}_9. \end{aligned} \tag{88}$$

9 Conclusions

1. The eigenvalue problems of the symmetric TBM of any even rank and sizes $m \times m, m \geq 1$ (sizes 2×2) are formulated (studied).
2. Some definitions and theorems concerning the TBM are formulated.
3. The formulas expressing the classical invariants of the TBM of any even rank and sizes 2×2 through the first invariants of powers of this TBM are given. The formulas which are inverse to the latter are obtained.
4. The complete orthonormal system of eigentensor columns for the TBM of any even rank and sizes 2×2 are constructed. The generalized eigenvalue problem of the TBM is formulated.
5. As a special case, the TBM of the elastic modulus tensors, the modulus tensor and the micropolar material with a center of symmetry are considered. The canonical representations of the TBM, the elastic strain energy, and the constitutive relations are given.
6. The classification of micropolar linearly-elastic anisotropic materials without a center of symmetry is given. The set of these anisotropic materials is divided into 18 classes, and the total number of anisotropic materials is equal to 131072.
7. The existence of a material with a negative Poisson's ratio is proved.

Acknowledgments The research was supported by the Russian Foundation for Basic Research, projects nos. 15–01–00848-a and 14–01–00317-a.

References

Alexandrov KS (1967) Elastic properties of anisotropic media. Abstract of DSc Dissertation, Institute of Crystallography Ac. Sc. USSR, Moscow
Annin BD, Ostrosablin NI (2008) Anisotropy of elastic properties of materials. J Appl Mech Tech Phys 49(6):998–1014
Chanyshev AI (1984a) Plasticity of anisotropic media. J Appl Mech Tech Phys 25(2):311–314
Chanyshev AI (1984b) Solution of limit load problems for a rigid-plastic anisotropic body. J Appl Mech Tech Phys 25(5):806–809
Chen S (1984) New concepts of elasticity theory and an application (in Chin.). Acta Mech Sin 16(3):259–274
Eremeyev VA, Pietraszkiewicz W (2012) Material symmetry group of the non-linear polar-elastic continuum. Int J Solids Struct 49(14):1993–2005
Eringen AC (1999) Microcontinuum field theories. Foundation and solids, vol 1. Springer, New York
Faddeev DK, Sominckii IS (1999) Problems in higher algebra. St. Petersburg State University, St. Petersburg
Gantmacher FR (1959) The theory of matrices, vol 1 and 2. Chelsea Publishing Company, New York
Korn GA, Korn TM (2000) Mathematical handbook for scientists and engineers: definitions, theorems, and formulas for reference and review. Dover Publications, New York
Kupradze VD, Gegelia TG, Basheleishvili MO, Burchuladze TV (1976) Three-dimensional problems of the mathematical theory of elasticity and thermoelasticity (in Russ.). Nauka, Moscow

Love AEH (2013) A treatise on the mathematical theory of elasticity, 4th edn. Cambridge University Press, Cambridge
Lur'e KA (1979) Some problems of optimal bending and tension for elastic plates (in Russ.). Izv Akad Nauk SSSR Mekh Tverd Tela 14(6):86–93
Mehrabadi MM, Cowin SC (1990) Eigentensors of linear anisotropic elastic materials. Q J MechAppl Math 43(1):15–41
Mehrabadi MM, Cowin SC (1991) Eigentensors of linear anisotropic elastic materials. Quart J Mech and Appl Math 44:333 Corrigendum
Mindlin RD (1964) Micro-structure in linear elasticity. Arch Ration Mech Anal 16:51–78
Minkevich LM (1973) Representation of the elastic modulus tensor and elastic compliance tensor through its eigentensors. In: Questions of mechanical systems of vibroimpact action. Novosibirsk, pp 107–110
Nikabadze MU (2008) On the eigenvalue and eigentensor problem for a tensor of even rank. Mech Solids 43(4):586–596
Nikabadze MU (2009a) On some problems of tensor calculus I. J Math Sci 161(5):668–697
Nikabadze MU (2009b) On some problems of tensor calculus II. J Math Sci 161(5):698–733
Nikabadze MU (2009c) On the construction of linearly independent tensors. Mech Solids 44(1):14–30
Nikabadze MU (2014) Construction of eigentensor columns in the linear micropolar theory of elasticity. Vestn Mosk Univ, Matem Mekhan 1:30–39
Nikabadze MU (2015) On some questions of tensor calculus with applications to mechanics. Tensor Analysis, PFUR, M, CMFD 55:3–194
Novozhilov VV (1961) Theory of elasticity. Pergamon Press, New York
Ostrosablin NI (1984) Structure of the elastic modulus tensor. Dynamics of a continuum, vol 66. Hydromechanics Inst., Novosibirsk, pp 113–125
Ostrosablin NI (1986a) Elastic eigenmoduli and eigenstates for the materials of crystallographic systems. In: Dynamics of a continuum, vol 75. Hydromechanics Inst., Novosibirsk, pp 110–127
Ostrosablin NI (1986b) On the structure of the elastic tensor and the classification of anisotropic materials. J Appl Mech Techn Phys 27(4):600–607
Ostrosablin NI (2000) Anisotropy and general solutions of equations in the linear theory of elasticity. Doct. diss. in math. and physics, Hydromechanics Inst., Novosibirsk
Pobedrya BE (1984) The plasticity theory of anisotropic materials. Application Problems of Plasticity, vol 26. Lobachevsky State Univ., Gor'kii, pp 110–115
Pobedrya BE (1986) The flow theory of anisotropic media. Plasticity and Viscoplasticity of Materials and Structures. Akad. Nauk SSSR, Sverdlovsk, pp 101–108
Pobedrya BE (1990) On the theory of plasticity of transversely isotropic materials. Izv Ross Akad Nauk Mekh Tverd Tela 25(3):96–101
Revuzhenko AF, Chanyshev AI, Shemyakin EI (1985) Mathematical models of elastic-plastic bodies. Actual problems of computational mathematics and mathematical modeling. Nauka, Novosibirsk, pp 108–119
Rychlewski J (1983) Mathematical structure of elastic solids. Preprint No. 217, Institute for Problems in Mechanics, Moscow
Rychlewski J (1984) On Hooke's law. Prikl Mekh Math 48(3):303–314
Sutcliffe S (1992) Spectral decomposition of the elasticity tensor. Trans ASME J Appl Mech 59(4):762–773
Theocaris PS (1989) The compliance fourth-rank tensor for the transtropic material and its spectral decomposition. Proc Nat Acad Athens 61(1):80–100
Theocaris PS, Philippidis TP (1989) Elastic eigenstates of a medium with transverse isotropy. Arch Mech Stosov 41(5):717–724
Theocaris PS, Philippidis TP (1990) Variational bounds on the eigenangle orthonormal of transversely isotropic materials. Acta Mech 85(1–2):13–26
Theocaris PS, Philippidis TP (1991) Spectral decomposition of compliance and stiffness fourth-rank tensors suitable for orthotropic materials. Z Angew Math und Mech 71(3):161–171

Todhunter I, Pearson K (1960) A history of the theory of elasticity and of the strength of materials from Galilei to Lord Kelvin, Part II, vol II. Saint-Venant to Lord Kelvin. Dover Publications Inc, New York

Tolokonnikov LA, Matchenko NM (1974) On presentations of the limiting conditions for initially isotropic solids. Strength Mater 6(3):319–322

Toupin RA (1962) Elastic materials with couple-stresses. Arch Ration Mech Anal 11(1):385–414

Vekua IN (1978) Foundations of tensor analysis and the theory of covariants (in Russ.). Nauka, Moscow

Watkins DS (2002) Fundamentals of matrix computations, 2nd edn. Wiley, New York

Zheng QS, Spencer AJM (1993) On the canonical representations for Kronecker powers of orthogonal tensors with application to material symmetry problems. Int J Eng Sci 31(4):617–635

Analytical Solutions in the Theory of Thin Bodies

Mikhail U. Nikabadze and Armine R. Ulukhanyan

Abstract Some questions about the parametrization of three-dimensional thin body with one small size under an arbitrary base surface and the changing of transverse coordinate from –1 to 1 are considered. The vector parametric equation of the thin body domain is given. In particular, we have defined the various families of bases and geometric characteristics generated by them. Expressions for the components of the second rank isotropic tensor are obtained. The representations of some differential operators, the equations of motion, and the constitutive relations of micropolar elasticity theory under the considered parametrization of the thin body domain are given. The inverse tensor operators to a tensor operator of the equations of motion in terms of displacements for an isotropic homogeneous material and to a stress operator are found. They allow decomposing equations and boundary conditions. The inverse matrix differential tensor operator to the matrix differential tensor operator of the equations of motion in displacements and rotations of the micropolar theory of elasticity is constructed for isotropic homogeneous materials with a symmetry center as well as for materials without a symmetry center. We obtain the equations with respect to displacement vector and rotation vector individually. As a special case, a reduced continuum is considered. Cases in which it is easy to invert the stress and the couple stress operator are found out. From the decomposed equations of classical (micropolar) theory of elasticity, the corresponding decomposed equations of quasistatic problems of theory of prismatic bodies with constant thickness in displacements (in displacements and rotations) are obtained. From these systems of equations, we derive the equations in moments of unknown vector functions with respect to any system of orthogonal polynomials. We obtain the systems of equations of various approximations (from zero to eighth order) in moments with respect to the systems of Legendre and second kind Chebyshev polynomials. The system splits and for each moment of unknown vector function we, obtain a high order

M.U. Nikabadze (✉)
Lomonosov Moscow State University, Moscow, Russia
e-mail: munikabadze@yandex.ru

A.R. Ulukhanyan
Bauman State Technical University, Moscow, Russia
e-mail: armine_msu@mail.ru

H. Altenbach and S. Forest (eds.), *Generalized Continua as Models for Classical and Advanced Materials*, Advanced Structured Materials 42,
DOI 10.1007/978-3-319-31721-2_15

elliptic type equation (the system order depends on the order of approximation), the characteristic roots of which can be easily found. Using the method of Vekua, their analytical solution is obtained. For micropolar theory of thin prismatic bodies with two small sizes and a the rectangular cross-section, the decomposed equations in moments of displacement and rotation vectors via an arbitrary system of polynomials (Legendre, Chebyshev) are obtained. Similar equations are also deduced for the reduced medium containing classical equation. The decomposed systems of equations of eight approximations for micropolar theory of multilayer prismatic bodies of constant thickness in moments of displacement and rotation vectors are obtained. Using Vekua method, we can find the analytical solutions for this system and for equations for the reduced medium.

Keywords Micropolar theory · Thin body · Tensor operator · Couple stress · Legendre polynomials · Couple stress-tensor-operator · Reduced medium

1 Parametrization of Thin Domain with One Small Size Under an Arbitrary Base Surface

Let V be a domain ff three-dimensional Euclidean space occupying of thin body.

Definition 15.1 A three-dimensional body (two-dimensional domain) is called thin body (domain) if its one or two dimensions (one dimension) are significantly smaller than the others.

Definition 15.2 Changing for some domain the coordinate system is called the parametrization of this domain.

Here, if the opposite will not be specified, we mainly consider a three-dimensional body, one size of which is less than the other ones, i.e. a three-dimensional thin body with one small size. Also, we will specify the concept of thin body.

The position vector of an arbitrary point of the domain of a thin body is represented as follows (in Fig. 1 the normal section of the thin body is shown)

$$\hat{\mathbf{r}}(x', x^3) = \mathbf{r}(x') + \left[\bar{h}(x') + x^3 h(x')\right]\mathbf{n}(x'), \quad -1 \leq x^3 \leq 1, \tag{1}$$

where $\mathbf{r} = \mathbf{r}(x')$ is the vector parametric equation of base surface S (in Fig. 1 the cross section of the base surface with a normal body section is shown), $x' = (x^1, x^2)$ is an arbitrary point on S, i.e. x^1 and x^2 are curvilinear coordinates on the base surface S,

$$\bar{h}(x') = \left[\overset{(+)}{h}(x') - \overset{(-)}{h}(x')\right]/2, \quad h(x') = \left[\overset{(+)}{h}(x') + \overset{(-)}{h}(x')\right]/2,$$

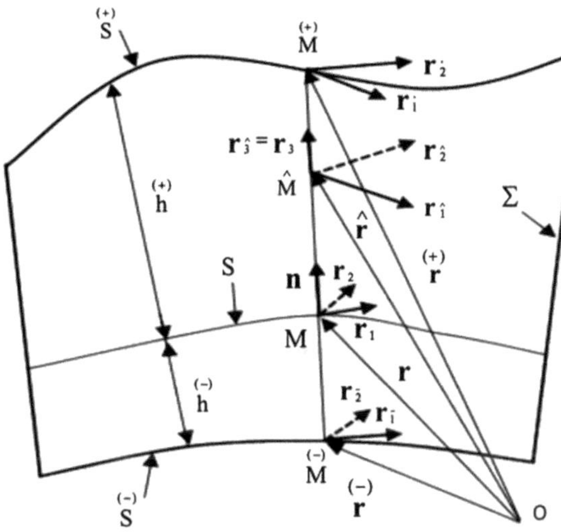

Fig. 1 Normal section of thin body

$\mathbf{n}(x')$ is the unit normal vector to S at the point x', $\overset{(-)}{h}(x')$ is the distance from the point x' to the corresponding point on the surface $\overset{(-)}{S}$, $\overset{(+)}{h}(x')$ is the distance from the same point x' to the corresponding point on the surface $\overset{(+)}{S}$ (in Fig. 1 the intersections of these surfaces with normal section of the body are shown),

$$2h(x') = \overset{(+)}{h}(x') + \overset{(-)}{h}(x')$$

is the thin body thickness at the point x'. Note that the point O, in general, does not belong to the plane of figure. It is easy to see that Eq. (1) determines the interior $\overset{(-)}{S}$ and exterior $\overset{(+)}{S}$ surfaces of the thin body for $\forall x', x^3 = -1$ and $\forall x', x^3 = 1$, respectively. The vector relationships are

$$\overset{(-)}{\mathbf{r}}(x') = \hat{\mathbf{r}}(x', x^3)\Big|_{x^3=-1} = \mathbf{r}(x') - \overset{(-)}{h}(x')\mathbf{n}(x'), \quad \forall x', \ x^3 = -1, \tag{2}$$

$$\overset{(+)}{\mathbf{r}}(x') = \hat{\mathbf{r}}(x', x^3)\Big|_{x^3=1} = \mathbf{r}(x') + \overset{(+)}{h}(x')\mathbf{n}(x'), \quad \forall x', \ x^3 = 1. \tag{3}$$

For $\forall x', x^3 = const$, where $x^3 \in (-1, 1)$, Eq. (1) determines the equidistant surface $\hat{S}$ from base surface S. Equation (1) for $x' \in S$, $x^3 \in [-1, 1]$ is a vector parametric equation of a domain of a thin body.

For the derivatives of $\hat{\mathbf{r}}(x', x^3)$ and $\overset{(\sim)}{\mathbf{r}}$, $\sim\in\{-, \emptyset, +\}$, in x^P let us introduce the notations[1] $\mathbf{r}_{\hat{P}} = \partial_P \hat{\mathbf{r}}$ and $\mathbf{r}_{\tilde{P}} = \partial_P \overset{(\sim)}{\mathbf{r}}$, $\sim\in\{-, \emptyset, +\}$, respectively. It is easy to see that due to (1)–(3) we have

$$\begin{aligned} &\mathbf{r}_{\hat{P}} = \left[g_P^Q - (\bar{h} + x^3 h) b_P^Q\right]\mathbf{r}_Q + (\partial_P \bar{h} + x^3 \partial_P h)\mathbf{n}, \\ &\mathbf{r}_{\underset{P}{-}} = (g_P^Q + \overset{(-)}{h} b_P^Q)\mathbf{r}_Q - \partial_P \overset{(-)}{h}\mathbf{n}, \quad \mathbf{r}_{\underset{P}{+}} = (g_P^Q + \overset{(+)}{h} b_P^Q)\mathbf{r}_Q - \partial_P \overset{(+)}{h}\mathbf{n}. \end{aligned} \tag{4}$$

To write Eq. (4) we used Weingarten derivation equations

$$\mathbf{n}_P = \partial_P \mathbf{n} = -b_P^Q \mathbf{r}_Q = -\underset{\sim}{\mathbf{b}} \cdot \mathbf{r}_P,$$

where $\underset{\sim}{\mathbf{b}}$ is the second tensor of surface S.

The vector pair $\mathbf{r}_{\underset{1}{*}}, \mathbf{r}_{\underset{2}{*}}$, $* \in \{-, \emptyset, \wedge, +\}$, defined at the points $\overset{(*)}{M} \in \overset{(*)}{S}$, $* \in \{-, \emptyset, \wedge, +\}$, gives us two-dimensional covariant surface bases. Using these bases, we can construct the corresponding contravariant bases $\overset{*}{\mathbf{r}}{}^1, \overset{*}{\mathbf{r}}{}^2$, $* \in \{-, \emptyset, \wedge, +\}$. Naturally, the covariant and contravariant bases generate their inherent geometric characteristics. In particular, we can define the following matrices:

$$g_{\tilde{I}\breve{J}} = \mathbf{r}_{\tilde{I}} \cdot \mathbf{r}_{\breve{J}}, \quad g_{\tilde{I}}^{\breve{J}} = \mathbf{r}_{\tilde{I}} \cdot \mathbf{r}^{\breve{J}}, \quad g^{\tilde{I}\breve{J}} = \mathbf{r}^{\tilde{I}} \cdot \mathbf{r}^{\breve{J}}, \quad \sim, \smile \in * \in \{-, \emptyset, \wedge, +\}. \tag{5}$$

Differentiating (1) w.r.t. x^3, we have

$$\mathbf{r}_{\hat{3}} = \partial_3 \hat{\mathbf{r}} = h(x')\mathbf{n}(x'), \quad \forall x^3 \in [-1, 1]. \tag{6}$$

According to (6), we can assume that

$$\mathbf{r}_{\underset{3}{-}} = \mathbf{r}_3 = \mathbf{r}_{\hat{3}} = \mathbf{r}_{\underset{3}{+}} = h(x')\mathbf{n}(x'), \quad \forall x^3 \in [-1, 1]. \tag{7}$$

Relations (4) and (7) allow us to define the spatial covariant bases $\mathbf{r}_{\underset{p}{*}}$, $* \in \{-, \emptyset, +\}$, at the points $\overset{(*)}{M} \in \overset{(*)}{S}$, $* \in \{-, \emptyset, +\}$. Therefore, the third basis vector of the spatial covariant bases at the points $\overset{(*)}{M} \in \overset{(*)}{S}$, $* \in \{-, \emptyset, \wedge, +\}$, is the same vector $\mathbf{r}_3 = h(x')\mathbf{n}(x')$.

Introducing the notation $z(x', x^3) = \bar{h}(x') + x^3 h(x')$, due to (7) the first relation (4) and (6) can be presented in the form

---

[1]In the following brief notes as $\overset{(\sim)}{M} \in \overset{(\sim)}{S}$, $\sim \in \{-, \emptyset, \wedge, +\}$ or $\mathbf{r}_{\tilde{p}} = g_{\tilde{p}}^{\breve{q}} \mathbf{r}_{\breve{q}}$, $\sim, \smile \{-, \emptyset, \wedge, +\}$, where $\emptyset$ is empty set, are applied. The first record means: if $\sim = -$ then $\overset{(-)}{M} \in \overset{(-)}{S}$; if $\sim = \emptyset$ then $M \in S$; if $\sim = \wedge$ than $\hat{M} \in \hat{S}$; if $\sim = +$ then $\overset{(+)}{M} \in \overset{(+)}{S}$. The second record means that if, for example, $\sim = \emptyset$, $\smile = -$ then $\mathbf{r}_p = g_p^{\bar{q}} \mathbf{r}_{\bar{q}}$; if $\sim = \wedge$, $\smile = \emptyset$ then $\mathbf{r}_{\hat{p}} = g_{\hat{p}}^q \mathbf{r}_q$ and soon. Going through all the values, we get all the relations.

$$\mathbf{r}_{\hat{p}} = \left(g_p^q - zb_p^q + h^{-1}\partial_M z g_p^M g_3^q\right)\mathbf{r}_q. \tag{8}$$

By analogy to Eq. (8) using the last two formulas of (4) and (7), we have

$$\mathbf{r}_{\bar{p}} = \left(g_p^q + \overset{(-)}{h} b_p^q - h^{-1}\partial_M \overset{(-)}{h} g_p^M g_3^q\right)\mathbf{r}_q, \quad \mathbf{r}_{\overset{+}{p}} = \left(g_p^q - \overset{(+)}{h} b_p^q + h^{-1}\partial_M \overset{(+)}{h} g_p^M g_3^q\right)\mathbf{r}_q. \tag{9}$$

In relations (8) and (9) $b_p^q = g_p^M g_N^q b_M^N$ are the components of the extended second tensor of surface S (Vekua 1978, 1985; Nikabadze 2015, 2007b). Obviously, formulas (9) can be obtained from (8) when $x^3 = -1$ and $x^3 = 1$, respectively.

The triples of vectors $\mathbf{r}_{\overset{*}{1}}, \mathbf{r}_{\overset{*}{2}}, \mathbf{r}_{\overset{*}{3}}$, $* \in \{-, \emptyset, \wedge, +\}$, defined at the points $\overset{(*)}{M} \in \overset{(*)}{S}$, $* \in \{-, \emptyset, \wedge, +\}$, give us three-dimensional covariant bases. Using these bases we can construct the corresponding contravariant bases $\mathbf{r}^{\overset{*}{1}}, \mathbf{r}^{\overset{*}{2}}, \mathbf{r}^{\overset{*}{3}}$, $* \in \{-, \emptyset, \wedge, +\}$. In fact, based on the definition of the contravariant bases (Vekua 1978; Lurie 1990; Pobedrya 1986; Nikabadze 2015, 2007b) we have

$$\mathbf{r}^{\tilde{k}} = \frac{1}{2} C^{\tilde{k}\tilde{p}\tilde{q}} \mathbf{r}_{\tilde{p}} \times \mathbf{r}_{\tilde{q}}, \quad \sim \in \{-, \emptyset, \wedge, +\}, \tag{10}$$

where $C^{\tilde{k}\tilde{p}\tilde{q}} = \left(\mathbf{r}^{\tilde{k}} \times \mathbf{r}^{\tilde{p}}\right) \cdot \mathbf{r}^{\tilde{q}}$, $\sim \in \{-, \emptyset, \wedge, +\}$ are the contravariant components of the discriminant tensors (Vekua 1978) at the points $\overset{(*)}{M} \in \overset{(*)}{S}$, $* \in \{-, \emptyset, \wedge, +\}$, respectively.

Analogously to Eq. (5), we introduce the following matrices:

$$g_{\tilde{p}\breve{q}} = \mathbf{r}_{\tilde{p}} \cdot \mathbf{r}_{\breve{q}}, \quad g_{\tilde{p}}^{\breve{q}} = \mathbf{r}_{\tilde{p}} \cdot \mathbf{r}^{\breve{q}}, \quad g^{\tilde{p}\breve{q}} = \mathbf{r}^{\tilde{p}} \cdot \mathbf{r}^{\breve{q}}, \quad \sim, \smile \in \{-, \emptyset, \wedge, +\}. \tag{11}$$

Because of the first two relations of (11) and (8), we have

$$g_{\hat{p}q} = g_{pq} - zb_{pq} + h^{-1}\partial_M z g_p^M g_{3q}, \quad g_{\hat{p}}^q = g_p^q - zb_p^q + h^{-1}\partial_M z g_p^M g_3^q. \tag{12}$$

Similarly to (12) with the help of (9) and the first two relations of (11), we find

$$\begin{aligned} g_{\bar{p}q} &= g_{pq} + \overset{(-)}{h} b_{pq} - h^{-1}\partial_M \overset{(-)}{h} g_p^M g_{3q}, \quad g_{\bar{p}}^q = g_p^q + \overset{(-)}{h} b_p^q - h^{-1}\partial_M \overset{(-)}{h} g_p^M g_3^q, \\ g_{\overset{+}{p}q} &= g_{pq} - \overset{(+)}{h} b_{pq} + h^{-1}\partial_M \overset{(+)}{h} g_p^M g_{3q}, \quad g_{\overset{+}{p}}^q = g_p^q - \overset{(+)}{h} b_p^q + h^{-1}\partial_M \overset{(+)}{h} g_p^M g_3^q. \end{aligned} \tag{13}$$

It is easy to see that from (12) and (13) we obtain

$$g_{\hat{p}3} = h\left(\partial_P \bar{h} + x^3 \partial_P h\right), \quad g_{\hat{p}}^3 = h^{-1}\left(\partial_P \bar{h} + x^3 \partial_P h\right), \quad \mathbf{r}^3 = h^{-1}(x')\mathbf{n}(x'), \tag{14}$$

$$g_{\bar{P}3} = -h\partial_P \overset{(-)}{h}, \quad g_{\bar{P}}^3 = -h^{-1}\partial_P \overset{(-)}{h}, \quad g_{\overset{+}{P}3} = h\partial_P \overset{(+)}{h}, \quad g_{\overset{+}{P}}^3 = h^{-1}\partial_P \overset{(+)}{h}. \tag{15}$$

It is seen that by the second formula (14) equalities (12) can be written as

$$\begin{aligned} g_{\hat{p}q} &= g_{pq} - (\bar{h} + x^3 h) b_{pq} + g^3_{\hat{M}} g^M_p g_{3q}, \\ g^q_{\hat{p}} &= g^q_p - (\bar{h} + x^3 h) b^q_p + g^3_{\hat{M}} g^M_p g^q_3. \end{aligned} \tag{16}$$

Next using a second and fourth formulas of (15) from (13) we have

$$\begin{aligned} g_{\underset{-}{p}q} &= g_{pq} + \overset{(-)}{h} b_{pq} + g^3_{\underset{-}{M}} g^M_p g_{3q}, \quad g^q_{\underset{-}{p}} = g^q_p + \overset{(-)}{h} b^q_p + g^3_{\underset{-}{M}} g^M_p g^q_3, \\ g_{\underset{+}{p}q} &= g_{pq} - \overset{(+)}{h} b_{pq} + g^3_{\underset{+}{M}} g^M_p g_{3q}, \quad g^q_{\underset{+}{p}} = g^q_p - \overset{(+)}{h} b^q_p + g^3_{\underset{+}{M}} g^M_p g^q_3. \end{aligned} \tag{17}$$

Note also that based on (15) the formulas (14) take the form

$$g_{\hat{P}3} = \frac{1}{2}\big[g_{\underset{+}{P}3} + g_{\underset{-}{P}3} + x^3(g_{\underset{+}{P}3} - g_{\underset{-}{P}3})\big], \quad g^3_{\hat{P}} = \frac{1}{2}\big[g^3_{\underset{+}{P}} + g^3_{\underset{-}{P}} + x^3(g^3_{\underset{+}{P}} - g^3_{\underset{-}{P}})\big]. \tag{18}$$

It is seen that according to (11) the is following relationship between basis vectors holds true

$$\mathbf{r}_{\tilde{p}} = g^{\breve{q}}_{\tilde{p}} \mathbf{r}_{\breve{q}}, \quad \sim, \smile \in \{-, \emptyset, \wedge, +\}, \tag{19}$$

which remains valid when the indices are raised or lowered. By (19), we can prove the validity of relation

$$g^{\breve{q}}_{\tilde{p}} = g^{\overset{*}{n}}_{\tilde{p}} g^{\breve{q}}_{\overset{*}{n}}, \quad \sim, \smile, * \in \{-, \emptyset, \wedge, +\}, \tag{20}$$

which remains valid when the indices are raised or lowered.

It is not difficult to find an expression for $g_{\hat{p}\hat{q}}$. According to (12), from (20) we have

$$g_{\hat{p}\hat{q}} = g^n_{\hat{p}} g_{n\hat{q}} = g_{pq} - 2z b_{pq} + z^2 b^M_p b_{Mq} + g^3_{\hat{S}} g^S_p g_{3q} + g^3_{\hat{T}} g^T_q g_{3q} + g_{33} g^3_{\hat{S}} g^3_{\hat{T}} g^S_p g^T_q. \tag{21}$$

We obtain an expression for $\sqrt{\hat{g}} = (\mathbf{r}_{\hat{1}} \times \mathbf{r}_{\hat{2}}) \cdot \mathbf{r}_{\hat{3}}$. Taking into account (19) when $\sim = \wedge$ and $\smile \in \{-, \emptyset, +\}$, we get

$$\sqrt{\hat{g}} = \sqrt{\overset{(\smile)}{g}} \det\big(g^{\breve{K}}_{\hat{I}}\big), \quad \det\big(g^{\breve{K}}_{\hat{I}}\big) = \frac{1}{2} \epsilon^{IJ} \epsilon_{KL} g^{\breve{K}}_{\hat{I}} g^{\breve{L}}_{\hat{J}}, \quad \smile \in \{-, \emptyset, +\}. \tag{22}$$

Here ϵ^{IJ}, ϵ_{KL} are two-dimensional Levi–Civita symbols, and

$$\sqrt{\overset{(\smile)}{g}} = (\mathbf{r}_{\breve{1}} \times \mathbf{r}_{\breve{2}}) \cdot \mathbf{r}_{\breve{3}}, \quad \smile \in \{-, \emptyset, +\}, \quad \overset{(-)}{g} = \hat{g}|_{x^3=-1}, \quad g = \hat{g}|_{x^3=0}, \quad \overset{(+)}{g} = \hat{g}|_{x^3=1}. \tag{23}$$

From (22), when $\smile = \emptyset$, we obtain

$$\hat{\vartheta} \equiv \sqrt{\hat{g} g^{-1}} = \det\left(g_{\hat{I}}^{K}\right) = \frac{1}{2}\,\epsilon^{IJ}\epsilon_{KL}g_{\hat{I}}^{K}g_{\hat{J}}^{L}. \tag{24}$$

In the following, we assume that $\hat{\vartheta} \not\approx 0$, i.e. the thin body has such a shape that the considered parametrization of domain is valid.

It should be noted that there is a more general relation than (22)

$$\sqrt{\overset{(\sim)}{g}} = \frac{1}{2}\sqrt{\overset{(\smile)}{g}}\,\epsilon^{IJ}\epsilon_{KL}g_{\tilde{I}}^{\breve{K}}g_{\tilde{J}}^{\breve{L}} = \sqrt{\overset{(\smile)}{g}}\,\det\left(g_{\tilde{P}}^{\breve{Q}}\right), \quad \sim, \smile \in \{-, \emptyset, \wedge, +\}. \tag{25}$$

From Eq. (25), we find

$$\det\left(g_{\tilde{P}}^{\breve{Q}}\right) = \sqrt{\overset{(\sim)}{g}\,\overset{(\smile)}{g}^{-1}} = \frac{1}{2}\,\epsilon^{IJ}\epsilon_{KL}g_{\tilde{I}}^{\breve{K}}g_{\tilde{J}}^{\breve{L}}, \quad \sim, \smile \in \{-, \emptyset, \wedge, +\}.$$

Now we can find the expression for $\mathbf{r}^{\hat{k}}$. According to (19) from (10), we have

$$\mathbf{r}^{\breve{k}} = \frac{1}{2}\sqrt{\overset{(\smile)}{g}^{-1}\overset{(\sim)}{g}}\,\epsilon^{kpq}\epsilon_{lmn}g_{\breve{p}}^{\tilde{m}}g_{\breve{q}}^{\tilde{n}}\mathbf{r}^{\tilde{l}} = \frac{1}{2}\sqrt{\overset{(\smile)}{g}^{-1}g}\,\epsilon^{kpq}\epsilon_{lmn}g_{\breve{p}}^{m}g_{\breve{q}}^{n}\mathbf{r}^{l}, \quad \sim, \smile \in \{-, \emptyset, \wedge, +\}. \tag{26}$$

Hence, when $\smile = \wedge$ we obtain

$$\mathbf{r}^{\hat{k}} = \frac{1}{2}\sqrt{\hat{g}^{-1}\overset{(\sim)}{g}}\,\epsilon^{kpq}\epsilon_{lmn}g_{\hat{p}}^{\tilde{m}}g_{\hat{q}}^{\tilde{n}}\mathbf{r}^{\tilde{l}} = \frac{1}{2}\,\hat{\vartheta}^{-1}\epsilon^{kpq}\epsilon_{lmn}g_{\hat{p}}^{m}g_{\hat{q}}^{n}\mathbf{r}^{l}, \quad \sim \in \{-, \emptyset, \wedge, +\}. \tag{27}$$

Here ϵ^{kpq}, ϵ_{lmn} are the Levi–Civita symbols. It is easy to notice that due to (26) we have

$$g_{\tilde{l}}^{\breve{k}} = \mathbf{r}^{\breve{k}} \cdot \mathbf{r}_{\tilde{l}} = \frac{1}{2}\sqrt{\overset{(\smile)}{g}^{-1}\overset{(\sim)}{g}}\,\epsilon^{kpq}\epsilon_{lmn}g_{\breve{p}}^{\tilde{m}}g_{\breve{q}}^{\tilde{n}},$$
$$g^{\breve{k}\tilde{l}} = \mathbf{r}^{\breve{k}} \cdot \mathbf{r}^{\tilde{l}} = \frac{1}{2}\sqrt{\overset{(\smile)}{g}^{-1}\overset{(\sim)}{g}}\,\epsilon^{kpq}\epsilon_{smn}g_{\breve{p}}^{\tilde{m}}g_{\breve{q}}^{\tilde{n}}g^{\tilde{s}\tilde{l}}, \quad \sim, \smile \in \{-, \emptyset, \wedge, +\}.$$

From here or using (27) we can find

$$g_{l}^{\hat{k}} = \frac{1}{2}\hat{\vartheta}^{-1}\epsilon^{kpq}\epsilon_{lmn}g_{\hat{p}}^{m}g_{\hat{q}}^{n}, \quad g^{\hat{k}l} = \frac{1}{2}\hat{\vartheta}^{-1}\epsilon^{kpq}\epsilon_{smn}g_{\hat{p}}^{m}g_{\hat{q}}^{n}g^{sl}.$$

We also find the relationship for $\mathbf{r}^{\hat{P}}$ and $\mathbf{r}^{\hat{3}}$. Based on (27) after simple calculations we have

$$\mathbf{r}^{\hat{P}} = g_{M}^{\hat{P}}\mathbf{r}^{M} = \hat{\vartheta}^{-1}A_{M}^{\hat{P}}\mathbf{r}^{M}, \quad g_{M}^{\hat{P}} = \hat{\vartheta}^{-1}A_{M}^{\hat{P}}, \quad A_{M}^{\hat{P}} = \epsilon^{PK}\epsilon_{ML}g_{\hat{K}}^{L}. \tag{28}$$

According to (19) we obtain

$$\mathbf{r}^{3} = g_{\hat{p}}^{3}\mathbf{r}^{\hat{p}} = \mathbf{r}^{\hat{3}} + g_{\hat{P}}^{3}\mathbf{r}^{\hat{P}} = \mathbf{r}^{\hat{3}} + g_{\hat{P}}^{3}g_{M}^{\hat{P}}\mathbf{r}^{M}.$$

Hence,

$$\mathbf{r}^{\hat{3}} = \mathbf{r}^3 - g^3_{\hat{P}} g^{\hat{P}}_M \mathbf{r}^M, \quad g^{\hat{3}}_M = -g^3_{\hat{P}} g^{\hat{P}}_M. \tag{29}$$

It is easy to see that from (29) we obtain

$$\begin{aligned} \mathbf{r}^{\bar{3}} &= \mathbf{r}^{\hat{3}}|_{x^3=-1} = \mathbf{r}^3 - g^3_{\bar{P}} g^{\bar{P}}_M \mathbf{r}^M, \quad g^{\bar{3}}_M = -g^3_{\bar{P}} g^{\bar{P}}_M, \\ \mathbf{r}^{\overset{+}{3}} &= \mathbf{r}^{\hat{3}}|_{x^3=1} = \mathbf{r}^3 - g^3_{\overset{+}{P}} g^{\overset{+}{P}}_M \mathbf{r}^M, \quad g^{\overset{+}{3}}_M = -g^3_{\overset{+}{P}} g^{\overset{+}{P}}_M. \end{aligned} \tag{30}$$

2 Representation of the Second Rank Isotropic Tensor and Its Components

Proceeding from the usual expression of the second rank isotropic tensor (SRIT) (Vekua 1978; Nikabadze 2015; Lurie 1990; Pobedrya 1986), according to (19) and (20) we get the representation

$$\underset{\sim}{\mathbf{E}} = g^{\breve{n}}_{\tilde{m}} \mathbf{r}^{\tilde{m}} \mathbf{r}_{\breve{n}}, \quad \sim, \smile \in \{-, \emptyset, \wedge, +\}, \tag{31}$$

which remains valid when the indices are raised or lowered.[2] As is seen from (31), the elements of matrices (11), introduced above, are components of the SRIT.

Definition 15.3 The components $g^{\breve{q}}_{\tilde{p}}$, $\sim, \smile \in \{-, \emptyset, \wedge, +\}$, $\sim \neq \smile$, and also their images, obtained from them by raising or lowering indices, are called the translation components of the SRIT under the considered parametrization of the thin body domain.

In what follows we will be interested in expressions of the translation components $g^{\hat{P}}_M$, $g^{\hat{3}}_M$ and components $g^{\hat{P}\hat{Q}}$, $g^{\hat{P}\hat{3}}$, $g^{\hat{3}\hat{3}}$ of the SRIT in the form of a uniformly convergent power series with respect to x^3. Analogous representations of basis vectors and components of the SRIT under the classical parametrization were given in Vekua (1978) and under the new parametrization presented in the second chapter of Nikabadze (2007b) (see also Nikabadze 2008a, 2014b, c, 2015).

Assuming that the considered body is thin, i.e. the norm $\| z\underset{\sim}{\mathbf{b}} \| < 1$, where $z(x', x^3) = \bar{h}(x') + x^3 h(x')$, for $\mathbf{r}^{\hat{P}}$ we obtain the representation

$$\begin{aligned} \mathbf{r}^{\hat{P}} = \mathbf{r}^P \cdot (\underset{\sim}{\mathbf{E}} - z\underset{\sim}{\mathbf{b}})^{-1} &= \mathbf{r}^P \cdot (\underset{\sim}{\mathbf{E}} + z\underset{\sim}{\mathbf{b}} + z^2 \underset{\sim}{\mathbf{b}}^2 + \cdots) \\ &= (g^P_M + z\, b^P_M + z^2\, b^P_N b^N_M + \cdots)\mathbf{r}^M. \end{aligned} \tag{32}$$

Note that

[2] It means if one of the dummy indices is lowered, then the corresponding index rises and vice versa.

$$(\underset{\sim}{\mathbf{E}} - z\underset{\sim}{\mathbf{b}})^{-1} = \sum_{s=0}^{\infty} z^s \underset{\sim}{\mathbf{b}}^s, \quad (\underset{\sim}{\mathbf{E}} - z\underset{\sim}{\mathbf{b}})^{-2} = \sum_{s=0}^{\infty} (1+s) z^s \underset{\sim}{\mathbf{b}}^s. \tag{33}$$

By (32) we have

$$\begin{aligned} &g_M^{\hat{P}} = \mathbf{r}^P \cdot \left[\underset{\sim}{\mathbf{E}} - (\bar{h} + x^3 h)\underset{\sim}{\mathbf{b}}\right]^{-1} \cdot \mathbf{r}_M = \sum_{s=0}^{\infty} \underset{(s)}{A}{}_M^P (\bar{h} + x^3 h)^s, \quad \underset{(0)}{A}{}_M^P = g_M^P, \\ &\underset{(1)}{A}{}_M^P = b_M^P, \quad \underset{(2)}{A}{}_M^P = b_N^P b_M^N, \ldots, \underset{(s)}{A}{}_M^P = b_{N_1}^P b_{N_2}^{N_1} \ldots b_{N_{s-1}}^{N_{s-2}} b_M^{N_{s-1}}. \end{aligned} \tag{34}$$

Knowing (33) and (34), it is easy to find representations of the other components. In fact, we have

$$\begin{aligned} &g_M^{\hat{3}} = -g_{\hat{P}}^3 g_M^{\hat{P}} = -h^{-1} \partial_P z \sum_{s=0}^{\infty} \underset{(s)}{A}{}_M^P z^s, \\ &g^{\hat{P}\hat{Q}} = g_M^{\hat{P}} g^{M\hat{Q}} = \mathbf{r}^P \cdot (\underset{\sim}{\mathbf{E}} - z\underset{\sim}{\mathbf{b}})^{-2} \cdot \mathbf{r}^Q = g^{QM} \sum_{s=0}^{\infty} (1+s) \underset{(s)}{A}{}_M^P z^s, \\ &g^{\hat{P}\hat{3}} = -g_{\hat{Q}}^3 g_M^{\hat{P}} g^{M\hat{Q}} = -g^{QM} g_{\hat{Q}}^3 \sum_{s=0}^{\infty} (1+s) \underset{(s)}{A}{}_M^P z^s, \\ &g^{\hat{3}\hat{3}} = g^{33} + g^{MN} g_M^{\hat{3}} g_N^{\hat{3}} = g^{33} + g_{\hat{P}}^3 g_{\hat{Q}}^3 g^{\hat{P}\hat{Q}} = g^{33} + g^{QM} g_{\hat{P}}^3 g_{\hat{Q}}^3 \sum_{s=0}^{\infty} (1+s) \underset{(s)}{A}{}_M^P z^s. \end{aligned} \tag{35}$$

Taking into account the representation $g_M^{\hat{P}}$ (34) and using the rule of multiplication of series in the form of Cauchy, we obtain

$$g_M^{\hat{P}} g_N^{\hat{Q}} = \sum_{s=0}^{\infty} \underset{(s)}{B}{}_{MN}^{PQ} z^s, \quad \underset{(s)}{B}{}_{MN}^{PQ} = \sum_{r=0}^{s} \underset{(s-r)}{A}{}_M^P \underset{(r)}{A}{}_N^Q. \tag{36}$$

Note that the components of the SRIT take part in representations of equations and constitutive equations of mechanics of a deformable thin body under a given parametrization. Therefore, the relations (34)–(36) play an important role in the construction of different variants of mathematical theories of thin bodies with the use of expansion with respect to orthogonal polynomials. The number of summands retained in the right-hand sides (34)–(36), depends on the character of the solved problem and the required accuracy of approximation. Note that all questions outlined in the second chapter of Nikabadze (2007b) (see also Nikabadze 2008a, 2014b, c, 2015), concerning the new parametrization and not included in this paper, can be easily transferred to the case of this parametrization.

It should be noted also that it is easy to obtain other types of parameterizations from the considered above parametrization of the thin body domain as special cases. For example,

(1) if we assume that $\overset{(-)}{h} = \overset{(+)}{h} \equiv h, x^3 h \equiv z$, then we can get classical parametrization where the base surface is considered the middle surface;

(2) if we assume that $\overset{(-)}{h} = \overset{(+)}{h} \equiv h$, then we will have a parametrization at which the base is regarded as a middle surface, and transverse coordinate $x^3 \in [-1, 1]$;

(3) if we assume that $\overset{(-)}{h} = 0$ and rename $\hat{\mathbf{r}}(x', x^3) = \mathbf{r}(x', z)$, $1 + x^3 = z$, $\mathbf{r}(x') = \overset{(-)}{\mathbf{r}}(x')$, then we obtain a new parametrization (Nikabadze 2007b, 2014b, c, 2015) (in this case, for convenience, it will be better to rename z through x^3 where $x^3 \in [0, 1]$).

3 Representations of Gradient, Divergence, Repeated Gradient and Laplacian

In this case with the help of (28) and (29) we have the next representation for the gradient of a tensor field $\hat{\mathbb{F}}(x', x^3)$

$$\hat{\nabla}\mathbb{F} = \mathbf{r}^{\hat{p}}\partial_{p}\mathbb{F} = \mathbf{N}\mathbb{F} + \mathbf{r}^3\nabla_3\mathbb{F} = \mathbf{r}^{\hat{P}}N_P\mathbb{F} + \mathbf{r}^3\nabla_3\mathbb{F} = \mathbf{r}^M g_M^{\hat{P}} N_P\mathbb{F} + \mathbf{r}^3\partial_3\mathbb{F}, \quad (37)$$

where we have used the differential operators

$$N_P = \partial_P - g_{\hat{P}}^3\partial_3, \quad \mathbf{N} = \mathbf{r}^{\hat{P}}N_P = \mathbf{r}^{\hat{P}}(\partial_P - g_{\hat{P}}^3\partial_3) = \mathbf{r}^M g_M^{\hat{P}}(\partial_P - g_{\hat{P}}^3\partial_3).$$

Here the SRIT components $g_{\hat{P}}^3$ (14) characterize the change of thickness.

It is easy to see that the divergence, say, of the second rank tensor $\underset{\sim}{\mathbf{P}}$ according to (37) can be represented in the form

$$\hat{\nabla}\cdot\underset{\sim}{\mathbf{P}} = \nabla_p\mathbf{P}^{\hat{p}} = g_M^{\hat{P}}N_P\mathbf{P}^M + \partial_3\mathbf{P}^3. \quad (38)$$

By (37) the repeated gradient of a tensor field $\hat{\mathbb{F}}(x', x^3)$ under the considered parametrization can be represented as follows

$$\begin{aligned}\hat{\nabla}\hat{\nabla}\mathbb{F} &= \mathbf{r}^M\mathbf{r}^N g_M^{\hat{P}}N_P\big(g_N^{\hat{Q}}N_Q\mathbb{F}\big) + \mathbf{r}^M\mathbf{r}^3 g_M^{\hat{P}}N_P\partial_3\mathbb{F} + \mathbf{r}^3\mathbf{r}^N\partial_3\big(g_N^{\hat{Q}}N_Q\mathbb{F}\big) + \mathbf{r}^3\mathbf{r}^3\partial_3^2\mathbb{F}\\ &= \mathbf{r}^M\mathbf{r}^N g_M^{\hat{P}}g_N^{\hat{Q}}N_PN_Q\mathbb{F} + \mathbf{r}^M\mathbf{r}^3 g_M^{\hat{P}}N_P\partial_3\mathbb{F} + \mathbf{r}^3\mathbf{r}^N g_N^{\hat{Q}}\nabla_3N_Q\mathbb{F} + \mathbf{r}^3\mathbf{r}^3\partial_3^2\mathbb{F}.\end{aligned} \quad (39)$$

Here

$$N_PN_Q = \nabla_P\nabla_Q - \big(g_{\hat{P}}^3\nabla_3\nabla_Q + g_{\hat{Q}}^3\nabla_P\nabla_3\big) + g_{\hat{P}}^3g_{\hat{Q}}^3\nabla_3^2.$$

It is easy to notice that based on (39) we obtain representations for the Laplacian

$$\begin{aligned}\hat{\Delta}\mathbb{F} &= \hat{\nabla}\cdot\hat{\nabla}\mathbb{F} = g^{MN}g_M^{\hat{P}}N_P\big(g_N^{\hat{Q}}N_Q\mathbb{F}\big) + g^{33}\partial_3^2\mathbb{F}\\ &= g^{MN}g_M^{\hat{P}}g_N^{\hat{Q}}N_PN_Q\mathbb{F} + g^{33}\partial_3^2\mathbb{F}.\end{aligned} \quad (40)$$

Next, we note that having the representations of the gradient (37) and repeated gradient (39), it is easy to get representations of other differential operators (rotor, repeated divergence, divergence gradient). For brevity, we do not consider this in detail.

4 Equations of Motion and Constitutive Relations in the Micropolar Theory of Thin Bodies

As is known (Eringen 1999), three-dimensional equations of motion of micropolar deformable rigid bodies, related to the current configuration, are represented in the form[3]

$$\nabla \cdot \underset{\sim}{\mathbf{P}} + \rho \mathbf{F} = \rho \dot{\mathbf{v}}, \quad \nabla \cdot \underset{\sim}{\boldsymbol{\mu}} + \underset{\approx}{\mathbf{C}} \overset{2}{\otimes} \underset{\sim}{\mathbf{P}} + \rho \mathbf{m} = \rho \dot{\mathbf{k}}, \tag{41}$$

and the equations of motion, related to the reference configuration, can be written as follows

$$\overset{\circ}{\nabla} \cdot \overset{\circ}{\underset{\sim}{\mathbf{P}}} + \overset{\circ}{\rho} \mathbf{F} = \overset{\circ}{\rho} \dot{\mathbf{v}}, \quad \overset{\circ}{\nabla} \cdot \overset{\circ}{\underset{\sim}{\boldsymbol{\mu}}} + \overset{\circ}{\underset{\approx}{\mathbf{C}}} \overset{2}{\otimes} \overset{\circ}{\underset{\sim}{\mathbf{P}}} + \overset{\circ}{\rho} \mathbf{m} = \overset{\circ}{\rho} \mathrm{d}\overset{\circ}{\mathbf{k}}/\mathrm{d}t. \tag{42}$$

Here, $\underset{\sim}{\mathbf{P}}$ and $\underset{\sim}{\boldsymbol{\mu}}$ are tensors of stresses and couple stresses; $\overset{\circ}{\underset{\sim}{\mathbf{P}}} = \sqrt{g/\overset{\circ}{g}}\overset{\circ}{\nabla}\overset{\circ}{\mathbf{r}}^T \cdot \underset{\sim}{\mathbf{P}}$ and $\overset{\circ}{\underset{\sim}{\boldsymbol{\mu}}} = \sqrt{g/\overset{\circ}{g}}\overset{\circ}{\nabla}\overset{\circ}{\mathbf{r}}^T \cdot \underset{\sim}{\boldsymbol{\mu}}$ are the tensors of stresses and couple stresses, related to the current configuration; $\sqrt{g} = (\mathbf{r}_1 \times \mathbf{r}_2) \cdot \mathbf{r}_3$, $\sqrt{\overset{\circ}{g}} = (\overset{\circ}{\mathbf{r}}_1 \times \overset{\circ}{\mathbf{r}}_2) \cdot \overset{\circ}{\mathbf{r}}_3$, $\overset{\circ}{\nabla}$ and ∇ are the Hamiltonians in reference and current configurations, respectively; ρ is the material density; $\mathbf{F}$ is the mass force density; $\mathbf{m}$ is the mass moment density; $\mathbf{v} = \dot{\mathbf{u}}$ is the velocity vector; $\mathbf{u}$ is the vector of displacement; $\boldsymbol{\varphi}$ is the vector of inner rotation; t is the time; the dot above the quantities denotes the time derivative; $\mathbf{k} = \underset{\sim}{\mathbf{I}} \cdot \boldsymbol{\omega}$ is the angular momentum; $\underset{\sim}{\mathbf{I}}$ is the microinertia tensor per unit mass; $\boldsymbol{\omega}$ is the angular velocity vector; $\overset{2}{\otimes}$ is the inner 2-product (Nikabadze 2008a, b; Nikabadze and Ulukhanyan 2008; Vekua 1978; Nikabadze et al. 2008; Nikabadze 2009a, b, 2014a, b, c); $\underset{\approx}{\mathbf{C}}$ is the discriminant third-rank tensor. It should be noted that the values marked with a circle on top, refer to the reference configuration. Note also that the superscript T in the upper right corner of the quantities denotes transposition.

It can be seen that the Eqs. (41) and (42) have the same form, so here we deal with the equations of the current configuration (41). It is easy to see that the equations of

[3]Second-rank tensors we mark from the bottom with the wave ($\underset{\sim}{\mathbf{P}}$), third-rank tensors we mark from below with the wave and hyphen ($\underset{\approx}{\mathbf{C}}$) and fourth-rank tensors we mark from below with two waves ($\underset{\approx}{\mathbf{A}}$).

motion of micropolar theory (41) according to (38) can be written in the form

$$g_M^{\hat{P}} N_P \mathbf{P}^M + \partial_3 \mathbf{P}^3 + \rho \mathbf{F} = \rho \dot{\mathbf{v}}, \quad g_M^{\hat{P}} N_P \boldsymbol{\mu}^M + \partial_3 \boldsymbol{\mu}^3 + \underset{\approx}{\mathbf{C}} \overset{2}{\otimes} \underset{\sim}{\mathbf{P}} + \rho \mathbf{m} = \rho \dot{\mathbf{k}}. \quad (43)$$

Equations (43) represent the system of equations of motion of micropolar mechanics of deformable solids under the considered parametrization of the thin body domain. Therefore, it is advisable to call it the system of equations of motion of micropolar mechanics of thin deformable solids.

In linear micropolar theory of elasticity, the constitutive relations (CR) in the case of isothermal processes (Kupradze et al 1976; Pobedrya 1995) can be written in the form (next we consider linear theory for very small $\mathbf{u}$ and $\boldsymbol{\varphi}$)

$$\underset{\sim}{\mathbf{P}} = \underset{\approx}{\mathbf{A}} \overset{2}{\otimes} \underset{\sim}{\boldsymbol{\gamma}} + \underset{\approx}{\mathbf{B}} \overset{2}{\otimes} \underset{\sim}{\boldsymbol{\varkappa}}, \quad \underset{\sim}{\boldsymbol{\mu}} = \underset{\approx}{\mathbf{C}} \overset{2}{\otimes} \underset{\sim}{\boldsymbol{\gamma}} + \underset{\approx}{\mathbf{D}} \overset{2}{\otimes} \underset{\sim}{\boldsymbol{\varkappa}}, \quad (44)$$

where $\underset{\sim}{\boldsymbol{\gamma}} = \nabla \mathbf{u} - \underset{\simeq}{\mathbf{C}} \cdot \boldsymbol{\varphi}$ is the strain tensor; $\underset{\sim}{\boldsymbol{\varkappa}} = \nabla \boldsymbol{\varphi}$ is the bending-torsion tensor in micropolar theory of elasticity (Nowacki 1970; Kupradze et al 1976); $\underset{\approx}{\mathbf{A}}$, $\underset{\approx}{\mathbf{C}} = \underset{\approx}{\mathbf{B}}^T$, $\underset{\approx}{\mathbf{D}}$ are the material tensors of the fourth rank. Taking into account the expressions for $\underset{\sim}{\boldsymbol{\gamma}}$ and $\underset{\sim}{\boldsymbol{\varkappa}}$, the CR (44), we can write

$$\underset{\sim}{\mathbf{P}} = \underset{\approx}{\mathbf{A}} \overset{2}{\otimes} \nabla \mathbf{u} + \underset{\approx}{\mathbf{B}} \overset{2}{\otimes} \nabla \boldsymbol{\varphi} - \underset{\approx}{\mathbf{A}} \overset{2}{\otimes} \underset{\simeq}{\mathbf{C}} \cdot \boldsymbol{\varphi}, \quad \underset{\sim}{\boldsymbol{\mu}} = \underset{\approx}{\mathbf{C}} \overset{2}{\otimes} \nabla \mathbf{u} + \underset{\approx}{\mathbf{D}} \overset{2}{\otimes} \nabla \boldsymbol{\varphi} - \underset{\approx}{\mathbf{C}} \overset{2}{\otimes} \underset{\simeq}{\mathbf{C}} \cdot \boldsymbol{\varphi}. \quad (45)$$

Now it is easy to find the desired representations of the CR under the considered parametrization of the thin body domain. Indeed, from (45) because of operator (37) we have

$$\begin{aligned}
\underset{\sim}{\mathbf{P}} &= g_M^{\hat{P}} \underset{\simeq}{\mathbf{A}}^{M\cdot} \cdot N_P \mathbf{u} + \underset{\simeq}{\mathbf{A}}^{3\cdot} \cdot \partial_3 \mathbf{u} + g_M^{\hat{P}} \underset{\simeq}{\mathbf{B}}^{M\cdot} \cdot N_P \boldsymbol{\varphi} + \underset{\simeq}{\mathbf{B}}^{3\cdot} \cdot \partial_3 \boldsymbol{\varphi} - \underset{\approx}{\mathbf{A}} \overset{2}{\otimes} \underset{\simeq}{\mathbf{C}} \cdot \boldsymbol{\varphi}, \\
\underset{\sim}{\boldsymbol{\mu}} &= g_M^{\hat{P}} \underset{\simeq}{\mathbf{C}}^{M\cdot} \cdot N_P \mathbf{u} + \underset{\simeq}{\mathbf{C}}^{3\cdot} \cdot \partial_3 \mathbf{u} + gM^{\hat{P}} \underset{\simeq}{\mathbf{D}}^{M\cdot} \cdot N_P \boldsymbol{\varphi} + \underset{\simeq}{\mathbf{D}}^{3\cdot} \cdot \partial_3 \boldsymbol{\varphi} - \underset{\approx}{\mathbf{C}} \overset{2}{\otimes} \underset{\simeq}{\mathbf{C}} \cdot \boldsymbol{\varphi}.
\end{aligned} \quad (46)$$

where we have introduced the notation

$$\underset{\simeq}{\mathbf{C}}^{m\cdot} = C^{ijmn} \mathbf{r}_i \mathbf{r}_j \mathbf{r}_n = \underset{\approx}{\mathbf{C}} \overset{2}{\otimes} \mathbf{r}^m \underset{\sim}{\mathbf{E}}, \ \underset{\simeq}{\mathbf{A}}^{m\cdot} = \underset{\approx}{\mathbf{A}} \overset{2}{\otimes} \mathbf{r}^m \underset{\sim}{\mathbf{E}}, \ \underset{\simeq}{\mathbf{D}}^{m\cdot} = \underset{\approx}{\mathbf{D}} \overset{2}{\otimes} \mathbf{r}^m \underset{\sim}{\mathbf{E}}, \ \underset{\simeq}{\mathbf{B}}^{m\cdot} = \underset{\approx}{\mathbf{B}} \overset{2}{\otimes} \mathbf{r}^m \underset{\sim}{\mathbf{E}}.$$

Hence, we can represent the relations (45) as follows

$$\begin{aligned}
\underset{\sim}{\mathbf{P}} &= \underset{\approx}{\mathbf{A}} \overset{2}{\otimes} \mathbf{r}^{\hat{P}} \partial_P \mathbf{u} + \underset{\approx}{\mathbf{A}} \overset{2}{\otimes} \mathbf{r}^{\hat{3}} \partial_3 \mathbf{u} + \underset{\approx}{\mathbf{B}} \overset{2}{\otimes} \mathbf{r}^{\hat{P}} \partial_P \boldsymbol{\varphi} + \underset{\approx}{\mathbf{B}} \overset{2}{\otimes} \mathbf{r}^{\hat{3}} \partial_3 \boldsymbol{\varphi} - \underset{\approx}{\mathbf{A}} \overset{2}{\otimes} \underset{\simeq}{\mathbf{C}} \cdot \boldsymbol{\varphi}, \\
\underset{\sim}{\boldsymbol{\mu}} &= \underset{\approx}{\mathbf{C}} \overset{2}{\otimes} \mathbf{r}^{\hat{P}} \partial_P \mathbf{u} + \underset{\approx}{\mathbf{C}} \overset{2}{\otimes} \mathbf{r}^{\hat{3}} \partial_3 \mathbf{u} + \underset{\approx}{\mathbf{D}} \overset{2}{\otimes} \mathbf{r}^{\hat{P}} \partial_P \boldsymbol{\varphi} + \underset{\approx}{\mathbf{D}} \overset{2}{\otimes} \mathbf{r}^{\hat{3}} \partial_3 \boldsymbol{\varphi} - \underset{\approx}{\mathbf{C}} \overset{2}{\otimes} \underset{\simeq}{\mathbf{C}} \cdot \boldsymbol{\varphi}.
\end{aligned} \quad (47)$$

According to (32) and (34), it is seen that Eq. (43), the CR (46) and (47) contain an infinite number of summands. Therefore, we have not to use them in this form. Only the relationships that are represented by a finite number of summands will be used in the application. In this context, we introduce a definition.

Definition 15.4 Relations (equations, CR and others), which are obtained from the corresponding representations under the considered parametrization of the thin body domain, if in the expansion $g_M^{\hat{P}}$ (see (34)) the first $r+1$ terms are preserved, are called the relations (equations, CR and others) of the rth order approximation (rth approximation).

It should be noted that the conservation of the first $r+1$ summands in the expansion $g_M^{\hat{P}}$ is equal to saving the same number of first terms in the expansion $\mathbf{r}^{\hat{P}}$, see (32).

Introducing the notation

$$\underset{(r)}{g}{}_M^{\hat{P}} = \sum_{s=0}^{r} \underset{(s)}{A}{}_M^P (\bar{h} + x^3 h)^s, \tag{48}$$

for example, from (43) and (46), and if we replace $g_M^{\hat{P}}$ to $\underset{(r)}{g}{}_M^{\hat{P}}$, then we obtain the equations of motion and the CR of the micropolar theory of thin elastic bodies of the rth approximation.

$$\begin{aligned}
&\underset{(r)}{g}{}_M^{\hat{P}} N_P \mathbf{P}^M_{(r)} + \partial_3 \mathbf{P}^3_{(r)} + \rho \mathbf{F} = \rho \partial_t^2 \mathbf{u}, \quad \underset{(r)}{g}{}_M^{\hat{P}} N_P \boldsymbol{\mu}^M_{(r)} + \partial_3 \boldsymbol{\mu}^3_{(r)} + \underset{\approx}{\mathbf{C}} \overset{2}{\otimes} \underset{\sim}{\mathbf{P}}_{(r)} + \rho \mathbf{m} = \underset{\sim}{\mathbf{J}} \cdot \partial_t^2 \boldsymbol{\varphi},\\
&\underset{\sim}{\mathbf{P}}_{(r)} = \underset{(r)}{g}{}_M^{\hat{P}} \underset{\approx}{\mathbf{A}}^{M\cdot} \cdot N_P \mathbf{u} + \underset{\approx}{\mathbf{A}}^{3\cdot} \cdot \partial_3 \mathbf{u} + \underset{(r)}{g}{}_M^{\hat{P}} \underset{\approx}{\mathbf{B}}^{M\cdot} \cdot N_P \boldsymbol{\varphi} + \underset{\approx}{\mathbf{B}}^{3\cdot} \cdot \partial_3 \boldsymbol{\varphi} - \underset{\approx}{\mathbf{A}} \overset{2}{\otimes} \underset{\approx}{\mathbf{C}} \cdot \boldsymbol{\varphi}, \quad \underset{\sim}{\mathbf{J}} = \rho \underset{\sim}{\mathbf{I}},\\
&\underset{\sim}{\boldsymbol{\mu}}_{(r)} = \underset{(r)}{g}{}_M^{\hat{P}} \underset{\approx}{\mathbf{C}}^{M\cdot} \cdot N_P \mathbf{u} + \underset{\approx}{\mathbf{C}}^{3\cdot} \cdot \partial_3 \mathbf{u} + \underset{(r)}{g}{}_M^{\hat{P}} \underset{\approx}{\mathbf{D}}^{M\cdot} \cdot N_P \boldsymbol{\varphi} + \underset{\approx}{\mathbf{D}}^{3\cdot} \cdot \partial_3 \boldsymbol{\varphi} - \underset{\approx}{\mathbf{C}} \overset{2}{\otimes} \underset{\approx}{\mathbf{C}} \cdot \boldsymbol{\varphi},\\
&\partial_t^2 = \partial^2 / \partial t^2.
\end{aligned} \tag{49}$$

It is easy to see that from (49) for $r=0$ we get the equations of motion and the CR of the zeroth approximation, and for $r=\infty$ we assume that $\underset{(\infty)}{g}{}_M^{\hat{P}} = g_M^{\hat{P}}$.

5 Classical Theory of Elasticity in Displacements

5.1 Equations of the Classical Theory of Elasticity in Displacements

The equations of motion in displacements of the classical theory of elasticity in the case of a linear homogeneous isotropic material can be written in the form

$$\underset{\sim}{\mathbf{L}} \cdot \mathbf{u} + \rho \mathbf{F} = \mathbf{0}, \tag{50}$$

where we introduce the following differential tensor-operator:

$$\underset{\sim}{\mathbf{L}} = \underset{\sim}{\mathbf{E}} Q_2 + a\nabla\nabla, \quad Q_2 = \mu\Delta - \rho\partial_t^2, \quad a = \lambda + \mu. \tag{51}$$

Here $\underset{\sim}{\mathbf{E}}$ is the second-rank isotropic tensor.

Let us denote the tensor operator of cofactors for the tensor-operator $\underset{\sim}{\mathbf{L}}$ with $\underset{\sim}{\mathbf{L}}_*$. Then, after elementary calculations for $\underset{\sim}{\mathbf{L}}_*$, we obtain the expression

$$\underset{\sim}{\mathbf{L}}_* = Q_2(\underset{\sim}{\mathbf{E}} Q_1 - a\nabla\nabla), \quad Q_1 = Q_2 + a\Delta = (\lambda + 2\mu)\Delta - \rho\partial_t^2. \tag{52}$$

Having the expression for $\underset{\sim}{\mathbf{L}}$ and $\underset{\sim}{\mathbf{L}}_*$, it is easy to prove the following relation

$$\underset{\sim}{\mathbf{L}} \cdot \underset{\sim}{\mathbf{L}}_* = \underset{\sim}{\mathbf{L}}_* \cdot \underset{\sim}{\mathbf{L}} = \underset{\sim}{\mathbf{E}} \det \underset{\sim}{\mathbf{L}}, \tag{53}$$

where the determinant $\det \underset{\sim}{\mathbf{L}}$ of the differential tensor operator $\underset{\sim}{\mathbf{L}}$ has the expression

$$L \equiv |\underset{\sim}{\mathbf{L}}| \equiv \det \underset{\sim}{\mathbf{L}} = Q_1 Q_2^2. \tag{54}$$

Now, we introduce the differential tensor operator

$$\underset{\sim}{\mathbf{N}} = \underset{\sim}{\mathbf{E}} Q_1 - a\nabla\nabla. \tag{55}$$

Then, obviously, the operator $\underset{\sim}{\mathbf{L}}_*$ can be written as

$$\underset{\sim}{\mathbf{L}}_* = Q_2 \underset{\sim}{\mathbf{N}}. \tag{56}$$

By virtue (51) and (55), it is not difficult to prove that

$$\underset{\sim}{\mathbf{L}} \cdot \underset{\sim}{\mathbf{N}} = \underset{\sim}{\mathbf{N}} \cdot \underset{\sim}{\mathbf{L}} = \underset{\sim}{\mathbf{E}} Q_1 Q_2. \tag{57}$$

Applying the operator $\underset{\sim}{\mathbf{L}}_*$, see (52) and (56), to (50) with a single multiplication and taking into account (53) and (54), we get

$$Q_2\big(Q_1 Q_2 \mathbf{u} + \mathbf{G}\big) = 0, \quad \mathbf{G} = \underset{\sim}{\mathbf{N}} \cdot (\rho\mathbf{F}). \tag{58}$$

We now seek the vector $\mathbf{u}$ in the form (representation of Galerkin 1930, 1931; Iacovache 1949)

$$\mathbf{u} = \underset{\sim}{\mathbf{N}} \cdot \mathbf{v} = (\underset{\sim}{\mathbf{E}} Q_1 - a\nabla\nabla) \cdot \mathbf{v}, \tag{59}$$

where $\mathbf{v}$ is an arbitrary vector field. Then substituting (59) in (50) and taking into account (57), we have

$$Q_1 Q_2 \mathbf{v} + \rho\mathbf{F} = 0. \tag{60}$$

Finally, applying $\underset{\sim}{\mathbf{N}}$ to (50) with a single multiplication, according to (57) we have

$$Q_1 Q_2 \mathbf{u} + \mathbf{G} = 0. \tag{61}$$

Thus, we got the decomposed systems of equations of elasticity theory for a linear elastic homogeneous material in displacements in the form of (60) and (61). Obviously, every equation of each of these systems is of fourth order, and each of these systems is of 12th order. Different representations of the general solution of the Lamé's equations can be found, for example, in Nowacki (1970), Pobedrya (1995). It can be seen that Eq. (61) compared with Eq. (60) has an advantage. The boundary conditions for Eq. (61) are the boundary conditions of the original problem, while the boundary conditions for Eq. (60) become more complicated due to the introduction of an additional vector field $\mathbf{v}$.

5.2 *On Boundary Conditions in the Linear Theory of Elasticity. Stress Tensor Operator*

The stress boundary conditions for a linear elastic inhomogeneous anisotropic material in an isothermal process can be represented as

$$\underset{\sim}{\mathbf{T}} \cdot \mathbf{u} = \mathbf{P}, \quad \underset{\sim}{\mathbf{T}} = \mathbf{r}_j \mathbf{r}_l n_i C^{ijkl} \nabla_k = \mathbf{n} \cdot \underset{\approx}{\mathbf{C}} \overset{2}{\otimes} \mathbf{r}^k \underset{\sim}{\mathbf{E}} \nabla_k = \mathbf{n} \cdot \underset{\approx}{\mathbf{C}} \cdot \nabla, \tag{62}$$

where $\mathbf{P}$ is a vector function given on the boundary; $\underset{\sim}{\mathbf{T}}$ is a stress tensor-operator. In the case of an isotropic body with the piecewise-plane boundary, we obtain after simple calculations

$$\begin{aligned}
\underset{\sim}{\mathbf{T}} &= \lambda \mathbf{n}\nabla + \mu[\underset{\sim}{\mathbf{E}}\mathbf{n} \cdot \nabla + (\mathbf{n}\nabla)^T] = [\lambda \underset{\approx}{\mathbf{C}}_{(2)} + \mu(\underset{\approx}{\mathbf{C}}_{(1)} + \underset{\approx}{\mathbf{C}}_{(3)})] \overset{2}{\otimes} \mathbf{n}\nabla, \\
\underset{\sim}{\mathbf{T}}_* &= \mu\{-(\lambda+\mu)\mathbf{n}\nabla + 2[(\lambda+\mu)\underset{\sim}{\mathbf{E}}\mathbf{n} \cdot \nabla - \lambda(\mathbf{n}\nabla)^T]\}\mathbf{n} \cdot \nabla \\
&\quad + \lambda\mu[(\mathbf{nn} - \underset{\sim}{\mathbf{E}})\Delta + \nabla\nabla] \\
&= \Big\{\mu\{-(\lambda+\mu)\underset{\approx}{\mathbf{C}}_{(2)} + 2[(\lambda+\mu)\underset{\approx}{\mathbf{C}}_{(1)} - \lambda \underset{\approx}{\mathbf{C}}_{(3)}]\} \overset{2}{\otimes} \mathbf{n}\nabla\Big\}\mathbf{n} \cdot \nabla \\
&\quad + \lambda\mu[(\mathbf{nn} - \underset{\sim}{\mathbf{E}})\Delta + \nabla\nabla], \\
\det \underset{\sim}{\mathbf{T}} &= |\underset{\sim}{\mathbf{T}}| = \mu^2[2(\lambda+\mu)\mathbf{nn} \overset{2}{\otimes} \nabla\nabla - \lambda\Delta]\mathbf{n} \cdot \nabla.
\end{aligned} \tag{63}$$

Here $\underset{\sim}{\mathbf{T}}_*$ is a tensor operator of the cofactors for $\underset{\sim}{\mathbf{T}}$; $\det \underset{\sim}{\mathbf{T}} = |\underset{\sim}{\mathbf{T}}|$ is a determinant of $\underset{\sim}{\mathbf{T}}$; $\underset{\approx}{\mathbf{C}}_{(i)}$, $i = 1, 2, 3$, are the linearly independent isotropic fourth-rank tensors. Applying tensor operator $\underset{\sim}{\mathbf{T}}_*^T$ from left (see the corresponding formula (63) with a single multiplication to (62), in virtue of relation $\underset{\sim}{\mathbf{T}}_*^T \cdot \underset{\sim}{\mathbf{T}} = \underset{\sim}{\mathbf{T}} \cdot \underset{\sim}{\mathbf{T}}_*^T = \underset{\sim}{\mathbf{E}}|\underset{\sim}{\mathbf{T}}|$ we obtain

$$|\underset{\sim}{\mathbf{T}}|\mathbf{u} = \underset{\sim}{\mathbf{T}}_*^T \cdot \mathbf{P}. \tag{64}$$

Thus, we got a decomposed stress boundary condition in the form (64). From the above it can be seen that the equations in displacements and the stress boundary conditions of the theory of elasticity of linear elastic homogeneous body with the piecewise-plane boundary are split (below we will always assume that the body has a piecewise-plane boundary when it comes to splitting of the boundary conditions).

It should be noted that taking into account the representations of the nabla operator and Laplacian under the different parameterizations of the thin body domain, it is not difficult to obtain the corresponding decomposed equations and stress boundary conditions for thin bodies from the above decomposed equations and the stress boundary conditions. For brevity, we do not consider this in detail. Below we write down the equations for prismatic bodies of constant thickness.

5.3 Quasi-Static Problems of the Classical Theory of Elasticity in Displacements

In the static or quasi-static case based on (61) and (64) we have the following equations and boundary conditions:

$$\Delta^2\mathbf{u} + \mathbf{G} = 0, \quad |\mathbf{T}|\mathbf{u} = \mathbf{T}_*^T \cdot \mathbf{P}. \tag{65}$$

5.4 Quasi-Static Problems of the Theory of Prismatic Bodies

Let us consider a prismatic body of constant thickness $2h$. We take the middle plane as the base plane. In this case $g_M^{\hat{P}} = \delta_M^P$, $g_{\hat{P}}^3 = 0$, $g^{33} = h^{-2}$, the nabla operator (37) and the Laplacian (40) are represented in the form (Nikabadze and Ulukhanyan 2008; Nikabadze 2008a, 2014b see also)

$$\begin{gathered} \hat{\nabla}\mathbb{F} = (\mathbf{r}^P\partial_P + \mathbf{r}^3\partial_3)\mathbb{F} = (\mathbf{r}^P\partial_P + h^{-1}\mathbf{n}\partial_3)\mathbb{F}, \quad -1 \le x^3 \le 1, \\ \Delta\mathbb{F} = \nabla^2\mathbb{F} = (g^{PQ}\partial_P\partial_Q + g^{33}\partial_3^2)\mathbb{F} = (\bar{\Delta} + h^{-2}\partial_3^2)\mathbb{F}, \quad \bar{\Delta} = g^{PQ}\partial_P\partial_Q. \end{gathered} \tag{66}$$

Taking into account the representation of the Laplacian, see the third relation (66), the equation for the prismatic bodies, see the first equality (65), can be represented as

$$(\bar{\Delta}^2 + 2h^{-2}\bar{\Delta}\partial_3^2 + h^{-4}\partial_3^4)\mathbf{u} + \mathbf{G} = 0. \tag{67}$$

Now, applying the kth moment operator of any system of orthogonal polynomials (Legendre, Chebyshev)[4] to Eq. (67), we obtain the following equations in moments of displacement vector

[4] The theory of moments with respect to the systems of orthogonal polynomials is presented in Nikabadze (2007a, 2008a, b, 2014b, c), Nikabadze and Ulukhanyan (2008), Nikabadze et al. (2008).

$$\bar{\Delta}^2\overset{(k)}{\mathbf{u}} + 2h^{-2}\bar{\Delta}\overset{(k)}{\mathbf{u}}'' + h^{-4}\overset{(k)}{\mathbf{u}}^{IV} + \overset{(k)}{\mathbf{G}} = 0, \quad k \in \mathbb{N}_0 \ (\mathbb{N}_0 = \{0, 1, 2, \ldots\}). \tag{68}$$

It should be noted that the stress boundary conditions for the systems of equations (68) are obtained from the second relation (65) after applying the kth moment operator to the corresponding polynomial system. In the general case, we shall not dwell on the boundary conditions. However, below we consider some special cases. Note also that the decomposed equations obtained above in the case of the quasi-static problem in the absence of body forces do not depend on material properties.

Let us now consider the system of equations in moments (68) more carefully. From this system changing $k = \overline{1, N}$ and neglecting the moments, order of which is more than N, we obtain a system of equation of Nth approximation. Giving N different values (starting from zero) we obtain the systems of equations of various approximation for prismatic bodies of constant thickness. From each of the deduced approximate system we can obtain setting into it moments of the displacement vector the equations of elliptic type (high-order), separately. Using new methods for solving the elliptic equations (method of (Vekua 1948)), we can write an analytical solution for each equation. Let us write down the first few approximations of Eq. (68) when applying Legendre polynomials.

5.5 *Equations of Quasi-Static Problems of the Theory of Prismatic Bodies*

To obtain the desired systems of equations, we need to find expressions for $\overset{(k)}{\mathbf{u}}''$ and $\overset{(k)}{\mathbf{u}}^{IV}$ when $-1 \leq x^3 \leq 1$. By the definition of the kth moment of some quantities for $\overset{(k)}{\mathbf{u}}''$ with respect to Legendre polynomials, we obtain (Nikabadze 2008a, b, 2014b, c)

$$\begin{aligned}
\overset{(k)}{\mathbf{u}}'' &= \overset{(k)}{\mathbf{M}}(\partial_3^2\mathbf{u}) = \frac{2k+1}{2}\int_{-1}^{1} \partial_3^2\mathbf{u}(x', x^3)P_k(x^3)dx^3 \\
&= (2k+1)\sum_{p=1}^{\infty} p(2k+2p+1)\overset{(k+2p)}{\mathbf{u}} \\
&= \frac{2k+1}{2}\{[(\partial_3\mathbf{u})^+ - (-1)^k(\partial_3\mathbf{u})^-]P_k(1) - [\overset{(+)}{\mathbf{u}} + (-1)^k\overset{(-)}{\mathbf{u}}]P_k'(1)\} + \underline{\overset{(k)}{\mathbf{u}}}'',
\end{aligned} \tag{69}$$

where $(\partial_3\mathbf{u})^+ = (\partial_3\mathbf{u})\big|_{x^3=1}$, $(\partial_3\mathbf{u})^- = (\partial_3\mathbf{u})\big|_{x^3=-1}$, and

$$\underline{\overset{(k)}{\mathbf{u}}}'' = (2k+1)\sum_{s=1} C_s^1(2k-2s+1)\overset{(k-2s)}{\mathbf{u}}. \tag{70}$$

Analogously to (69) and (70) for $\overset{(k)}{\mathbf{u}}^{IV}$ and $\underline{\overset{(k)}{\mathbf{u}}}^{IV}$, we have

$$
\begin{aligned}
\overset{(k)}{\mathbf{u}}{}^{IV} &= \overset{(k)}{\mathbf{M}}(\partial_3^4\mathbf{u}) = \frac{2k+1}{2}\int_{-1}^{1}\partial_3^4\mathbf{u}(x', x^3)P_k(x^3)dx^3 \\
&= (2k+1)\sum_{s=1}^{\infty} C_{s+2}^3(2k+2s+1)(2k+2s+3)(2k+2s+5)\overset{(k+2s+2)}{\mathbf{u}} \\
&= \frac{2k+1}{2}\{[(\partial_3^3\mathbf{u})^+ - (-1)^k(\partial_3^3\mathbf{u})^-]P_k(1) - [(\partial_3^2\mathbf{u})^+ \\
&\quad - (-1)^{k-1}(\partial_3^2\mathbf{u})^-]P_k'(1) + [(\partial_3\mathbf{u})^+ - (-1)^{k-2}(\partial_3\mathbf{u})^-]P_k''(1) \\
&\quad - [\overset{(+)}{\mathbf{u}} + (-1)^k\overset{(-)}{\mathbf{u}}]P_k'''(1)\} + \underline{\overset{(k)}{\mathbf{u}}}{}^{IV},
\end{aligned}
\tag{71}
$$

$$
\underline{\overset{(k)}{\mathbf{u}}}{}^{IV} = (2k+1)\sum_{s=1} C_{s+2}^3(2k-2s+1)(2k-2s-1)(2k-2s-3)\overset{(k-2s-2)}{\mathbf{u}}, \tag{72}
$$

where C_{s+2}^3 are binomial coefficients; $\overset{(k)}{\mathbf{M}}$ is the kth moment operator.

It is seen from (69) and (71) that $\overset{(k)}{\mathbf{u}}{}''$ and $\overset{(k)}{\mathbf{u}}{}^{IV}$ are represented as an infinite sum of moments of displacement vector, or as the final sum of moments of displacement vector, or as the sum of values of displacement vector and their partial derivatives in x^3 on the face surface, i.e. when $x^3 = -1$ and $x^3 = 1$. Therefore, taking into account (69) and (71), from (68) we get different representations of the systems of equations for prismatic thin bodies of constant thickness in moments of displacement vector with respect to the system of Legendre polynomials. In particular, the system of equations (68) is represented by the kinematic boundary conditions on the face surface. Consequently, the system of equations (68) can be represented with the help of the static boundary conditions on the face surface. In fact, it is easy to show that, based on Hooke's law or from the boundary conditions (62) for an isotropic medium, we obtain

$$
(\partial_3 u_J)^\pm = \frac{\pm 1}{\mu}\overset{(\pm)}{P}_J - \partial_J\overset{(\pm)}{u}_3, \quad (\partial_3 u_3)^\pm = \frac{1}{\lambda+2\mu}(\pm\overset{(\pm)}{P}_3 - \lambda\partial_K\overset{(\pm)}{u}_K),
$$

and from here we will have

$$
\begin{aligned}
(\partial_3 u_J)^+ \pm (-1)^k(\partial_3 u_J)^- &= \frac{1}{\mu}[\overset{(+)}{P}_J \pm (-1)^{k+1}\overset{(-)}{P}_J] - \partial_J[\overset{(+)}{u}_3 \pm (-1)^k\overset{(-)}{u}_3], \\
(\partial_3 u_3)^+ \pm (-1)^k(\partial_3 u_3)^- &= \frac{1}{\lambda+2\mu}[\overset{(+)}{P}_3 \pm (-1)^{k+1}\overset{(-)}{P}_3] \\
&\quad - \frac{\lambda}{\lambda+2\mu}\partial_K[\overset{(+)}{u}_K \pm (-1)^k\overset{(-)}{u}_K],
\end{aligned}
\tag{73}
$$

where $\overset{(+)}{P}_k$ and $\overset{(-)}{P}_k$ are components of the given stresses $\overset{(+)}{\mathbf{P}}$ and $\overset{(-)}{\mathbf{P}}$; $\overset{(+)}{u}_k$ and $\overset{(-)}{u}_k$ are components of the given displacement vectors $\overset{(+)}{\mathbf{u}}$ and $\overset{(-)}{\mathbf{u}}$ on the face surface $\overset{(+)}{S}$ and $\overset{(-)}{S}$, respectively.

Substituting (73) into (69) and (71), then by deriving relations from (68) we obtain the desired system of equations with allowance for static and kinematic boundary conditions, as well as the values of the partial derivatives of second and third order of displacement vector in x^3 on the face surface. For brevity, it is not be written out. However, we note that the meanings of expressions for the partial derivatives of the displacement vector with respect to x^3 when $x^3 = -1$ and $x^3 = 1$ in (69) and (71) can be represented in terms of moments of displacement vector in the form

$$(\partial_3^s \mathbf{u})^+ \pm (-1)^k (\partial_3^s \mathbf{u})^- = \sum_{n=s}^{\infty} [1 \pm (-1)^{k+n-s}] \overset{(n)}{\mathbf{u}} P_n^{(s)}(1), \quad s = 0, 1, 2, 3, \ k \in \mathbb{N}_0 \tag{74}$$

and when writing the system of equations of Nth approximation, Eq. (74) can be replaced by the approximate relations

$$(\partial_3^s \mathbf{u})^+ \pm (-1)^k (\partial_3^s \mathbf{u})^- \approx \sum_{n=s}^{N} [1 \pm (-1)^{k+n-s}] \overset{(n)}{\mathbf{u}} P_n^{(s)}(1), \ s = 0, 1, 2, 3, \ k = \overline{0, N}. \tag{75}$$

In this case, if we want the boundary conditions on the face surface to be reflected in the equations, it is necessary to use relations (73) and (75) simultaneously. In particular, if the static boundary conditions are given on the face surfaces, you must use (73) and (75), when $s = 0, 2, 3$. If the kinematic boundary conditions are known on the face surfaces, you should use relations (75), when $s = 1, 2, 3$. Of course, we can consider different variants of the static and kinematic boundary conditions on the face surface. For brevity, we will not stop on these. Below, using a simplified method for reducing the infinite system to a finite, we obtain the system of equations of the first few approximations of the static problem of theory of prismatic bodies with constant thickness in moments of displacement vector with respect to the system of Legendre polynomials.

5.5.1 Equations of Nth Approximation of the Quasi-Static Problems of the Theory of Prismatic Bodies in Moments with Respect to the Legendre Polynomials Without Boundary Conditions on the Face Surface

In this case from (69) and (71) we have the following approximate relations:

$$\overset{(k)}{\mathbf{u}}{}'' \approx \overset{(k)}{\mathbf{u}}{}''_{(N)} = \frac{2k+1}{4} \sum_{p=k+2}^{N} (p-k)(p+k+1)[1+(-1)^{k+p}] \overset{(p)}{\mathbf{u}}, \quad 0 \le k \le N-2;$$

$$\overset{(k)IV}{\mathbf{u}} \approx \overset{(k)IV}{\mathbf{u}}_{(N)} = \frac{2k+1}{2}\sum_{s=k+4}^{N} b_{[\frac{s-k-2}{2}]}(k+s-1)(k+s+1)(k+s+3)[1+(-1)^{k+s}]\overset{(s)}{\mathbf{u}} \tag{76}$$

with

$$0 \le k \le N-4, \quad b_n = C_{n+2}^3 = \frac{1}{3!}n(n+1)(n+2), \quad n \in \mathbb{N} \ (\mathbb{N} = \{1, 2, 3, \dots\}).$$

Here, the index $[(s-k+2)/2]$ denotes the integer part of the number $(s-k+2)/2$.

Taking into account (76) from (68) we obtain the desired system of equations in the form

$$\bar{\Delta}^2\overset{(k)}{\mathbf{u}} + 2h^{-2}\bar{\Delta}\overset{(k)''}{\mathbf{u}}_{(N)} + h^{-4}\overset{(k)IV}{\mathbf{u}}_{(N)} + \overset{(k)}{\mathbf{G}} = 0, \quad k = \overline{0, N}. \tag{77}$$

Hence, it is easy to obtain a system of equations of any order of approximation. Let us derive, for example, a system of equations of the fifth approximation. The first and second approximations are given in Ulukhanyan (2010, 2011).

The system of equations of fifth ($N = 5$) approximation. In this case $k = \overline{0, 5}$ and with (77) we obtain

$$\begin{aligned} &\bar{\Delta}^2\overset{(0)}{\mathbf{u}} + 2h^{-2}\bar{\Delta}(3\overset{(2)}{\mathbf{u}} + 10\overset{(4)}{\mathbf{u}}) + 105h^{-4}\overset{(4)}{\mathbf{u}} + \overset{(0)}{\mathbf{G}} = 0,\\ &\bar{\Delta}^2\overset{(1)}{\mathbf{u}} + 2h^{-2}\bar{\Delta}(15\overset{(3)}{\mathbf{u}} + 42\overset{(5)}{\mathbf{u}}) + 945h^{-4}\overset{(5)}{\mathbf{u}} + \overset{(1)}{\mathbf{G}} = 0,\\ &\bar{\Delta}^2\overset{(2)}{\mathbf{u}} + 70h^{-2}\bar{\Delta}\overset{(4)}{\mathbf{u}} + \overset{(2)}{\mathbf{G}} = 0, \quad \bar{\Delta}^2\overset{(3)}{\mathbf{u}} + 126h^{-2}\bar{\Delta}\overset{(5)}{\mathbf{u}} + \overset{(3)}{\mathbf{G}} = 0,\\ &\bar{\Delta}^2\overset{(4)}{\mathbf{u}} + \overset{(4)}{\mathbf{G}} = 0, \quad \bar{\Delta}^2\overset{(5)}{\mathbf{u}} + \overset{(5)}{\mathbf{G}} = 0. \end{aligned} \tag{78}$$

Displacement vector has the form

$$\mathbf{u} \approx \overset{(0)}{\mathbf{u}} + \overset{(1)}{\mathbf{u}}P_1(x^3) + \overset{(2)}{\mathbf{u}}P_2(x^3) + \overset{(3)}{\mathbf{u}}P_3(x^3) + \overset{(4)}{\mathbf{u}}P_4(x^3) + \overset{(5)}{\mathbf{u}}P_5(x^3).$$

Note that the system of equations (78) decomposes into two system of equations. The first system consists of the first, third and fifth equations of system, and the second system consists of the second, fourth and sixth equations. Let us write down these systems separately. We have

$$\begin{aligned} &\bar{\Delta}^2\overset{(0)}{\mathbf{u}} + 2h^{-2}\bar{\Delta}(3\overset{(2)}{\mathbf{u}} + 10\overset{(4)}{\mathbf{u}}) + 105h^{-4}\overset{(4)}{\mathbf{u}} + \overset{(0)}{\mathbf{G}} = 0,\\ &\bar{\Delta}^2\overset{(2)}{\mathbf{u}} + 70h^{-2}\bar{\Delta}\overset{(4)}{\mathbf{u}} + \overset{(2)}{\mathbf{G}} = 0, \quad \bar{\Delta}^2\overset{(4)}{\mathbf{u}} + \overset{(4)}{\mathbf{G}} = 0;\\ &\bar{\Delta}^2\overset{(1)}{\mathbf{u}} + 2h^{-2}\bar{\Delta}(15\overset{(3)}{\mathbf{u}} + 42\overset{(5)}{\mathbf{u}}) + 945h^{-4}\overset{(5)}{\mathbf{u}} + \overset{(1)}{\mathbf{G}} = 0,\\ &\bar{\Delta}^2\overset{(3)}{\mathbf{u}} + 126h^{-2}\bar{\Delta}\overset{(5)}{\mathbf{u}} + \overset{(3)}{\mathbf{G}} = 0, \quad \bar{\Delta}^2\overset{(5)}{\mathbf{u}} + \overset{(5)}{\mathbf{G}} = 0. \end{aligned} \tag{79}$$

Introducing the notation

$$L = \begin{pmatrix} \bar{\Delta}^2 & 6h^{-2}\bar{\Delta} & 20h^{-2}\bar{\Delta} + 105h^{-4} \\ 0 & \bar{\Delta}^2 & 70h^{-2}\bar{\Delta} \\ 0 & 0 & \bar{\Delta}^2 \end{pmatrix}, \quad U = \begin{pmatrix} \overset{(0)}{\mathbf{u}} \\ \overset{(2)}{\mathbf{u}} \\ \overset{(4)}{\mathbf{u}} \end{pmatrix}, \quad G = \begin{pmatrix} \overset{(0)}{\mathbf{G}} \\ \overset{(2)}{\mathbf{G}} \\ \overset{(4)}{\mathbf{G}} \end{pmatrix},$$

the system of the first three equations (79) can be written as a matrix equation

$$LU + G = 0. \tag{80}$$

It is easy to calculate that the matrix of the cofactors L_* and the determinant $|L|$ for differential matrix L will have the form

$$L_* = \begin{pmatrix} \bar{\Delta}^4 & 0 & 0 \\ 6h^{-2}\bar{\Delta}^3 & \bar{\Delta}^4 & 0 \\ 20h^{-2}\bar{\Delta}^3 + 315h^{-4}\bar{\Delta}^2 & -70h^{-2}\bar{\Delta}^3 & \bar{\Delta}^4 \end{pmatrix}, \quad |L| = \bar{\Delta}^6.$$

It is not difficult to see that L_* can be written as

$$L_* = \Delta^2 N, \quad N = \begin{pmatrix} \bar{\Delta}^2 & 0 & 0 \\ 6h^{-2}\bar{\Delta} & \bar{\Delta}^2 & 0 \\ 20h^{-2}\bar{\Delta} + 315h^{-4} & -70h^{-2}\bar{\Delta} & \bar{\Delta}^2 \end{pmatrix}.$$

Applying the differential operator N^T to Eq. (80) from the left, in virtue of relation $N^T L = E\bar{\Delta}^4$, where E is the third order identity matrix, we have

$$\bar{\Delta}^4 U + N^T G = 0,$$

and from here, obviously, we find

$$\begin{aligned} &\bar{\Delta}^4 \overset{(0)}{\mathbf{u}} + \bar{\Delta}^2 \overset{(0)}{G} - 6h^{-2}\bar{\Delta}\overset{(2)}{G} - (20h^{-2}\bar{\Delta} + 315h^{-4})\overset{(4)}{G} = 0, \\ &\bar{\Delta}^3 \overset{(2)}{\mathbf{u}} + \bar{\Delta}\overset{(2)}{G} - 70h^{-2}\overset{(4)}{G} = 0, \quad \bar{\Delta}^2 \overset{(4)}{\mathbf{u}} + \overset{(0)}{G} = 0. \end{aligned} \tag{81}$$

From (81), it can be seen that we have got an eighth-order inhomogeneous equation with respect to $\overset{(0)}{\mathbf{u}}$ and its general solution is expressed using four analytic functions. We have the sixth and fourth order equations with respect to $\overset{(2)}{\mathbf{u}}$ and $\overset{(4)}{\mathbf{u}}$, and its general solutions are given by means of three and two analytic functions, respectively (Vekua 1948). For brevity, we will not write general solution of these equations. Important is the fact that it is possible to obtain analytical solutions. Of course, analogously to (80) we can consider a system, consisting of the second, fourth and sixth equations of (79), and for them to obtain the analytical solutions. Note, that the analytical solutions can be obtained for system of equations of a higher order approximation (see below the micropolar theory case).

5.5.2 Equations of Nth Approximation of the Quasi-Static Problems of Theory of Prismatic Bodies in Moments with Respect to the Legendre Polynomials with the Static Boundary Conditions on the Face Surface

To get this system of equations analogously to (76) due to (75), we will present $\overset{(k)}{\mathbf{u}}{}''_{(N)}$ and $\overset{(k)}{\mathbf{u}}{}^{IV}_{(N)}$ (see (69) and (71)) in the form of

$$
\begin{aligned}
\overset{(k)}{\mathbf{u}}{}'' &\approx \overset{(k)}{\mathbf{u}}{}''_{(N)}\\
&= \frac{2k+1}{2}\Big\{[(\partial_3\mathbf{u})^+ - (-1)^k(\partial_3\mathbf{u})^-] - \sum_{n=0}^{N}[1+(-1)^{k+n}]\overset{(n)}{\mathbf{u}}P_n'(1)\Big\} + \overset{(k)}{\underline{\mathbf{u}}}{}''_{(N)},\\
\overset{(k)}{\mathbf{u}}{}^{IV} &\approx \overset{(k)}{\mathbf{u}}{}^{IV}_{(N)}\\
&= \frac{2k+1}{2}\Big\{\sum_{n=3}^{N}[1+(-1)^{k+n}]\overset{(n)}{\mathbf{u}}P_n^{(3)}(1) - \sum_{n=2}^{N}[1+(-1)^{k+n}]\overset{(n)}{\mathbf{u}}P_n^{(2)}(1)P_k'(1)\\
&\quad + [(\partial_3\mathbf{u})^+ - (-1)^k(\partial_3\mathbf{u})^-]P_k''(1) - \sum_{n=0}^{N}[1+(-1)^{k+n}]\overset{(n)}{\mathbf{u}}P_n^{(3)}(1)\Big\} + \overset{(k)}{\underline{\mathbf{u}}}{}^{IV}_{(N)}.
\end{aligned}
\tag{82}
$$

It is easy to see that based on (82) we get

$$
2h^{-2}\bar{\Delta}\overset{(k)}{\mathbf{u}}{}''_{(N)} + h^{-4}\overset{(k)}{\mathbf{u}}{}^{IV}_{(N)} = 2h^{-2}\bar{\Delta}\overset{(k)}{\underline{\mathbf{u}}}{}''_{(N)} + h^{-4}\overset{(k)}{\underline{\mathbf{u}}}{}^{IV}_{(N)}
$$

$$
\begin{aligned}
&-\frac{2k+1}{2}\Big\{\sum_{n=0}^{N}2h^{-2}P_k'(1)[1+(-1)^{k+n}]\bar{\Delta}\overset{(n)}{\mathbf{u}} + \sum_{n=0}^{1}P_k'''(1)[1+(-1)^{k+n}]\overset{(n)}{\mathbf{u}}\\
&\quad + [P_2''(1)P_k'(1) + P_k'''(1)][1+(-1)^k]\overset{(2)}{\mathbf{u}}\\
&\quad + \sum_{n=3}^{N}[P_n''(1)P_k'(1) - P_k'''(1) - P_n'''(1)][1+(-1)^{k+n}]\overset{(n)}{\mathbf{u}}\Big\}\\
&\quad + \frac{2k+1}{2}[2h^{-2}\bar{\Delta} + h^{-4}P_k''(1)][(\partial_3\mathbf{u})^+ - (-1)^k(\partial_3\mathbf{u})^-].
\end{aligned}
\tag{83}
$$

From (73) using (75) when $s=0$, we have

$$
\begin{aligned}
&(\partial_3\mathbf{u})^+ \pm (-1)^k(\partial_3\mathbf{u})^-]\\
&= \frac{1}{\mu}[\overset{(+)}{P}_J \pm (-1)^{k+1}\overset{(-)}{P}_J]\mathbf{e}_J + \frac{1}{\lambda+2\mu}[\overset{(+)}{P}_3 \pm (-1)^{k+1}\overset{(-)}{P}_3]\mathbf{n}\\
&\quad - \sum_{n=0}^{N}[1\pm(-1)^{k+n}]\partial_J\overset{(n)}{u}_3\mathbf{e}_J - \frac{\lambda}{\lambda+2\mu}\sum_{n=0}^{N}[1\pm(-1)^{k+n}]\partial_L\overset{(n)}{u}_L\mathbf{n}.
\end{aligned}
\tag{84}
$$

Taking into account (84) in (83), and then substituting the relation into (77), we obtain the desired system of equations. For brevity, we will not reproduce it. From

(83) and (84) it is seen that the system of equation with the boundary conditions on the face surfaces is rather cumbersome. Generally speaking, to deal with the system of equations with the boundary conditions is a much more difficult problem than the system of equations without them. However, this problem can be a little simplified. In fact, in this case due to (84) it is better to express the $\overset{(k)}{\mathbf{u}}{}''_{(N)}$ (see the first formula (82)) through the boundary conditions on the front surface, and for $\overset{(k)}{\mathbf{u}}{}^{IV}_{(N)}$ to use the second formula (76). Note that when applying the system of Chebyshev polynomials, we obtain analogous to (78) and (79) equations under any parametrization of the thin body domain. The difference is only in the coefficients (Nikabadze 2008a, b, 2014b, c).

6 Micropolar Theory of Elasticity in Displacements and Rotations

6.1 Equations of Motion of 3D Micropolar Theory of Elasticity in Displacements and Rotation Vectors

Let us consider the micropolar material with a center of symmetry. In this case $\underset{\approx}{\mathbf{C}}^T = \underset{\approx}{\mathbf{B}} = 0$ and constitutive relations (44) will get the form

$$\underset{\sim}{\mathbf{P}} = \underset{\approx}{\mathbf{A}} \overset{2}{\otimes} \underset{\sim}{\gamma}, \quad \underset{\sim}{\mu} = \underset{\approx}{\mathbf{D}} \overset{2}{\otimes} \underset{\sim}{\varkappa} \quad (\underset{\sim}{\gamma} = \nabla \mathbf{u} - \underset{\simeq}{\mathbf{C}} \cdot \boldsymbol{\varphi}, \quad \underset{\sim}{\varkappa} = \nabla \boldsymbol{\varphi}). \tag{85}$$

Taking into account (85), from the Eq. (41) after simple transformations we obtain

$$\underset{\sim}{\mathbf{A}}^{(1)} \cdot \mathbf{u} + \underset{\sim}{\mathbf{A}}^{(2)} \cdot \boldsymbol{\varphi} + \rho \mathbf{F} = 0, \quad \underset{\sim}{\mathbf{A}}^{(3)} \cdot \mathbf{u} + \underset{\sim}{\mathbf{A}}^{(4)} \cdot \boldsymbol{\varphi} + \rho \mathbf{m} = 0, \tag{86}$$

where for differential tensor operators $\underset{\sim}{\mathbf{A}}^{(k)}$, $k = 1, 2, 3, 4$, we have the following expression:

$$\begin{aligned}
\underset{\sim}{\mathbf{A}}^{(1)} &= \mathbf{r}_j \mathbf{r}_l (A^{ijkl} \nabla_i \nabla_k + \nabla_i A^{ijkl} \nabla_k) - \underset{\sim}{\mathbf{E}} \rho \partial_t^2, \\
\underset{\sim}{\mathbf{A}}^{(2)} &= -\mathbf{r}_j \mathbf{r}^m C_{mkl} (A^{klij} \nabla_i + \nabla_i A^{klij}), \\
\underset{\sim}{\mathbf{A}}^{(3)} &= \mathbf{r}^m \mathbf{r}_j C_{mkl} A^{klij} \nabla_i, \\
\underset{\sim}{\mathbf{A}}^{(4)} &= \mathbf{r}_j \mathbf{r}_l (D^{ijkl} \nabla_i \nabla_k + \nabla_i D^{ijkl} \nabla_k) - \underset{\simeq}{\mathbf{C}} \overset{2}{\otimes} \underset{\approx}{\mathbf{A}} \overset{2}{\otimes} \underset{\simeq}{\mathbf{C}} - \underset{\sim}{\mathbf{J}} \partial_t^2.
\end{aligned}$$

If we introduce the matrix differential tensor operator and vector-columns

$$\underset{\sim}{\mathbf{M}} = \begin{pmatrix} \underset{\sim}{\mathbf{A}}^{(1)} & \underset{\sim}{\mathbf{A}}^{(2)} \\ \underset{\sim}{\mathbf{A}}^{(3)} & \underset{\sim}{\mathbf{A}}^{(4)} \end{pmatrix}, \quad \mathbf{U} = \begin{pmatrix} \mathbf{u} \\ \boldsymbol{\varphi} \end{pmatrix}, \quad \mathbf{X} = \begin{pmatrix} \rho \mathbf{F} \\ \rho \mathbf{m} \end{pmatrix}, \tag{87}$$

then Eq. (86) can be shortly represented as follows:

$$\underset{\sim}{\mathbf{M}} \cdot \mathbf{U} + \mathbf{X} = \mathbf{0}. \tag{88}$$

In case of homogeneous isotropic material, the equations of motion in displacement and rotation vectors have the form (Kupradze et al 1976; Nowacki 1970; Eringen 1999)

$$\begin{aligned} &(\mu + \alpha)\Delta\mathbf{u} + (\lambda + \mu - \alpha)\,\mathrm{grad\,div}\,\mathbf{u} + 2\alpha\,\mathrm{rot}\boldsymbol{\varphi} + \rho\mathbf{F} = \rho\partial_t^2\mathbf{u},\\ &(\delta + \beta)\Delta\boldsymbol{\varphi} + (\gamma + \delta - \beta)\,\mathrm{grad\,div}\boldsymbol{\varphi} + 2\alpha\,\mathrm{rot}\mathbf{u} - 4\alpha\,\boldsymbol{\varphi} + \rho\mathbf{m} = \underset{\sim}{\mathbf{J}} \cdot \partial_t^2\boldsymbol{\varphi}, \end{aligned} \tag{89}$$

and the differential tensor operators $\underset{\sim}{\mathbf{A}}^{(k)}$, $k = 1, 2, 3, 4$, are represented as

$$\begin{aligned} &\underset{\sim}{\mathbf{A}} \equiv \underset{\sim}{\mathbf{A}}^{(1)} = \underset{\sim}{\mathbf{E}}Q_2 + d\nabla\nabla,\\ &\underset{\sim}{\mathbf{B}} \equiv \underset{\sim}{\mathbf{A}}^{(2)} = \underset{\sim}{\mathbf{A}}^{(3)} = -2\alpha\underset{\simeq}{\mathbf{C}} \cdot \nabla,\\ &\underset{\sim}{\mathbf{C}} \equiv \underset{\sim}{\mathbf{A}}^{(4)} = \underset{\sim}{\mathbf{E}}Q_4 + m\nabla\nabla, \end{aligned} \tag{90}$$

$$\begin{aligned} &d = \lambda + \mu - \alpha,\quad l = 4\alpha,\quad b = \mu + \alpha,\quad g = \delta + \beta,\quad m = \gamma + \delta - \beta,\\ &\underset{\sim}{\mathbf{J}} = J\underset{\sim}{\mathbf{E}},\quad J = \rho I,\\ &Q_2 = b\Box_2 = b\Delta - \rho\partial_t^2,\quad \Box_2 = \Delta - \frac{\rho}{b}\partial_t^2,\\ &Q_4 = g\Box_4 = g\Delta - l - J\partial_t^2,\quad \Box_4 = \Delta - \frac{l}{g} - \frac{J}{g}\partial_t^2. \end{aligned} \tag{91}$$

Note that besides the operators indicated in (91), operators

$$\begin{aligned} &Q_1 = Q_2 + d\Delta = (b + d)\Box_1,\quad Q_3 = Q_4 + m\Delta = (g + m)\Box_3,\\ &\Box_1 = \Delta - \frac{\rho}{b + d}\partial_t^2,\quad \Box_3 = \Delta - \frac{l}{g + m} - \frac{J}{g + m}\partial_t^2. \end{aligned} \tag{92}$$

are introduced in consideration. It is seen that for a homogeneous isotropic body the system of equations (86) has a form

$$\underset{\sim}{\mathbf{A}} \cdot \mathbf{u} + \underset{\sim}{\mathbf{B}} \cdot \boldsymbol{\varphi} + \rho\mathbf{F} = 0,\quad \underset{\sim}{\mathbf{B}} \cdot \mathbf{u} + \underset{\sim}{\mathbf{C}} \cdot \boldsymbol{\varphi} + \rho\mathbf{m} = 0, \tag{93}$$

$$\underset{\sim}{\mathbf{A}} = \underset{\sim}{\mathbf{E}}Q_2 + d\nabla\nabla,\quad \underset{\sim}{\mathbf{B}} = -2\alpha\underset{\simeq}{\mathbf{C}} \cdot \nabla,\quad \underset{\sim}{\mathbf{C}} = \underset{\sim}{\mathbf{E}}Q_4 + m\nabla\nabla.$$

It is easy to prove that the operators $\underset{\sim}{\mathbf{A}}$, $\underset{\sim}{\mathbf{B}}$, $\underset{\sim}{\mathbf{C}}$ commute in pairs with respect to a single multiplication, i.e. $\underset{\sim}{\mathbf{A}} \cdot \underset{\sim}{\mathbf{B}} = \underset{\sim}{\mathbf{B}} \cdot \underset{\sim}{\mathbf{A}}$, $\underset{\sim}{\mathbf{A}} \cdot \underset{\sim}{\mathbf{C}} = \underset{\sim}{\mathbf{C}} \cdot \underset{\sim}{\mathbf{A}}$, $\underset{\sim}{\mathbf{B}} \cdot \underset{\sim}{\mathbf{C}} = \underset{\sim}{\mathbf{C}} \cdot \underset{\sim}{\mathbf{B}}$.

Starting from (93), we obtain the equations with respect to the vectors $\mathbf{u}$ and $\boldsymbol{\varphi}$ separately. Therefore, we apply the operator $\underset{\sim}{\mathbf{B}}$ to the first equation (93) with the following single multiplication, and we apply the operator $\underset{\sim}{\mathbf{A}}$ to the second equation. As a result, we find

$$(\underset{\sim}{\mathbf{B}}\cdot\underset{\sim}{\mathbf{A}})\cdot\mathbf{u}+(\underset{\sim}{\mathbf{B}}\cdot\underset{\sim}{\mathbf{B}})\cdot\boldsymbol{\varphi}+\underset{\sim}{\mathbf{B}}\cdot(\rho\mathbf{F})=0,\ (\underset{\sim}{\mathbf{A}}\cdot\underset{\sim}{\mathbf{B}})\cdot\mathbf{u}+(\underset{\sim}{\mathbf{A}}\cdot\underset{\sim}{\mathbf{C}})\cdot\boldsymbol{\varphi}+\underset{\sim}{\mathbf{A}}\cdot(\rho\mathbf{m})=0. \quad (94)$$

Taking into account the commutativity of operators $\underset{\sim}{\mathbf{A}}$ and $\underset{\sim}{\mathbf{B}}$ and subtracting the second equation from the first, we obtain

$$(\underset{\sim}{\mathbf{A}}\cdot\underset{\sim}{\mathbf{C}}-\underset{\sim}{\mathbf{B}}^2)\cdot\boldsymbol{\varphi}+\underset{\sim}{\mathbf{A}}\cdot(\rho\mathbf{m})-\underset{\sim}{\mathbf{B}}\cdot(\rho\mathbf{F})=0. \quad (95)$$

Now we apply the operator $\underset{\sim}{\mathbf{C}}$ to the first equation (93) with the single multiplication, and we apply the operator $\underset{\sim}{\mathbf{B}}$ to the second equation. Obviously, we have

$$\begin{aligned}(\underset{\sim}{\mathbf{C}}\cdot\underset{\sim}{\mathbf{A}})\cdot\mathbf{u}+(\underset{\sim}{\mathbf{C}}\cdot\underset{\sim}{\mathbf{B}})\cdot\boldsymbol{\varphi}+\underset{\sim}{\mathbf{C}}\cdot(\rho\mathbf{F})=0,\\(\underset{\sim}{\mathbf{B}}\cdot\underset{\sim}{\mathbf{B}})\cdot\mathbf{u}+(\underset{\sim}{\mathbf{B}}\cdot\underset{\sim}{\mathbf{C}})\cdot\boldsymbol{\varphi}+\underset{\sim}{\mathbf{B}}\cdot(\rho\mathbf{m})=0.\end{aligned} \quad (96)$$

Taking into account the communicative property of operators $\underset{\sim}{\mathbf{A}}$ and $\underset{\sim}{\mathbf{C}}$, as well as $\underset{\sim}{\mathbf{B}}$ and $\underset{\sim}{\mathbf{C}}$, with respect to a single multiplication and subtracting the second equation from the first, we obtain the equation

$$(\underset{\sim}{\mathbf{A}}\cdot\underset{\sim}{\mathbf{C}}-\underset{\sim}{\mathbf{B}}^2)\cdot\mathbf{u}+\underset{\sim}{\mathbf{C}}\cdot(\rho\mathbf{F})-\underset{\sim}{\mathbf{B}}\cdot(\rho\mathbf{m})=0. \quad (97)$$

Introducing the notation $\underset{\sim}{\mathbf{D}}=\underset{\sim}{\mathbf{A}}\cdot\underset{\sim}{\mathbf{C}}-\underset{\sim}{\mathbf{B}}^2$, Eqs. (95) and (97) can be written as

$$\underset{\sim}{\mathbf{D}}\cdot\mathbf{u}+\underset{\sim}{\mathbf{C}}\cdot(\rho\mathbf{F})-\underset{\sim}{\mathbf{B}}\cdot(\rho\mathbf{m})=0,\quad \underset{\sim}{\mathbf{D}}\cdot\boldsymbol{\varphi}+\underset{\sim}{\mathbf{A}}\cdot(\rho\mathbf{m})-\underset{\sim}{\mathbf{B}}\cdot(\rho\mathbf{F})=0. \quad (98)$$

Then after some simple calculations we find

$$\begin{aligned}&\underset{\sim}{\mathbf{A}}\cdot\underset{\sim}{\mathbf{C}}=\underset{\sim}{\mathbf{E}}Q_2Q_4+(mQ_1+dQ_4)\nabla\nabla,\quad \underset{\sim}{\mathbf{B}}^2=-4\alpha^2(\underset{\sim}{\mathbf{E}}\Delta-\nabla\nabla),\\&\underset{\sim}{\mathbf{D}}=\underset{\sim}{\mathbf{E}}(Q_2Q_4+4\alpha^2\Delta)+(mQ_1+dQ_4-4\alpha^2)\nabla\nabla,\\&\underset{\sim}{\mathbf{D}}_*=(Q_2Q_4+4\alpha^2\Delta)[\underset{\sim}{\mathbf{E}}Q_1Q_3-(mQ_1+dQ_4-4\alpha^2)\nabla\nabla],\\&|\underset{\sim}{\mathbf{D}}|=Q_1Q_3(Q_2Q_4+4\alpha^2\Delta)^2,\quad \underset{\sim}{\mathbf{D}}\cdot\underset{\sim}{\mathbf{D}}_*=\underset{\sim}{\mathbf{D}}_*\cdot\underset{\sim}{\mathbf{D}}=\underset{\sim}{\mathbf{E}}|\underset{\sim}{\mathbf{D}}|\ \ (\underset{\sim}{\mathbf{D}}_*^T=\underset{\sim}{\mathbf{D}}_*),\end{aligned} \quad (99)$$

where $\underset{\sim}{\mathbf{D}}_*$ is the differential tensor- operator of the cofactors to the operator $\underset{\sim}{\mathbf{D}}$. Let us introduce the differential tensor operator

$$\underset{\sim}{\mathbf{N}}=\underset{\sim}{\mathbf{E}}Q_1Q_3-(mQ_1+dQ_4-4\alpha^2)\nabla\nabla. \quad (100)$$

Then, with (99) and (100) we have

$$\begin{aligned}\underset{\sim}{\mathbf{D}}_*\ &=(Q_2Q_4+4\alpha^2\Delta)\underset{\sim}{\mathbf{N}},\\\underset{\sim}{\mathbf{D}}\cdot\underset{\sim}{\mathbf{N}}&=\underset{\sim}{\mathbf{N}}\cdot\underset{\sim}{\mathbf{D}}=\underset{\sim}{\mathbf{E}}Q_1Q_3(Q_2Q_4+4\alpha^2\Delta),\\\underset{\sim}{\mathbf{N}}\cdot\underset{\sim}{\mathbf{B}}&=-\underset{\sim}{\mathbf{B}}\cdot\underset{\sim}{\mathbf{N}}=-2\alpha Q_1Q_3\underset{\approx}{\mathbf{C}}\cdot\nabla,\\\underset{\sim}{\mathbf{N}}\cdot\underset{\sim}{\mathbf{C}}&=\underset{\sim}{\mathbf{C}}\cdot\underset{\sim}{\mathbf{N}}=Q_3[\underset{\sim}{\mathbf{E}}Q_1Q_4-(dQ_4-4\alpha^2)\nabla\nabla],\\\underset{\sim}{\mathbf{N}}\cdot\underset{\sim}{\mathbf{A}}&=\underset{\sim}{\mathbf{A}}\cdot\underset{\sim}{\mathbf{N}}=Q_1[\underset{\sim}{\mathbf{E}}Q_2Q_3-(mQ_2-4\alpha^2)\nabla\nabla].\end{aligned} \quad (101)$$

If we look for the solution of the Eq. (98) as (analog to Galerkin)

$$\mathbf{u} = \underset{\sim}{\mathbf{N}} \cdot \mathbf{v}, \quad \boldsymbol{\varphi} = \underset{\sim}{\mathbf{N}} \cdot \boldsymbol{\psi}, \tag{102}$$

then taking into account the corresponding relations (99) and (101), due to (100) from Eq. (98), we obtain the following equations with respect to $\mathbf{v}$ and $\boldsymbol{\psi}$ separately:

$$\begin{aligned} &Q_1 Q_3(Q_2 Q_4 + 4\alpha^2 \Delta)\mathbf{v} + (\underset{\sim}{\mathbf{E}} Q_4 + m\nabla\nabla) \cdot (\rho \mathbf{F}) + 2\alpha(\underset{\approx}{\mathbf{C}} \cdot \nabla) \cdot (\rho \mathbf{m}) = 0, \\ &Q_1 Q_3(Q_2 Q_4 + 4\alpha^2 \Delta)\boldsymbol{\psi} + (\underset{\sim}{\mathbf{E}} Q_2 + d\nabla\nabla) \cdot (\rho \mathbf{m}) + 2\alpha(\underset{\approx}{\mathbf{C}} \cdot \nabla) \cdot (\rho \mathbf{F}) = 0. \end{aligned} \tag{103}$$

Now applying the tensor operator $\underset{\sim}{\mathbf{N}}$ to the Eq. (98) with the following single multiplication and taking into account (101) we obtain

$$\begin{aligned} &Q_3\{Q_1[(Q_2 Q_4 + 4\alpha^2 \Delta)\mathbf{u} + 2\alpha(\underset{\approx}{\mathbf{C}} \cdot \nabla) \cdot (\rho \mathbf{m})] \\ &\qquad + [\underset{\sim}{\mathbf{E}} Q_1 Q_4 - (d Q_4 - 4\alpha^2)\nabla\nabla] \cdot (\rho \mathbf{F})\} = 0, \\ &Q_1\{Q_3[(Q_2 Q_4 + 4\alpha^2 \Delta)\boldsymbol{\varphi} + 2\alpha(\underset{\approx}{\mathbf{C}} \cdot \nabla) \cdot (\rho \mathbf{F})] \\ &\qquad + [\underset{\sim}{\mathbf{E}} Q_2 Q_3 - (m Q_2 - 4\alpha^2)\nabla\nabla] \cdot (\rho \mathbf{m})\} = 0. \end{aligned} \tag{104}$$

Consider now the equations of a micropolar homogeneous isotropic medium in the form (88). Denote by

$$\underset{\sim}{\mathbf{M}}_* = \begin{pmatrix} \hat{\mathbf{A}} & \hat{\mathbf{B}}^{(1)} \\ \hat{\mathbf{B}}^{(2)} & \hat{\mathbf{C}} \end{pmatrix} \tag{105}$$

the matrix differential tensor operator of the cofactors to the differential operator $\underset{\sim}{\mathbf{M}}$ of Eq. (88). After simple, but very cumbersome computations we get

$$\begin{aligned} &\hat{\mathbf{A}} = Q_3(Q_2 Q_4 + 4\alpha^2 \Delta)[\underset{\sim}{\mathbf{E}} Q_1 Q_4 - (d Q_4 - 4\alpha^2)\nabla\nabla] \quad (\hat{\mathbf{A}}^T = \hat{\mathbf{A}}), \\ &\hat{\mathbf{B}} = \hat{\mathbf{B}}^{(1)} = \hat{\mathbf{B}}^{(2)} = -2\alpha Q_1 Q_3(Q_2 Q_4 + 4\alpha^2 \Delta)\underset{\approx}{\mathbf{C}} \cdot \nabla) \quad (\hat{\mathbf{B}}^T = -\hat{\mathbf{B}}), \\ &\hat{\mathbf{C}} = Q_1(Q_2 Q_4 + 4\alpha^2 \Delta)[\mathbf{E} Q_2 Q_3 - (m Q_2 - 4\alpha^2)\nabla\nabla] \quad (\hat{\mathbf{C}}^T = \hat{\mathbf{C}}). \end{aligned} \tag{106}$$

Note that the expression for the cofactors of the elements of the determinant of the matrix differential operator of the homogeneous equation of stationary oscillations of micropolar theory of elasticity are given in Kupradze et al (1976), which are obtained from the expressions for the differential tensor-operators components (106), if we replace the second time derivative (∂_t^2) by the square of the oscillation frequency (σ^2).

Let us introduce the matrix differential tensor operators (differential tensor block matrices)

$$\underset{\sim}{\mathbf{N}}^{(1)} = \begin{pmatrix} \underset{\sim}{\mathbf{R}} & \underset{\sim}{\mathbf{S}}^{(2)} \\ \underset{\sim}{\mathbf{S}}^{(1)} & \underset{\sim}{\mathbf{T}} \end{pmatrix}, \quad \underset{\sim}{\mathbf{N}}^{(2)} = \begin{pmatrix} \underset{\sim}{\mathbf{R}} & \underset{\sim}{\mathbf{S}}^{(1)} \\ \underset{\sim}{\mathbf{S}}^{(2)} & \underset{\sim}{\mathbf{T}} \end{pmatrix}, \tag{107}$$

$$\begin{aligned} &\underset{\sim}{\mathbf{R}} = \underset{\sim}{\mathbf{E}} Q_1 Q_4 - (d Q_4 - 4\alpha^2)\nabla\nabla, \quad \underset{\sim}{\mathbf{S}}^{(1)} = Q_3 \underset{\sim}{\mathbf{B}}, \quad \underset{\sim}{\mathbf{S}}^{(2)} = Q_1 \underset{\sim}{\mathbf{B}}, \\ &\underset{\sim}{\mathbf{B}} = -2\alpha \underset{\approx}{\mathbf{C}} \cdot \nabla, \quad \underset{\sim}{\mathbf{T}} = \underset{\sim}{\mathbf{E}} Q_2 Q_3 - (m Q_2 - 4\alpha^2)\nabla\nabla. \end{aligned}$$

Then, obviously,

$$\underset{\sim}{\mathbf{N}}^{(1)T} = \begin{pmatrix} \underset{\sim}{\mathbf{R}} & -\underset{\sim}{\mathbf{S}}^{(1)} \\ -\underset{\sim}{\mathbf{S}}^{(2)} & \underset{\sim}{\mathbf{T}} \end{pmatrix}, \quad \underset{\sim}{\mathbf{N}}^{(2)T} = \begin{pmatrix} \underset{\sim}{\mathbf{R}} & -\underset{\sim}{\mathbf{S}}^{(2)} \\ -\underset{\sim}{\mathbf{S}}^{(1)} & \underset{\sim}{\mathbf{T}} \end{pmatrix}. \tag{108}$$

If we will search for the solution of the Eq. (88) as (analog of Galerkin)

$$\mathbf{U} = \underset{\sim}{\mathbf{N}}^{(1)T} \cdot \mathbf{V}, \quad \mathbf{U} = \begin{pmatrix} \mathbf{u} \\ \boldsymbol{\varphi} \end{pmatrix}, \quad \mathbf{V} = \begin{pmatrix} \mathbf{v} \\ \boldsymbol{\psi} \end{pmatrix}, \tag{109}$$

then we get the following equations:

$$Q_1(Q_2Q_4 + 4\alpha^2\Delta)\mathbf{v} + \rho\mathbf{F} = 0, \quad Q_3(Q_2Q_4 + 4\alpha^2\Delta)\boldsymbol{\psi} + \rho\mathbf{m} = 0. \tag{110}$$

If we apply the operator $\underset{\sim}{\mathbf{N}}^{(2)T}$ from the left to Eq. (88) with the single multiplication then we have equations

$$\begin{aligned} &Q_1[(Q_2Q_4 + 4\alpha^2\Delta)\mathbf{u} + 2\alpha(\underset{\approx}{\mathbf{C}} \cdot \nabla) \cdot (\rho\mathbf{m})] \\ &+ [\underset{\sim}{\mathbf{E}}Q_1Q_4 - (dQ_4 - 4\alpha^2)\nabla\nabla] \cdot (\rho\mathbf{F}) = 0, \\ &Q_3[(Q_2Q_4 + 4\alpha^2\Delta)\boldsymbol{\varphi} + 2\alpha(\underset{\approx}{\mathbf{C}} \cdot \nabla) \cdot (\rho\mathbf{F})] \\ &+ [\underset{\sim}{\mathbf{E}}Q_2Q_3 - (mQ_2 - 4\alpha^2)\nabla\nabla] \cdot (\rho\mathbf{m}) = 0. \end{aligned} \tag{111}$$

We see that when $\alpha = 0$, i.e. in the case of a reduced medium, from the first equation (111) follows the classical equation (61), and the second equation has a similar view. Furthermore, it is clear that if we have Eq. (111), then the Eq. (104) are also valid.

Note that similar equations were obtained in Sandru (1966). Similar equations in another way were obtained by Nowacki (1970). Finally, Sandru and Nowacki gave the same representations of the displacement and rotation vectors and they are reduced to (109). Aero and Kuvshinskii (1964) deserve great attention because the system of equilibrium equations for the isotropic elastic body without a center of symmetry and in the absence of mass loads is decomposed into two independent systems of equations.

Note that Eq. (111) have an advantage compared with Eq. (110), because the stress boundary conditions for Eq. (111) are the boundary conditions of the initial-boundary value problem. The boundary conditions for Eq. (110) have more complex expressions with respect to the introduced vector fields $\mathbf{v}$ and $\boldsymbol{\psi}$. Note also that the equations can be split if the material does not have a center of symmetry (Nikabadze 2008b, 2014b, c). For brevity, we do not consider this in detail.

6.2 On Stress Boundary Conditions. Tensor Operators of Stress and Couple Stress

The boundary conditions for a linear elastic inhomogeneous anisotropic body with no symmetry center in an isothermal process can be represented as

$$\underset{\sim}{\mathbf{T}}^{(1)} \cdot \mathbf{u} + \underset{\sim}{\mathbf{T}}^{(2)} \cdot \boldsymbol{\varphi} = \mathbf{P}_{(n)}, \quad \underset{\sim}{\mathbf{T}}^{(3)} \cdot \mathbf{u} + \underset{\sim}{\mathbf{T}}^{(4)} \cdot \boldsymbol{\varphi} = \boldsymbol{\mu}_{(n)}, \tag{112}$$

where we have introduced the following differential operators:

$$\begin{aligned} \underset{\sim}{\mathbf{T}}^{(1)} &= \mathbf{r}_j \mathbf{r}_l n_i A^{ijkl} \nabla_k, \quad \underset{\sim}{\mathbf{T}}^{(2)} = \mathbf{r}_j \mathbf{r}_l n_i B^{ijkl} \nabla_k - \mathbf{n} \cdot \underset{\approx}{\mathbf{A}} \overset{2}{\otimes} \underset{\simeq}{\mathbf{C}}, \\ \underset{\sim}{\mathbf{T}}^{(3)} &= \mathbf{r}_j \mathbf{r}_l n_i C^{ijkl} \nabla_k, \quad \underset{\sim}{\mathbf{T}}^{(4)} = \mathbf{r}_j \mathbf{r}_l n_i D^{ijkl} \nabla_k - \mathbf{n} \cdot \underset{\approx}{\mathbf{C}} \overset{2}{\otimes} \underset{\simeq}{\mathbf{C}}. \end{aligned} \tag{113}$$

Let us introduce the matrix differential tensor operator (tensor operator of stress and couple stress) and the vector columns:

$$\mathbb{T} = \begin{pmatrix} \underset{\sim}{\mathbf{T}}^{(1)} & \underset{\sim}{\mathbf{T}}^{(2)} \\ \underset{\sim}{\mathbf{T}}^{(3)} & \underset{\sim}{\mathbf{T}}^{(4)} \end{pmatrix}, \quad \mathbb{U} = \begin{pmatrix} \mathbf{u} \\ \boldsymbol{\varphi} \end{pmatrix}, \quad \mathbb{Q}_{(n)} = \begin{pmatrix} \mathbf{P}_{(n)} \\ \boldsymbol{\mu}_{(n)} \end{pmatrix}. \tag{114}$$

Then, the boundary conditions can be represented in form

$$\begin{pmatrix} \underset{\sim}{\mathbf{T}}^{(1)} & \underset{\sim}{\mathbf{T}}^{(2)} \\ \underset{\sim}{\mathbf{T}}^{(3)} & \underset{\sim}{\mathbf{T}}^{(4)} \end{pmatrix} \cdot \begin{pmatrix} \mathbf{u} \\ \boldsymbol{\varphi} \end{pmatrix} = \begin{pmatrix} \mathbf{P}_{(n)} \\ \boldsymbol{\mu}_{(n)} \end{pmatrix} \quad \text{or shortly} \quad \mathbb{T} \cdot \mathbb{U} = \mathbb{Q}_{(n)}. \tag{115}$$

In case of isotropic material with no symmetry center, we have

$$\begin{aligned} \underset{\sim}{\mathbf{T}}^{(1)} &= a_2 \underset{\sim}{\mathbf{E}} \mathbf{n} \cdot \nabla + a_1 \mathbf{n} \nabla + a_3 (\mathbf{n} \nabla)^T, \\ \underset{\sim}{\mathbf{T}}^{(2)} &= b_2 \underset{\sim}{\mathbf{E}} \mathbf{n} \cdot \nabla + b_1 \mathbf{n} \nabla + b_3 (\mathbf{n} \nabla)^T - (a_2 - a_3) \mathbf{n} \cdot \underset{\simeq}{\mathbf{C}}, \\ \underset{\sim}{\mathbf{T}}^{(3)} &= b_2 \underset{\sim}{\mathbf{E}} \mathbf{n} \cdot \nabla + b_1 \mathbf{n} \nabla + b_3 (\mathbf{n} \nabla)^T, \\ \underset{\sim}{\mathbf{T}}^{(4)} &= d_2 \underset{\sim}{\mathbf{E}} \mathbf{n} \cdot \nabla + d_1 \mathbf{n} \nabla + d_3 (\mathbf{n} \nabla)^T - (b_2 - b_3) \mathbf{n} \cdot \underset{\simeq}{\mathbf{C}}. \end{aligned} \tag{116}$$

Let us introduce also the differential tensor operator

$$\underset{\sim}{\mathbf{T}}^{'(4)} = d_2 \underset{\sim}{\mathbf{E}} \mathbf{n} \cdot \nabla + d_1 \mathbf{n} \nabla + d_3 (\mathbf{n} \nabla)^T. \tag{117}$$

It is easy to see that

$$\underset{\sim}{\mathbf{T}}^{(2)} = \underset{\sim}{\mathbf{T}}^{(3)} - (a_2 - a_3) \mathbf{n} \cdot \underset{\simeq}{\mathbf{C}}, \quad \underset{\sim}{\mathbf{T}}^{(4)} = \underset{\sim}{\mathbf{T}}^{'(4)} - (b_2 - b_3) \mathbf{n} \cdot \underset{\simeq}{\mathbf{C}}. \tag{118}$$

We assume that the body has a piecewise-plane boundary. Then, denoting by $\underset{\sim}{\mathbf{T}}^{(1)}_*$, $|\underset{\sim}{\mathbf{T}}^{(1)}|$, $\underset{\sim}{\mathbf{T}}^{(3)}_*$, $|\underset{\sim}{\mathbf{T}}^{(3)}|$, $\underset{\sim}{\mathbf{T}}^{'(4)}_*$, $|\underset{\sim}{\mathbf{T}}^{'(4)}|$ the differential tensor-operators of cofactors and the determinants of tensor operators $\underset{\sim}{\mathbf{T}}^{(1)}$, $\underset{\sim}{\mathbf{T}}^{(3)}$ and $\underset{\sim}{\mathbf{T}}^{'(4)}$ respectively, after some calculations we get

$$
\begin{aligned}
\underset{*}{\mathbf{T}}^{(1)} &= [(a_1 + a_2)(a_2 + a_3)\underset{\sim}{\mathbf{E}}\mathbf{n} \cdot \nabla - a_3(a_1 + a_2)\mathbf{n}\nabla \\
&\quad - a_1(a_2 + a_3)(\mathbf{n}\nabla)^T]\mathbf{n} \cdot \nabla + a_1a_3[\nabla\nabla + (\mathbf{nn} - \underset{\sim}{\mathbf{E}})\Delta], \\
|\underset{\sim}{\mathbf{T}}^{(1)}| &= a_2[(a_1 + a_2)(a_2 + a_3)\mathbf{nnn}\overset{3}{\otimes}\nabla\nabla\nabla - a_1a_3\Delta\mathbf{n} \cdot \nabla] \\
&= a_2[(a_1 + a_2)(a_2 + a_3)\mathbf{nn}\overset{2}{\otimes}\nabla\nabla - a_1a_3\Delta]\mathbf{n} \cdot \nabla, \\
\underset{*}{\mathbf{T}}^{(3)} &= [(b_1 + b_2)(b_2 + b_3)\underset{\sim}{\mathbf{E}}\mathbf{n} \cdot \nabla - b_3(b_1 + b_2)\mathbf{n}\nabla \\
&\quad - b_1(b_2 + b_3)(\mathbf{n}\nabla)^T]\mathbf{n} \cdot \nabla + b_1b_3[\nabla\nabla + (\mathbf{nn} - \underset{\sim}{\mathbf{E}})\Delta], \\
|\underset{\sim}{\mathbf{T}}^{(3)}| &= b_2[(b_1 + b_2)(b_2 + b_3)\mathbf{nn}\overset{2}{\otimes}\nabla\nabla - b_1b_3\Delta]\mathbf{n} \cdot \nabla, \\
\underset{*}{\mathbf{T}}^{'(4)} &= [(d_1 + d_2)(d_2 + d_3)\underset{\sim}{\mathbf{E}}\mathbf{n} \cdot \nabla - d_3(d_1 + d_2)\mathbf{n}\nabla \\
&\quad - d_1(d_2 + d_3)(\mathbf{n}\nabla)^T]\mathbf{n} \cdot \nabla + d_1d_3[\nabla\nabla + (\mathbf{nn} - \underset{\sim}{\mathbf{E}})\Delta], \\
|\underset{\sim}{\mathbf{T}}^{'(4)}| &= d_2[(d_1 + d_2)(d_2 + d_3)\mathbf{nn}\overset{2}{\otimes}\nabla\nabla - d_1d_3\Delta]\mathbf{n} \cdot \nabla.
\end{aligned}
\tag{119}
$$

Note that we want to decompose the boundary conditions, i.e. we want to get the boundary conditions for $\mathbf{u}$ and $\boldsymbol{\varphi}$ separately. For brevity, we consider the case when $b_2 = b_3, a_2 = a_3$. Then $\underset{\sim}{\mathbf{T}}^{(2)} = \underset{\sim}{\mathbf{T}}^{(3)}, \underset{\sim}{\mathbf{T}}^{(4)} = \underset{\sim}{\mathbf{T}}^{'(4)}$ and the boundary conditions (115) can be written as

$$
\underset{\sim}{\mathbf{T}}^{(1)} \cdot \mathbf{u} + \underset{\sim}{\mathbf{T}}^{(3)} \cdot \boldsymbol{\varphi} = \mathbf{P}_{(n)}, \quad \underset{\sim}{\mathbf{T}}^{(3)} \cdot \mathbf{u} + \underset{\sim}{\mathbf{T}}^{'(4)} \cdot \boldsymbol{\varphi} = \boldsymbol{\mu}_{(n)}. \tag{120}
$$

In this case it is easy to obtain the boundary conditions with respect to $\mathbf{u}$ and $\boldsymbol{\varphi}$ separately. Indeed, applying the operator $|\underset{\sim}{\mathbf{T}}^{(3)}|\underset{*}{\mathbf{T}}^{(1)T}$ to the first relation (120) with a single multiplication, and the operator $|\underset{\sim}{\mathbf{T}}^{(1)}|\underset{*}{\mathbf{T}}^{(3)T}$ to the second relation and taking into account

$$
\underset{*}{\mathbf{T}}^{(1)T} \cdot \underset{\sim}{\mathbf{T}}^{(1)} = \underset{\sim}{\mathbf{E}}|\underset{\sim}{\mathbf{T}}^{(1)}|, \quad \underset{*}{\mathbf{T}}^{(3)T} \cdot \underset{\sim}{\mathbf{T}}^{(3)} = \underset{\sim}{\mathbf{E}}|\underset{\sim}{\mathbf{T}}^{(3)}|,
$$

and then subtracting the second relation from the first obtained relation, we have

$$
\left(|\underset{\sim}{\mathbf{T}}^{(3)}|\underset{*}{\mathbf{T}}^{(1)T} \cdot \underset{\sim}{\mathbf{T}}^{(3)} - |\underset{\sim}{\mathbf{T}}^{(1)}|\underset{*}{\mathbf{T}}^{(3)T} \cdot \underset{\sim}{\mathbf{T}}^{'(4)}\right) \cdot \boldsymbol{\varphi} = |\underset{\sim}{\mathbf{T}}^{(3)}|\underset{*}{\mathbf{T}}^{(1)T} \cdot \mathbf{P}_{(n)} - |\underset{\sim}{\mathbf{T}}^{(1)}|\underset{*}{\mathbf{T}}^{(3)T} \cdot \boldsymbol{\mu}_{(n)}. \tag{121}
$$

Analogous to (121), applying the operator $|\underset{\sim}{\mathbf{T}}^{'(4)}|\underset{*}{\mathbf{T}}^{(3)T}$ to Eq. $(120)_1$ with a single multiplication, and the operator $|\underset{\sim}{\mathbf{T}}^{(3)}|\underset{*}{\mathbf{T}}^{'(4)T}$ to the second relation and taking into account

$$
\underset{*}{\mathbf{T}}^{(3)T} \cdot \underset{\sim}{\mathbf{T}}^{(3)} = \underset{\sim}{\mathbf{E}}|\underset{\sim}{\mathbf{T}}^{(3)}|, \quad \underset{*}{\mathbf{T}}^{'(4)T} \cdot \underset{\sim}{\mathbf{T}}^{'(4)} = \underset{\sim}{\mathbf{E}}|\underset{\sim}{\mathbf{T}}^{'(4)}|,
$$

and then subtracting the second relation from the first obtained relation, we have

$$
\left(|\underset{\sim}{\mathbf{T}}^{'(4)}|\underset{*}{\mathbf{T}}^{(3)T} \cdot \underset{\sim}{\mathbf{T}}^{(1)} - |\underset{\sim}{\mathbf{T}}^{(3)}|\underset{*}{\mathbf{T}}^{'(4)T} \cdot \underset{\sim}{\mathbf{T}}^{(3)}\right) \cdot \mathbf{u} = |\underset{\sim}{\mathbf{T}}^{'(4)}|\underset{*}{\mathbf{T}}^{(3)T} \cdot \mathbf{P}_{(n)} - |\underset{\sim}{\mathbf{T}}^{(3)}|\underset{*}{\mathbf{T}}^{'(4)T} \cdot \boldsymbol{\mu}_{(n)}. \tag{122}
$$

The relations (121) and (122) are the desired boundary conditions.

Note that the boundary conditions can be decomposed in a more general case. For brevity, we will not stop on t. Next, we note only that the boundary conditions are relatively easy to split in the following cases:

(1) $a_2 = a_3$,
(2) $b_2 = b_3$,
(3) $d_2 = d_3$,
(4) $\left(\underset{\sim}{\mathbf{T}}^{(1)} \cdot \underset{\sim}{\mathbf{T}}^{(3)} = \underset{\sim}{\mathbf{T}}^{(3)} \cdot \underset{\sim}{\mathbf{T}}^{(1)}\right) \Leftrightarrow \left(\xi = \dfrac{a_1}{a_3} = \dfrac{b_1}{b_3}\right)$,
(5) $\left(\underset{\sim}{\mathbf{T}}^{(1)} \cdot \underset{\sim}{\mathbf{T}}^{'(4)} = \underset{\sim}{\mathbf{T}}^{'(4)} \cdot \underset{\sim}{\mathbf{T}}^{(1)}\right) \Leftrightarrow \left(\eta = \dfrac{a_1}{a_3} = \dfrac{d_1}{d_3}\right)$,
(6) $\left(\underset{\sim}{\mathbf{T}}^{(1)} \cdot \underset{\sim}{\mathbf{T}}^{'(4)} = \underset{\sim}{\mathbf{T}}^{'(4)} \cdot \underset{\sim}{\mathbf{T}}^{(1)}\right) \Leftrightarrow \left(\zeta = \dfrac{b_1}{b_3} = \dfrac{d_1}{d_3}\right)$,
(7) $\left(\underset{\sim}{\mathbf{T}}^{(1)},\ \underset{\sim}{\mathbf{T}}^{(3)},\ \underset{\sim}{\mathbf{T}}^{'(4)} \text{ commute in pairs}\right) \Leftrightarrow \left(\xi = \dfrac{a_1}{a_3} = \dfrac{b_1}{b_3} = \dfrac{d_1}{d_3}\right)$,
(8) for materials with a center of symmetry, including the reduced media. In this case for unreduced media $\underset{\sim}{\mathbf{T}}^{(2)} = (a_3 - a_2)\mathbf{n} \cdot \underset{\approx}{\mathbf{C}}$, $\underset{\sim}{\mathbf{T}}^{(3)} = 0$ and the tensor operator of stress and couple stress is a differential triangular tensor block matrix and it is easy to dismember the boundary conditions.

For the reduced medium $\underset{\sim}{\mathbf{T}}^{(2)} = 0$ ($a_2 = a_3$), $\underset{\sim}{\mathbf{T}}^{(3)} = 0$ and the tensor operator of stress and of couple stress is a differential tensor block diagonal matrix. This matrix can be easily inverted. Therefore it is easy to dismember the boundary conditions.

Let the second, third and seventh conditions be fulfilled simultaneously, i.e.

$$a_2 = a_3, \quad b_2 = b_3, \quad d_2 = d_3, \quad \xi = \frac{a_1}{a_3} = \frac{b_1}{b_3} = \frac{d_1}{d_3}, \tag{123}$$

Taking into account (123), operators (116) will take the form

$$\begin{aligned} \underset{\sim}{\mathbf{T}}^{(1)} &= a_2[\underset{\sim}{\mathbf{E}}\mathbf{n} \cdot \nabla + \xi \mathbf{n}\nabla + (\mathbf{n}\nabla)^T], \\ \underset{\sim}{\mathbf{T}}^{(2)} &= \underset{\sim}{\mathbf{T}}^{(3)} = b_2[\underset{\sim}{\mathbf{E}}\mathbf{n} \cdot \nabla + \xi \mathbf{n}\nabla + (\mathbf{n}\nabla)^T], \\ \underset{\sim}{\mathbf{T}}^{(4)} &= d_2[\underset{\sim}{\mathbf{E}}\mathbf{n} \cdot \nabla + \xi \mathbf{n}\nabla + (\mathbf{n}\nabla)^T]. \end{aligned} \tag{124}$$

Introducing the notation

$$\underset{\sim}{\mathbf{T}} = \left(\underset{\approx}{\mathbf{C}}_{(1)} + \xi \underset{\approx}{\mathbf{C}}_{(2)} + \underset{\approx}{\mathbf{C}}_{(3)}\right) \overset{2}{\otimes} \mathbf{n}\nabla = [\underset{\sim}{\mathbf{E}}\mathbf{n} \cdot \nabla + \xi \mathbf{n}\nabla + (\mathbf{n}\nabla)^T], \tag{125}$$

the differential tensor-operators (124) can be written as

$$\underset{\sim}{\mathbf{T}}^{(1)} = a_2 \underset{\sim}{\mathbf{T}}, \quad \underset{\sim}{\mathbf{T}}^{(2)} = \underset{\sim}{\mathbf{T}}^{(3)} = b_2 \underset{\sim}{\mathbf{T}}, \quad \underset{\sim}{\mathbf{T}}^{(4)} = d_2 \underset{\sim}{\mathbf{T}}, \tag{126}$$

and the boundary conditions (120) can be written as

$$a_2 \underset{\sim}{\mathbf{T}} \cdot \mathbf{u} + b_2 \underset{\sim}{\mathbf{T}} \cdot \boldsymbol{\varphi} = \mathbf{P}_{(n)}, \quad b_2 \underset{\sim}{\mathbf{T}} \cdot \mathbf{u} + d_2 \underset{\sim}{\mathbf{T}} \cdot \boldsymbol{\varphi} = \boldsymbol{\mu}_{(n)}. \tag{127}$$

It is easy to find an expression for $\underset{\sim}{\mathbf{T}}_*$ and $|\underset{\sim}{\mathbf{T}}|$. They have the form

$$\begin{aligned}\underset{\sim}{\mathbf{T}}_* &= [2(1+\xi)\underset{\sim}{\mathbf{E}}\mathbf{n}\cdot\nabla - (1+\xi)\mathbf{n}\nabla - 2\xi(\mathbf{n}\nabla)^T]\mathbf{n}\cdot\nabla + \xi[\nabla\nabla + (\mathbf{nn} - \underset{\sim}{\mathbf{E}})\Delta],\\ |\underset{\sim}{\mathbf{T}}| &= [2(1+\xi)\mathbf{nn}\overset{2}{\otimes}\nabla\nabla - \xi\Delta]\mathbf{n}\cdot\nabla.\end{aligned} \tag{128}$$

Assuming that $a_2d_2 - b_2^2 \neq 0$ and solving the system of equations (127) with respect to $\underset{\sim}{\mathbf{T}}\cdot\mathbf{u}$ and $\underset{\sim}{\mathbf{T}}\cdot\boldsymbol{\varphi}$, we obtain

$$\begin{aligned}\underset{\sim}{\mathbf{T}}\cdot\mathbf{u} &= (a_2d_2 - b_2^2)^{-1}(d_2\mathbf{P}_{(n)} - b_2\boldsymbol{\mu}_{(n)}),\\ \underset{\sim}{\mathbf{T}}\cdot\boldsymbol{\varphi} &= (a_2d_2 - b_2^2)^{-1}(a_2\boldsymbol{\mu}_{(n)} - b_2\mathbf{P}_{(n)}).\end{aligned} \tag{129}$$

Multiplying each Eq. (129) from the left by $\underset{\sim}{\mathbf{T}}_*^T$ $(\underset{\sim}{\mathbf{T}}_*^T\cdot\underset{\sim}{\mathbf{T}} = \underset{\sim}{\mathbf{E}}|\underset{\sim}{\mathbf{T}}|)$, we have

$$\begin{aligned}|\underset{\sim}{\mathbf{T}}|\mathbf{u} &= (a_2d_2 - b_2^2)^{-1}\underset{\sim}{\mathbf{T}}_*^T\cdot(d_2\mathbf{P}_{(n)} - b_2\boldsymbol{\mu}_{(n)}),\\ |\underset{\sim}{\mathbf{T}}|\boldsymbol{\varphi} &= (a_2d_2 - b_2^2)^{-1}\underset{\sim}{\mathbf{T}}_*^T\cdot(a_2\boldsymbol{\mu}_{(n)} - b_2\mathbf{P}_{(n)}).\end{aligned}$$

6.3 *Quasi-Static Problems of Micropolar Elasticity Theory in Displacements and Rotations*

For brevity, we consider the material with a center of symmetry. Then, in the case of quasi-static, for example, from (111) we have the equations

$$Q_1^*(Q_2^*Q_4^* + 4\alpha^2\Delta)\mathbf{u} + \mathbf{S}^* = 0, \quad Q_3^*(Q_2^*Q_4^* + 4\alpha^2\Delta)\boldsymbol{\varphi} + \mathbf{H}^* = 0, \tag{130}$$

where we have introduced the following notations:

$$\begin{aligned}\mathbf{S}^* &= 2\alpha Q_1^*(\underset{\approx}{\mathbf{C}}\cdot\nabla)\cdot(\rho\mathbf{m}) + [\underset{\sim}{\mathbf{E}}Q_1^*Q_4^* - (dQ_4^* - 4\alpha^2)\nabla\nabla]\cdot(\rho\mathbf{F}),\\ \mathbf{H}^* &= 2\alpha Q_3^*(\underset{\approx}{\mathbf{C}}\cdot\nabla)\cdot(\rho\mathbf{F}) + [\underset{\sim}{\mathbf{E}}Q_2^*Q_3^* - (mQ_2^* - 4\alpha^2)\nabla\nabla]\cdot(\rho\mathbf{m}),\\ Q_1^* &= (b+d)\Delta, \quad Q_2^* = b\Delta, \quad Q_3^* = (g+m)\Delta - l, \quad Q_4^* = g\Delta - l,\\ d &= \lambda + \mu - \alpha, \quad l = 4\alpha, \quad b = \mu + \alpha, \quad m = \gamma + \delta - \beta, \quad g = \delta + \beta.\end{aligned} \tag{131}$$

Next, by notations (131), the Eq. (130) can be written as

$$\begin{aligned}&[(\lambda+2\mu)(\mu+\alpha)(\delta+\beta)\Delta^3 - 4\alpha\mu(\lambda+2\mu)\Delta^2]\mathbf{u} + \mathbf{S}^* = 0,\\ &\big\{(\gamma+2\delta)(\mu+\alpha)(\delta+\beta)\Delta^3 - 4\alpha[\mu(\gamma+2\delta) + (\mu+\alpha)(\delta+\beta)]\Delta^2\\ &\qquad + 16\alpha^2\mu\Delta\big\}\boldsymbol{\varphi} + \mathbf{H}^* = 0.\end{aligned} \tag{132}$$

It is easy to see that when $\alpha = 0$, i.e. in the case of the reduced medium, from (130) (or from (132)) we obtain the following equation:

$$\Delta^2\mathbf{u} + \mathbf{G} = 0, \quad \Delta^2\boldsymbol{\varphi} + \mathbf{H} = 0, \tag{133}$$

where we have introduced the notations

$$\begin{aligned} \mathbf{G} &= \frac{1}{\mu(\lambda + 2\mu)}[\mathbf{E}(\lambda + 2\mu)\Delta - (\lambda + \mu)\nabla\nabla] \cdot (\rho\mathbf{F}), \\ \mathbf{H} &= \frac{1}{(\delta + \beta)(\gamma + 2\delta)}[\mathbf{E}(\gamma + 2\delta)\Delta - (\gamma + \delta - \beta)\nabla\nabla] \cdot (\rho\mathbf{m}). \end{aligned} \tag{134}$$

Note that the first equation (133) is the classical equation and the second equation has a similar form. The difference is only in the coefficients. So, since we can find an analytical solution of the classical equation (see the above cases of prismatic bodies), then we can also find the analytical solutions of the Eq. (133) for the reduced medium because in this case they are written out by analogy. However, we note that in case of absence of volume loads the equations of reduced medium do not depend on the properties of the material (although their dependence on the material constants in the case of presence of volume loads, is not difficult). This fact suggests that these equations can be used to identify the material constants of the reduced medium. To solve this problem, most likely, it is sufficient to write the general solution of these equations, and then consider suitable simple boundary value problems. Moreover, it is possible to experiment on suitable samples made from any material, which is convenient in terms of experimentation.

6.4 *Prismatic Bodies with Constant Thickness in Displacements and Rotations and in Moments of Displacement and Rotation Vectors*

Consider a prismatic body with constant thickness $2h$. As a base plane, as above, we take the median plane. Then, using the representation of Laplacian (see the second formula (66)) for Δ^2 and Δ^3, we have the expression

$$\begin{aligned} \Delta^2 &= \bar{\Delta}^2 + 2h^{-2}\bar{\Delta}\partial_3^2 + h^{-4}\partial_3^4, \quad \bar{\Delta} = g^{PQ}\partial_P\partial_Q, \\ \Delta^3 &= \bar{\Delta}^3 + 3h^{-2}\bar{\Delta}^2\partial_3^2 + 3h^{-4}\bar{\Delta}\partial_3^4 + h^{-6}\partial_3^6. \end{aligned}$$

By virtue of the latest formulas and representation of Laplacian (see the second formula (66)), Eq. (132) for the theory of prismatic bodies of constant thickness in displacements and rotations can be written as

$$\begin{aligned} [\bar{\Delta}^3 + A\bar{\Delta}^2 + h^{-2}(3\bar{\Delta} + 2A)\bar{\Delta}\partial_3^2 + h^{-4}(3\bar{\Delta} + A)\partial_3^4 + h^{-6}\partial_3^6]\mathbf{u} + \mathbf{S}^{**} &= 0, \\ [\bar{\Delta}^3 + (B\bar{\Delta} + A)\bar{\Delta} + h^{-2}[(3\bar{\Delta} + 2B)\bar{\Delta} + C]\partial_3^2 + h^{-4}(3\bar{\Delta} + B)\partial_3^4 + h^{-6}\partial_3^6]\boldsymbol{\varphi} & \\ +\mathbf{H}^{**} &= 0; \end{aligned} \tag{135}$$

$$\mathbf{S}^{**} = \frac{\mathbf{S}^*}{(\lambda+2\mu)(\mu+\alpha)(\delta+\beta)}, \quad \mathbf{H}^{**} = \frac{\mathbf{H}^*}{(\gamma+2\delta)(\mu+\alpha)(\delta+\beta)},$$
$$A = -\frac{4\alpha\mu}{(\mu+\alpha)(\delta+\beta)},$$
$$B = -\frac{4\alpha[\mu(\gamma+2\delta)+(\mu+\alpha)(\delta+\beta)]}{(\gamma+2\delta)(\mu+\alpha)(\delta+\beta)}, \quad C = \frac{16\alpha^2\mu}{(\gamma+2\delta)(\mu+\alpha)(\delta+\beta)}. \tag{136}$$

Applying the kth moment operator of any system of orthogonal polynomials (Legendre, Chebyshev) to Eq. (135), we find the following equations in moments of displacement and rotation vectors for the micropolar theory of prismatic bodies with constant thickness:

$$\begin{aligned}
&[\bar{\Delta}^3 + (B\bar{\Delta} + A)\bar{\Delta}]\overset{(k)}{\boldsymbol{\varphi}} + h^{-2}[(3\bar{\Delta} + 2B)\bar{\Delta} + C]\overset{(k)}{\boldsymbol{\varphi}}{}'' + h^{-4}(3\bar{\Delta} + B)\overset{(k)}{\boldsymbol{\varphi}}{}^{IV} + h^{-6}\overset{(k)}{\boldsymbol{\varphi}}{}^{VI}\\
&+ \overset{(k)}{\mathbf{H}}{}^{**} = 0,\\
&[\bar{\Delta}^3 + A\bar{\Delta}^2]\overset{(k)}{\mathbf{u}} + h^{-2}(3\bar{\Delta} + 2A)\bar{\Delta}\overset{(k)}{\mathbf{u}}{}'' + h^{-4}(3\bar{\Delta} + A)\overset{(k)}{\mathbf{u}}{}^{IV} + h^{-6}\overset{(k)}{\mathbf{u}}{}^{VI} + \overset{(k)}{\mathbf{S}}{}^{**} = 0,\\
&k \in \mathbb{N}_0.
\end{aligned} \tag{137}$$

Having Eq. (137), we can easily obtain the system of equations for various approximations. From the structure of Eq. (137), it can be seen that the first equation can be considered absolutely similar to the second. Therefore, for brevity we consider only the second equation (137), from which we obtain a system of equations of the eighth approximation. Then, from the obtained equations we derive equations for each moment of the displacement vector separately (from the zeroth to the eighth order).

Assuming that the moments in (137) are considered with respect to a system of Legendre polynomials, by (69), (71) and the formula (see Nikabadze 2008a, b, 2014b, c)

$$\overset{(n)}{\mathbf{u}}{}^{VI}(x') = (2n+1)\sum_{k=1}^{\infty} C_{k+4}^5 \prod_{s=1}^{5}(2n+2k+2s-1)\overset{(n+2k+4)}{\mathbf{u}}, \quad n \in \mathbb{N}_0, \tag{138}$$

say, from the second equation (137) we have a system of equations of eighth ($N = 8$) approximation, splitting into two systems

$$L^{(\alpha)}\mathbf{U}^{(\alpha)} + \boldsymbol{\Phi}^{(\alpha)} = \mathbf{O}^{(\alpha)}, \quad \alpha = 1, 2. \tag{139}$$

Here, we have introduced the following notation:

$$\mathbf{U}^{(1)} = \begin{pmatrix} \overset{(0)}{\mathbf{u}} \\ \overset{(2)}{\mathbf{u}} \\ \overset{(4)}{\mathbf{u}} \\ \overset{(6)}{\mathbf{u}} \\ \overset{(8)}{\mathbf{u}} \end{pmatrix}, \quad \mathbf{U}^{(2)} = \begin{pmatrix} \overset{(1)}{\mathbf{u}} \\ \overset{(3)}{\mathbf{u}} \\ \overset{(5)}{\mathbf{u}} \\ \overset{(7)}{\mathbf{u}} \end{pmatrix}, \quad \boldsymbol{\Phi}^{(1)} = \begin{pmatrix} \overset{(0)}{\mathbf{S}}{}^{**} \\ \overset{(2)}{\mathbf{S}}{}^{**} \\ \overset{(4)}{\mathbf{S}}{}^{**} \\ \overset{(6)}{\mathbf{S}}{}^{**} \\ \overset{(8)}{\mathbf{S}}{}^{**} \end{pmatrix}, \quad \boldsymbol{\Phi}^{(2)} = \begin{pmatrix} \overset{(1)}{\mathbf{S}}{}^{**} \\ \overset{(3)}{\mathbf{S}}{}^{**} \\ \overset{(5)}{\mathbf{S}}{}^{**} \\ \overset{(7)}{\mathbf{S}}{}^{**} \end{pmatrix},$$

$$\mathbf{O}^{(1)} = \begin{pmatrix} \mathbf{0} \\ \mathbf{0} \\ \mathbf{0} \\ \mathbf{0} \\ \mathbf{0} \end{pmatrix}, \quad \mathbf{O}^{(2)} = \begin{pmatrix} \mathbf{0} \\ \mathbf{0} \\ \mathbf{0} \\ \mathbf{0} \end{pmatrix},$$

$$L^{(1)} = \begin{pmatrix} L_{11}^{(1)} & L_{12}^{(1)} & L_{13}^{(1)} & L_{14}^{(1)} & L_{15}^{(1)} \\ 0 & L_{22}^{(1)} & L_{23}^{(1)} & L_{24}^{(1)} & L_{25}^{(1)} \\ 0 & 0 & L_{33}^{(1)} & L_{34}^{(1)} & L_{35}^{(1)} \\ 0 & 0 & 0 & L_{44}^{(1)} & L_{45}^{(1)} \\ 0 & 0 & 0 & 0 & L_{55}^{(1)} \end{pmatrix},$$

$$\begin{gathered} L_{11}^{(1)} = \bar{\Delta}^3 + A\bar{\Delta}^2, \quad L_{12}^{(1)} = a_{03}\bar{\Delta}^2 + a_{04}\bar{\Delta}, \quad L_{13}^{(1)} = a_{05}\bar{\Delta}^2 + a_{06}\bar{\Delta} + a_{07}, \\ L_{14}^{(1)} = a_{08}\bar{\Delta}^2 + a_{09}\bar{\Delta} + a_{010}, \quad L_{15}^{(1)} = a_{011}\bar{\Delta}^2 + a_{012}\bar{\Delta} + a_{013}, \quad L_{22}^{(1)} = \bar{\Delta}^3 + A\bar{\Delta}^2, \\ L_{23}^{(1)} = a_{26}\bar{\Delta}^2 + a_{27}\bar{\Delta}, \quad L_{24}^{(1)} = a_{28}\bar{\Delta}^2 + a_{29}\bar{\Delta} + a_{210}, \\ L_{25}^{(1)} = a_{211}\bar{\Delta}^2 + a_{212}\bar{\Delta} + a_{213}, \quad L_{33}^{(1)} = \bar{\Delta}^3 + A\bar{\Delta}^2, \quad L_{34}^{(1)} = a_{49}\bar{\Delta}^2 + a_{410}\bar{\Delta}, \\ L_{35}^{(1)} = a_{411}\bar{\Delta}^2 + a_{412}\bar{\Delta} + a_{413}, \quad L_{44}^{(1)} = \bar{\Delta}^3 + A\bar{\Delta}^2, \\ L_{45}^{(1)} = a_{612}\bar{\Delta}^2 + a_{613}\bar{\Delta}, \quad L_{55}^{(1)} = \bar{\Delta}^3 + A\bar{\Delta}^2; \end{gathered}$$

$$\begin{gathered} a_{03} = 9h^{-2}, \quad a_{04} = 6Ah^{-2}, \quad a_{05} = 30h^{-2}, \quad a_{06} = 20Ah^{-2} + 315h^{-4}, \\ a_{07} = 105Ah^{-4}, \quad a_{08} = 63h^{-2}, \quad a_{09} = 42Ah^{-2} + 3780h^{-4}, \\ a_{010} = 1260Ah^{-4} + 10395h^{-6}, \quad a_{011} = 108h^{-2}, \\ a_{012} = 72Ah^{-2} + 20790h^{-4}, \quad a_{013} = 6930Ah^{-4} + 270270h^{-6}, \quad a_{26} = 105h^{-2}, \\ a_{27} = 70Ah^{-2}, \quad a_{28} = 270h^{-2}, \quad a_{29} = 180Ah^{-2} + 10395h^{-4}, \quad a_{210} = 3465Ah^{-4}, \\ a_{211} = 495h^{-2}, \quad a_{212} = 330Ah^{-2} + 77220h^{-4}, \\ a_{213} = 25740Ah^{-4} + 675675h^{-6}, \\ a_{49} = 297h^{-2}, \quad a_{410} = 198Ah^{-2}, \quad a_{411} = 702h^{-2}, \\ a_{412} = 468Ah^{-2} + 57915h^{-4}, \\ a_{413} = 19305Ah^{-4}, \quad a_{612} = 585h^{-2}, \quad a_{613} = 390Ah^{-2}; \end{gathered}$$

$$L^{(2)} = \begin{pmatrix} \bar{\Delta}^3 + A\bar{\Delta}^2 & a_{13}\bar{\Delta}^2 + a_{14}\bar{\Delta} & a_{15}\bar{\Delta}^2 + a_{16}\bar{\Delta} + a_{17} & a_{18}\bar{\Delta}^2 + a_{19}\bar{\Delta} + a_{110} \\ 0 & \bar{\Delta}^3 + A\bar{\Delta}^2 & a_{36}\bar{\Delta}^2 + a_{37}\bar{\Delta} & a_{38}\bar{\Delta}^2 + a_{39}\bar{\Delta} + a_{310} \\ 0 & 0 & \bar{\Delta}^3 + A\bar{\Delta}^2 & a_{59}\bar{\Delta}^2 + a_{510}\bar{\Delta} \\ 0 & 0 & 0 & \bar{\Delta}^3 + A\bar{\Delta}^2 \end{pmatrix},$$

$$
\begin{aligned}
& a_{13}=45h^{-2},\ a_{14}=30Ah^{-2},\ a_{15}=126h^{-2},\ a_{16}=84Ah^{-2}+2835h^{-4},\\
& a_{17}=945Ah^{-4},\ a_{18}=243h^{-2},\ a_{19}=162Ah^{-2}+24948h^{-4},\\
& a_{110}=8316Ah^{-4}+135135h^{-6},\ a_{36}=189h^{-2},\ a_{37}=126Ah^{-2},\ a_{38}=462h^{-2},\\
& a_{39}=308Ah^{-2}+27027h^{-4},\ a_{310}=9009Ah^{-4},\ a_{59}=429h^{-2},\\
& a_{510}=286Ah^{-2};
\end{aligned}
$$

$\mathbf{0}$ is a three-component zero vector.

The displacement vector has the following expression

$$
\begin{aligned}
\mathbf{u} \approx\ & \overset{(0)}{\mathbf{u}}+\overset{(1)}{\mathbf{u}}P_1(x^3)+\overset{(2)}{\mathbf{u}}P_2(x^3)+\overset{(3)}{\mathbf{u}}P_3(x^3)+\overset{(4)}{\mathbf{u}}P_4(x^3)+\overset{(5)}{\mathbf{u}}P_5(x^3)+\overset{(6)}{\mathbf{u}}P_6(x^3)\\
& +\overset{(7)}{\mathbf{u}}P_7(x^3)+\overset{(8)}{\mathbf{u}}P_8(x^3).
\end{aligned}
$$

It is seen that the matrices $L^{(\alpha)}$, $\alpha = 1, 2$, are the upper triangular differential matrix operators and their determinants $|L^{(\alpha)}|$, $\alpha = 1, 2$, have the following expressions

$$
|L^{(1)}| = \bar{\Delta}^{10}(\bar{\Delta}+A)^5, \quad |L^{(2)}| = \bar{\Delta}^{8}(\bar{\Delta}+A)^4. \tag{140}
$$

From (140) it is clear that $|L^{(\alpha)}|$, $\alpha = 1, 2$, are different from zero. Therefore, we can find their cofactors $L_*^{(\alpha)}$, $\alpha = 1, 2$. Indeed, by simple calculations we have

$$
L_*^{(1)} = \begin{pmatrix}
L_{*11}^{(1)} & L_{*12}^{(1)} & L_{*13}^{(1)} & L_{*14}^{(1)} & L_{*15}^{(1)}\\
0 & L_{*22}^{(1)} & L_{*23}^{(1)} & L_{*24}^{(1)} & L_{*25}^{(1)}\\
0 & 0 & L_{*33}^{(1)} & L_{*34}^{(1)} & L_{*35}^{(1)}\\
0 & 0 & 0 & L_{*44}^{(1)} & L_{*45}^{(1)}\\
0 & 0 & 0 & 0 & L_{*55}^{(1)}
\end{pmatrix},
$$

$$
\begin{aligned}
L_{*11}^{(1)} &= L_{*22}^{(1)} = L_{*33}^{(1)} = L_{*44}^{(1)} = L_{*55}^{(1)} = \Delta^{12}+4A\Delta^{11}+6A^2\Delta^{10}+4A^3\Delta^9+A^4\Delta^8,\\
L_{*12}^{(1)} &= -9/h^2\Delta^{11}-33A/h^2\Delta^{10}-45A^2/h^2\Delta^9-27A^3/h^2\Delta^8-6A^4/h^2\Delta^7,\\
L_{*13}^{(1)} &= -30/h^2\Delta^{11}+a_1/h^4\Delta^{10}+a_2A/h^4\Delta^9+a_3A^2/h^4\Delta^8+a_4A^3/h^4\Delta^7\\
&\quad +a_5A^4/h^4\Delta^6,\\
L_{*14}^{(1)} &= -63/h^2\Delta^{11}+a_6/h^4\Delta^{10}+a_7/h^6\Delta^9+a_8A/h^6\Delta^8+a_9A^2/h^6\Delta^7\\
&\quad +a_{10}A^3/h^6\Delta^6+a_{11}A^4/h^6\Delta^5,\\
L_{*15}^{(1)} &= -108/h^2\Delta^{11}+a_{12}/h^4\Delta^{10}+a_{13}/h^6\Delta^9+a_{14}/h^8\Delta^8+a_{15}A/h^8\Delta^7\\
&\quad +a_{16}A^2/h^8\Delta^6+a_{17}A^3/h^8\Delta^5+a_{18}A^4/h^8\Delta^4,\\
L_{*23}^{(1)} &= -105/h^2\Delta^{11}-385A/h^2\Delta^{10}-525A^2/h^2\Delta^9-315A^3/h^2\Delta^8-70A^4/h^2\Delta^7,
\end{aligned}
$$

$$
\begin{aligned}
L^{(1)}_{*24} &= -270/h^2\Delta^{11} + a_{19}/h^4\Delta^{10} + a_{20}A/h^4\Delta^9 + a_{21}/A^2h^4\Delta^8 + a_{22}A^3/h^4\Delta^7 \\
&\quad + 10395A^4/h^4\Delta^6, \\
L^{(1)}_{*25} &= -495/h^2\Delta^{11} + a_{23}/h^4\Delta^{10} + a_{24}/h^6\Delta^9 + a_{25}A/h^6\Delta^8 + a_{26}A^2/h^6\Delta^7 \\
&\quad + a_{27}A^3/h^6\Delta^6 - 2702700A^4/h^6\Delta^5, \\
L^{(1)}_{*34} &= -297/h^2\Delta^{11} - 1089A/h^2\Delta^{10} - 1485A^2/h^2\Delta^9 - 891A^3/h^2\Delta^8 \\
&\quad - 198A^4/h^2\Delta^7, \\
L^{(1)}_{*35} &= -702/h^2\Delta^{11} + a_{28}/h^4\Delta^{10} + a_{29}A/h^4\Delta^9 + a_{30}A^2/h^4\Delta^8 + a_{31}A^3/h^4\Delta^7 \\
&\quad + 57915A^4/h^4\Delta^6, \\
L^{(1)}_{*45} &= -585/h^2\Delta^{11} - 2145A/h^2\Delta^{10} - 2925A^2/h^2\Delta^9 - 1755A^3/h^2\Delta^8 \\
&\quad - 390A^4/h^2\Delta^7,
\end{aligned}
$$

$$
\begin{aligned}
a_1 &= 630 - 110Ah^2, \quad a_2 = -(150A + 105)h^2 + 2205, \\
a_3 &= -(90A + 315)h^2 + 2940, \\
a_4 &= -(20A + 315)h^2 + 1785, \quad a_5 = -105h^2 + 420, \quad a_6 = 7560 - 231Ah^2, \\
a_7 &= -315A^2h^4 + 25200Ah^2 - 10395B - 93555, \\
a_8 &= -189A^2h^4 + (31500A + 31185)h^2 - 31185B - 311850, \\
a_9 &= -42A^2h^4 + (17640A + 83160)h^2 - 31185B - 415800, \\
a_{10} &= (3780A + 72765)h^2 - 10395B - 259875, \\
a_{11} &= 20790h^2 - 62370, \quad a_{12} = 41580 - 396Ah^2, \\
a_{13} &= -540A^2h^4 + 138600Ah^2 - 2702700, \\
a_{14} &= -324A^3h^6 + (173250A^2 + 73710A)h^4 - 8181810Ah^2 + 6081075B \\
&\quad + 24324300, \\
a_{15} &= -72A^3h^6 + (97020A^2 + 196560A)h^4 - (9656010A + 12162150)h^2 \\
&\quad + 16216200B + 77026950, \\
a_{16} &= (20790A^2 + 171990A)h^4 - (5307120A + 28378350)h^2 + 14189175B \\
&\quad + 105405300, \\
a_{17} &= 49140Ah^4 - (1130220A + 22297275)h^2 + 4054050B + 66891825, \\
a_{18} &= -6081075h^2 + 16216200, \quad a_{19} = 20790 - 990Ah^2, \\
a_{20} &= -1350Ah^2 + 69300, \\
a_{21} &= -810Ah^2 + 86625, \quad a_{22} = -180Ah^2 + 48510, \\
a_{23} &= 154440 - 1815Ah^2, \quad a_{24} = -2475A^2h^4 + 514800Ah^2 - 6756750, \\
a_{25} &= -1485A^2h^4 + 643500Ah^2 - 20270250, \\
a_{26} &= -330A^2h^4 + 360360Ah^2 - 23648625, \quad a_{27} = 77220Ah^2 - 12837825, \\
a_{28} &= 115830 - 2574Ah^2, \quad a_{29} = -3510Ah^2 + 386100, \\
a_{30} &= -2106Ah^2 + 482625, \quad a_{31} = -468Ah^2 + 270270;
\end{aligned}
$$

$$
L^{(2)}_* = \begin{pmatrix} L^{(2)}_{*11} & L^{(2)}_{*12} & L^{(2)}_{*13} & L^{(2)}_{*14} \\ 0 & L^{(2)}_{*22} & L^{(2)}_{*23} & L^{(2)}_{*24} \\ 0 & 0 & L^{(2)}_{*33} & L^{(2)}_{*34} \\ 0 & 0 & 0 & L^{(2)}_{*44} \end{pmatrix},
$$

$$
\begin{aligned}
L_{*11}^{(2)} &= L_{*22}^{(2)} = L_{*33}^{(2)} = L_{*44}^{(2)} = \Delta^9 + 3A\Delta^8 + 3A^2\Delta^7 + A^3\Delta^6, \\
L_{*12}^{(2)} &= -45/h^2\Delta^8 - 120A/h^2\Delta^7 - 105A^2/h^2\Delta^6 - 30A^3/h^2\Delta^5, \\
L_{*13}^{(2)} &= -126/h^2\Delta^8 + b_1/h^4\Delta^7 + b_2A/h^4\Delta^6 + b_3A^2/h^4\Delta^5 + 2835A^3/h^4\Delta^4, \\
L_{*14}^{(2)} &= -243/h^2\Delta^8 + b_4/h^4\Delta^7 + b_5/h^6\Delta^6 + b_6A/h^6\Delta^5 + b_7A^2/h^6\Delta^4 \\
&\quad - 540540A^3/h^6\Delta^3, \\
L_{*23}^{(2)} &= -189/h^2\Delta^8 - 504A/h^2\Delta^7 - 441A^2/h^2\Delta^6 - 126A^3/h^2\Delta^5, \\
L_{*24}^{(2)} &= -462/h^2\Delta^8 + b_8/h^4\Delta^7 + b_9A/h^4\Delta^6 + b_{10}A^2/h^4\Delta^5 + 27027A^3/h^4\Delta^4, \\
L_{*34}^{(2)} &= -429/h^2\Delta^8 - 1144A/h^2\Delta^7 - 1001A^2/h^2\Delta^6 - 286A^3/h^2\Delta^5,
\end{aligned}
$$

$$
\begin{aligned}
&b_1 = 5670 - 336Ah^2, \quad b_2 = -294Ah^2 + 13230, \quad b_3 = -84Ah^2 + 10395, \\
&b_4 = 49896 - 648Ah^2, \quad b_5 = -567A^2h^4 + 116424Ah^2 - 1351350, \\
&b_6 = -162A^2h^4 + 91476Ah^2 - 2702700, \quad b_7 = 24948Ah^2 - 2027025, \\
&b_8 = 54054 - 1232Ah^2, \quad b_9 = -1078Ah^2 + 126126, \quad b_{10} = -308Ah^2 + 99099.
\end{aligned}
$$

Then, obviously, we will have the relations

$$
L_*^{(\alpha)T} L^{(\alpha)} = L^{(\alpha)} L_*^{(\alpha)T} = E^{(\alpha)}|L^{(\alpha)}|, \alpha = 1, 2, \tag{141}
$$

where $E^{(1)}$ and $E^{(2)}$ are the identity matrices of the fifth and fourth order, respectively.

Applying the operator $L_*^{(\alpha)T}$ from the left to Eq. (139) and taking into account (141), we have the following decomposed system of equations:

$$
|L^{(\alpha)}|\mathbf{U}^{(\alpha)} + L_*^{(\alpha)T}\boldsymbol{\Phi}^{(\alpha)} = \mathbf{O}^{(\alpha)}, \quad \alpha = 1, 2. \tag{142}
$$

From here for each of the moments $\overset{(k)}{\mathbf{u}}$, $k = \overline{0, 8}$, of displacement vectors separately analogous to (81) we obtain the equation of elliptic type of high order (the equations with the differential operators $|L^{(1)}|$ have the 30th order, and the equations with the differential operators $|L^{(2)}|$ are 24th order). If the operators $|L^{(\alpha)}|$ and $L_*^{(\alpha)}$ have a common factor, it is possible to reduce the order of the equation. For each of Eq. (142), using the method of Vekua (1948), we can write an analytical solution. But we will not dwell on this. If necessary, it is not hard to do so.

It should be noted that the appropriate corrective term must be added to the analytical solution depending on the given boundary conditions in both the classical and the micropolar cases. This corrective term provides the fulfillment of the boundary conditions on the face surfaces. They are constructed in Nikabadze (2008a, b, 2014b, c), so we will not dwell on them.

6.5 *Multilayer Prismatic Bodies*

Using the rule, set out in Nikabadze (2008a, b, 2014b, c), to obtain the desired relation of the multilayer thin body from the corresponding relation of monolayer thin body under the new parametrization, the system of equations of the micropolar theory of multilayer prismatic bodies with constant thickness (each layer has a constant thickness) in displacements and rotations, analogous to (135), can be written as

$$
\begin{aligned}
&[\bar{\Delta}^3 + (\underset{s}{B}\bar{\Delta} + \underset{s}{A})\bar{\Delta} + \underset{s}{h}^{-2}[(3\bar{\Delta} + 2\underset{s}{B})\bar{\Delta} + \underset{s}{C}]\partial_3^2 + \underset{s}{h}^{-4}(3\bar{\Delta} + \underset{s}{B})\partial_3^4 + \underset{s}{h}^{-6}\partial_3^6]\underset{s}{\boldsymbol{\varphi}} \\
&+ \underset{s}{\mathbf{H}}^{**} = 0, \\
&[\bar{\Delta}^3 + \underset{s}{A}\bar{\Delta}^2 + \underset{s}{h}^{-2}(3\bar{\Delta} + 2\underset{s}{A})\bar{\Delta}\partial_3^2 + \underset{s}{h}^{-4}(3\bar{\Delta} + \underset{s}{A})\partial_3^4 + \underset{s}{h}^{-6}\partial_3^6]\underset{s}{\mathbf{u}} + \underset{s}{\mathbf{S}}^{**} = 0, \\
&s = \overline{1, K},
\end{aligned}
\tag{143}
$$

where $\underset{s}{h} = \text{const}$ is the sth layer thickness ($s = \overline{1, K}$), $\bar{\Delta} = g^{\bar{I}\bar{J}}\partial_I\partial_J$. Here we have introduced the notations

$$
\begin{aligned}
&\underset{s}{\mathbf{S}}^{**} = \frac{\underset{s}{\mathbf{S}}^*}{(\underset{s}{\lambda} + 2\underset{s}{\mu})(\underset{s}{\mu} + \underset{s}{\alpha})(\underset{s}{\delta} + \underset{s}{\beta})}, \quad \underset{s}{\mathbf{H}}^{**} = \frac{\underset{s}{\mathbf{H}}^*}{(\underset{s}{\gamma} + 2\underset{s}{\delta})(\underset{s}{\mu} + \underset{s}{\alpha})(\underset{s}{\delta} + \underset{s}{\beta})}, \\
&\underset{s}{A} = -\frac{4\underset{s}{\alpha}\underset{s}{\mu}}{(\underset{s}{\mu} + \underset{s}{\alpha})(\underset{s}{\delta} + \underset{s}{\beta})}, \\
&\underset{s}{B} = -\frac{4\underset{s}{\alpha}[\underset{s}{\mu}(\underset{s}{\gamma} + 2\underset{s}{\delta}) + (\underset{s}{\mu} + \underset{s}{\alpha})(\underset{s}{\delta} + \underset{s}{\beta})]}{(\underset{s}{\gamma} + 2\underset{s}{\delta})(\underset{s}{\mu} + \underset{s}{\alpha})(\underset{s}{\delta} + \underset{s}{\beta})}, \quad \underset{s}{C} = \frac{16\underset{s}{\alpha}^2\underset{s}{\mu}}{(\underset{s}{\gamma} + 2\underset{s}{\delta})(\underset{s}{\mu} + \underset{s}{\alpha})(\underset{s}{\delta} + \underset{s}{\beta})}, \\
&s = \overline{1, K}.
\end{aligned}
$$

Applying the kth moment operator of any system of orthogonal polynomials (Legendre, Chebyshev) (or based on (137) to Eq. (143) and using the above rule), we obtain the following equations in moments of displacement and rotation vectors for the micropolar theory of prismatic bodies of constant thickness:

$$
\begin{aligned}
&[\bar{\Delta}^3 + (\underset{s}{B}\bar{\Delta} + \underset{s}{A})\bar{\Delta}]\overset{(k)}{\underset{s}{\boldsymbol{\varphi}}} + \underset{s}{h}^{-2}[(3\bar{\Delta} + 2\underset{s}{B})\bar{\Delta} + \underset{s}{C}]\overset{(k)}{\underset{s}{\boldsymbol{\varphi}}}{}'' + \underset{s}{h}^{-4}(3\bar{\Delta} + \underset{s}{B})\overset{(k)}{\underset{s}{\boldsymbol{\varphi}}}{}^{IV} + \underset{s}{h}^{-6}\overset{(k)}{\underset{s}{\boldsymbol{\varphi}}}{}^{VI} \\
&+ \overset{(k)}{\underset{s}{\mathbf{H}}}{}^{**} = 0, \\
&[\bar{\Delta}^3 + \underset{s}{A}\bar{\Delta}^2]\overset{(k)}{\underset{s}{\mathbf{u}}} + \underset{s}{h}^{-2}(3\bar{\Delta} + 2\underset{s}{A})\bar{\Delta}\overset{(k)}{\underset{s}{\mathbf{u}}}{}'' + \underset{s}{h}^{-4}(3\bar{\Delta} + \underset{s}{A})\overset{(k)}{\underset{s}{\mathbf{u}}}{}^{IV} + \underset{s}{h}^{-6}\overset{(k)}{\underset{s}{\mathbf{u}}}{}^{VI} + \overset{(k)}{\underset{s}{\mathbf{S}}}{}^{**} = 0, \\
&k \in \mathbb{N}_0, \ s = \overline{1, K}.
\end{aligned}
\tag{144}
$$

Finally, by (142) from the second equation (144), we will have the following decomposed system of equations:

$$
|\underset{s}{L}^{(k)}|\underset{s}{\mathbf{U}}^{(k)} + \underset{s*}{L}^{(k)T}\underset{s}{\boldsymbol{\Phi}}^{(k)} = \underset{s}{\mathbf{O}}^{(k)}, \quad k = 1, 2, \quad s = \overline{1, K}. \tag{145}
$$

Note that a similar (145) system of equations is obtained from the first equation (144), which we do not write here.

In the above relations s is an index of layers, K is a number of layers. It should be noted that, as in the case of a single layer prismatic body, so in the case of multilayer prismatic body for each of Eq. (144), using the method of Vekua (1948), we can write an analytical solution. Consequently, for correct statement of problems the boundary conditions in moments and the interlayer contact conditions must be added to the Eqs. (144) and (145), see Nikabadze (2008a, b, 2014b, c). The problem of corrective term in this case is solved analogously to the above. At the same time the analytical solution of each layer (except the first and last) with the corrective terms can be written so that it satisfies the interlayer contact conditions. For the first (last) layer can be represented the analytical solution by means of corrective terms in such a way that it satisfies the boundary conditions on the inner (outer) surface and the interlayer contact conditions on the outer (inner) surface. Therefore, we suppose that the interlayer contact conditions would be taken into account better if the order of approximation is higher. This is very important in the theory of multilayer structures.

6.6 *Prismatic Bodies with Two Small Sizes*

Let us consider the prismatic homogeneous body with two small sizes and a rectangular cross-section with sides $2h_1$ and $2h_2$. In this case we use the parametrization under an arbitrary baseline. As a baseline, we take a middle line (straight line). It is easy to notice that in this case we have, see Nikabadze (2008b, 2014b, c)

$$\begin{aligned} &h_I = \overset{(-)}{h}_I = \overset{(+)}{h}_I = \text{const}, \quad k_1 = 0, \quad k_2 = 0, \quad g_{\hat{3}}^{I} = 0, \quad g_{\hat{3}3} = g_{\hat{3}}^{3} = \hat{\vartheta} = 1, \\ &g_3^{\hat{3}} = \hat{\vartheta}^{-1} = 1, \quad g^{11} = h_1^{-2}, \quad g^{22} = h_2^{-2}, \quad g^{12} = 0, \qquad g^{33} = 1, \quad N_3 = \partial_3. \end{aligned} \tag{146}$$

By (146) and $\hat{\nabla}\mathbb{F} = g_3^{\hat{3}}\mathbf{r}^3 N_3\mathbb{F} + \mathbf{r}^P\partial_P\mathbb{F}$ $(N_3 = \partial_3 - g_{\hat{3}}^{P}\partial_P)$, see Nikabadze (2008b, 2014b), operators $\hat{\Delta}$, $\hat{\Delta}^2$ and $\hat{\Delta}^3$ will have the expressions

$$\begin{aligned} \hat{\Delta} &= \partial_3^2 + \tilde{\Delta}, \quad \tilde{\Delta} = g^{PQ}\nabla_P\nabla_Q = h_1^{-2}\partial_1^2 + h_2^{-2}\partial_2^2, \\ \hat{\Delta}^2 &= \partial_3^4 + 2\partial_3^2\tilde{\Delta} + \tilde{\Delta}^2, \quad \hat{\Delta}^3 = \partial_3^6 + 3\partial_3^4\tilde{\Delta} + 3\partial_3^2\tilde{\Delta}^2 + \tilde{\Delta}^3. \end{aligned} \tag{147}$$

Based on the second formula (147), operators $\tilde{\Delta}^2$ and $\tilde{\Delta}^3$ can be written as

$$\begin{aligned} \tilde{\Delta}^2 &= h_1^{-4}\partial_1^4 + 2h_1^{-2}h_2^{-2}\partial_1^2\partial_2^2 + h_2^{-4}\partial_2^4, \\ \tilde{\Delta}^3 &= h_1^{-6}\partial_1^6 + 3h_1^{-2}h_2^{-2}(h_1^{-2}\partial_1^2 + h_2^{-2}\partial_2^2)\partial_1^2\partial_2^2 + h_2^{-6}\partial_2^6. \end{aligned} \tag{148}$$

If $h = h_1 = h_2$ (the cross-section is a square), then from second relation (147) and (148) we get

$$\begin{aligned} \tilde{\Delta} &= h^{-2}(\partial_1^2 + \partial_2^2) = h^{-2}\Delta, \quad \tilde{\Delta}^2 = h^{-4}(\partial_1^4 + 2\partial_1^2\partial_2^2 + \partial_2^4) = h^{-4}\Delta^2, \\ \tilde{\Delta}^3 &= h^{-6}[\partial_1^6 + 3(\partial_1^2 + \partial_2^2)\partial_1^2\partial_2^2 + \partial_2^6] = h^{-6}\Delta^3, \quad \Delta = \partial_1^2 + \partial_2^2. \end{aligned} \tag{149}$$

Now we write Eq. (132) for the considered prismatic body with two small sizes. For this purpose, it is sufficient to replace Laplacian Δ and the nabla operator ∇, existing in Eq. (132), on $\hat{\Delta}$ and $\hat{\nabla} = \mathbf{r}^3\partial_3 + \mathbf{r}^P\partial_P$, respectively. After such changes, we will have

$$(\hat{\Delta}^3 + A\hat{\Delta}^2)\mathbf{u} + \hat{\mathbf{S}}^{**} = 0, \quad (\hat{\Delta}^3 + B\hat{\Delta}^2 + C\hat{\Delta})\boldsymbol{\varphi} + \hat{\mathbf{H}}^{**} = 0, \tag{150}$$

where $A, B, C, \hat{\mathbf{S}}^{**}$ and $\hat{\mathbf{H}}^{**}$ are given by (136). The expressions for $\hat{\mathbf{S}}^{**}$ and $\hat{\mathbf{H}}^{**}$ are obtained from $\mathbf{S}^{**}$ and $\mathbf{H}^{**}$ if in $\mathbf{S}^{*}$ and $\mathbf{H}^{*}$ (see (131)) we replace Δ and ∇ by $\hat{\Delta}$ and $\hat{\nabla}$, respectively.

Taking into account Eq. (147), (150) can be written as:

$$\begin{aligned} &[\partial_3^6 + A\partial_3^4 + \partial_3^2(3\partial_3^2 + 2A)\tilde{\Delta} + (3\partial_3^2 + A)\tilde{\Delta}^2 + \tilde{\Delta}^3]\mathbf{u} + \hat{\mathbf{S}}^{**} &&= 0,\\ &\{\partial_3^6 + B\partial_3^4 + [\partial_3^2(3\partial_3^2 + 2B) + C]\tilde{\Delta} + (3\partial_3^2 + B)\tilde{\Delta}^2 + \tilde{\Delta}^3\}\varphi + \hat{\mathbf{H}}^{**} &&= 0. \end{aligned} \tag{151}$$

It is easy to notice that in this case we obtain the following equations for the reduced medium ($\alpha = 0$) based on (133):

$$(\partial_3^4 + 2\partial_3^2\tilde{\Delta} + \tilde{\Delta}^2)\mathbf{u} + \hat{\mathbf{G}} = 0, \quad (\partial_3^4 + 2\partial_3^2\tilde{\Delta} + \tilde{\Delta}^2)\varphi + \hat{\mathbf{H}} = 0, \tag{152}$$

where expressions for $\hat{\mathbf{G}}$ and $\hat{\mathbf{H}}$ are obtained from the expressions for $\mathbf{G}$ and $\mathbf{H}$ (see (134)), if the operators Δ and ∇ are replaced by $\hat{\Delta}$ and $\hat{\nabla}$, respectively. Note that the first equation (152) is the equation of the classical theory of prismatic thin bodies with two small sizes and with a rectangular cross-section. Note also that Eq. (152) can be obtained from (151) if $\alpha = 0$. Furthermore, it is seen that Eq. (152) do not depend on the material properties in the absence of volume loads.

Next, before getting equations in moments of displacement and rotation vectors, we define the moment of (m, n)th order ((m, n)th moment) of any tensor field. If $\{u_k\}_{k=0}^{\infty}$ is an orthogonal system of polynomials on the segment $[a, b]$, and $\mathbb{F}(x', x^3)$ is any tensor field, then the moment of (m, n)th order of the tensor field $\mathbb{F}(x', x^3)$ with respect to the system of polynomials $\{u_k\}_{k=0}^{\infty}$ is defined as follows.

Definition 15.5 The (m, n)th moment of the tensor field $\mathbb{F}(x', x^3)$ with respect to the system of polynomials $\{u_k\}_{k=0}^{\infty}$, denoted by $\overset{(m,n)}{\mathbb{M}}(\mathbb{F})$, is defined to be the integral

$$\overset{(m,n)}{\mathbb{M}}(\mathbb{F}) = ||u_m||^{-2}||u_n||^{-2}\int_a^b\int_a^b \mathbb{F}(x^1, x^2, x^3)u_m(x^1)u_n(x^2)h(x^1)h(x^2)dx^1dx^2. \tag{153}$$

Here $||u_k||$ is the norm of a polynomial u_k, and h is the weighting function. Note that issues related to the theory of thin bodies with two small sizes are desribed in Nikabadze et al. (2008), Nikabadze (2008b, 2014b, c).

Now it is not difficult to obtain equations for the micropolar theory of prismatic thin bodies with two small sizes, having a rectangle cross-section, in moments of displacement and rotation vectors with respect to any system of polynomials (Legendre, Chebyshev). Indeed, it is easy to see that by (153), definition of "prime" operator (Nikabadze 2008a, b, 2014b, c), the second formula (147) and (148) for any tensor field $\mathbb{F}(x', x^3)$ we have the formulas (Nikabadze 2008a, b, 2014b, c)

$$
\begin{aligned}
&\overset{(m,n)}{\mathbf{M}}(\tilde{\Delta}\mathbb{F}) = h_1^{-2}\overset{(m'',n)}{\mathbb{F}} + h_2^{-2}\overset{(m,n'')}{\mathbb{F}},\\
&\overset{(m,n)}{\mathbf{M}}(\tilde{\Delta}^2\mathbb{F}) = h_1^{-4}\overset{(m^{IV},n)}{\mathbb{F}} + 2h_1^{-2}h_2^{-2}\overset{(m'',n'')}{\mathbb{F}} + h_2^{-4}\overset{(m,n^{IV})}{\mathbb{F}},\\
&\overset{(m,n)}{\mathbf{M}}(\tilde{\Delta}^3\mathbb{F}) = h_1^{-6}\overset{(m^{VI},n)}{\mathbb{F}} + 3h_1^{-2}h_2^{-2}\Big(h_1^{-2}\overset{(m^{IV},n'')}{\mathbb{F}} + h_2^{-2}\overset{(m'',n^{IV})}{\mathbb{F}}\Big) + h_2^{-6}\overset{(m,n^{VI})}{\mathbb{F}},\\
&m,n \in \mathbb{N}_0,
\end{aligned}
\tag{154}
$$

which are true for any system of polynomials (Legendre, Chebyshev).

Applying to (151) the (m,n)th moment operator of some systems of polynomials, due to (154) we obtain the desired equations in the form

$$
\begin{aligned}
&\partial_3^6\overset{(m,n)}{\mathbf{u}} + A\partial_3^4\overset{(m,n)}{\mathbf{u}} + \partial_3^2(3\partial_3^2+2A)\Big(h_1^{-2}\overset{(m'',n)}{\mathbf{u}} + h_2^{-2}\overset{(m,n'')}{\mathbf{u}}\Big)\\
&\quad + (3\partial_3^2 + A)\Big(h_1^{-4}\overset{(m^{IV},n)}{\mathbf{u}} + 2h_1^{-2}h_2^{-2}\overset{(m'',n'')}{\mathbf{u}} + h_2^{-4}\overset{(m,n^{IV})}{\mathbf{u}}\Big)\\
&\quad + h_1^{-6}\overset{(m^{VI},n)}{\mathbf{u}} + 3h_1^{-2}h_2^{-2}\Big(h_1^{-2}\overset{(m^{IV},n'')}{\mathbf{u}} + h_2^{-2}\overset{(m'',n^{IV})}{\mathbf{u}}\Big) + h_2^{-6}\overset{(m,n^{VI})}{\mathbf{u}} + \overset{(m,n)}{\mathbf{M}}(\hat{\mathbf{S}}^{**}) = 0,\\
&\partial_3^6\overset{(m,n)}{\boldsymbol{\varphi}} + B\partial_3^4\overset{(m,n)}{\boldsymbol{\varphi}} + [\partial_3^2(3\partial_3^2+2B)+C]\Big(h_1^{-2}\overset{(m'',n)}{\boldsymbol{\varphi}} + h_2^{-2}\overset{(m,n'')}{\boldsymbol{\varphi}}\Big)\\
&\quad + (3\partial_3^2 + B)\Big(h_1^{-4}\overset{(m^{IV},n)}{\boldsymbol{\varphi}} + 2h_1^{-2}h_2^{-2}\overset{(m'',n'')}{\boldsymbol{\varphi}} + h_2^{-4}\overset{(m,n^{IV})}{\boldsymbol{\varphi}}\Big)\\
&\quad + h_1^{-6}\overset{(m^{VI},n)}{\boldsymbol{\varphi}} + 3h_1^{-2}h_2^{-2}\Big(h_1^{-2}\overset{(m^{IV},n'')}{\boldsymbol{\varphi}} + h_2^{-2}\overset{(m'',n^{IV})}{\boldsymbol{\varphi}}\Big) + h_2^{-6}\overset{(m,n^{VI})}{\boldsymbol{\varphi}} + \overset{(m,n)}{\mathbf{M}}(\hat{\mathbf{H}}^{**}) = 0,\\
&m,n\in\mathbb{N}_0.
\end{aligned}
\tag{155}
$$

Analogous to (155) in the case of the reduced medium ($\alpha = 0$) based on (152) we obtain the following equation in displacement and rotation vectors with respect to any systems of polynomials (Legendre, Chebyshev) for the theory of prismatic thin bodies with two small sizes and with the rectangle cross-section:

$$
\begin{aligned}
&\partial_3^4\overset{(m,n)}{\mathbf{u}} + 2\partial_3^2\Big(h_1^{-2}\overset{(m'',n)}{\mathbf{u}} + h_2^{-2}\overset{(m,n'')}{\mathbf{u}}\Big) + h_1^{-4}\overset{(m^{IV},n)}{\mathbf{u}} + 2h_1^{-2}h_2^{-2}\overset{(m'',n'')}{\mathbf{u}} + h_2^{-4}\overset{(m,n^{IV})}{\mathbf{u}}\\
&+ \overset{(m,n)}{\mathbf{M}}(\hat{\mathbf{G}}) = 0,\\
&\partial_3^4\overset{(m,n)}{\boldsymbol{\varphi}} + 2\partial_3^2\Big(h_1^{-2}\overset{(m'',n)}{\boldsymbol{\varphi}} + h_2^{-2}\overset{(m,n'')}{\boldsymbol{\varphi}}\Big) + h_1^{-4}\overset{(m^{IV},n)}{\boldsymbol{\varphi}} + 2h_1^{-2}h_2^{-2}\overset{(m'',n'')}{\boldsymbol{\varphi}} + h_2^{-4}\overset{(m,n^{IV})}{\boldsymbol{\varphi}}\\
&+ \overset{(m,n)}{\mathbf{M}}(\hat{\mathbf{H}}) = 0, \quad m,n\in\mathbb{N}_0.
\end{aligned}
\tag{156}
$$

Note that the first relation in (156) is the equation in moments of displacement vector with respect to any systems of polynomials (Legendre, Chebyshev) for the classical theory of prismatic thin bodies with two small sizes and with a rectangular cross-section.

In order to write the systems of equations (155) and (156) in moments with respect to any system of orthogonal polynomials, it is enough to express the $\overset{(m'',n)}{\mathbf{F}}$, $\overset{(m,n'')}{\mathbf{F}}$, $\overset{(m^{IV},n)}{\mathbf{F}}$, $\overset{(m,n^{IV})}{\mathbf{F}}$, $\overset{(m'',n'')}{\mathbf{F}}$, $\overset{(m^{VI},n)}{\mathbf{F}}$, $\overset{(m,n^{VI})}{\mathbf{F}}$, $\overset{(m^{IV},n'')}{\mathbf{F}}$ and $\overset{(m'',n^{IV})}{\mathbf{F}}$, where $\mathbf{F} = \mathbf{u}$ or $\mathbf{F} = \boldsymbol{\varphi}$, to express in terms of moments $\mathbf{F}$ with respect to the considered system of polynomials (Nikabadze 2008b, 2014b, c). Here, for brevity, we will not dwell on it. However, we note that all that has been said above about the monolayer and multilayer thin bodies with one small size is true in this case.

Acknowledgments The research was supported by the Russian Foundation for Basic Research, projects nos. 15–01–00848-a and 14–01–00317-a.

References

Aero EL, Kuvshinskii EV (1964) Continual theory of asymmetric elasticity. Equilibrium of an isotropic body. Solid State Phys 6(9):2689–2699

Eringen AC (1999) Microcontinuum field theories, vol 1. Foundation and solids, Springer, New York

Galerkin BG (1930) Contribution à la solution générale du problème de la théorie de l'élasticité dans le cas de trois dimensions. CR Acad Sci 190:1047–1048

Galerkin BG (1931) Contribution à la solution générale du problème de la théorie de l'élasticité dans le cas de trois dimensions. CR Acad Sci 193:568–571

Iacovache M (1949) O extindere a metodei lui galerkin pentru sistemul ecuatiilor elasticitvelocity eii. Bull Acad Sci RPR, Ser A 1:593

Kupradze VD, Gegelia TG, Basheleishvili MO, Burchuladze TV (1976) Three-dimensional problems of the mathematical theory of elasticity and thermoelasticity. Nauka, Moscow (in Russia)

Lurie AI (1990) Nonlinear theory of elasticity. North-Holland, Dordrecht

Nikabadze MU (2007a) Some issues concerning a version of the theory of thin solids based on expansions in a system of chebyshev polynomials of the second kind. Mech Solids 42(3):391–421

Nikabadze MU (2007b) Some problems of tensor calculus, vol II. Moscow State University, Moscow (in Russia)

Nikabadze MU (2008a) Method of orthogonal polynomials in mechanics of micropolar and classical elasticity thin bodies. i., available from VINITI, 135 – B2014 (20.05.2014)

Nikabadze MU (2008b) Method of orthogonal polynomials in mechanics of micropolar and classical elasticity thin bodies. ii., available from VINITI, 136 – B2014 (20.05.2014)

Nikabadze MU (2009a) On some problems of tensor calculus. i. J Math Sci 161(5):668–697

Nikabadze MU (2009b) On some problems of tensor calculus. ii. J Math Sci 161(5):698–733

Nikabadze MU (2014a) Construction of eigentensor columns in the linear micropolar theory of elasticity. Mosc Univ Mech Bull 69(1):1–9

Nikabadze MU (2014b) Development of the method of orthogonal polynomials in mechanics of micropolar and classical elasticity thin bodies. Lomonosov Moscow State University, Moscow (in Russia)

Nikabadze MU (2014c) Method of orthogonal polynomials in mechanics of micropolar and classical elasticity thin bodies. Dsc thesis, Moscow Aviation Institute (National Research University), Moscow

Nikabadze MU (2015) On some questions of tensor calculus with applications to mechanics. In: Tensor analysis, vol 55. PFUR, Moscow, pp 3–194 (in Russia)

Nikabadze MU, Ulukhanyan AR (2008) Mathematical modeling of elastic thin bodies with one small dimension with the use of systems of orthogonal polynomials, available from VINITI, 723 – B2008 (21.08.2008)

Nikabadze MU, Kantor MM, Ulukhanyan AR (2008) On mathematical modeling of elastic thin bodies and numerical realization of some tasks of strip, available from VINITI, 204 – B2011 (29.04.2011)

Nowacki W (1970) Teoria Sprężystości. Państwowe Wydawnictwo Naukowe, Warsaw

Pobedrya BE (1986) Lectures in tensor analysis. Moscow State University, Moscow (in Russia)

Pobedrya BE (1995) Numerical methods in the theory of elasticity and plasticity. Moscow State University, Moscow (in Russia)

Sandru N (1966) On some problems of the linear theory of the asymmetric elasticity. Int J Eng Sci 4(1):81–94
Ulukhanyan AR (2010) Representation of solutions to equations of hyperbolic type. Mosc Univ Mech Bull 65(2):47–50
Ulukhanyan AR (2011) Dynamic equations of the theory of thin prismatic bodies with expansion in the system of Legendre polynomials. Mech Solids 46(3):467–479
Vekua IN (1948) New methods for solving elliptic equations. OGIZ, Moscow (in Russia)
Vekua IN (1978) Fundamentals of tensor analysis and covariant theory. Nauka, Moscow (in Russia)
Vekua IN (1985) Shell theory: general methods of construction. Pitman, Boston

Method for Calculating the Characteristics of Elastic State Media with Internal Degrees of Freedom

Sergey N. Romashin, Victoria Yu. Presnetsova, Larisa Yu. Frolenkova and Vladimir S. Shorkin

Abstract In practical use of non-traditional models of Cosserat, Leroux, Toupin, Mindlin, Aero, the problem of determining the elastic constants arise. This work is devoted to solving this problem. The solution is based on a comparison of the governing relations of the non-traditional model with its counterpart. It is obtained from the conversion of a specially constructed for this purpose variant of nonlocal elastic medium in the local model.

Keywords Adhesion · Elastic material · Lennard–Jones potential · Morse potential · Stockmayer potential · Nonlocal surface and volume forces and moments · Cosserat · Leroux · Toupin · Mindlin · Aero

1 Introduction

When using non-traditional models of Cosserat, Leroux, Toupin, Mindlin, Aero (Lurie et al. 2003a) for the calculation of mechanical processes with specific elastic constructional materials, the problem of determining the elastic constants arises. It is difficult or impossible to apply the methods of solid-state physics (Partenskii 1979; Vakilov et al. 1997) for such materials. The mechanical processes in such materials are conveniently described on the basis of the phenomenological approach, which is based on continuum thermodynamics. Therefore, methods for calculating the

S.N. Romashin · V.Yu. Presnetsova · L.Yu. Frolenkova · V.S. Shorkin (✉)
State University - Education-Science-Production Complex, 29 Naugorskoe Shosse, 302020 Orel, Russia
email: VShorkin@yandex.ru

S.N. Romashin
email: Sromashin@yandex.ru

V.Yu. Presnetsova
email: alluvian@mail.ru

L.Yu. Frolenkova
email: Larafrolenkova@yandex.ru

H. Altenbach and S. Forest (eds.), *Generalized Continua as Models for Classical and Advanced Materials*, Advanced Structured Materials 42,
DOI 10.1007/978-3-319-31721-2_16

characteristics of the elastic media with internal degrees of freedom should be based on the phenomenological macroexperiment.

One of the phenomena, which is described by the models of Cosserat, Leroux, Tupin, Mindlin, Aero, is the adhesion (Johnson 1997; Maugis 2000; Johnson and Greenwood 1997; Johnson et al. 1971; Maugis 1991; Goryacheva and Makhovskaya 2001; Yu and Polycarpou 2004). The characteristics of adhesion are its energy and adhesive force. In this paper, the elastic constants are determined numerically based on a nonlocal theory for an elastic medium.

2 Theoretical Statements

The interaction of bodies $B \equiv B_{(k)}$ ($k = 1, 2, \ldots$—number of bodies) is considered. The bodies are bounded by smooth surfaces $A \equiv A_{(k)}$. The surfaces have outward unit normals $\vec{n} \equiv \vec{n}_{(k)}$. The bodies $B \equiv B_{(k)}$ are composed of homogeneous, isotropic, linear elastic materials. Each of them is considered to be dedicated from the infinite medium $\Omega \equiv \Omega_{(k)}$. The material $\Omega \equiv \Omega_{(k)}$ and $B \equiv B_{(k)}$ is the same. This assumption excludes the influence of its boundary region on the material properties of the body $B \equiv B_{(k)}$. The state $B \equiv B_{(k)}$ inside $\Omega \equiv \Omega_{(k)}$ is the reference. It corresponds to time $t = 0$.

Every body $B \equiv B_{(k)}$ can be represented as a junction of non-intersecting parts $\Delta B_n \equiv \Delta B_{(k)n}$ ($n = 1, 2, \ldots, N$): $B \equiv B_{(k)} = \bigcup_{n=1}^{n=N} \Delta B_n$. When the diameter of the portion $\Delta B_{(k)n}$ tends to zero and $N \to \infty$ the following relations are valid:

$$\Delta B_n \equiv \Delta B_{(k)n} \to \mathrm{d}B \equiv dB_{(k)}, B \equiv B_{(k)} = \int \mathrm{d}B \equiv \int \mathrm{d}B_{(k)}$$

The density of the materials are $\rho \equiv \rho_{(k)}$ and their temperatures $T \equiv T_{(k)}$ are distributed uniformly and do not change over time. Herewith: $T_{(1)} \equiv T_{(2)}$.

Two cases are considered. In the first case, the proper rotations of the particles are not taken into account. The particles are material points. In the second case, these rotations are taken into account. The model of particles is a dumbbell. The first case is considered in detail. The second case may be analyzed similarly to the first case.

In the reference configuration the arbitrary body B occupies the region V and the centers of inertia of its particles $\mathrm{d}B$ have the radius vectors $\vec{r} \in V$. The position of an arbitrary particle $\mathrm{d}B_2$ relative to another arbitrary particle $\mathrm{d}B_1 \equiv \mathrm{d}B$ is defined by the relative radius vector $\vec{l}_{12} = \vec{r}_2 - \vec{r}_1$ with length $l_{12} = |\vec{l}_{12}| = |\vec{r}_2 - \vec{r}_1|$.

Under the influence of external mechanical impacts, including the allocation of $B \equiv B_{(k)}$ from $\Omega \equiv \Omega_{(k)}$, particles $dB \equiv dB_{(k)}$ acquire new locations, which are characterized by the radius vectors $\vec{R} \equiv \vec{R}_{(k)} \in V_{t(k)}$ and the displacement vectors $\vec{u}(\vec{r}, t) = \vec{R}(\vec{r}, t) - \vec{r}$. Area $V_{t(k)}$ is the area occupied by the body $B \equiv B_{(k)}$ in the current configuration. The position of a particle $\mathrm{d}B_2$ relative to particle $\mathrm{d}B_1 \equiv dB$ will change and will be determined by the radius vector

$$\vec{L}_{12} = \vec{R}_2 - \vec{R}_1 = (\vec{r}_2 - \vec{r}_1) + (\vec{u}_2 - \vec{u}_1) = \vec{l}_{12} + \Delta\vec{u}_{12}$$

and the length $L_{12} = |\vec{L}_{12}| = |\vec{R}_2 - \vec{R}_1|$. If body $B \equiv B_{(k)}$ is deformed, then $L_{12} \neq l_{12}$.

It is assumed that the deformations are small: $|L - l| / l \ll 1$. Therefore, the material density and volumes of elementary particles in the reference and current status are equal.

Vectors $\Delta\vec{u}_{1j}$ can be represented as series according to exterior powers $\vec{l}_{1j}$

$$\Delta\vec{u}_{1j} = \sum_{n=1}^{\infty} \frac{1}{n!} \left(\nabla_{1j}^{n}\vec{u}\right) \overbrace{\cdots\cdot}^{n\text{ times}} \vec{l}_{1j}^{\,n} = \sum_{n=1}^{\infty} \frac{(-1)^n}{n!} \left(\nabla^{n}\vec{u}\right) \overbrace{\cdots\cdot}^{n\text{ times}} \left(\vec{l}_{1j}\right)^n, j = 2, 3, \ldots, \quad (1)$$

where $\nabla = \mathrm{d}\ldots/\mathrm{d}\vec{r}$ is the differential *del* operator on a vector $\vec{r}$ and $\nabla_{1j} = \mathrm{d}\ldots/\mathrm{d}\vec{l}_{1j}$ on a vector $\vec{l}_{1j}$.

It is believed that the vector $\vec{r}_{(1)}$ receives the increment $\mathrm{d}\vec{r}_{(1)}$, then the vector $\vec{l}_{1j}$ receives the increment $\mathrm{d}\vec{l}_{1j} = -\mathrm{d}\vec{r}$. It means: $\nabla_{1j}^{n} = (-1)^n \nabla^n$. When changing the relative positions of the particles $\mathrm{d}B_1 \equiv \mathrm{d}B$ and $\mathrm{d}B_2$, they are deformed and rotated. For a particle $\mathrm{d}B_1 \equiv \mathrm{d}B$, deformations and rotations are characterized by strain deviator $\gamma_{ij}\left(\vec{e}_i\vec{e}_j\right)$, tensor of cubic strains $\theta\delta_{ij}\left(\vec{e}_i\vec{e}_j\right)$ and the rotation tensor $\omega_k\varepsilon_{ijk}\left(\vec{e}_i\vec{e}_j\right)$. Here $\vec{e}_i$ $(i = 1, 2, 3)$ is the orthonormal basis of the Cartesian coordinate system x_i $(\vec{e}_i x_i = \vec{r})$; δ_{ij} is the Kronecker delta; ε_{ijk} is the Levi-Civita symbol.

These tensors form a distortion tensor $d_{ij}\vec{e}_i\vec{e}_j$. It is expressed by the first gradient of the displacement vector $\nabla\vec{u} = u_{i},_{j}\,\vec{e}_i\vec{e}_j$.

$$\nabla\vec{u} = u_{i},_{j}\,\vec{e}_i\vec{e}_j = d_{ij}\vec{e}_i\vec{e}_j = \gamma_{ij}\left(\vec{e}_i\vec{e}_j\right) + \theta\delta_{ij}\left(\vec{e}_i\vec{e}_j\right) + \omega_k\varepsilon_{ijk}\left(\vec{e}_i\vec{e}_j\right) = d_{ij}\vec{e}_i\vec{e}_j. \quad (2)$$

Equality $\nabla\vec{u} = u_{i},_{j}\,\vec{e}_i\vec{e}_j = d_{ij}\vec{e}_i\vec{e}_j$ is the condition for the existence of the vector potential $\vec{u}$ for distortion tensor $d_{ij}\vec{e}_i\vec{e}_j$ (Lurie et al. 2003a). Equality

$$d_{ij},_{k}\,\varepsilon_{ijk} = 0 \quad (3)$$

is the condition for the existence of this potential. If (2) holds, then rounding any closed circuit does not detect a gap field of displacement vector. Medium is defect-free. If the condition (3) is not fulfilled, then

$$d_{in},_{m}\,\varepsilon_{nmj} = \left[\gamma_{in} + \frac{1}{3}\theta\delta_{in} - \omega_k\varepsilon_{ink}\right],_{m}\varepsilon_{nmj} = \Xi_{ij} \neq 0.$$

In this case, the displacement vector field will be discontinuous. There are dislocations in the material (Lurie et al. 2003a, b; Lurie and Belov 2008).

The distortion tensor is the gradient of the curvature tensor. It is expressed by the second gradient of the displacement vector

$$\nabla\nabla\vec{u} = \nabla^2\vec{u} = u_{i},_{jn}\,\vec{e}_i\vec{e}_j\vec{e}_n = D_{ijn}\vec{e}_i\vec{e}_j\vec{e}_n = d_{ij},_{n}\,\vec{e}_i\vec{e}_j\vec{e}_n. \quad (4)$$

If the second gradient of displacement is necessary to use to characterize the kinematics of the continuum, then it may appear dislocations.

In Frolenkova and Shorkin (2013) on the basis of Gibbs (1928) kinematic sign of adhesion of the two bodies is formulated. Any material fiber, which intersects the contact surface $A_{(12)}$, should preserve the smoothness of strain distribution there along under deformation of the combined body $B = B_1 \cup B_2$. In Frolenkova and Shorkin (2013) it is suggested that this characteristic may be fulfilled by using the curvature tensor for a description of deformations (4). This means that dislocation fields must arise during the adhesion of elastic materials. This characteristic is enough for adhesion of two materials.

The occurrence of dislocations in the surface layer leads to a rotation of the particles of the medium, which is independent of its rotation (Aero and Kuvshinskii 1960). When there is the adhesion of a dielectric and a metal, the dielectric enters the electric field created by the electrical double layer of the metal. The similar situation arises in the dielectric, which is the gasket of the capacitor. Due to this, the surface layer of the dielectric is polarized. Its particles acquire a particular orientation. Under external mechanical influences, they turn independently of the rotation of the medium. We assume, that the polar dielectric is placed into a homogeneous electrostatic field with strength $\vec{E} = \vec{e}E$, $|\vec{e}| = 1$. The polarization vector $\vec{P}$ of the volume unit of dielectric is proportional to the strength $\vec{E}$, which caused it: $\vec{P} = \varepsilon_0 \kappa_e \vec{E}$ (ε_0 is an electric constant; κ_e is the polarizability of the volume unit).

It is believed, that all the particles dB are dipoles-dumbbells with opposite charges at the ends. They have the same polarization vector $\vec{P}dV = (\varepsilon_0 \kappa_e E)\vec{e}dV = \chi \vec{e}dV$. The relative position of the interacting particles in new condition is characterized by a vector $\vec{L}_{12}$ instead $\vec{l}_{12}$—the vector of their relative position in the reference state. Rotation from $\vec{l}_{12}$ to $\vec{L}_{12}$ is characterized by a vector $\frac{1}{2}(\nabla \times \vec{u}) = \frac{1}{2} rot\vec{u}$. The rotary reaction of particle dB_1 to external influences is characterized by its rotation relative to its initial position by an angle $\vec{\varphi} = \vec{\varphi}_j$. In the reference state, all particles have the same orientation vector $\vec{e}$. This vector coincides with the direction of polarization. In the current state, their orientation vectors are different. So, we can talk about the function $\vec{\varphi} = \vec{\varphi}(\vec{r})$ and the notation

$$\vec{\varphi}_m = \vec{\varphi} + \sum_{n=1}^{\infty} \frac{1}{n!} (\nabla^n \vec{\varphi}) \cdot \ldots \cdot \vec{l}^{\,n}_{1m},$$

where $m = 1, 2, \ldots$ is number of dielectric particle. In the subsequent arguments, we restrict ourselves to the notation:

$$\vec{\varphi}_m = \vec{\varphi} + (\nabla \vec{\varphi}) \cdot \vec{l}_{1m}$$

If there is material adhesion, it is assumed (by analogy with Ruelle 1969), that the total potential energy of the combined body $B = B_{(1)} \cup B_{(2)}$ is the sum of the potential energies of many-particle interactions inside each of the bodies $B_{(1)}$ and $B_{(2)}$, and between them. The quantities

$$\Phi^{(2)}_{(kp)}\left(\vec{R}_{(k)}, \vec{R}_{(p)}\right) \mathrm{d}V_{(k)}\mathrm{d}V_{(p)}, \Phi^{(3)}_{(kpq)}\left(\vec{R}_{(k)}, \vec{R}_{(p)}, \vec{R}_{(q)}\right) \mathrm{d}V_{(k)}\mathrm{d}V_{(p)}\mathrm{d}V_{(q)}, \ldots$$

are the potentials of pair, triple, etc. interactions of paired particles $\mathrm{d}B_{(k)}$, $\mathrm{d}B_{(p)}$, $\mathrm{d}B_{(q)}$ bodies $B_{(k)}$, $B_{(p)}$, $B_{(q)}$ $(k, p, q = 1, 2)$. In this case, $dV_{(k)}$, $dV_{(p)}$, $dV_{(q)}$—the volumes of the interacting particles in the reference state. Functions (hereafter the potentials) $\Phi^{(2)}_{(kp)}\left(\vec{R}_{(k)}, \vec{R}_{(p)}\right)$, $\Phi^{(3)}_{(kpq)}\left(\vec{R}_{(k)}, \vec{R}_{(p)}, \vec{R}_{(q)}\right)$ for a homogeneous isotropic material depend only on the distance between the interacting particles in the current state.

The energy $\mathrm{d}W_{(1)}\left(\vec{R}_{(1)}\right) = w_{(1)}\left(\vec{R}_{(1)}\right) \mathrm{d}V_{(1)}$ of infinitesimal particle, e.g., $\mathrm{d}B_{(1)}$ with the volume $\mathrm{d}V_{(1)}$ and center of inertia $\vec{R}_{(1)}$ is represented as

$$\begin{aligned} w_{(1)} &= \left(\vec{R}_{(1)}\right) \mathrm{d}V_{(1)} = \left(w_{(11)} + w_{(12)}\right) dV_{(1)} \\ &= \left[\int\limits_{V_{(1)}} \Phi^{(2)}_{(11)} \mathrm{d}V_{(1)} + \frac{1}{2!} \int\limits_{V_{(1)}} \int\limits_{V_{(1)}} \Phi^{(3)}_{(111)} \mathrm{d}V_{(1)} \mathrm{d}V_{(1)} + \ldots\right] \mathrm{d}V_{(1)} \\ &+ \left[\int\limits_{V_{(2)}} \Phi^{(2)}_{(12)} \mathrm{d}V_{(2)} + \frac{1}{2!} \sum_{k=1}^{2} \int\limits_{V_{(2)}} \int\limits_{V_{(k)}} \Phi^{(3)}_{(12k)} \mathrm{d}V_{(2)} \mathrm{d}V_{(k)} + \ldots\right] \mathrm{d}V_{(1)}. \end{aligned}$$

In this equality $w_{(11)}$ is the cubic density of the potential energy, which arises due to the interaction of the particles of the body $B_{(1)}$ among themselves; $w_{(12)}$ is addition to the quantity of $w_{(11)}$, which arises from the interaction of particles of the body $B_{(1)}$ with the particles of the body $B_{(2)}$. Each particle $\mathrm{d}B_{(1)} \subset B_{(1)}$ is affected by the forces from the other particles $dB_{(1)}$ of the same body $B_{(1)}$, particles $\mathrm{d}B_{(2)}$ body $B_{(2)}$ and the medium, which surrounds both bodies. The first forces are called forces of cohesive interaction of parts of the body (Ashcroft and Mermin 1976). Their cubic density is:

$$\vec{f}_{(11)} = -\nabla w_{(11)}.$$

The second forces are adhesive forces (Johnson 1997). Their cubic density is:

$$\vec{f}_{(12)} = -\nabla w_{(12)} = -\vec{f}_{(21)}. \tag{5}$$

During the deformation of the material interacting particles $\mathrm{d}B_{1(k)}$ and $\mathrm{d}B_{j(p)}$ experienced relative displacements $\Delta\vec{u}_{1j}$. Decomposition (1) is valid for the particles. At the same time, it's permissible to represent the potentials of pair, triple, etc. interactions as second-order polynomials relatively $\Delta\vec{u}_{1j}$ (Shorkin 2011). Absolute term of the polynomial and its coefficients are expressed through the potentials of many-particle interactions in the reference state.

Changing the cubic density $\Delta w_{(11)}$ of the potential energy of body $B_{(1)}$ is a function of the sequence $\{\nabla^n \vec{u}\}$ of displacement gradients. If we differentiate dependence $\Delta w_{(11)}\left(\nabla\vec{u}, \nabla^2\vec{u}, \ldots\right)$ by the gradients $\nabla^n\vec{u}$, we obtain the expressions for the stress tensors, which are developed in the material of body $B_{(1)}$.

$$P^{(n)} = \frac{\partial \Delta w_{(11)}}{\partial \left(\nabla^n \vec{u}\right)} = P^{0(n)} + \sum_{m=1}^{\infty} \left(\nabla^m \vec{u}\right) \overbrace{\cdot \ldots \cdot}^{n\ times} C^{(m,n)}, \tag{6}$$

where $P^{0(n)}$ are tensors of initial stress; $C^{(m,n)}$ are tensors, which characterize the mechanical properties of the material. Taking into account only pair and triple interactions, the defining relations have the form:

$$P^{0(m)} = \frac{1}{2!}\int\limits_V \frac{1}{m!}\left(\nabla_{12}\Phi^{(2)}_{(11)}\right)\vec{l}^{\,m}_{12}\mathrm{d}V_2 + \frac{1}{3!}\sum_{j=2}^{3}\int\limits_V\left[\int\limits_V \frac{1}{m!}\left(\nabla_{1j}\Phi^{(3)}_{(111)}\right)\vec{l}^{\,m}_{1j}\mathrm{d}V_2\right]\mathrm{d}V_3, \tag{7}$$

$$C^{n,m} = \frac{1}{2!}\int\limits_V \frac{1}{m!n!}\vec{l}^{\,n}_{12}\left(\nabla_1^2\Phi^{(2)}_{(11)}\right)\vec{l}^{\,m}_{12}\mathrm{d}V_2 + \frac{1}{3!}\sum_{p,q=2}^{3}\int\limits_V\left[\int\limits_V \frac{1}{m!n!}\vec{l}^{\,n}_{1p}\left(\nabla_p\nabla_q\Phi^{(3)}_{(111)}\right)\vec{l}^{\,m}_{1q}\mathrm{d}V_2\right]\mathrm{d}V_3. \tag{8}$$

Jump to a specific local model is the replacement of (1) for the sum of one, two, etc. terms. Herewith, sequence $\{P^{(n)}\}$ stores a corresponding number of elements. The equation of motion for interacting bodies $B_{(1)}$ and $B_{(2)}$ (Shorkin 2011) in stresses for the local model has the form:

$$\rho_{(k)}\frac{\partial^2\vec{u}_{(k)}}{\partial t^2} = \nabla\cdot\left(P^{(1)}_{(k)} - \nabla\cdot\left(P^{(2)}_{(k)} - \nabla\cdot\left(P^{(3)}_{(k)} - \ldots\right)\right)\right) + \vec{f}_{(kp)} + \vec{\psi}_{(k)};\ k,p = 1,2;\ k \neq p; \tag{9}$$

where $P^{(m)}_{(k)}$, $m = 1, 2, 3, \ldots$ is the internal stress tensor.

The vector fields $\vec{\psi}_{(k)} = \vec{\psi}_{(k)}\left(\vec{r}_{(k)}\right)$ are defined. The fields $\vec{f}_{(kp)} = \vec{f}_{(kp)}\left(\vec{r}_{(k)}, \vec{r}_{(p)}\right)$ are defined by (5). The value $w_{(12)}$ is calculated through the interaction potentials of particles in the assumption of the absence of deformations in them. The interaction potentials must be known.

At time $t = t_{0(k)}$ we set the initial conditions of the displacements distribution and velocities of the particles of the body $B_{(k)}$, which occupies an area $V_{(k)}$:

$$\vec{u}_{(k)}\left(\vec{r}, t_{0(k)}\right) = \vec{u}_{0(k)}\left(\vec{r}\right), \quad \left.\frac{\partial \vec{u}_{(k)}\left(\vec{r}, t\right)}{\partial t}\right|_{t=t_{0(k)}} = \vec{v}_{0(k)}\left(\vec{r}\right). \tag{10}$$

At any time on a free from the contact surface portion $A_{(k)}$ of each of the bodies $B_{(k)}$ ($A_{(k)}$ is the complete surface in the absence of contact) the boundary conditions are set:

$$\vec{n}_{(k)} \cdot \left[P_{(k)}^{(1)} - \nabla \cdot \left(P_{(k)}^{(2)} - \ldots\right)\right] - \nabla_A \cdot \left[\vec{n}_{(k)} \cdot \left(P_{(k)}^{(2)} - \ldots\right)\right] = \Pi_{(k)}^{(0)} \equiv \vec{\tau}_{(k)}, \tag{11}$$

or

$$\vec{u}_{(k)} = \vec{u}_{A_{(k)}};$$

$$\vec{n}_{(k)} \cdot \left[P_{(k)}^{(2)} - \nabla \cdot \left(P_{(k)}^{(3)} - \ldots\right)\right] - \nabla_A \cdot \left[\vec{n}_{(k)} \cdot \left(P_{(k)}^{(3)} - \ldots\right)\right] = \Pi_{(k)}^{(1)}, \tag{12}$$

or

$$\nabla_{\vec{n}} \vec{u}_{(k)} = \Gamma_{A_{(k)}};$$

where $\Pi_{(k)}^{(0)} \equiv \vec{\tau}_{(k)}$—classical surface forces; $\Pi_{(k)}^{(m)}$, $m = 1, 2, \ldots$ are tensors, which characterize non-classical surface interactions, that will be able to perform the work on tensor characteristics $\nabla_{\vec{n}} \vec{u}_{(k)} = \Gamma_{A_{(k)}}, \ldots$ unevenness of displacements distribution.

In the equations of motion vectors $\vec{f}_{(kk)}$ are defined by the equalities:

$$\vec{f}_{(kk)} = \nabla \cdot \left(P_{(k)}^{(1)} - \nabla \cdot \left(P_{(k)}^{(2)} - \nabla \cdot \left(P_{(k)}^{(3)} - \ldots\right)\right)\right), \quad k = 1, 2.$$

Therefore, the use of expressions (9)–(12) and (6)–(9) makes a conjugate problem of the contact interaction of elastic bodies with regard to their adhesion. In the reference state, the potentials of all the many-particle interactions should be known.

When the independent rotations are considered, the free energy consists of two terms. The first term is the energy. The energy arises when there are translational relative displacements. The second term arises because of relative rotation. It is considered further. The interaction of a real pair of dipoles is described by the Stockmayer potential (Hirschfelder et al. 1966):

$$\phi_{(1,2)} = 4\varepsilon\left[\left(\frac{\sigma}{r}\right)^{12} - \left(\frac{\sigma}{r}\right)^{6}\right] - \frac{\mu_1\mu_2}{r^3} g(\Theta_1, \Theta_2, \varphi),$$

$$g(\Theta_1, \Theta_2, \varphi) = 2\cos\Theta_1 \cos\Theta_2 - 2\sin\Theta_1 \sin\Theta_2 \cos\varphi,$$

where ε, σ are parameters of the Lennard–Jones potential (Kittle 1966); μ_1, μ_2 are the values of the dipole moments of the interacting particles; r is the distance between their centers of mass; $\varphi = \varphi_1 - \varphi_2$; Θ_1, Θ_2, φ_1, φ_2 are vector angles $\vec{\mu}_1$, $\vec{\mu}_2$ of the dipole moments in a spherical coordinate system with the axis. This axis is oriented from the first particle to second and passes through their centers of mass. For particle interaction of continuum, the Lennard–Jones potential is replaced by many-particle interaction potential.

In the current state for the potential energy of relative rotation, expression can be obtained:

$$F^* = \Phi_{(1,2)} dV_1^* dV_2^* = \frac{(P_1^* dV_1^*)(P_2^* dV_2^*)}{l_{12}^{*3}} g(\vec{e}_1^{\,*}, \vec{e}_2^{\,*}, \vec{e}_{12}^{\,*}),$$
$$g(\vec{e}_1^{\,*}, \vec{e}_2^{\,*}, \vec{e}_{12}^{\,*}) = 3(\vec{e}_1^{\,*} \cdot \vec{e}_{12}^{\,*})(\vec{e}_2^{\,*} \cdot \vec{e}_{12}^{\,*}) - (\vec{e}_1^{\,*} \cdot \vec{e}_2^{\,*})$$

Next, we assume:

$$P_m^* = P_m = P, \quad m = 1, 2.$$

In its turn:

$$\vec{e}_p^{\,*} = \vec{e}_p + \Delta\vec{e}_p = \vec{e}_p + \vec{\varphi}_p \times \vec{e}_p \quad p = 1, 2, 12, \quad \vec{e}_1 = \vec{e}_2 = \vec{e}.$$

In order to know the rotational part of the cubic density of the elastic strain energy, it is necessary to calculate the sum of paired rotational interactions of particles B_1 and B_2, which are located in the single volume. Integration gives the expression:

$$\left[\vec{\varphi} - \frac{1}{2}(\nabla \times \vec{u})\right] \cdot \beta_{11} \cdot \left[\vec{\varphi} - \frac{1}{2}(\nabla \times \vec{u})\right] + (\nabla\vec{\varphi}) \cdot\cdot \beta_{22} \cdot\cdot (\nabla\vec{\varphi}).$$

The mechanical properties of the material, exhibited during independent rotation of its particles, are characterized by tensors

$$\beta_{11} \approx 1.1 \times 10^{-1} \varepsilon_0 \left(\kappa_e^2/\varepsilon\right) E^2 \left[\left(\vec{i}\,\vec{i}\right) + \left(\vec{j}\,\vec{j}\right)\right], \tag{13}$$

$$\beta_{22} \approx 1.3 \times 10^{-3} \varepsilon_0 \left(\kappa_e^2/\varepsilon\right) E^2 \left[\left(\vec{j}\,\vec{i}\right) - \left(\vec{i}\,\vec{j}\right)\right]\left[\left(\vec{j}\,\vec{i}\right) - \left(\vec{i}\,\vec{j}\right)\right] \tag{14}$$

where $\vec{i}, \vec{j}$ are unit vectors of orthonormal basis $(\vec{i}, \vec{j}, \vec{k})$ of Cartesian coordinate system, where the third vector $\vec{k}$ coincides with the vector $\vec{e}$. When there is adhesion, the vector is perpendicular to the contact surface. Expressions (13) and (14) are obtained under representation of the free energy of the elastic rotations as a quadratic form.

3 Numerical Results

Expressions (7), (8), (13) and (14) show, that the characteristics of the elastic state of the material are calculated by the potentials of nonlocal interaction of its particles. To account for the dielectric polarization, form of the potential is selected in the Stockmayer form. The singularity of metals—pressure of the electron gas is taken into account in Shorkin et al. (2011). The potentials of nonlocal interaction of particles—material points are requested to identify with a nonlinear dispersion law (Shorkin 2011; Kittle 1966)

$$\omega^2 = f_{0(kk)}K^2 - f_{1(kk)}K^4 + \ldots,$$

where ω is the oscillation frequency, K is the wave number and $f_{0(kk)}, f_{1(kk)}$ are coefficients obtained in the experiment. The dispersion law is approximated by a polynomial of degree n. The value of the degree n is determined by the condition of current task. Geometrical condition of adhesion (Frolenkova and Shorkin 2013) is the continuity and smoothness of field variations for those displacements of contacting bodies, which they obtained by adhesion. To fulfill the conditions it is sufficient to apply only the first two displacement gradients in the description of the deformations, which occur under adhesion of the two bodies. It is enough to take $n = 2$.

In this case, dependencies of the potentials of pair and triple interactions are approximated by functions

$$\Phi^{(2)}_{(kp)} = \Phi^{(2)}_{0(kp)} \left(e^{-2\beta_{(kp)} l_{12(kp)}} - 2e^{-\beta_{(kp)} l_{12(kp)}}\right), \tag{15}$$

$$\Phi^{(3)}_{(kpq)} = \Phi^{(3)}_{0(kpq)} \left(e^{-2\beta_{(kp)} l_{12(kp)}} - 2e^{-\beta_{(kp)} l_{12(kp)}}\right) \left(e^{-2\beta_{(kp)} l_{13(kp)}} - 2e^{-\beta_{(kp)} l_{13(kp)}}\right), \tag{16}$$

where $\beta_{(kk)}, \beta_{(kp)}, \Phi^{(2)}_{0(kp)}, \Phi^{(3)}_{0(kpq)}$ are parameters. These functions are equal to zero at an infinite distance. These particles may belong to the body $B_{(k)}$ $(k = p = q)$ or another $(k \neq p \vee k \neq q)$.

In the first case $(k = p = q)$, for the parameters $\Phi^{(2)}_{0(kp)}, \Phi^{(3)}_{0(kpq)}, \beta_{(kp)}$ the calculating formulas are obtained

$$\left(\frac{\Phi^{(2)}_{0(kk)}}{\beta^3_{(kk)}}\right) = \frac{4\mu_{(k)} + 45\lambda_{(k)}}{294\pi} = \frac{E_{(k)}}{294\pi} \frac{2 + 41\nu_{(k)}}{(1 + \nu_{(k)})(1 - 2\nu_{(k)})}, \tag{17}$$

$$\left(\frac{\Phi^{(3)}_{0(k)}}{\beta^6_{(kk)}}\right) = \frac{27(\lambda_{(k)} - \mu_{(k)})}{1764\pi^2} = \frac{27E_{(k)}}{1764\pi^2} \frac{4\nu_{(k)} - 1}{2(1 + \nu_{(k)})(1 - 2\nu_{(k)})}, \tag{18}$$

$$\beta_{(kk)} = \frac{1}{2}\sqrt{3\pi \left(\frac{f_{0(kk)}}{f_{1(kk)}}\right) \frac{15\left(\frac{\Phi^{(2)}_{0(kk)}}{\beta^3_{(kk)}}\right) + \left(\frac{1563\pi}{4}\right)\left(\frac{\Phi^{(3)}_{0(kk)}}{\beta^6_{(kk)}}\right)}{2\mu_{(k)} + \lambda_{(k)}}}, \tag{19}$$

where λ, μ are Lamé's parameters, E is the Young's modulus and ν is the Poisson's ratio. Formulas (17) and (18) are the result of the comparison of Voigt notation for tensor traditional characteristics of the elastic state of the material (Nowacki 1975) with the first term of the polynomial $\omega^2 = f(K^2)$. Equations (17) and (18) are constructed with the help of Eqs. (7), (15) and (16):

$$[C^{(1,1)}]_{\mathrm{Cl}} = \begin{bmatrix} \lambda+2\mu & \lambda & \lambda & 0 & 0 & 0 \\ \lambda & \lambda+2\mu & \lambda & 0 & 0 & 0 \\ \lambda & \lambda & \lambda+2\mu & 0 & 0 & 0 \\ 0 & 0 & 0 & \mu & 0 & 0 \\ 0 & 0 & 0 & 0 & \mu & 0 \\ 0 & 0 & 0 & 0 & 0 & \mu \end{bmatrix}.$$

The general presentation of the matrix of elastic parameters in the linear elasticity theory is:

$$\begin{bmatrix} C_{1111} & C_{1122} & C_{1133} & C_{1123} & C_{1131} & C_{1112} \\ C_{2211} & C_{2222} & C_{2233} & C_{2223} & C_{2231} & C_{2212} \\ C_{3311} & C_{3322} & C_{3333} & C_{3323} & C_{3331} & C_{3312} \\ C_{3211} & C_{3222} & C_{3233} & C_{3223} & C_{3231} & C_{3212} \\ C_{1311} & C_{1322} & C_{1333} & C_{1323} & C_{1331} & C_{1312} \\ C_{2111} & C_{2122} & C_{2133} & C_{2123} & C_{2131} & C_{2112} \end{bmatrix}.$$

The matrix of elastic parameters, which are expressed in terms of the interaction potential pairs and triplets of particles has the form:

$$[C^{(1,1)}]_{ijkl} = \begin{bmatrix} \eta+\chi+\varepsilon & \frac{\eta+\chi}{3}+\varepsilon & \frac{\eta+\chi}{3}+\varepsilon & 0 & 0 & 0 \\ \frac{\eta+\chi}{3}+\varepsilon & \eta+\chi+\varepsilon & \frac{\eta+\chi}{3}+\varepsilon & 0 & 0 & 0 \\ \frac{\eta+\chi}{3}+\varepsilon & \frac{\eta+\chi}{3}+\varepsilon & \eta+\chi+\varepsilon & 0 & 0 & 0 \\ 0 & 0 & 0 & \frac{\eta+\chi}{3} & 0 & 0 \\ 0 & 0 & 0 & 0 & \frac{\eta+\chi}{3} & 0 \\ 0 & 0 & 0 & 0 & 0 & \frac{\eta+\chi}{3} \end{bmatrix}$$

with

$$\eta = \frac{\pi}{2\cdot 2!}\left[-\int_0^\infty l_1^4 \frac{\mathrm{d}^2\Phi^{(2)}(l_1)}{\mathrm{d}l_1^2}\mathrm{d}l_1\right]\cdot\frac{4}{5}, \tag{20}$$

$$\chi = 2\frac{\pi}{2\cdot 3!}\cdot 4\pi\left[\int_0^\infty l_2^2\mathrm{d}l\int_0^\infty l_1^2\frac{\partial^2\Phi^{(3)}(l_1,l_2)}{\partial l_1^2}\mathrm{d}l_1\right]\cdot\frac{4}{5}, \tag{21}$$

$$\varepsilon = 2\frac{16\pi^2}{9\cdot 3!}\left[\int_0^\infty l_2^2\mathrm{d}l_2\int_0^\infty l_1^2\frac{\partial^2\Phi^{(3)}(l_1,l_2)}{\partial l_1\partial l_2}\mathrm{d}l_1\right]. \tag{22}$$

The first and the third matrix are written for isotropic materials. They are equal.

Equation (19) is obtained by comparing the first and second terms of submission of the dispersion law $\omega^2 = f(K^2)$ for plane longitudinal waves in the form of a polynomial of the second degree. In the second case $k \neq p \vee k \neq q$, when the bodies of different materials interact, for determining the parameters $\Phi^{(2)}_{0(kp)}$, $\Phi^{(3)}_{0(kpq)}$, $\beta_{(kp)}$ are used dependencies of characteristics of elastic state of two-component solid solutions on the concentration of their components. The element of volume $\mathrm{d}\langle V\rangle$ of the solid solution is expressed through elements of volumes $\mathrm{d}V_{(1)}$ of $\mathrm{d}V_{(2)}$ and components of its materials $B_{(1)}$ and $B_{(2)}$.

$$\mathrm{d}\langle V\rangle = (1-c)\mathrm{d}V_{(1)} + c\mathrm{d}V_{(2)},$$

where c and $(1-c)$—the volume concentrations of the materials of bodies $B_{(2)}$ and $B_{(1)}$. Each part $\mathrm{d}\langle V\rangle$ of each of the interacting particles interacts with each part of the other interacting particles. Therefore, we have the equalities

$$\langle \Phi^{(2)}\rangle = (1-c)^2\Phi^{(2)}_{(11)} + 2c(1-c)\Phi^{(2)}_{(12)} + c^2\Phi^{(2)}_{(22)},$$
$$\langle \Phi^{(3)}\rangle = (1-c)^3\Phi^{(3)}_{(111)} + 3c(1-c)^2\Phi^{(3)}_{(112)} + 3c^2(1-c)\Phi^{(3)}_{(122)} + c^3\Phi^{(3)}_{(222)}.$$

Taking this into account we can get:

$$\frac{4\langle\mu\rangle + 45\langle\lambda\rangle}{294\pi} = \frac{\langle E\rangle}{294\pi}\frac{2+\langle\nu\rangle}{(1+\langle\nu\rangle)(1-2\langle\nu\rangle)} = \left(\frac{\langle\Phi^{(2)}_0\rangle}{\langle\beta\rangle^3}\right)$$
$$= (1-c)^2\left(\frac{\Phi^{(2)}_{0(11)}}{\beta^3_{(11)}}\right) + 2c(1-c)\left(\frac{\Phi^{(2)}_{0(12)}}{\beta^3_{(12)}}\right) + c^2\left(\frac{\Phi^{(2)}_{0(22)}}{\beta^3_{(22)}}\right),$$
$$\frac{27\left(\langle\lambda\rangle - \langle\mu\rangle\right)}{1764\pi^2} = \frac{27\langle E\rangle}{1764\pi^2}\frac{4\langle\nu\rangle - 1}{2(1+\langle\nu\rangle)(1-2\langle\nu\rangle)} = \left(\frac{\langle\Phi^{(3)}_0\rangle}{\langle\beta\rangle^6}\right)$$
$$= (1-c)^3\left(\frac{\Phi^{(3)}_{0(111)}}{\beta^6_{(11)}}\right) + 3c(1-c)^2\left(\frac{\Phi^{(3)}_{0(112)}}{\beta^3_{(11)}\beta^3_{(22)}}\right)$$
$$+ 3c^2(1-c)\left(\frac{\Phi^{(3)}_{0(122)}}{\beta^3_{(12)}\beta^3_{(22)}}\right) + c^3\left(\frac{\Phi^{(3)}_{0(222)}}{\beta^6_{(22)}}\right)$$

with

$$\langle\beta\rangle = \frac{1}{2}\sqrt{3\pi\left(\frac{\langle f_0\rangle}{\langle f_1\rangle}\right)\frac{15\left(\frac{\langle\Phi^{(2)}_0\rangle}{\langle\beta\rangle^3}\right) + \left(\frac{1563\pi}{4}\right)\left(\frac{\langle\Phi^{(3)}_0\rangle}{\langle\beta\rangle^6}\right)}{2\langle\mu\rangle + \langle\lambda\rangle}}$$

In the calculations, we need to use two values $c = c_1$ and $c = c_2$. The validity of the proposed methodology of calculating the characteristics of the elastic state has indirect confirmation.

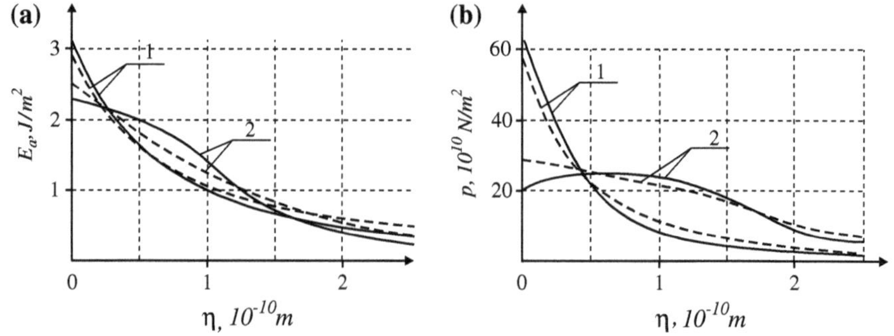

Fig. 1 The dependence of the potential E_a and the force of attraction p from the distance η for combinations *Cu-Al—curves 1* and *Al-Al—curves 2*

The methodology proposed in this paper, allows to calculate the interaction potential of semi-infinite bodies $B_{(1)}$ and $B_{(2)}$ (Fig. 1a), the force of attraction p (Fig. 1b), which depends on the distance η between the units of area boundary planes $A_{(1)}$ and $A_{(1)}$. The results of calculation (Fig. 1, dotted line) are compared with the works (Vakilov et al. 1997) and (Johnson 1997) (Fig. 1, solid line). Conformity is satisfactory. The values of adhesive forces, which are calculated at zero distance between the boundaries of the semi-infinite bodies (Table 1) are also compared.

Table 1 The comparison of the results of calculation of adhesive forces

Materials $B_{(1)}$–$B_{(2)}$	σ_t, 10^4 MPa	σ_t, 10^4 MPa settlement	$(E_{(1)} - E_{(2)})$, 10^3 MPa[3]	$\nu_{(1)} - \nu_{(2)}$[3]
Ag–Ag	3.18[1]	3.14	83 – 83	0.37 – 0.37
Al–Al	3.27[1]	2.21	75 – 75	0.34 – 0.34
Au–Au	3.22[1]	4.07	78 – 78	0.40 – 0.40
Cu–Cu	4.76[1]	4.85	120 – 120	0.38 – 0.38
V–V	6.94[1]	5.94	170 – 170	0.36 – 0.36
W–W	7.34[1]	7.64	350 – 350	0.30 – 0.30
Si–Si	3.16[1]	2.84	160 – 160	0.27 – 0.27
Fe–Cr	7.40[2]	7.12	210 – 300	0.31 – 0.31
Fe–Cu	6.00[2]	5.33	210 – 120	0.31 – 0.38
Cu–Al	4.40[2]	3.27	120 – 75	0.38 – 0.34

[1]Nevolin and Fazylov (2011)
[2]Vakilov et al. (1997)
[3]Babichev et al. (1991)

4 Conclusion

In the paper we proposed a method of modeling the behavior of elastic media. This method is based on obtaining the explicit form of the free energy dependence on the characteristics of the thermodynamic state of the medium. This allows to calculate the characteristics of the elastic state. As a result, the possibilities of a quantitative description of the non-classical behavior of elastic media are expanded, the amount of experimental studies to determine the parameters, which characterize the properties of materials, is reduced.

The work of each of the bodies is considered to be continously. The particles, that compose the body, complitly fill the volumes. When there is adhesion of the two bodies, it turns a solid third body. Thus, there is no equilibrium distances between the particle of bodies and the bodies themselves, and it is not necessary to define them. Excluding the equilibrium distances between the particles and the bodies is one of the purposes of the paper. If there are equilibrium distances, this contradicts the hypothesis of the continuity of the material.

The interaction potentials of infinitely small particles do not coincide with the potentials of the interaction of atoms and do not contain equilibrium distances. Their parameters are calculated using the Lamé parameters. The formulas for these calculations were obtined by comparing the tensor of elastic deformations of the linear elasticity theory with the same tensor, which components are expressed in terms of the interaction potential of the particles.

Acknowledgments The work was performed in the framework of the base part of the state task for 2014–2016 years (project code 286).

References

Aero EL, Kuvshinskii EV (1960) Basic equations of the theory of elasticity of media with rotational interaction of particles. Fizika Tverdogo Tela [Phys Solids] 2(7):1399–1409

Ashcroft NW, Mermin ND (1976) Solid state physics. Saunders College, Philadelphia

Babichev AP, Babushkina NA, Bratkovsky AM (1991) In: Grigoriev IS, Malikova EH (eds) Physical quantities: a handbook [Russian translation]. Energoatomizdat, Moscow

Frolenkova LY, Shorkin VS (2013) Method of calculating the surface and adhesion energies of elastic bodies. PNRPU Mech Bull 50(1):235–259

Gibbs JW (1928) The Collected works, vol 1. Thermodynamics. Longmans, Green and Co., New York

Goryacheva IG, Makhovskaya YY (2001) Adhesive interaction of elastic bodies. J Appl Math Mech 6(2):273–282

Hirschfelder JO, Curtiss CF, Bird RB (1966) Molecular theory of gases and liquids. Wiley, New York

Johnson KL (1997) Adhesion and friction between a smooth elastic spherical asperity and plane surface. Proc Royal Soc Lond Ser A 453:163–179

Johnson KL, Greenwood JA (1997) An adhesion map for the contact of elastic spheres. J Colloid Interface Sci 192(2):326–333

Johnson KL, Kendall K, Roberts AD (1971) Surface energy and the contact of elastic solids. Proc Royal Soc Lond Ser A 324:301–313
Kittle C (1966) Introduction to solid state physics. Wiley, New York
Lurie SA, Belov PA (2008) Cohesion field: Barenblatt's hypothesis as formal corollary of theory of continuous media with conserved dislocations. Int J Fract 50(1–2):181–194
Lurie S, Belov P, Volkov-Bogorodsky D, Tuchkova N (2003a) Nanomechanical modeling of the nanostructures and dispersed composites. Comput Mater Sci 28(3–4):529–539
Lurie SA, Belov PA, Volkov-Bogorodsky DB (2003b) Multiscale modeling in the mechanics of materials: Cohesion, interfacial interactions, inclusions and defects. In: Wendland WL, Efendiev M (eds) Analysis and simulation of multifield problems, vol 15. Lecture notes in applied and computational mechanics. Springer, Berlin, pp 101–110
Maugis D (1991) Adhesion of spheres: the JKR-DMT transition using a Dugdale model. J Colloid Interface Sci 150:243–269
Maugis D (2000) Adhesion and rupture of elastic solids. Solid-state sciences. Springer, Berlin
Nevolin VK, Fazylov FR (2011) On the adhesion theory of solids in terms of the dielectric formalism. Phys Solid State 53(3):634–637
Nowacki W (1975) Theory of elasticity [Russian translation]. Mir, Moscow
Partenskii MB (1979) Self-consistent electron theory of a metallic surface. Sov Phys Usp 22(5):330–351
Ruelle D (1969) Statistical mechanics: rigorous results. Benjamin-Cummings, New York
Shorkin VS (2011) Nonlinear dispersion properties of high-frequency waves in the gradient theory of elasticity. Mech Solids 6:898–912
Shorkin VS, Frolenkova LY, Azarov AS (2011) Accounting the triple interaction influence of environment particles on superficial and adhesive properties of solids. Materialovedenie (Materials Science) 2:2–8
Vakilov AN, Mamonova MV, Prudnikov VV (1997) Adhesion of metals and semiconductors analyzed by a dielectric formalism. Phys Solid State 39(6):864–867
Yu N, Polycarpou AA (2004) Adhesive contact based on the Lennard–Jones potential: a correction to the value of the equilibrium distance as used in the potential. J Colloid Interface Sci 278:428–435

Variational Theories of Two-Phase Continuum Poroelastic Mixtures: A Short Survey

Roberto Serpieri, Alessandro Della Corte, Francesco Travascio and Luciano Rosati

Abstract A comprehensive survey is presented on two-phase and multi-phase continuum poroelasticity theories whose governing equations at a macroscopic level are based, to different extents, either on the application of classical variational principles or on variants of Hamilton's least Action principle. As a focal discussion, the 'closure problem' is recalled, since it is widespread opinion in the multiphase poroelasticity community that even the simpler two-phase purely-mechanical problem of poroelasticity has to be regarded as a still-open problem of applied continuum mechanics. This contribution integrates a previous review by Bedford and Drumheller, and covers the period from the early use of variational concepts by Biot, together with the originary employment of porosity-enriched kinematics by Cowin and co-workers, up to variational theories of multiphase poroelasticity proposed in the most recent years.

Keywords Variational poroelasticity · Compressible phases · Generalized continua · Effective stress · VMTPM

R. Serpieri (✉)
Dipartimento di Ingegneria, Università degli Studi del Sannio, Piazza Roma, 21, 82100 Benevento, Italy
e-mail: roberto.serpieri@unisannio.it

A. Della Corte
Dipartimento di Ingegneria Meccanica e Aerospaziale, Università di Roma La Sapienza, via Eudossiana 18, 00184 Rome, Italy
e-mail: alessandro.dellacorte@uniroma1.it

F. Travascio
Biomechanics Research Laboratory, Department of Industrial Engineering, University of Miami, 1251 Memorial Drive, MEB 276, Coral Gables, FL 33146, USA
e-mail: f.travascio@miami.edu

L. Rosati
Dipartimento di Strutture per l'Ingegneria e l'Architettura (DIST), Universitá degli Studi di Napoli Federico II, Via Claudio, 21, 80125 Naples, Italy
e-mail: rosati@unina.it

H. Altenbach and S. Forest (eds.), *Generalized Continua as Models for Classical and Advanced Materials*, Advanced Structured Materials 42,
DOI 10.1007/978-3-319-31721-2_17

1 Introduction

Theoretical poroelasticity has a very wide range of applications. Besides the well known applications to soil mechanics (Fillunger 1936; Terzaghi 1936; Biot 1956), poroelastic models have been increasingly applied for the description of complex biological phenomenology such as biological tissue mechanics and remodeling processes (see e.g., Cowin 1999; Giorgio et al. 2014; Andreaus et al. 2014; Madeo et al. 2011; Ateshian and Ricken 2010; Ehlers and Bluhm 2013; Mow et al. 1980). However, most problems of geomechanics and biomechanics require a multiphase continuum description for a proper understanding and prevision of several intertwined mechanical phenomena. As well-known in geomechanics, this is the case of saturated and partially saturated soils (Schrefler 2002; Madeo et al. 2013). Similarly, in biomechanics, cartilaginous tissues have been described as mixtures of a solid phase made up of structural macromolecules plus an interstitial fluid phase consisting of water and solutes (Lai et al. 1991; Gu et al. 1998; Huyghe and Janssen 1997; Travascio et al. 2014).

However, the achievement of a general consistent theory for two-phase continuum poroelasticity, capable of addressing systems with any degree and range of compressibility of the constituent phases, represents a long dated challenge of theoretical and applied continuum mechanics.

The insightful historical retrospective survey by de Boer (1996) provides evidence of the complexity of the construction of a *standard* continuum theory of this type, undertaken over the last century. In particular, by the term *standard* we refer to the formulation of a generally agreed minimal set of mathematically consistent and physically plausible governing equations of two-phase poroelasticity, deducible from the classical principles of physics with assessed predictive capabilities. The review by de Boer covers a large part of the approaches extending from the early Terzaghi–Fillunger dispute (Fillunger 1936; Terzaghi 1936), including the fundamental theoretical contributions by Biot (1956) and the fundamental experimental evidences from geomechanics (Skempton 1954; Nur and Byerlee 1971), up to the more recent group of theories frequently gathered under the term *Theories of Immiscible Mixtures* (TIM).

A comprehensive survey on TIMs proposed until 1983 has been provided by Bedford and Drumheller (1983). Such a review includes mixture theories derived from continuous models more general than classical Cauchy one which are based on continuum mechanics frameworks employing enhanced microstructural descriptions for the solid phase: in particular the theory of linear elasticity with microstructure by Mindlin (1964), Eringen's micromorphic theory (Eringen 1968), as well as Goodman and Cowin's theories which employ an additional equation of motion in either postulated form (Goodman and Cowin 1972) or developed proceeding from a postulated variational principle (Cowin and Goodman 1976).

After the middle of the eighties, driven by the increase of applications in geomechanics, biomechanics, environmental engineering and material engineering, theoretical research efforts have been aimed at developing general and comprehensive multiphase flow theories. Nevertheless, investigators kept searching for a fundamental set of governing equations achieving general consensus. Research in this area has accordingly experienced a proliferation of porous media frameworks which have proceeded quite independently by stressing different arguments in order to achieve the formulation of a standard macroscopic governing set of continuum equations.

To logically organize the research efforts driven by such a *multiplication of languages* from the eighties until current times, classifications of the mainstream approaches can be attempted, without any claim of completeness and of clean-cut separation.

A first classification can be considered by identifying two approaches:

1. *Purely Macroscale Theories* (PMT) which are based on the introduction of kinematic descriptors or constitutive features expressly at the macroscale level;
2. *Upscaling/Averaging Theories* (AT) which proceed from considering a detailed representation of the geometry and flow processes at the microscale.

A general agreement on the superiority of either PMT and AT has not been reached yet (see e.g. the debates in Gray et al. 2013a, b; Baveye 2013). In any case, PMT are exposed to the criticism of lacking a strong connection with the pore scale physics and of performing implicit approximations, while AT can be criticized for introducing assumptions justifiable only on a heuristic basis, and also for lacking a clear link with the macroscopic measurement processes.

Among PM theories, a further classification can be performed according to the setting employed for the definition of energy potentials of the constituent phases (Gajo 2010). Two families can be identified: the first one includes PMT approaches where a single macroscopic energy potential of the whole saturated mixture gives rise to stresses for both solid and fluid phases (see, e.g., Coussy et al. 1998); the other one includes approaches where the two phases can be treated as superposed continua, each one endowed with a separate energy potential. As observed by several authors, this second and more general approach requires, alongside of linear momentum and mass balances, an additional governing equation (Svendsen and Hutter 1995), usually referred to as *closure equation*.

In this respect, several candidate closure equations have been proposed for the identification of the additional equation (or of the set of additional equations), capable of providing the minimal set of governing balance equations necessary to achieve a general consistent formulation of compressible poroelasticity. Within a wide family of more general formulations (see for instance Bowen 1982; Hassanizadeh and Gray 1990; Schrefler 2002) the closure of the poroelastic problem has been sought by supplementing momentum and mass balance equations with the second law of thermodynamics, in agreement with an early indication by Truesdell (quoted by Bedford and Drumheller) according to which "*the 'missing principle', surely, is a proper generalization of the Clausius–Duhem inequality*".

Similar to Drumheller (1978), Bowen (1982) includes the evolutions equations of volume fractions, as well as momentum of momentum balances, among the governing equations, tracing back to Cosserat's theory (Cosserat and Cosserat 1909;

Eringen and Kafadar 1976). A further closure of the biphasic problem has been proposed by incorporating of the saturation constraint in the entropy inequality, using an incompressibility hypothesis and a Lagrangian multiplier (Svendsen and Hutter 1995; de Boer 1996). Moment of momentum balance is also considered together with a multiplicative decomposition of the strain tensor (Diebels 1999).

Alternatively, porosity was added as an additional independent state field by Albers and Wilmański (2006) and by Wilmański (1998), who investigated several additional balance equations in the form of porosity balance or integrability condition for the deformation of the solid skeleton. A geometric saturation constraint has also been employed, combined with a multiplicative decomposition of the deformation gradient, as a closure equation of the formulation (de Boer 2005).

It is therefore evident that even the simpler two-phase purely-mechanical problem of poroelasticity has to be regarded as a still-open problem of continuum mechanics. This opinion is widely spread in the multiphase poroelasticity community. As stated by de Boer: *"the necessity to attack the problem of developing a consistent general poroelasticity theory is still existent"* (de Boer 2005), as well as by Lopatnikov and Gillespie (2010) *"... in spite of a tremendous number of publications in this field, the discussion continues about physical background of the poroelastic theory. Even the form of basic governing equations are sufficiently different [...] in frame of different approaches that one can find in literature. It seems that there is no final agreement about consistency of proposed different approaches."*

Turning to the objective of the present survey, attention is herein focused on the subclass of two-phase continuum poroelasticity theories which can be classified to be of *variational* type and ascribable to the PMT group. Thus, ruling out the major effort of an accurate updated review of the currently available porous media frameworks, this contribution is aimed at providing an updated survey on the variational subclass of poroelastic multiphase theories, since the authors share the convincement that variational statements are privileged means for the continuum description of physical phenomena ensuring *"a natural and rigorously correct way to think of [...] continuum physics"* (Oden and Reddy 2012).

A further, highly relevant feature of the variational approach is that the principle of minimization is very convenient as a basis for numerical simulations. Indeed, it is well-known that Finite Element (FE) methods are the natural discretization of theories presented in weak formulation. More specifically, numerical investigation of poroelastic continua has greatly benefited from the development of high-regularity FE schemes such as isogeometric analysis (see Hughes et al. 2005, as one of the stem works), a technique that is particularly suitable for generalized continuum theories as recently shown in different contexts (Greco and Cuomo 2014, 2016; Cazzani et al. 2014; Cuomo et al. 2014).

For the purpose of the present contribution, we include under the term *variational theories of poroelasticity* all continuum theories of porous multiphase materials (including single phase theories) which found the derivation of governing equations upon the application of classical variational principles: Hamilton Least Action Principle of mechanics (Landau and Lifshitz 1976; Moiseiwitsch 2013; Berdichevsky 2009), Principle of Virtual Powers (for functionals admitting first differentials), and

Principle of Virtual Works (see also the retrospective by dell'Isola and Placidi 2012 on the application of variational principles to continuum mechanics).

Concerning the focus on PMT, it has to be added that the already mentioned microstructured continua theory can be seen as a general framework in which one can find higher gradient continuum theories as a particular case. Researches on microstructured and higher gradient continua are experiencing a significant intensification, especially in connection with the development of computer-aided manufacturing techniques (as high-precision and multi-material 3D-printing) which allow the designing and the fabrication of micro- and even nanostructures characterized by a high degree of complexity, whose effects at the macroscale cannot be captured by standard continuum theories (dell'Isola et al. 2015).

2 Variational Theories of the 70's and the 80's

An important remark should be made concerning the notation adopted hereafter. As always happens when a research field gradually produces a considerable amount of literature, there is a general tendency towards a relative uniformity in the notation and conventions which is agreed within the scientific community of specialists in the topic. This fact has obvious advantages, allowing for the immediate understanding of the equations in a paper without requiring a detailed reading of the discursive parts. On the other hand, the literature usually converges only asymptotically towards this status, and there are many cases (especially among pioneering works) in which substantially different notations and conventions are used. Thus, it appeared illogical to the authors not to exploit the advantages of a coherent and uniform notation even though this has meant, in some cases, the modification of the original format of some equations. Therefore, all equations and formulas in the present work have to be intended as conceptually identical (but not philologically accurate) rendering of the original ones in the cited papers.

The first use of some of the ideas of variational approaches in the derivation of a theory of mixtures has been claimed by Truesdell and Toupin (1960) (p. 567) to trace back to Duhem (1893).

It is also important to recall that the seminal and influential poroelastic theory by Biot (1941, 1956, 1962), while originarily obtained on a somewhat intuitive definition of stress measures and elastic relations based on concise mechanical considerations (Bear and Corapcioglu 2012), was subsequently framed by Biot (1972) into a variational theoretical framework in the context of quasi-static and isothermal deformations. Therein, the equilibrium equations are obtained proceeding from the statement of a principle of virtual work, and later extended to account for nonisothermal deformations and to include dynamical forces (Biot 1977).

In particular, variational concepts are applied by Biot (1972), proceeding from the introduction of an 'isothermal free energy density function' which depends on the finite strain of the solid and on a quantity m defined therein as the total mass of fluid added in the pores of the sample during deformation. This particular choice for the descriptors, referred to open mechanical systems where mass can enter or leave, has been criticized by more than one author. For instance, it has been observed that it is not possible to construct a true variational principle as the Biot model contains nonequilibrium variables, see, e.g., Wilmański (2006).

Further works in the seventies, containing some applications of variational concepts to the derivation of multiphase porous media theories, are the papers by Kenyon (1976) where a variational postulate is proposed to justify the linear momentum balance equations introduced by Truesdell (1969) in postulated form. The kinematic descriptors considered are the densities and the deformation gradients of each phase; however, this theory makes no use of volume fractions. Also, an application of variational concepts to formulate a theory of two-phase mixtures is reported by Aizicovici and Aron (1977), although this study proceeds from postulated equations of motion.

2.1 Cowin's Theories Including Porosity

Theories of ideal multiphase mixtures which, in some respects, can be stated to have a variational character are those by Nunziato and Walsh (1980) and Passman (1977). These frameworks consist of extensions of the continuum theory for granular materials (Goodman and Cowin 1972), and rely on the key idea to add the volume fraction of the solid phase (ϕ) as an additional kinematic continuum scalar descriptor. Moreover the frameworks by Nunziato and Walsh (1980) and Passman (1977) use an additional balance scalar equation, proposed by Goodman and Cowin (1972) and denominated therein *equation of balance of equilibrated force*. This additional equation pairs the number of unknown fields, incremented by one as a result of the introduction of ϕ among the kinematic descriptors, with the number of momentum balance scalar PDEs.

Although the theory by Goodman and Cowin (1972) cannot be termed variational, since it is based on thermodynamic arguments and ad-hoc modified forms of the momentum and energy balances, the balance of equilibrated force is motivated by a variational analysis. Later, Cowin and Goodman (1976) have shown that the so called balance of equilibrated forces can be derived proceeding from a postulated variational principle. In particular, such variational theory is derived by addressing the dependence of a *density of stored energy function* upon the solid volume fraction, the true density of the solid porous phase ρ, the solid volume fraction ϕ and its space gradient $\phi\nabla$.

It should be remarked that the variational theory by Cowin and Goodman (1976) is not a standard variational theory in several respects. Actually, Eq. (13) therein presents a postulated condition, directly expressed in the form of first-variations containing two postulated quantities: a quantity H, stated to be a *self-equilibrated stress system*, and a second quantity l, stated to be a *self-equilibrated body force*. There is a potential misunderstanding in this last choice. Indeed, talking about *stress* seems to focus just on classical (Cauchy-type) external actions, while the proposed model entails the presence of more general external actions. Thus, *generalized stress* would probably have been a more appropriate wording in this case. A further uncommon feature of this theory is that the stress tensor of the solid phase is originally defined as a quantity work-associated with the solid true density ρ, instead of being defined as a quantity work-associated with the symmetric part of the displacement gradient in a standard way.

2.2 *Mindlin's Variational Single-Phase Theory of Materials with Microstructure*

Although not directly applied to multiphase problems, the (single phase) continuum theory of materials with microstructure by Mindlin (1964) has provided a useful (and in some respects 'canonical') background for the subsequent development, on a variational basis, of multiphase poroelastic continuum frameworks which, exploiting ideas similar to those in Cowin and Goodman (1976), Passman (1977), Nunziato and Walsh (1980), employ enhanced kinematics with additional descriptors such as the porosity.

In Mindlin's theory the equations of motion are derived by using Hamilton's principle which can be conveniently written as follows:

$$\delta \int_{t_1}^{t_2} (T - V)\mathrm{d}t + \int_{t_1}^{t_2} \delta W \mathrm{d}t = 0, \tag{1}$$

where t_1 and t_2 are two arbitrarily assigned time instants, T is the kinetic energy, V is the internal potential energy while the term δW is the virtual work of external body forces, external traction vectors, generalized body forces and generalized surface forces (termed *double forces* by Mindlin).

A quite general framework is considered in which a macroscopic second-order tensor field $\boldsymbol{\psi}$ is added as a further kinematic descriptor, termed *microdeformation*, complementing the displacement field $\mathbf{u}$.

As a consequence of such choice of kinematic descriptors, it is derived from (1) a vector linear momentum balance, expressing the stationarity of (1) with respect to $\mathbf{u}$ plus additional stationarity scalar conditions expressing stationarity with respect to the independent components ψ_{ij}.

The strain measures of this theory are the standard strain tensor $\boldsymbol{\varepsilon} = \mathrm{sym}(\nabla \mathbf{u})$ plus two additional strain measure fields related to $\boldsymbol{\psi}$: the so-called *relative deformation field*, defined as $\boldsymbol{\gamma} = \nabla \mathbf{u} - \boldsymbol{\psi}$, and a *microdeformation gradient* field $\boldsymbol{\kappa} = \nabla \boldsymbol{\psi}$. On this basis, the strain energy is a homogeneous quadratic function of $\boldsymbol{\varepsilon}$, $\boldsymbol{\gamma}$ and $\boldsymbol{\kappa}$.

2.3 The Variational Theory of Immiscible and Structured Mixtures by Bedford and Drumheller

A fundamental advancement in the derivation of variational theories of multiphase porous media and structured mixtures has been provided by Bedford and Drumheller (1978, 1979, 1983). These authors have extended the ideas laying the basis of single-continuum framework of microstructured continua by Mindlin (1964) and the approaches for the variational treatment of a single continuum in solid and fluid mechanics (Lanczos 1970; Herivel 1955; Eckart 1960; Finlayson 2013; Leech 1977; Oden and Reddy 2012) to derive the balance equations for porous multiphase problems by means of Hamilton's principle. Accordingly, momentum balance equations are derived from a stationarity condition representing a variant of Eq. (1).

A multiphase framework is considered with index of the generic phase ξ hereby indicated by script $(\cdot)^{(\xi)}$. From a constitutive point of view, denoting by $\phi^{(\xi)}$ the volume fraction of the generic ξth phase and by $\hat{\rho}^{(\xi)}$ its 'true density', related to the relevant apparent density $\bar{\rho}^{(\xi)}$ by the usual relation:

$$\hat{\rho}^{(\xi)} = \frac{\bar{\rho}^{(\xi)}}{\phi^{(\xi)}}, \tag{2}$$

it is assumed (Bedford and Drumheller 1979) that each phase ξ has a strain energy density ψ which is only dependent on $\hat{\rho}^{(\xi)}$ while in (Bedford and Drumheller 1978) a dependence of upon $\hat{\rho}^{(\xi)}$ and the (infinitesimal) strain tensor ε is considered.

The primary descriptors of such formulation are fields $\phi^{(\xi)}$ and $\hat{\rho}^{(\xi)}$, together with the placement field $\boldsymbol{\chi}^{(\xi)}$ which operates the association $\mathbf{x}^{(\xi)} = \boldsymbol{\chi}^{(\xi)}\left(\mathbf{X}^{(\xi)}\right)$ between the current position of phase ξ and its material position $\mathbf{X}^{(\xi)}$. This choice of fields amounts to a total of five fields per each phase. In agreement with Leech (1977), the least-action condition is written integrating over a fixed reference volume domain containing a fixed mass of mixture.

It is important to remark that, in the formulation stated by Bedford and Drumheller, the primary descriptors are not unconstrained fields. Actually, fields $\phi^{(\xi)}$, $\hat{\rho}^{(\xi)}$, and $\boldsymbol{\chi}^{(\xi)}$ are constrained by the equations of mass balance:

$$J^{(\xi)}\bar{\rho}^{(\xi)} = \bar{\rho}_0^{(\xi)} \tag{3}$$

and by the volume fraction constraint stating that space is completely saturated by the phases so that the sum of volume fractions equals unity, viz.:

$$\sum_{\xi=1}^{N} \phi^{(\xi)} = 1, \tag{4}$$

where N is the number of phases. In order to respect (3) and (4), the variations $\delta\phi^{(\xi)}$, $\delta\hat{\rho}^{(\xi)}$ and $\delta\mathbf{x}^{(\xi)}$ are also constrained each other. Such constraints are included by Bedford and Drumheller through the addition of (3) and (4) into (1) with the aid of Lagrange multipliers λ and μ_ξ. The resulting equation has the format:

$$\delta\int_{t_1}^{t_2}(T-V)\mathrm{d}t+\int_{t_1}^{t_2}\delta W\mathrm{d}t$$
$$+\int_{t_1}^{t_2}\left[\sum_{\xi=1}^{N}\int_{\Omega}\mu_\xi\delta\left(J^{(\xi)}-\frac{\phi_0^{(\xi)}\hat{\rho}_0^{(\xi)}}{\phi^{(\xi)}\hat{\rho}^{(\xi)}}\right)\mathrm{d}V_0-\int_{\Omega}\lambda\delta\left(\sum_{\xi=1}^{N}\phi^{(\xi)}\right)\Bigg|_{\mathbf{x}}\mathrm{d}V_0\right]\mathrm{d}t=0. \tag{5}$$

The physical interpretation of λ and μ_ξ is also discussed by Bedford and Drumheller (1978). Resorting to the standard notion of Lagrange multipliers as generalized forces ensuring the constraints to be satisfied, and to some considerations on pressure force balances, the authors justify the interpretation of λ as an *interface pressure* between constituents, and infer for μ_ξ the relationship $\mu_\xi=\dfrac{p^{(\xi)}\phi^{(\xi)}}{J^{(\xi)}}$, where $p^{(\xi)}$ indicates the pressure of the ξth constituent.

In this respect, it is important to remark that the mechanical consistency of the choice of incorporating of the effect of constraints in a variational framework has been subjected to debate and objections between researchers. In their valuable review, Bedford and Drumheller (1983) recall a criticism by Truesdell and Toupin (1960) (pp. 594, 595) who have indeed observed that incorporating the effect of constraints in variational principles "... is a somewhat dubious blessing". Bedford and Drumheller have rebutted that the volume fraction constraint does not entail ill-posedness issues and have remarked that the admissibility and usefulness of the volume fraction constraint in multiphase theories can be standardly accepted as a continuum mechanical analogue to the treatment of connections between rigid parts in the variational description of the mechanics of rigid bodies.

3 Most Recent Theories

In more recent years, researches on multiphase theories on a variational macroscopic basis have still continued to appear in the specialized literature. Referring the readers to the original papers for further details, a brief account of these theories and of the key ideas is given in this subsection, proceeding in chronological order.

3.1 Variational Theories by Lopatnikov and Co-workers

The Least Action principle has been employed by Lopatnikov and coworkers to obtain continuum governing equations for binary poroelastic mixtures (Lopatnikov and Cheng 2004; Lopatnikov and Gillespie 2010). Important differences with the framework by Bedford and Drumheller are the following:

- As a peculiar feature of this formulation, distinction is characteristically made between a notion of *internal strain tensor* and a notion of *external strain tensor*. For the definition of these quantities, the reader is referred to the original papers where these concepts are introduced (Lopatnikov and Cheng 2002, 2004; Lopatnikov and Gillespie 2010). Relationships between variations of external and internal parameters of the material are introduced and referred to as *material structural equations*. Such relations have a constitutive nature and, hence, appear to be medium-dependent. Lopatnikov and Gillespie (2010) discuss several options for their definition (see p. 482 therein).
- This theory is essentially formulated in infinitesimal displacements.
- The least Action condition is formulated without making explicit statement of the recourse to Lagrange multipliers, even if the theory contains constraints for the variation fields. Specifically, the system of governing equations contains mass conservation equations. Most importantly, the authors infer from mass conservation relationships between variation of porosity $\delta\phi^{(f)}$ and $\delta\hat{\rho}^{(f)}$ involving also the gradient of porosity $\phi^{(f)}\nabla$.

Lopatnikov and Gillespie remark that the presence of a dependence upon $\phi^{(f)}\nabla$ in their mass conservation relation is an important difference with respect to other previously proposed multiphase variational frameworks such as the one by Bedford and Drumheller. Actually, they show that, in nonhomogenous media, an additional volume force interaction between solid and fluid phases appears in the governing equations. This force, which is proportional to $\phi^{(f)}\nabla$, is traced back by the authors to an interaction force term deduced earlier by Nikolaevskiy (see Nikolaevskiy 2005), based on phenomenological reasoning.

This theory is next deployed to analyze the equilibrium state of a fluid and elastic penetrable material, encapsulated in a rigid volume (Lopatnikov and Gillespie 2011). In Lopatnikov and Gillespie (2012) the derivation of interfacial conditions, compatible with the governing differential equations of the theory, is presented.

3.2 Variational Higher Gradient Theories by dell'Isola and Co-workers

An investigation of porous media following a consistent variational approach is the one pursued by Madeo et al. (2013), dell'Isola et al. (2009), Sciarra et al. (2007), dell'Isola et al. (2005a, b, 1998).

In dell'Isola et al. (1998), a micro–macro identification is indeed performed for a compaction of grounds with fluid inclusions with the fluid being confined into the pores. The work particularly focuses on the effect of a length-scale l characterizing pore size. The model is a microstructured continuum of the type introduced by Eringen, and the main result is the dependence of evolution equations on the length l.

In Sciarra et al. (2005), the behavior of a sponge under an increase of the outside fluid pressure is studied by using the Principle of Virtual Power, with second gradients of the displacement included as a further deformation measure. In particular, a simple idea is introduced: the boundary pressure is divided between the solid and fluid pressures, $p_f = d_f p^{ext}$, $p_s = d_s p^{ext}$ with $d_f + d_s = 1$; quantity, p_{ext} is the external pressure and d_f and d_s are coefficients which depend on the constituent apparent densities, regarded as state parameters, under the condition that the work performed by these tractions vanishes in every cyclic process over the parameter space. This condition restricts the permissible constitutive relations for the dividing coefficient, which turns out to be characterized by a single material parameter. Moreover, a stability analysis of the solutions is performed by Sciarra et al. (2005).

In dell'Isola et al. (2009), a (classical) solid fluid mixture is studied in the framework of an extended Hamilton–Rayleigh principle. A general set of boundary conditions at fluid-permeable interfaces between dissimilar fluid-filled porous matrices is established, including jump conditions, friction and inertia effects. In particular, solid and fluid domains $B_s \subset \Re^3$ and $B_f \subset \Re^3$ are introduced, as well as the maps

$$\boldsymbol{\chi}_s : B_s \times (0, T) \to \Re^3 \qquad \boldsymbol{\chi}_f : B_f \times (0, T) \to \Re^3, \tag{6}$$

which represent the (time dependent) placement of the solid and fluid constituent; the motion of the fluid inside the solid matrix is described by the function

$$\boldsymbol{\chi}_{sf} : B_s \times (0, T) \to B_f. \tag{7}$$

General motion equations relative to a representative elementary volume are then derived through lengthy computations:

$$\begin{gathered} -\left(\bar{\rho}^{(s)}\dot{\mathbf{v}}_s + \bar{\rho}^{(f)}\dot{\mathbf{v}}_f^{Ⓢ}\right) + div\left(\mathbf{F}_s^T \cdot \frac{\partial \Psi}{\partial \mathbf{E}}\right) - \frac{\partial \Psi}{\partial \boldsymbol{\chi}_s} = -div\left(J_s(\boldsymbol{\Pi}_f^{Ⓢ})^T \cdot \mathbf{F}_s^{-T}\right) \\ \bar{\rho}^{(f)}\left[\mathbf{F}_s^T \cdot \dot{\mathbf{v}}_f^{Ⓢ} + \nabla\left(\frac{\partial \Psi}{\partial \bar{\rho}^{(f)}}\right)\right] = \mathbf{F}_s^T \cdot \left[J_s \kappa^{Ⓢ} - div\left(J_s(\boldsymbol{\Pi}^{Ⓢ} - \boldsymbol{\Pi}_f^{Ⓢ})^T \cdot \mathbf{F}_s^{-T}\right)\right] \end{gathered} \tag{8}$$

with the following boundary conditions:

$$\begin{gathered} \left[\left|\mathbf{F}_s \cdot \frac{\partial \Psi}{\partial \mathbf{E}} - (\mathbf{v}_f^{Ⓢ} - \mathbf{v}_s) \otimes \mathbf{D} + J_s(\boldsymbol{\Pi}_f^{Ⓢ})^T \cdot \mathbf{F}_s^{-T}\right|\right] \cdot \mathbf{N}_s = \mathbf{o} \\ \left[\left|\mathbf{G}_s^{-T} \cdot \left(\bar{\rho}^{(f)} \frac{\partial \Psi}{\partial \bar{\rho}^{(f)}} \mathbf{N}_s - \frac{1}{2}\bar{\rho}^{(f)}\left(\mathbf{v}_f^{Ⓢ}\right)^2 \cdot \mathbf{N}_s + \mathbf{F}_s^T \cdot \mathbf{v}_f^{Ⓢ} \otimes \mathbf{D} \cdot \mathbf{N}_s\right) \cdot \boldsymbol{\Gamma}^{Ⓢ}\right|\right] \\ + \left[\left|\mathbf{G}_s^{-T} \cdot \mathbf{F}_s^T \cdot \left(J_s\left((\boldsymbol{\Pi}^{Ⓢ} - \boldsymbol{\Pi}_f^{Ⓢ})^T\right) \cdot \mathbf{F}_s^{-T} \cdot \mathbf{N}_s - ||J_s \mathbf{F}_s^{-T} \cdot \mathbf{N}_s|| \sigma^{Ⓢ}\right) \cdot \boldsymbol{\Gamma}^{Ⓢ}\right|\right] = \mathbf{o}. \end{gathered}$$

In the previous equations, the following definitions are used: $\mathbf{F}_i = \nabla \chi_i$, $\mathbf{G}_s = \nabla \chi_{sf}$, $\mathbf{v}_i = \frac{\partial \chi_i}{\partial t}$, $\mathbf{u}_i = \frac{\partial \chi_{sf}}{\partial t}$. Moreover, in the general case Ψ is the sum of a non homogeneous deformation energy potential $\Psi_i(\mathbf{E}, \bar{\rho}^{(f)}, \mathbf{X}_s)$ and a potential accounting for external body forces $\Psi_g = (\bar{\rho}^{(s)} + \bar{\rho}^{(f)})E_p(\chi_s, \mathbf{X}_s)$. Finally, $J_s = \det \mathbf{F}_s$, $\mathbf{E}$ is the Green-Lagrange strain tensor, $\boldsymbol{\Pi}$ is the Brinkman stress tensor, $\boldsymbol{\Pi}_f$ is the fluid viscous stress tensor and the acceleration fields $\dot{\mathbf{v}}_s$, $\dot{\mathbf{v}}_f$ are the time derivatives of $\mathbf{v}_s$ and $\mathbf{v}_f$ respectively, $\bar{\rho}^{(s)}$ and $\bar{\rho}^{(f)}$ are apparent mass densities for the solid and the fluid, and the symbol Ⓢ denotes the transport of a tensor field from the configuration where it is defined to B_s; for the meaning of the remaining symbols κ, σ, E_p, for the meaning of the $(\cdot)$ operations and for other details we refer the reader to the paper, and especially to the technical Appendices in (dell'Isola et al. 2009).

3.3 The VMTPM Framework and the Extrinsic/Intrinsic Treatment

Among more recent contributions, a two-phase poroelastic formulation, also based on the least-action principle, has been proposed in (Serpieri and Rosati 2011) and in (Serpieri 2011). Chronologically, such formulation is subsequent to the works of Lopatnikov and co-workers, and dell'Isola and co-workers. This theory, hereby abbreviated in VMTPM (Variational Macroscopic Theory of Porous Media), consists of an application to poroelastic problems of a generalized continuum formulation with additional kinematic descriptors, in the wake of the ideas of Mindlin and Beford and Drumheller. However, in contrast to previous applications of generalized continua theories to poroelastic problems, the kinematic of VMTPM is enriched with a so-called scalar field of *intrinsic* volumetric strain $\hat{J}^{(s)}$ in place of a porosity field. More specifically, $\hat{J}^{(s)}$ is an additional macroscopic scalar field, introduced on a purely kinematic rationale, which essentially corresponds to the ratio $\hat{\rho}^{(s)}/\hat{\rho}_0^{(s)}$ between 'true' densities before and after deformation. This field is independent from the primary macroscopic volumetric strain measure $\bar{J}^{(s)} = \det(\nabla\chi)$ which remains instead ordinarily defined as the determinant of the macroscopic deformation gradient, and termed *extrinsic* volumetric strain in order to remark its difference with $\hat{J}^{(s)}$. It should be noted that $\hat{J}^{(s)}$ has a direct relation with the porosity field. In a region of a porous medium undergoing a macroscopically homogeneous deformation, the value of $\hat{J}^{(s)}$ can be macroscopically measured by the following relation which links $\hat{J}^{(s)}$ to the porosities before ($\phi_0^{(f)}$) and after deformation ($\phi^{(f)}$):

$$\hat{J}^{(s)} = \bar{J}^{(s)}\left(1 - \phi^{(f)}\right) / \left(1 - \phi_0^{(f)}\right). \tag{9}$$

Hence if the medium is saturated and with completely interconnected pores, the measurement of $\hat{J}^{(s)}$ can be translated into the measurement of the fluid leaving or entering this region as a consequence of a loading-induced deformation.

Based on this choice of extrinsic/intrinsic kinematic descriptors, the associated stress measures consist of an extrinsic stress tensor, work-associated with the extrinsic strain ($\check{\boldsymbol{\sigma}}^{(s)}$), and an scalar intrinsic pressure $\hat{p}^{(s)}$ (Serpieri et al. 2013, 2015; Serpieri and Travascio 2015; Travascio et al. 2015). In such works it is shown that, in undrained conditions (i.e., when no relative solid-fluid motion takes place in a region of a biphasic mixture), VMTPM predicts that the external stress, the fluid pressure, and the stress tensor work-associated with the extrinsic strain of the solid phase are partitioned according to a relation which is formally strictly compliant with Terzaghi's law, irrespective of the microstructural and constitutive features of a given medium.

In (Serpieri and Travascio 2015) and (Travascio et al. 2015) the constitutive response for isotropic media is found to be strongly determined by an additional dimensionless parameter $\bar{k}_r$ with bounds $-1 \leq \bar{k}_r \leq 0$, in a way similar to the role played by the Poisson's coefficient in characterizing the isotropic response of a single continuum medium. In particular, $\bar{k}_r$ appears in the expression for linear isotropic media in undrained conditions (i.e., fast loading) of Skempton's coefficient B, defined as the ratio between the induced pressure p of the interstitial fluid and the applied stress, $t_x^{(ext)}$ (Skempton 1954). Specifically:

$$B = \frac{p}{t_x^{(ext)}} = -\frac{\left(1+\bar{k}_r\right)\hat{k}_{sf}}{\left[2\bar{\mu}+\bar{\lambda}+\left(1+\bar{k}_r\right)^2\hat{k}_{sf}\right]}, \tag{10}$$

where $\bar{\mu}$ and $\bar{\lambda}$ are Lamé moduli and $\hat{k}_{sf}$ is a coupling modulus of intrinsic stiffness in series. The reader is again referred to Serpieri and Travascio (2015) for an exhaustive definition of these parameters on a variational basis. Serpieri and Travascio (2015) have found a peculiar mechanical behavior predicted by VMTMP which is discriminated by $\bar{k}_r$: the extrinsic pressure can actually become *tensile* (negative) even in presence of compressive external stresses under specific values of $\bar{k}_r$. Although such behavior may appear counterintuitive at first sight (by drawing a straightforward parallel with the traditional Cauchy stress tensor in single continuum mechanics), it is shown in the referenced work that such condition entails no violation of positive definiteness of strain energy, so that compressive external tractions always induce compressive strains and the interstitial fluid pressures is always positive and compressive.

Also, Travascio et al. (2015) have reported another unique feature characterizing VMTPM mixtures and, once more, modulated by $\bar{k}_r$: during displacement-controlled static compression, the mixture can express either a *stress-relaxing* or a *stress-tensing* behavior. The stress relaxation is a well-known phenomenon in poroelasticity, whose description has been documented in several studies (Mow et al. 1980; Ehlers and Bluhm 2013): the solid stress increases as the compression is applied; subsequently, fluid redistribution within the mixture occurs, and the stress relaxes to an equilibrium value which is held indefinitely, as long as the system is compressed. The behavior of stress tensing mixtures is substantially different: during compression, the solid stress progressively tenses upon reaching, in its drained state, an equilibrium value once again depending on applied deformation and stiffness of the solid phase. As a further confirmation of the important role played by $\bar{k}_r$, Travascio et al. (2013) showed in a numerical study simulating uniaxial stress relaxation tests on bovine articular cartilage that the consolidation time of the tissue reduces three-fold when $\bar{k}_r$ varies from 0 to -0.25.

4 Conclusions

The present survey on variational macroscopic continuum approaches to multiphase poroelasticity highlighted the existence of fundamental features shared by the theories reviewed in this paper. Also, several aspects have been pointed out, where agreement between the surveyed theories is not found. As such, these aspects deserve further investigation by the generalized continua community.

A fundamental feature which almost all the theories herein presented have in common is the resort to kinematics with additional descriptors (i.e., porosity, intrinsic strain, etc.) for a proper formulation of the problem. In this respect, generalized continua models appear to be the natural setting to properly address the multiphase problem, even in absence of a specific focus on microstructural or multi-scale effects.

On the other hand, some important still-open issues can be identified, where further investigation is needed either to assess the higher degree of mechanical consistency and of predictive capabilities of any of the existing frameworks over the others, or to formulate more comprehensive theories. In particular, the following issues are considered to be relevant:

- The role of constraints in relation to the variational treatment, with special reference to mass balance; in particular the well-posedness of the variational statement of the problem in presence of mass balance constraints for the primary fields appears to be a relevant research issue.
- In variational theories making use of Lagrange multipliers, an assessment of the physical meaning of stress quantities in relation to boundary data and to the macroscopic measurement process could be a relevant research endeavor.
- Even if the set of Euler–Lagrange equations for multiphase problems appears to be very broad (as very broad are the possibilities of conceiving enriched kinematics in generalized continua frameworks) an important objective for the generalized continua community should be the agreement on a set of minimal medium-independent equilibrium equations. Moreover, any new theory should be downward compatible with such equations.
- The theory should be based on the minimum possible number of parameters, which should have a clear physical-mechanical meaning. In addition, their experimental characterization should be possible.
- The identification of a generally agreed set of governing balance equations necessary to achieve a consistent formulation of compressible poroelasticity in a vari-

ational multiphase framework could benefit from contributions coming from all disciplines (e.g., theoretical mechanics, geomechanics, biomechanics, etc.) with the aim of identifying appropriate benchmark programs for validation of continuum poroelasticity theories.

References

Aizicovici S, Aron M (1977) A variational theorem in the linear theory of mixtures of two elastic solids. The quasi-static case. Acta Mech 27(1):275–280

Albers B, Wilmański K (2006) Influence of coupling through porosity changes on the propagation of acoustic waves in linear poroelastic materials. Arch Mech 58(4–5):313–325

Andreaus U, Giorgio I, Lekszycki T (2014) A 2-d continuum model of a mixture of bone tissue and bio-resorbable material for simulating mass density redistribution under load slowly variable in time. Zeitschrift für Angewandte Mathematik und Mechanik 94(12):978–1000

Ateshian GA, Ricken T (2010) Multigenerational interstitial growth of biological tissues. Biomech Model Mechanobiol 9(6):689–702

Baveye PC (2013) Comment on "Averaging theory for description of environmental problems: What have we learned?" by William G. Gray, Cass T. Miller, and Bernhard A. Schrefler. Adv Water Resour 52:328–330

Bear J, Corapcioglu MY (2012) Fundamentals of transport phenomena in porous media, vol 82. Springer Science and Business Media, Berlin

Bedford A, Drumheller D (1978) A variational theory of immiscible mixtures. Arch Ration Mech Anal 68(1):37–51

Bedford A, Drumheller D (1979) A variational theory of porous media. Int J Solids Struct 15(12):967–980

Bedford A, Drumheller DS (1983) Theories of immiscible and structured mixtures. Int J Eng Sci 21(8):863–960

Berdichevsky V (2009) Variational principles of continuum mechanics. Springer, Berlin

Biot M (1972) Theory of finite deformations of porous solids. Indiana Univ Math J 21(7):597–620

Biot M (1977) Variational Lagrangian-thermodynamics of nonisothermal finite strain mechanics of porous solids and thermomolecular diffusion. Int J Solids Struct 13(6):579–597

Biot MA (1941) General theory of three-dimensional consolidation. J Appl Phys 12(2):155–164

Biot MA (1956) Theory of propagation of elastic waves in a fluid-saturated porous solid. i. low-frequency range. J Acoust Soc Am 28(2):168–178

Biot MA (1962) Mechanics of deformation and acoustic propagation in porous media. J Appl Phys 33(4):1482–1498

Bowen RM (1982) Compressible porous media models by use of the theory of mixtures. Int J Eng Sci 20(6):697–735

Cazzani A, Malagù M, Turco E (2014) Isogeometric analysis of plane-curved beams. Math Mech Solids. doi:10.1177/1081286514531265

Cosserat E, Cosserat F (1909) Théorie des Corps Déformables (Theory of deformable structures). Hermann and Fils, Paris

Coussy O, Dormieux L, Detournay E (1998) From mixture theory to Biot's approach for porous media. Int J Solids Struct 35(34):4619–4635

Cowin S, Goodman M (1976) A variational principle for granular materials. ZAMM-J Appl Math Mech/Zeitschrift für Angewandte Mathematik und Mechanik 56(7):281–286

Cowin SC (1999) Bone poroelasticity. J Biomech 32(3):217–238

Cuomo M, Contrafatto L, Greco L (2014) A variational model based on isogeometric interpolation for the analysis of cracked bodies. Int J Eng Sci 80:173–188

de Boer R (1996) Highlights in the historical development of the porous media theory: toward a consistent macroscopic theory. Appl Mech Rev 49(4):201–262
de Boer R (2005) Theoretical poroelasticity—a new approach. Chaos, Solitons Fractals 25(4):861–878
dell'Isola F, Placidi L (2012) Variational principles are a powerful tool also for formulating field theories. CISM Courses and Lectures, vol 535. Springer, Berlin
dell'Isola F, Rosa L, Wozniak C (1998) A micro-structured continuum modelling compacting fluid-saturated grounds: the effects of pore-size scale parameter. Acta Mech 127(1–4):165–182
dell'Isola F, Sciarra G, Coussy O (2005a) A second gradient theory for deformable fluid-saturated porous media. In: Poromechanics III: Biot Centennial (1905-2005)—Proceedings of the 3rd Biot conference on poromechanics, pp 135–140
dell'Isola F, Sciarra G, Romesh B (2005b) A second gradient model for deformable porous matrices filled with an inviscid fluid. Solid Mech Appl 125:221–229
dell'Isola F, Madeo A, Seppecher P (2009) Boundary conditions at fluid-permeable interfaces in porous media: a variational approach. Int J Solids Struct 46(17):3150–3164
dell'Isola F, Steigmann D, Della Corte A (2015) Synthesis of complex structures. Designing micro-structure to deliver targeted macro-scale response. Appl Mech Rev. doi:10.1115/1.4032206
Diebels S (1999) A micropolar theory of porous media: constitutive modelling. Transp Porous Media 34(1–3):193–208
Drumheller DS (1978) The theoretical treatment of a porous solid using a mixture theory. Int J Solids Struct 14(6):441–456
Duhem P (1893) Dissolutions et mélanges. 2ème mémoire, Les propriétés physiques des dissolutions. Au siège des Facultés (Lille)
Eckart C (1960) Variation principles of hydrodynamics. Phys Fluids (1958-1988) 3(3):421–427
Ehlers W, Bluhm J (2013) Porous media: theory, experiments and numerical applications. Springer Science and Business Media, Berlin
Eringen AC (1968) Mechanics of micromorphic continua. Springer, Berlin
Eringen AC, Kafadar CB (1976) Polar field theories. Academic Press, Cambridge
Fillunger P (1936) Erdbaumechanik?. Selbstverl. d. Verf., Wien
Finlayson BA (2013) The method of weighted residuals and variational principles, vol 73. SIAM, Philadelphia
Gajo A (2010) A general approach to isothermal hyperelastic modelling of saturated porous media at finite strains with compressible solid constituents. In: Proceedings of the Royal Society of London A: mathematical, physical and engineering sciences, The Royal Society
Giorgio I, Andreaus U, Madeo A (2014) The influence of different loads on the remodeling process of a bone and bioresorbable material mixture with voids. Contin Mech Thermodyn 28(1–2):21–40
Goodman M, Cowin S (1972) A continuum theory for granular materials. Arch Ration Mech Anal 44(4):249–266
Gray WG, Miller CT, Schrefler BA (2013a) Averaging theory for description of environmental problems: what have we learned? Adv Water Resour 51:123–138
Gray WG, Miller CT, Schrefler BA (2013b) Response to comment on "averaging theory for description of environmental problems: what have we learned". Adv Water Resour 51:331–333
Greco L, Cuomo M (2014) An implicit G1 multi patch B-spline interpolation for Kirchhoff-Love space rod. Comput Methods Appl Mech Eng 269:173–197
Greco L, Cuomo M (2016) An isogeometric implicit G1 mixed finite element for Kirchhoff space rods. Comput Methods Appl Mech Eng 298:325–349
Gu W, Lai W, Mow V (1998) A mixture theory for charged-hydrated soft tissues containing multi-electrolytes: passive transport and swelling behaviors. J Biomech Eng 120(2):169–180
Hassanizadeh SM, Gray WG (1990) Mechanics and thermodynamics of multiphase flow in porous media including interphase boundaries. Adv Water Resour 13(4):169–186
Herivel JW (1955) The derivation of the equations of motion of an ideal fluid by Hamilton's principle. In: Mathematical Proceedings of the Cambridge Philosophical Society, Cambridge University Press 51(02):344–349

Hughes TJ, Cottrell JA, Bazilevs Y (2005) Isogeometric analysis: cad, finite elements, nurbs, exact geometry and mesh refinement. Comput Methods Appl Mech Eng 194(39):4135–4195
Huyghe JM, Janssen J (1997) Quadriphasic mechanics of swelling incompressible porous media. Int J Eng Sci 35(8):793–802
Kenyon DE (1976) Thermostatics of solid-fluid mixtures. Arch Ration Mech Anal 62(2):117–129
Lai W, Hou J, Mow V (1991) A triphasic theory for the swelling and deformation behaviors of articular cartilage. J Biomech Eng 113(3):245–258
Lanczos C (1970) The variational principles of mechanics, vol 4. Courier Corporation, North Chelmsford
Landau L, Lifshitz E (1976) Mechanics: vol 1 (Course of theoretical physics). Butterworth-Heinemann, Oxford
Leech C (1977) Hamilton's principle applied to fluid mechanics. Q J Mech Appl Math 30(1):107–130
Lopatnikov S, Cheng A (2002) Variational formulation of fluid infiltrated porous material in thermal and mechanical equilibrium. Mech Mater 34(11):685–704
Lopatnikov S, Cheng A (2004) Macroscopic Lagrangian formulation of poroelasticity with porosity dynamics. J Mech Phys Solids 52(12):2801–2839
Lopatnikov S, Gillespie J (2010) Poroelasticity-I: governing equations of the mechanics of fluid-saturated porous materials. Transp Porous Media 84(2):471–492
Lopatnikov S, Gillespie J (2011) Poroelasticity-II: on the equilibrium state of the fluid-filled penetrable poroelastic body. Transp Porous Media 89(3):475–486
Lopatnikov S, Gillespie J (2012) Poroelasticity-III: conditions on the interfaces. Transp Porous Media 93(3):597–607
Madeo A, Lekszycki T, dell'Isola F (2011) A continuum model for the bio-mechanical interactions between living tissue and bio-resorbable graft after bone reconstructive surgery. Comptes Rendus - Mecanique 339(10):625–640
Madeo A, dell'Isola F, Darve F (2013) A continuum model for deformable, second gradient porous media partially saturated with compressible fluids. J Mech Phys Solids 61(11):2196–2211
Mindlin R (1964) Micro-structure in linear elasticity. Arch Ration Mech Anal 16(1):51–78
Moiseiwitsch BL (2013) Variational principles. Courier Corporation, North Chelmsford
Mow V, Kuei S, Lai W, Armstrong C (1980) Biphasic creep and stress relaxation of articular cartilage in compression: theory and experiments. J Biomech Eng 102(1):73–84
Nikolaevskiy V (2005) Biot-Frenkel poromechanics in Russia (review). J Eng Mech 131(9):888–897
Nunziato JW, Walsh EK (1980) On ideal multiphase mixtures with chemical reactions and diffusion. Arch Ration Mech Anal 73(4):285–311
Nur A, Byerlee J (1971) An exact effective stress law for elastic deformation of rock with fluids. J Geophys Res 76(26):6414–6419
Oden JT, Reddy JN (2012) Variational methods in theoretical mechanics. Springer Science and Business Media, Berlin
Passman S (1977) Mixtures of granular materials. Int J Eng Sci 15(2):117–129
Schrefler B (2002) Mechanics and thermodynamics of saturated/unsaturated porous materials and quantitative solutions. Appl Mech Revi 55(4):351–388
Sciarra G, dell'Isola F, Hutter K (2005) Dilatancy and compaction around a cylindrical cavern leached-out in a fluid saturated salt rock. In: Poromechanics III: Biot Centennial (1905-2005) - Proceedings of the 3rd Biot Conference on Poromechanics, pp 681–687
Sciarra G, dell'Isola F, Coussy O (2007) Second gradient poromechanics. Int J Solids Struct 44(20):6607–6629
Serpieri R (2011) A rational procedure for the experimental evaluation of the elastic coefficients in a linearized formulation of biphasic media with compressible constituents. Transp Porous Media 90(2):479–508
Serpieri R, Rosati L (2011) Formulation of a finite deformation model for the dynamic response of open cell biphasic media. J Mech Phys Solids 59(4):841–862

Serpieri R, Travascio F (2015) General quantitative analysis of stress partitioning and boundary conditions in undrained biphasic porous media via a purely macroscopic and purely variational approach. Contin Mech Thermodyn 28(1–2):235–261

Serpieri R, Travascio F, Asfour S (2013) Fundamental solutions for a coupled formulation of porous biphasic media with compressible solid and fluid phases. In: Computational methods for coupled problems in Science and Engineering V -A Conference Celebrating the 60th Birthday of Eugenio Onate, COUPLED PROBLEMS, pp 1142–1153

Serpieri R, Travascio F, Asfour S, Rosati L (2015) Variationally consistent derivation of the stress partitioning law in saturated porous media. Int J Solids Struct 56–57:235–247

Skempton A (1954) The pore-pressure coefficients a and b. Geotechnique 4(4):143–147

Svendsen B, Hutter K (1995) On the thermodynamics of a mixture of isotropic materials with constraints. Int J Eng Sci 33(14):2021–2054

Terzaghi K (1936) The shearing resistance of saturated soils and the angle between the planes of shear. In: Proceedings of the international conference on soil mechanics and foundation engineering, Cambridge (MA), USA

Travascio F, Serpieri R, Asfour S (2013) Articular cartilage biomechanics modeled via an intrinsically compressible biphasic model: implications and deviations from an incompressible biphasic approach. In: Proceedings of the ASME 2013 summer bioengineering conference, American Society of Mechanical Engineers, pp V01BT55A004–V01BT55A004

Travascio F, Eltoukhy M, Cami S, Asfour S (2014) Altered mechano-chemical environment in hip articular cartilage: effect of obesity. Biomech Model Mechanobiol 13(5):945–959

Travascio F, Asfour S, Serpieri R, Rosati L (2015) Analysis of the consolidation problem of compressible porous media by a macroscopic variational continuum approach. Math Mech Solids. doi:10.1177/1081286515616049

Truesdell C (1969) Rational thermodynamics: a course of lectures on selected topics. McGraw-Hill, New York

Truesdell C, Toupin R (1960) The classical field theories. Springer, Berlin

Wilmański K (1998) A thermodynamic model of compressible porous materials with the balance equation of porosity. Transp Porous Media 32(1):21–47

Wilmański K (2006) A few remarks on Biot's model and linear acoustics of poroelastic saturated materials. Soil Dyn Earthq Eng 26(6):509–536

Buckling of Sandwich Tube with Foam Core Under Combined Loading

Denis N. Sheydakov and Nikolay E. Sheydakov

Abstract In the framework of a general stability theory for three-dimensional bodies the buckling analysis is carried out for the nonlinearly elastic three-layer cylindrical tube subjected to axial compression under internal or external pressure. It is assumed that the middle layer (core) of the tube is made of metal or polymer foam, and to describe its behavior the model of micropolar continuum is used. Such approach allows to study in detail the influence of foam microstructure on the deformation stability, which is especially important when the macroscopic dimensions of the tube are comparable with the average size of the foam cells. The inner and outer layers (coatings) of the tube are assumed to be made of the classic non-polar materials. Applying linearization the neutral equilibrium equations have been derived, which describe the perturbed state of the cylindrical sandwich tube. By solving these equations numerically for some specific materials, the critical curves and corresponding buckling modes have been found and the stability regions have been constructed in the planes of loading parameters (relative axial compression and internal or external pressure). Using the obtained results, the influence of coatings properties, as well as the overall size of the tube, on the loss of stability has been analyzed.

Keywords Nonlinear elasticity · Micropolar medium · Stability of deformable bodies · Sandwich tube · Foam

---

D.N. Sheydakov (✉)
South Scientific Center of Russian Academy of Sciences, Chekhova Ave. 41, 344006 Rostov-on-Don, Russia
e-mail: sheidakov@mail.ru

N.E. Sheydakov
Rostov State University of Economics, Bolshaya Sadovaya St. 69, 344002 Rostov-on-Don, Russia
e-mail: sheidakov@donpac.ru

H. Altenbach and S. Forest (eds.), *Generalized Continua as Models for Classical and Advanced Materials*, Advanced Structured Materials 42,
DOI 10.1007/978-3-319-31721-2_18

1 Introduction

The problem of equilibrium stability for the deformable bodies is of major importance both from theoretical and practical points of view, because the depletion of the load-bearing capacity and collapse of engineering structures often occurs due to the buckling under external loads. In the case of elastic medium, the stability theory is extensively developed for the classic non-polar materials. However, due to the increasing number of new construction materials, the problem of stability analysis for bodies with a microstructure becomes important. Examples of such new materials are metal and polymer foams (Ashby et al. 2000; Banhart 2000; Banhart and Fleck 1999; Degischer and Kriszt 2002; Gibson and Ashby 1997), which are widely used in the modern aerospace and automotive industries since they combine low weight, high specific strength and excellent possibilities to absorb energy. As a rule, constructions made of foams have a sandwich structure (foam core covered by hard and stiff shells). This is necessary for corrosion or thermal protection, and optimization of mechanical properties in the process of loading. Due to the microstructure influence, the behavior of foams cannot be adequately described within the framework of classic theory of elasticity. One approach to modeling them is to use the Cosserat continuum (Altenbach et al. 2010; Cosserat and Cosserat 1909; Eringen 1999; Kafadar and Eringen 1971; Maugin 1998; Toupin 1964), i.e. medium with couple stresses and rotational degrees of freedom. This allows, in particular, to describe the size effects observed experimentally for metal and polymer foams (Lakes 1995, 1986). Given the above, in the present paper we have carried out the stability analysis for a common construction element—a sandwich tube with foam core.

2 Initial Strain State of Sandwich Tube

We consider an elastic sandwich tube of length l, inner radius r_- and outer radius r_+ (see Fig. 1). The middle layer of the tube (core) of thickness h is made of metal or polymer foam, and to described its behavior the micropolar continuum model is used. The inner and outer layers (coatings) have thicknesses h_- and h_+, respectively, and are assumed to be made of the classic non-polar materials. Here and below, the indices "−" and "+" will denote quantities related to the inner and outer coatings. The quantities without these indices will relate to the foam core of the sandwich tube. In the case of axial compression of the tube under internal or external hydrostatic pressure, the position of a medium particle in the strained state is given by the radius vectors $\vec{R}_-$, $\vec{R}$ and $\vec{R}_+$ (Lurie 1990; Sheydakov 2010; Zubov 1997) $(c_- = r_- + h_-,\ c_+ = r_+ - h_+)$:

$$R = \begin{cases} f_-(r), & r_- \le r \le c_-, \\ f(r), & c_- \le r \le c_+, \\ f_+(r), & c_+ \le r \le r_+, \end{cases} \qquad \begin{array}{ll} \Phi = \varphi, & 0 \le \varphi \le 2\pi, \\ Z = \alpha z, & 0 \le z \le l, \end{array} \tag{1}$$

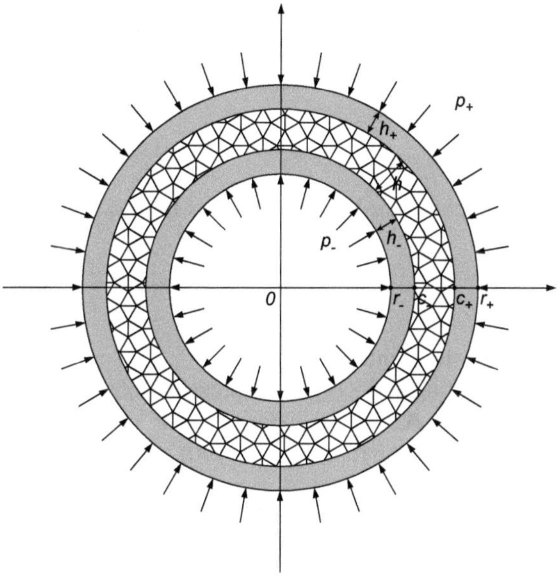

Fig. 1 Cross section of undeformed sandwich tube

$$
\begin{aligned}
&\vec{R}_{-} = f_{-}(r)\,\vec{e}_R + \alpha z\,\vec{e}_Z, && r_{-} \le r \le c_{-},\\
&\vec{R} = f(r)\,\vec{e}_R + \alpha z\,\vec{e}_Z, && c_{-} \le r \le c_{+},\\
&\vec{R}_{+} = f_{+}(r)\,\vec{e}_R + \alpha z\,\vec{e}_Z, && c_{+} \le r \le r_{+}.
\end{aligned} \tag{2}
$$

Here r, φ, z are the cylindrical coordinates in the reference configuration (Lagrangian coordinates), R, Φ, Z are the Eulerian cylindrical coordinates, $\{\vec{e}_r, \vec{e}_\varphi, \vec{e}_z\}$ and $\{\vec{e}_R, \vec{e}_\Phi, \vec{e}_Z\}$ are orthonormal vector bases of Lagrangian and Eulerian coordinates, respectively, α is the given compression ratio along the axis of the tube, $f(r)$ and $f_\pm(r)$ are the unknown functions which characterize the radial deformation of the sandwich tube.

A proper orthogonal tensor of microrotation $\mathbf{H}$, which determines the rotation of the medium particle, is given for the micropolar core ($c_{-} \le r \le c_{+}$):

$$
\mathbf{H} = \vec{e}_r \otimes \vec{e}_R + \vec{e}_\varphi \otimes \vec{e}_\Phi + \vec{e}_z \otimes \vec{e}_Z. \tag{3}
$$

According to the expressions (1) and (2), the deformation gradients $\mathbf{C}_{-}$, $\mathbf{C}$ and $\mathbf{C}_{+}$ are (hereinafter the $'$ denotes the derivative with respect to r):

$$
\begin{aligned}
\mathbf{C} &= \mathrm{Grad}\vec{\mathrm{R}} = f'\vec{e}_r \otimes \vec{e}_R + \frac{f}{r}\vec{e}_\varphi \otimes \vec{e}_\Phi + \alpha\vec{e}_z \otimes \vec{e}_Z,\\
\mathbf{C}_\pm &= \mathrm{Grad}\vec{\mathrm{R}}_\pm = f'_\pm\vec{e}_r \otimes \vec{e}_R + \frac{f_\pm}{r}\vec{e}_\varphi \otimes \vec{e}_\Phi + \alpha\vec{e}_z \otimes \vec{e}_Z,
\end{aligned} \tag{4}
$$

where Grad is the gradient operator in the Lagrangian coordinates.

It follows from relations (3) and (4) that for the micropolar core ($c_- \le r \le c_+$) the wryness tensor $\mathbf{L}$ is equal to zero (Eremeyev and Pietraszkiewicz 2012; Nikitin and Zubov 1998; Pietraszkiewicz and Eremeyev 2009a, b)

$$\mathbf{L} \times \mathbf{I} = -(\mathrm{Grad}\,\mathbf{H}) \cdot \mathbf{H}^{\mathrm{T}} = \mathbf{0}$$

and stretch tensor $\mathbf{Y}$ is expressed as follows

$$\mathbf{Y} = \mathbf{C} \cdot \mathbf{H}^{\mathrm{T}} = f' \vec{e}_r \otimes \vec{e}_r + \frac{f}{r} \vec{e}_\varphi \otimes \vec{e}_\varphi + \alpha \vec{e}_z \otimes \vec{e}_z. \tag{5}$$

Here $\mathbf{I}$ is the identity tensor.

According to Eq. (4), the expressions of stretch tensors $\mathbf{U}_-$, $\mathbf{U}_+$ and macrorotation tensors $\mathbf{A}_-$, $\mathbf{A}_+$ for the inner ($r_- \le r \le c_-$) and outer ($c_+ \le r \le r_+$) non-polar coatings have the form (Lurie 1990)

$$\begin{aligned} \mathbf{U}_\pm &= \left(\mathbf{C}_\pm \cdot \mathbf{C}_\pm^{\mathrm{T}}\right)^{\frac{1}{2}} = f'_\pm \vec{e}_r \otimes \vec{e}_r + \tfrac{f_\pm}{r} \vec{e}_\varphi \otimes \vec{e}_\varphi + \alpha \vec{e}_z \otimes \vec{e}_z, \\ \mathbf{A}_\pm &= \mathbf{U}_\pm^{-1} \cdot \mathbf{C}_\pm = \vec{e}_r \otimes \vec{e}_R + \vec{e}_\varphi \otimes \vec{e}_\Phi + \vec{e}_z \otimes \vec{e}_Z. \end{aligned} \tag{6}$$

We assume that the elastic properties of the sandwich tube are described by the model of physically linear material. In the case of micropolar core, the specific strain energy is a quadratic form of tensors $\mathbf{Y} - \mathbf{I}$ and $\mathbf{L}$ (Eremeyev and Zubov 1994; Lakes 1995):

$$\begin{aligned} W(\mathbf{Y}, \mathbf{L}) &= \frac{1}{2} \lambda \mathrm{tr}^2 (\mathbf{Y} - \mathbf{I}) + \frac{1}{2} [\mu + \kappa] \, \mathrm{tr} \left[(\mathbf{Y} - \mathbf{I}) \cdot (\mathbf{Y} - \mathbf{I})^{\mathrm{T}}\right] \\ &+ \frac{1}{2} \mu \mathrm{tr} (\mathbf{Y} - \mathbf{I})^2 + \frac{1}{2} \gamma_1 \mathrm{tr}^2 \mathbf{L} + \frac{1}{2} \gamma_2 \mathrm{tr} \left(\mathbf{L} \cdot \mathbf{L}^{\mathrm{T}}\right) + \frac{1}{2} \gamma_3 \mathrm{tr}\, \mathbf{L}^2, \end{aligned} \tag{7}$$

while in the case of non-polar coatings, it is a quadratic form of tensor $\mathbf{U}_- - \mathbf{I}$ or $\mathbf{U}_+ - \mathbf{I}$, respectively (Lurie 1990):

$$W_\pm(\mathbf{U}_\pm) = \frac{1}{2} \lambda_\pm \mathrm{tr}^2 (\mathbf{U}_\pm - \mathbf{I}) + \mu_\pm \mathrm{tr} (\mathbf{U}_\pm - \mathbf{I})^2. \tag{8}$$

Here λ, μ and $\lambda_\pm$, $\mu_\pm$ are the Lame parameters for the foam core and coatings, respectively, κ, γ_1, γ_2, γ_3 are micropolar elastic parameters of the foam.

It follows from the expressions (3), (5) and (7) that for the micropolar core the Piola-type couple stress tensor $\mathbf{G}$ is equal to zero at the considered deformation of sandwich tube

$$\mathbf{G} = \frac{\partial W}{\partial \mathbf{L}} \cdot \mathbf{H} = \left[\gamma_1 (\mathrm{tr}\, \mathbf{L}) \mathbf{I} + \gamma_2 \mathbf{L} + \gamma_3 \mathbf{L}^{\mathrm{T}}\right] \cdot \mathbf{H} = \mathbf{0}$$

and Piola-type stress tensor $\mathbf{D}$ is

$$\mathbf{D} = \frac{\partial W}{\partial \mathbf{Y}} \cdot \mathbf{H} = \left[\lambda \mathrm{tr}\,(\mathbf{Y} - \mathbf{I})\,\mathbf{I} + \mu\left(\mathbf{Y}^{\mathrm{T}} - \mathbf{I}\right) + (\mu + \kappa)(\mathbf{Y} - \mathbf{I})\right] \cdot \mathbf{H}$$
$$= \left[\lambda s + \chi\left(f' - 1\right)\right] \vec{e}_r \otimes \vec{e}_R + \left[\lambda s + \chi\left(\frac{f}{r} - 1\right)\right] \vec{e}_\varphi \otimes \vec{e}_\Phi$$
$$+ \left[\lambda s + \chi\,(\alpha - 1)\right] \vec{e}_z \otimes \vec{e}_Z; \quad s = f' + \frac{f}{r} + \alpha - 3, \quad \chi = 2\mu + \kappa. \tag{9}$$

According to the relations (6) and (8), the expressions of Piola stress tensors $\mathbf{D}_-$ and $\mathbf{D}_+$ for non-polar coatings have the form:

$$\mathbf{D}_\pm = \frac{\partial W_\pm}{\partial \mathbf{U}_\pm} \cdot \mathbf{A}_\pm = \left(\lambda_\pm \mathrm{tr}\,(\mathbf{U}_\pm - \mathbf{I})\,\mathbf{I} + 2\mu_\pm (\mathbf{U}_\pm - \mathbf{I})\right) \cdot \mathbf{A}_\pm$$
$$= \left(\lambda_\pm s_\pm + 2\mu_\pm \left(f'_\pm - 1\right)\right) \vec{e}_r \otimes \vec{e}_R + \left(\lambda_\pm s_\pm + 2\mu_\pm \left(\frac{f_\pm}{r} - 1\right)\right) \vec{e}_\varphi \otimes \vec{e}_\Phi$$
$$+ \left(\lambda_\pm s_\pm + 2\mu_\pm (\alpha - 1)\right) \vec{e}_z \otimes \vec{e}_Z; \qquad s_\pm = f'_\pm + \frac{f_\pm}{r} + \alpha - 3. \tag{10}$$

The equilibrium equations for the sandwich tube in the absence of mass forces and moments are written as follows (Eremeyev and Zubov 1994; Zubov 1997)

$$\begin{array}{lll} \mathrm{Div}\mathbf{D}_- = \vec{0}, & & r_- \le r \le c_-, \\ \mathrm{Div}\mathbf{D} = \vec{0}, & \mathrm{Div}\mathbf{G} + \left(\mathbf{C}^{\mathrm{T}} \cdot \mathbf{D}\right)_\times = \vec{0}, & c_- \le r \le c_+, \\ \mathrm{Div}\mathbf{D}_+ = \vec{0}, & & c_+ \le r \le r_+, \end{array} \tag{11}$$

where Div is the divergence in the Lagrangian coordinates. The symbol $_\times$ represents the vector invariant of a second-order tensor.

By solving Eqs. (11) while taking into account the relations (9) and (10), we find the form of unknown functions $f_-(r)$, $f(r)$ and $f_+(r)$:

$$f(r) = C_1 r + \frac{C_2}{r}, \qquad f_\pm(r) = C_1^\pm r + \frac{C_2^\pm}{r}. \tag{12}$$

The constants C_1^-, C_2^-, C_1, C_2, C_1^+, C_2^+ are determined from boundary conditions

$$\begin{array}{ll} \vec{e}_r \cdot \mathbf{D}_\pm\big|_{r=r_\pm} = -p_\pm J_\pm \vec{e}_r \cdot \mathbf{C}_\pm^{-\mathrm{T}}, & J_\pm = \det\mathbf{C}_\pm, \\ \vec{e}_r \cdot \mathbf{D}_\pm\big|_{r=c_\pm} = \vec{e}_r \cdot \mathbf{D}\big|_{r=c_\pm}, & f_\pm(c_\pm) = f(c_\pm), \end{array} \tag{13}$$

which express the action of hydrostatic pressure p_- and p_+ (referred to the unit area of the deformed configuration) on the inner ($r = r_-$) and outer ($r = r_+$) surfaces of the sandwich tube, respectively, as well as a rigid coupling of the foam core with the inner ($r = c_-$) and outer ($r = c_+$) coatings.

Using the relations (4), (9), (10) and (12), the boundary conditions (13) are written as a system of six linear algebraic equations

$$(2\lambda + \chi) C_1 - \frac{\chi}{c_\pm^2} C_2 - 2(\lambda_\pm + \mu_\pm) C_1^\pm + \frac{2\mu_\pm}{c_\pm^2} C_2^\pm = (\alpha - 3)(\lambda_\pm - \lambda) - 2\mu_\pm + \chi,$$

$$(2\lambda_\pm + 2\mu_\pm + \alpha p_\pm) C_1^\pm + \frac{\alpha p_\pm - 2\mu_\pm}{r_\pm^2} C_2^\pm = 2\mu_\pm - \lambda_\pm (\alpha - 3),$$

$$c_\pm C_1 + \frac{1}{c_\pm} C_2 - c_\pm C_1^\pm - \frac{1}{c_\pm} C_2^\pm = 0$$

by solving which we find the unknown constants. In this paper, the obtained general expressions for the constants C_1^-, C_2^-, C_1, C_2, C_1^+, C_2^+ are not presented due to their cumbersomeness.

3 Perturbed State of Sandwich Tube

We assume that in addition to the above-described state of equilibrium for the sandwich tube, there is an infinitely close equilibrium state under the same external loads, which is determined by the radius vector $\vec{R} + \eta\vec{v}$ and microrotation tensor $\mathbf{H} - \eta\mathbf{H} \times \vec{\boldsymbol{\omega}}$ for the micropolar core, and by the radius vectors $\vec{R}_- + \eta\vec{v}_-$ and $\vec{R}_+ + \eta\vec{v}_+$ for non-polar coatings. Here η is a small parameter, $\vec{v}_-$, $\vec{v}$ and $\vec{v}_+$ are vectors of additional displacements, $\vec{\boldsymbol{\omega}}$ is a linear incremental rotation vector, which characterizes the small rotation of the micropolar medium particles, measured from the initial strain state.

The perturbed state of equilibrium for the sandwich tube is described by the equations (Eremeyev and Zubov 1994; Green and Adkins 1960; Ogden 1997):

$$\begin{aligned} &\mathrm{Div}\mathbf{D}^\bullet_- = \vec{0}, \\ &\mathrm{Div}\mathbf{D}^\bullet = \vec{0}, \quad \mathrm{Div}\mathbf{G}^\bullet + \left[\mathrm{Grad}\vec{v}^{\mathrm{T}} \cdot \mathbf{D} + \mathbf{C}^{\mathrm{T}} \cdot \mathbf{D}^\bullet\right]_\times = \vec{0}, \\ &\mathrm{Div}\mathbf{D}^\bullet_+ = \vec{0}, \end{aligned} \tag{14}$$

where $\mathbf{D}^\bullet$ and $\mathbf{G}^\bullet$ are the linearized Piola-type stress and couple stress tensors for the foam core, $\mathbf{D}^\bullet_-$ and $\mathbf{D}^\bullet_+$ are the linearized Piola stress tensors for the coatings. In the case of physically linear micropolar material (7), the following relations are valid for the first two tensors (Eremeyev and Zubov 1994):

$$\begin{aligned} \mathbf{D}^\bullet = \left(\frac{\partial W}{\partial \mathbf{Y}}\right)^\bullet \cdot \mathbf{H} + \frac{\partial W}{\partial \mathbf{Y}} \cdot \mathbf{H}^\bullet = \left(\lambda \left(\mathrm{tr}\,\mathbf{Y}^\bullet\right)\mathbf{I} + (\mu + \kappa)\mathbf{Y}^\bullet + \mu \mathbf{Y}^{\bullet\mathrm{T}}\right) \cdot \mathbf{H} \\ - \left(\lambda \mathrm{tr}\,(\mathbf{Y} - \mathbf{I})\,\mathbf{I} + \mu\left(\mathbf{Y}^{\mathrm{T}} - \mathbf{I}\right) + (\mu + \kappa)(\mathbf{Y} - \mathbf{I})\right) \cdot \mathbf{H} \times \vec{\boldsymbol{\omega}}, \end{aligned} \tag{15}$$

$$\begin{aligned} \mathbf{G}^\bullet = \left(\frac{\partial W}{\partial \mathbf{L}}\right)^\bullet \cdot \mathbf{H} + \frac{\partial W}{\partial \mathbf{L}} \cdot \mathbf{H}^\bullet = \left(\gamma_1 \left(\mathrm{tr}\,\mathbf{L}^\bullet\right)\mathbf{I} + \gamma_2 \mathbf{L}^\bullet + \gamma_3 \mathbf{L}^{\bullet\mathrm{T}}\right) \cdot \mathbf{H} \\ - \left(\gamma_1 (\mathrm{tr}\,\mathbf{L})\,\mathbf{I} + \gamma_2 \mathbf{L} + \gamma_3 \mathbf{L}^{\mathrm{T}}\right) \cdot \mathbf{H} \times \vec{\boldsymbol{\omega}}, \end{aligned} \tag{16}$$

$$\mathbf{Y}^\bullet = (\mathrm{Grad}\vec{v} + \mathbf{C} \times \vec{\omega}) \cdot \mathbf{H}^{\mathrm{T}}, \qquad \mathbf{L}^\bullet = \mathrm{Grad}\, \vec{\omega} \cdot \mathbf{H}^{\mathrm{T}}$$

Here $\mathbf{Y}^\bullet$ is the linearized stretch tensor and $\mathbf{L}^\bullet$ is the linearized wryness tensor for the micropolar core.

Representations of the linearized Piola stress tensors $\mathbf{D}^\bullet_-$ and $\mathbf{D}^\bullet_+$ for the physically linear non-polar material (8) have the form (Sheydakov 2011):

$$\mathbf{D}^\bullet_\pm = \left(\frac{\partial W_\pm}{\partial \mathbf{U}_\pm}\right)^\bullet \cdot \mathbf{A}_\pm + \frac{\partial W_\pm}{\partial \mathbf{U}_\pm} \cdot \mathbf{A}^\bullet_\pm = \left(\lambda_\pm \left(\mathrm{tr}\mathbf{U}^\bullet_\pm\right)\mathbf{I} + 2\mu_\pm \mathbf{U}^\bullet_\pm\right) \cdot \mathbf{A}_\pm$$
$$+ \left(\lambda_\pm \mathrm{tr}\left(\mathbf{U}_\pm - \mathbf{I}\right)\mathbf{I} + 2\mu_\pm \left(\mathbf{U}_\pm - \mathbf{I}\right)\right) \cdot \mathbf{U}_\pm^{-1} \cdot \left(\mathrm{Grad}\, \vec{v}_\pm - \mathbf{U}^\bullet_\pm \cdot \mathbf{A}_\pm\right). \qquad (17)$$

Here $\mathbf{U}^\bullet_-$ and $\mathbf{U}^\bullet_+$ are the linearized stretch tensors for coatings, which can be expressed in terms of the linearized Cauchy-Green deformation tensors $\mathbf{G}^\bullet_-$ and $\mathbf{G}^\bullet_+$:

$$\mathbf{G}^\bullet_\pm = \left(\mathbf{U}_\pm \cdot \mathbf{U}_\pm\right)^\bullet = \mathbf{U}^\bullet_\pm \cdot \mathbf{U}_\pm + \mathbf{U}_\pm \cdot \mathbf{U}^\bullet_\pm,$$
$$\mathbf{G}^\bullet_\pm = \left(\mathbf{C}_\pm \cdot \mathbf{C}^{\mathrm{T}}_\pm\right)^\bullet = \mathrm{Grad}\vec{v}_\pm \cdot \mathbf{C}^{\mathrm{T}}_\pm + \mathbf{C}_\pm \cdot \left(\mathrm{Grad}\vec{v}_\pm\right)^{\mathrm{T}}.$$

The linearized boundary conditions on the inner ($r = r_-$) and outer ($r = r_+$) surfaces of the tube, as well as on the interfaces between the coatings and the foam core ($r = c_\pm$) are written as follows (Sheydakov 2010, 2011):

$$\vec{e}_r \cdot \mathbf{D}^\bullet_\pm\big|_{r=r_\pm} = -p_\pm J_\pm \vec{e}_r \cdot \mathbf{C}^{-\mathrm{T}}_\pm \cdot \left[(\mathrm{div}\, \vec{v}_\pm)\mathbf{I} - \mathrm{grad}\, \vec{v}^{\mathrm{T}}_\pm\right],$$
$$\vec{e}_r \cdot \mathbf{D}^\bullet_\pm\big|_{r=c_\pm} = \vec{e}_r \cdot \mathbf{D}^\bullet\big|_{r=c_\pm} \qquad \vec{e}_r \cdot \mathbf{G}^\bullet\big|_{r=c_\pm} = \vec{0}, \qquad \vec{v}_\pm|_{r=c_\pm} = \vec{v}|_{r=c_\pm}, \qquad (18)$$

where div and grad are the divergence and gradient in the Eulerian coordinates.

We assume that there is no friction at the ends of the sandwich tube ($z = 0, l$) and constant normal displacements are given. This leads to the following linearized end conditions:

- for the inner coating ($r_- \le r \le c_-$)

$$\vec{e}_z \cdot \mathbf{D}^\bullet_- \cdot \vec{e}_R\big|_{z=0,l} = \vec{e}_z \cdot \mathbf{D}^\bullet_- \cdot \vec{e}_\Phi\big|_{z=0,l} = \vec{e}_z \cdot \vec{v}_-|_{z=0,l} = 0, \qquad (19)$$

- for the micropolar core ($c_- \le r \le c_+$)

$$\vec{e}_z \cdot \mathbf{D}^\bullet \cdot \vec{e}_R\big|_{z=0,l} = \vec{e}_z \cdot \mathbf{D}^\bullet \cdot \vec{e}_\Phi\big|_{z=0,l} = \vec{e}_z \cdot \vec{v}|_{z=0,l} = 0,$$
$$\vec{e}_z \cdot \mathbf{G}^\bullet \cdot \vec{e}_Z\big|_{z=0,l} = \vec{e}_r \cdot \vec{\omega}|_{z=0,l} = \vec{e}_\varphi \cdot \vec{\omega}|_{z=0,l} = 0, \qquad (20)$$

- for the outer coating ($c_+ \le r \le r_+$)

$$\vec{e}_z \cdot \mathbf{D}^\bullet_+ \cdot \vec{e}_R\big|_{z=0,l} = \vec{e}_z \cdot \mathbf{D}^\bullet_+ \cdot \vec{e}_\Phi\big|_{z=0,l} = \vec{e}_z \cdot \vec{v}_+|_{z=0,l} = 0. \qquad (21)$$

The vectors of additional displacements $\vec{v}_-$, $\vec{v}$, $\vec{v}_+$ and vector of incremental rotation $\vec{\omega}$ in the basis of Eulerian cylindrical coordinates are written as:

$$\begin{aligned}\vec{v} &= v_R\vec{e}_R + v_\Phi\vec{e}_\Phi + v_Z\vec{e}_Z,\\ \vec{v}_\pm &= v_R^\pm\vec{e}_R + v_\Phi^\pm\vec{e}_\Phi + v_Z^\pm\vec{e}_Z,\\ \vec{\omega} &= \omega_R\vec{e}_R + \omega_\Phi\vec{e}_\Phi + \omega_Z\vec{e}_Z.\end{aligned} \tag{22}$$

To solve the linearized boundary-value problem (14), (18)–(21) for a system of twelve partial differential equations, we use the following substitutions

$$\begin{aligned}v_R &= V_R(r)\cos n\varphi\cos\beta z, & v_R^\pm &= V_R^\pm(r)\cos n\varphi\cos\beta z,\\ v_\Phi &= V_\Phi(r)\sin n\varphi\cos\beta z, & v_\Phi^\pm &= V_\Phi^\pm(r)\sin n\varphi\cos\beta z,\\ v_Z &= V_Z(r)\cos n\varphi\sin\beta z, & v_Z^\pm &= V_Z^\pm(r)\cos n\varphi\sin\beta z,\end{aligned} \tag{23}$$

$$\begin{aligned}\omega_R &= \Omega_R(r)\sin n\varphi\sin\beta z,\\ \omega_\Phi &= \Omega_\Phi(r)\cos n\varphi\sin\beta z,\\ \omega_Z &= \Omega_Z(r)\sin n\varphi\cos\beta z,\end{aligned} \tag{24}$$

$$\beta = \pi m/l, \qquad m, n = 0, 1, 2, \ldots$$

that lead to the separation of variables φ and z in this problem and allow to satisfy the linearized end conditions (19)–(21). As a result, the stability analysis of the sandwich tube with foam core is reduced to solving a linear homogeneous boundary-value problem for a system of twelve ordinary differential equations. A detailed derivation of this equations can be found in the Appendix.

4 Numerical Results

In the present paper, we have carried out the stability analysis for the sandwich tube with identical inner and outer coatings ($h_- = h_+,\ \lambda_- = \lambda_+,\ \mu_- = \mu_+$). At that it was assumed that the tube core is made of dense polyurethane foam. The micropolar elastic parameters for this material have been previously identified by Lakes (for details, see Lakes 1986, 1995):

$$\begin{aligned}\lambda &= 797.3\ \text{MPa}, & \mu &= 99.67\ \text{MPa}, & \kappa &= 8.67\ \text{MPa},\\ \gamma_1 &= -26.65\ \text{N}, & \gamma_2 &= 45.3\ \text{N}, & \gamma_3 &= 34.65\ \text{N}.\end{aligned}$$

Two coating materials were considered–a less stiff polycarbonate

$$\lambda_\pm = 2.3\times 10^3\ \text{MPa}, \quad \mu_\pm = 0.8\times 10^3\ \text{MPa}$$

and a more stiff aluminum

$$\lambda_{\pm} = 61.9 \times 10^3 \text{ MPa}, \quad \mu_{\pm} = 26.2 \times 10^3 \text{ MPa}.$$

For convenience, the following dimensionless parameters were introduced:

- length-to-radius ratio $l^* = l/r_+$;
- thickness-to-radius ratio $H^* = (r_+ - r_-)/r_+$;
- relative axial compression $\delta = 1 - \alpha$;
- relative internal pressure $p_-^* = p_-/\mu$;
- relative external pressure $p_+^* = p_+/\mu$;
- relative thickness of inner coating $h_-^* = h_-/(r_+ - r_-) \cdot 100\,\%$;
- relative thickness of outer coating $h_+^* = h_+/(r_+ - r_-) \cdot 100\,\%$;
- relative radius $r_+^* = r_+/l_b$.

Here, the characteristic length for bending l_b is the engineering constant of micropolar material (Eringen 1999; Lakes 1995). It is expressed through the elastic parameters of physically linear model (7) as follows

$$l_b = \sqrt{\frac{\gamma_2}{2\,(2\mu + \kappa)}}$$

and for the dense polyurethane foam $l_b = 0.33$ mm.

By numerical solution (Zubov and Sheidakov 2008) of the linearized boundary-value problem described in Sect. 3 we have found the critical curves, corresponding to the various buckling modes of the sandwich tube with micropolar core. Based on the analysis of these curves, the stability regions were constructed in the planes of loading parameters (relative axial compression δ and relative internal p_-^* or external p_+^* pressure) for tubes of various sizes. For all presented results, the length-to-radius ratio of the undeformed tube is 40 ($l^* = 40$), the wall thickness is 10 % of the tube radius ($H^* = 0.1$).

To study the size effect on the equilibrium bifurcation of sandwich tubes with foam core, we have carried out the stability analysis for tubes having the same proportions (i.e. thickness-to-radius ratio H^*, length-to-radius ratio l^* and relative thicknesses of inner h_-^* and outer h_+^* coatings are the same), but different overall size. In classic elasticity such tubes become unstable under the same strains. But, according to the obtained results, for the sandwich tubes with micropolar core the situation is different. In Figs. 2 and 3 the stability regions (boundaries of the stability regions) are presented for the sandwich tubes with very thin $\left(h_{\pm}^* = 1\,\%\right)$ polycarbonate and aluminum coatings in the case of internal ($p_+^* = 0$) and external ($p_-^* = 0$) pressure, respectively. The relative radius r_+^* has been used as the size (scale) parameter of the tubes at fixed aspect ratios H^*, l^*, h_-^* and h_+^*. It is evident from figures that tubes become more stable with a decrease in size. This size effect is very significant for small tubes $\left(r_+^* < 5\right)$, but negligible for large ones. For example, the stability boundaries for $r_+^* = 5$ and $r_+^* = 10$ differ very little on graphs. Moreover, according to the obtained results, the stability regions for the larger sandwich tubes $\left(r_+^* \geq 10\right)$

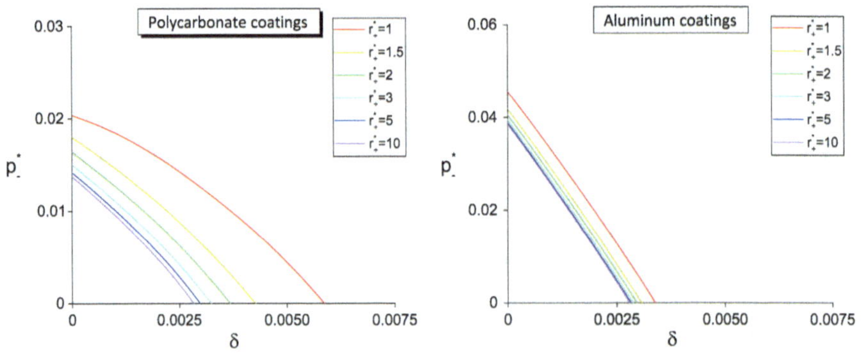

Fig. 2 Size effect on stability of sandwich tubes with very thin coatings in case of internal pressure

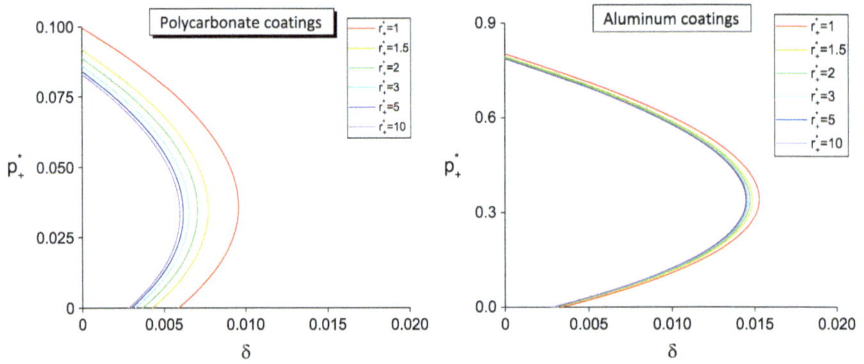

Fig. 3 Size effect on stability of sandwich tubes with very thin coatings in case of external pressure

are virtually the same, and coincide with the stability region for the sandwich tube with non-polar core, whose elastic properties are described by the model of physically linear material (8).

To analyze the influence of coatings on the loss of stability, we compared the stability regions for sandwich tubes with coatings of different thickness. In Figs. 4 and 5 the stability boundaries are constructed for the tubes with polycarbonate and aluminum coatings in the case of internal pressure ($p_+^* = 0$). The stability regions for the same tubes in the case of external pressure ($p_-^* = 0$) are shown in Figs. 6 and 7, respectively. The results are presented for two different tube sizes: $r_+^* = 1$ and $r_+^* = 5$. Through the comparison, we have determined that the sandwich tubes with thinner and less stiff coatings are more stable with respect to the axial compression, while tubes with thicker and stiffer coatings are more stable with respect to the internal and external pressure. This is more evident for small tubes ($r_+^* < 5$). For large tubes under simple axial compression ($p_-^* = 0$, $p_+^* = 0$) the influence of coatings thickness on the stability becomes less significant. Due to this, for example,

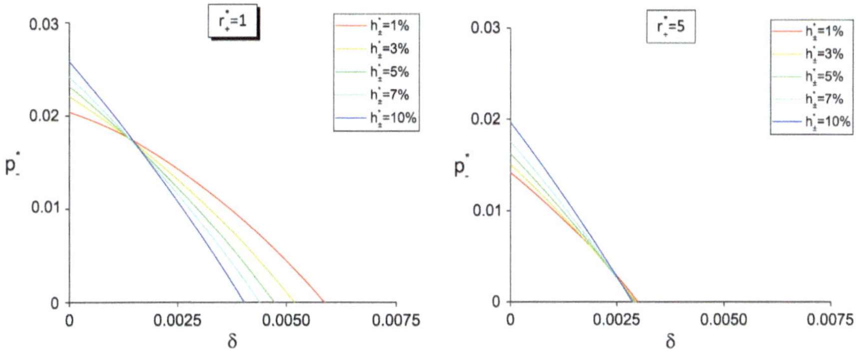

Fig. 4 Influence of polycarbonate coatings on sandwich tube stability in case of internal pressure

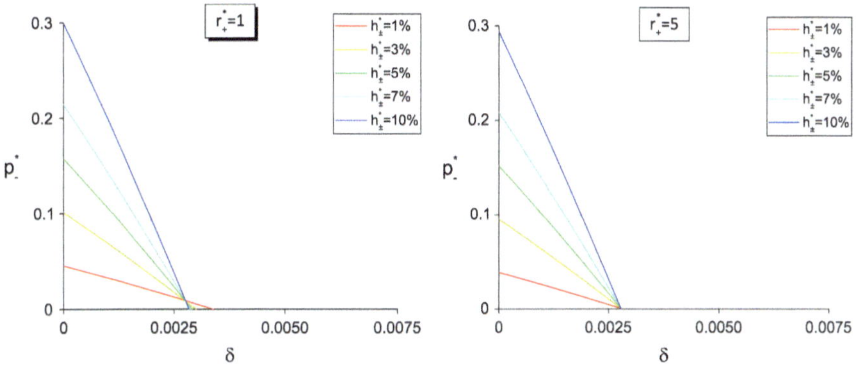

Fig. 5 Influence of aluminum coatings on sandwich tube stability in case of internal pressure

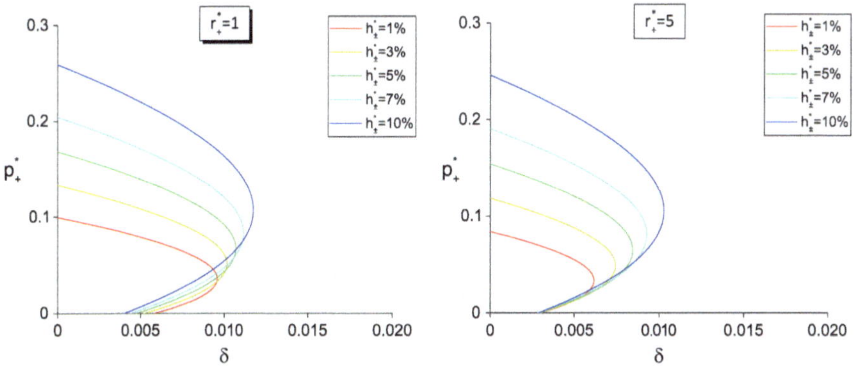

Fig. 6 Influence of polycarbonate coatings on sandwich tube stability in case of external pressure

the large sandwich tubes ($r_+^* \geq 5$) with thicker coatings are generally more stable in the case of internal pressure (see Figs. 4 and 5).

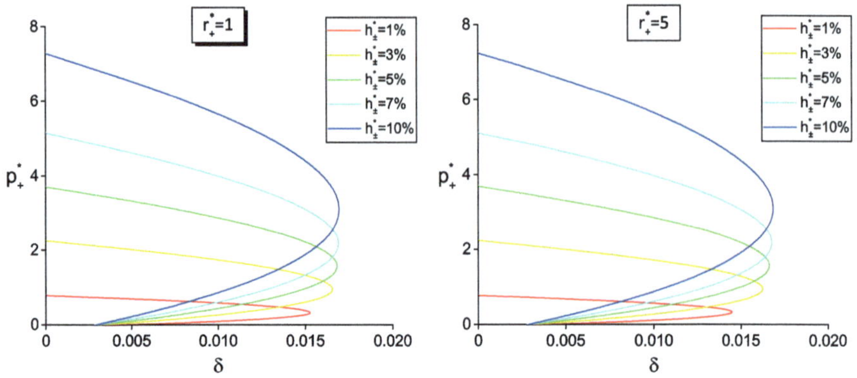

Fig. 7 Influence of aluminum coatings on sandwich tube stability in case of external pressure

5 Conclusion

In the framework of bifurcation approach, we have studied the stability of a sandwich tube with a micropolar core. For the model of physically linear material, a system of linearized equilibrium equations was derived, which describes the behavior of a sandwich tube in a perturbed state. In the case of a tube with polyurethane foam core and polycarbonate or aluminum coatings the stability regions were constructed in the planes of loading parameters. Based on these results, we have found that the sandwich tubes with foam core become more stable with a decrease in size. Additionally, it was determined that the tubes with thinner and less stiff coatings are more stable with respect to the axial compression, while tubes with thicker and stiffer coatings are more stable with respect to the internal and external pressure.

Acknowledgments This work was supported by the Russian Science Foundation (grant number 14-19-01676).

Appendix: Derivation of Neutral Equilibrium Equations

With respect to the representations (3), (4), (6) and (22), the expressions for the linearized stretch tensors $\mathbf{Y}^\bullet$, $\mathbf{U}_-^\bullet$ and $\mathbf{U}_+^\bullet$, and wryness tensor $\mathbf{L}^\bullet$ have the form:

$$
\begin{aligned}
\mathbf{Y}^{\bullet} &= \left(\frac{\partial v_{\Phi}}{\partial r} - f'\omega_{Z}\right)\vec{e}_r \otimes \vec{e}_{\varphi} + \frac{1}{r}\left(\frac{\partial v_R}{\partial \varphi} - v_{\Phi} + f\omega_Z\right)\vec{e}_{\varphi} \otimes \vec{e}_r \\
&+ \left(\frac{\partial v_Z}{\partial r} + f'\omega_{\Phi}\right)\vec{e}_r \otimes \vec{e}_z + \left(\frac{\partial v_R}{\partial z} - \alpha\omega_{\Phi}\right)\vec{e}_z \otimes \vec{e}_r \\
&+ \frac{1}{r}\left(\frac{\partial v_Z}{\partial \varphi} - f\omega_R\right)\vec{e}_{\varphi} \otimes \vec{e}_z + \left(\frac{\partial v_{\Phi}}{\partial z} + \alpha\omega_R\right)\vec{e}_z \otimes \vec{e}_{\varphi} \\
&+ \frac{\partial v_R}{\partial r}\vec{e}_r \otimes \vec{e}_r + \frac{1}{r}\left(\frac{\partial v_{\Phi}}{\partial \varphi} + v_R\right)\vec{e}_{\varphi} \otimes \vec{e}_{\varphi} + \frac{\partial v_Z}{\partial z}\vec{e}_z \otimes \vec{e}_z, \qquad (25)
\end{aligned}
$$

$$
\begin{aligned}
\mathbf{U}^{\bullet}_{\pm} &= \frac{\partial v_R^{\pm}}{\partial r}\vec{e}_r \otimes \vec{e}_r + \frac{1}{r}\left(\frac{\partial v_{\Phi}^{\pm}}{\partial \varphi} + v_R^{\pm}\right)\vec{e}_{\varphi} \otimes \vec{e}_{\varphi} + \frac{\partial v_Z^{\pm}}{\partial z}\vec{e}_z \otimes \vec{e}_z \\
&+ \frac{1}{rf'_{\pm} + f_{\pm}}\left(f'_{\pm}\left(\frac{\partial v_R^{\pm}}{\partial \varphi} - v_{\Phi}^{\pm}\right) + f_{\pm}\frac{\partial v_{\Phi}^{\pm}}{\partial r}\right)\left(\vec{e}_r \otimes \vec{e}_{\varphi} + \vec{e}_{\varphi} \otimes \vec{e}_r\right) \\
&+ \frac{1}{f'_{\pm} + \alpha}\left(f'_{\pm}\frac{\partial v_R^{\pm}}{\partial z} + \alpha\frac{\partial v_Z^{\pm}}{\partial r}\right)\left(\vec{e}_r \otimes \vec{e}_z + \vec{e}_z \otimes \vec{e}_r\right) \\
&+ \frac{1}{f_{\pm} + \alpha r}\left(f_{\pm}\frac{\partial v_{\Phi}^{\pm}}{\partial z} + \alpha\frac{\partial v_Z^{\pm}}{\partial \varphi}\right)\left(\vec{e}_z \otimes \vec{e}_{\varphi} + \vec{e}_{\varphi} \otimes \vec{e}_z\right), \qquad (26)
\end{aligned}
$$

$$
\begin{aligned}
\mathbf{L}^{\bullet} &= \frac{\partial \omega_R}{\partial r}\vec{e}_r \otimes \vec{e}_r + \frac{1}{r}\left(\frac{\partial \omega_{\Phi}}{\partial \varphi} + \omega_R\right)\vec{e}_{\varphi} \otimes \vec{e}_{\varphi} + \frac{\partial \omega_Z}{\partial z}\vec{e}_z \otimes \vec{e}_z \\
&+ \frac{\partial \omega_{\Phi}}{\partial r}\vec{e}_r \otimes \vec{e}_{\varphi} + \frac{1}{r}\left(\frac{\partial \omega_R}{\partial \varphi} - \omega_{\Phi}\right)\vec{e}_{\varphi} \otimes \vec{e}_r + \frac{\partial \omega_Z}{\partial r}\vec{e}_r \otimes \vec{e}_z \\
&+ \frac{\partial \omega_R}{\partial z}\vec{e}_z \otimes \vec{e}_r + \frac{1}{r}\frac{\partial \omega_Z}{\partial \varphi}\vec{e}_{\varphi} \otimes \vec{e}_z + \frac{\partial \omega_{\Phi}}{\partial z}\vec{e}_z \otimes \vec{e}_{\varphi}. \qquad (27)
\end{aligned}
$$

According to the relations (3), (5), (6), (15)–(17), (22), (25)–(27), the linearized Piola-type stress tensor $\mathbf{D}^{\bullet}$ and couple stress tensor $\mathbf{G}^{\bullet}$, and the linearized Piola stress tensors $\mathbf{D}^{\bullet}_{-}$ and $\mathbf{D}^{\bullet}_{+}$ are written as follows:

$$
\begin{aligned}
\mathbf{D}^{\bullet} &= \left[(\zeta + \tau)\frac{\partial v_R}{\partial r} + \frac{\lambda}{r}\left(\frac{\partial v_{\Phi}}{\partial \varphi} + v_R\right) + \lambda\frac{\partial v_Z}{\partial z}\right]\vec{e}_r \otimes \vec{e}_R \\
&+ \left[\tau\frac{\partial v_{\Phi}}{\partial r} + \frac{\mu}{r}\left(\frac{\partial v_R}{\partial \varphi} - v_{\Phi}\right) + B_3\omega_Z\right]\vec{e}_r \otimes \vec{e}_{\Phi} \\
&+ \left[\tau\frac{\partial v_Z}{\partial r} + \mu\frac{\partial v_R}{\partial z} - B_2\omega_{\Phi}\right]\vec{e}_r \otimes \vec{e}_Z + \left[\tau\frac{\partial v_R}{\partial z} + \mu\frac{\partial v_Z}{\partial r} + B_2\omega_{\Phi}\right]\vec{e}_z \otimes \vec{e}_R \\
&+ \left[\frac{\tau}{r}\left(\frac{\partial v_R}{\partial \varphi} - v_{\Phi}\right) + \mu\frac{\partial v_{\Phi}}{\partial r} - B_3\omega_Z\right]\vec{e}_{\varphi} \otimes \vec{e}_R
\end{aligned}
$$

$$
+\left[\lambda\frac{\partial v_R}{\partial r}+\frac{\zeta+\tau}{r}\left(\frac{\partial v_\Phi}{\partial\varphi}+v_R\right)+\lambda\frac{\partial v_Z}{\partial z}\right]\vec{e}_\varphi\otimes\vec{e}_\Phi
+\left[\frac{\tau}{r}\frac{\partial v_Z}{\partial\varphi}+\mu\frac{\partial v_\Phi}{\partial z}+B_1\omega_R\right]\vec{e}_\varphi\otimes\vec{e}_Z+\left[\tau\frac{\partial v_\Phi}{\partial z}+\frac{\mu}{r}\frac{\partial v_Z}{\partial\varphi}-B_1\omega_R\right]\vec{e}_z\otimes\vec{e}_\Phi
+\left[\lambda\frac{\partial v_R}{\partial r}+\frac{\lambda}{r}\left(\frac{\partial v_\Phi}{\partial\varphi}+v_R\right)+(\zeta+\tau)\frac{\partial v_Z}{\partial z}\right]\vec{e}_z\otimes\vec{e}_Z, \qquad (28)
$$

$$
\mathbf{G}^\bullet=\left[(\gamma+\gamma_2)\frac{\partial\omega_R}{\partial r}+\frac{\gamma_1}{r}\left(\frac{\partial\omega_\Phi}{\partial\varphi}+\omega_R\right)+\gamma_1\frac{\partial\omega_Z}{\partial z}\right]\vec{e}_r\otimes\vec{e}_R
+\left[\gamma_2\frac{\partial\omega_\Phi}{\partial r}+\frac{\gamma_3}{r}\left(\frac{\partial\omega_R}{\partial\varphi}-\omega_\Phi\right)\right]\vec{e}_r\otimes\vec{e}_\Phi
+\left[\gamma_2\frac{\partial\omega_Z}{\partial r}+\gamma_3\frac{\partial\omega_R}{\partial z}\right]\vec{e}_r\otimes\vec{e}_Z+\left[\gamma_2\frac{\partial\omega_R}{\partial z}+\gamma_3\frac{\partial\omega_Z}{\partial r}\right]\vec{e}_z\otimes\vec{e}_R
+\left[\frac{\gamma_2}{r}\left(\frac{\partial\omega_R}{\partial\varphi}-\omega_\Phi\right)+\gamma_3\frac{\partial\omega_\Phi}{\partial r}\right]\vec{e}_\varphi\otimes\vec{e}_R
+\left[\gamma_1\frac{\partial\omega_R}{\partial r}+\frac{\gamma+\gamma_2}{r}\left(\frac{\partial\omega_\Phi}{\partial\varphi}+\omega_R\right)+\gamma_1\frac{\partial\omega_Z}{\partial z}\right]\vec{e}_\varphi\otimes\vec{e}_\Phi
+\left[\frac{\gamma_2}{r}\frac{\partial\omega_Z}{\partial\varphi}+\gamma_3\frac{\partial\omega_\Phi}{\partial z}\right]\vec{e}_\varphi\otimes\vec{e}_Z+\left[\gamma_2\frac{\partial\omega_\Phi}{\partial z}+\frac{\gamma_3}{r}\frac{\partial\omega_Z}{\partial\varphi}\right]\vec{e}_z\otimes\vec{e}_\Phi
+\left[\gamma_1\frac{\partial\omega_R}{\partial r}+\frac{\gamma_1}{r}\left(\frac{\partial\omega_\Phi}{\partial\varphi}+\omega_R\right)+(\gamma+\gamma_2)\frac{\partial\omega_Z}{\partial z}\right]\vec{e}_z\otimes\vec{e}_Z, \qquad (29)
$$

$$
\mathbf{D}^\bullet_\pm=\left[(\zeta_\pm+\mu_\pm)\frac{\partial v^\pm_R}{\partial r}+\frac{\lambda_\pm}{r}\left(\frac{\partial v^\pm_\Phi}{\partial\varphi}+v^\pm_R\right)+\lambda_\pm\frac{\partial v^\pm_Z}{\partial z}\right]\vec{e}_r\otimes\vec{e}_R
+\left[\left(\mu_\pm+B^\pm_3\right)\frac{\partial v^\pm_\Phi}{\partial r}+\frac{\mu_\pm-B^\pm_3}{r}\left(\frac{\partial v^\pm_R}{\partial\varphi}-v^\pm_\Phi\right)\right]\vec{e}_r\otimes\vec{e}_\Phi
+\left[\left(\mu_\pm+B^\pm_2\right)\frac{\partial v^\pm_Z}{\partial r}+\left(\mu_\pm-B^\pm_2\right)\frac{\partial v^\pm_R}{\partial z}\right]\vec{e}_r\otimes\vec{e}_Z
+\left[\frac{\mu_\pm+B^\pm_3}{r}\left(\frac{\partial v^\pm_R}{\partial\varphi}-v^\pm_\Phi\right)+\left(\mu_\pm-B^\pm_3\right)\frac{\partial v^\pm_\Phi}{\partial r}\right]\vec{e}_\varphi\otimes\vec{e}_R
+\left[\lambda_\pm\frac{\partial v^\pm_R}{\partial r}+\frac{\zeta_\pm+\mu_\pm}{r}\left(\frac{\partial v^\pm_\Phi}{\partial\varphi}+v^\pm_R\right)+\lambda_\pm\frac{\partial v^\pm_Z}{\partial z}\right]\vec{e}_\varphi\otimes\vec{e}_\Phi
+\left[\frac{\mu_\pm+B^\pm_1}{r}\frac{\partial v^\pm_Z}{\partial\varphi}+\left(\mu_\pm-B^\pm_1\right)\frac{\partial v^\pm_\Phi}{\partial z}\right]\vec{e}_\varphi\otimes\vec{e}_Z
+\left[\left(\mu_\pm+B^\pm_2\right)\frac{\partial v^\pm_R}{\partial z}+\left(\mu_\pm-B^\pm_2\right)\frac{\partial v^\pm_Z}{\partial r}\right]\vec{e}_z\otimes\vec{e}_R
$$

$$+\left[\left(\mu_\pm + B_1^\pm\right)\frac{\partial v_\Phi^\pm}{\partial z} + \frac{\mu_\pm - B_1^\pm}{r}\frac{\partial v_Z^\pm}{\partial \varphi}\right]\vec{e}_z \otimes \vec{e}_\Phi$$
$$+\left[\lambda_\pm \frac{\partial v_R^\pm}{\partial r} + \frac{\lambda_\pm}{r}\left(\frac{\partial v_\Phi^\pm}{\partial \varphi} + v_R^\pm\right) + (\zeta_\pm + \mu_\pm)\frac{\partial v_Z^\pm}{\partial z}\right]\vec{e}_z \otimes \vec{e}_Z. \quad (30)$$

Here the following notations are used:

$$\zeta = \lambda + \mu, \quad \zeta_\pm = \lambda_\pm + \mu_\pm, \quad \tau = \mu + \kappa, \quad \gamma = \gamma_1 + \gamma_3, \quad \xi = n^2 + r^2\beta^2 + 1,$$

$$B_1 = \mu\left(\frac{f}{r} + \alpha\right) + \lambda s - \chi, \qquad B_1^\pm = \mu_\pm + \frac{(\lambda_\pm s_\pm - 2\mu_\pm)\, r}{f_\pm + \alpha r},$$
$$B_2 = \mu\left(f' + \alpha\right) + \lambda s - \chi, \qquad B_2^\pm = \mu_\pm + \frac{\lambda_\pm s_\pm - 2\mu_\pm}{f_\pm' + \alpha},$$
$$B_3 = \mu\left(f' + \frac{f}{r}\right) + \lambda s - \chi, \qquad B_3^\pm = \mu_\pm + \frac{(\lambda_\pm s_\pm - 2\mu_\pm)\, r}{r f_\pm' + f_\pm}.$$

By taking into account the expressions (4), (9), (22)–(24), (28)–(30), we derive a neutral equilibrium equations (14):

$$(\zeta + \tau)\, V_R'' + \frac{\zeta + \tau}{r} V_R' - \frac{\zeta + \tau\xi}{r^2} V_R + \frac{n\zeta}{r} V_\Phi' - \frac{n\,(\zeta + 2\tau)}{r^2} V_\Phi$$
$$+ \beta\zeta V_Z' + \beta B_2 \Omega_\Phi - \frac{n B_3}{r}\Omega_Z = 0,$$
$$\tau V_\Phi'' - \frac{n\zeta}{r} V_R' - \frac{n\,(\zeta + 2\tau)}{r^2} V_R + \frac{\tau}{r} V_\Phi' - \frac{\zeta n^2 + \tau\xi}{r^2} V_\Phi - \frac{n\beta\zeta}{r} V_Z$$
$$- \beta B_1 \Omega_R + B_3' \Omega_Z + B_3 \Omega_Z' = 0,$$
$$\tau V_Z'' - \beta\zeta V_R' - \frac{\beta\zeta}{r} V_R - \frac{n\beta\zeta}{r} V_\Phi + \frac{\tau}{r} V_Z' - \left(\zeta\beta^2 + \frac{\xi - 1}{r^2}\tau\right) V_Z$$
$$+ \frac{n B_1}{r}\Omega_R - B_2 \Omega_\Phi' - \left(B_2' + \frac{B_2}{r}\right)\Omega_\Phi = 0,$$
$$(\gamma + \gamma_2)\left(\Omega_R'' + \frac{\Omega_R'}{r}\right) - \left[\frac{\gamma + \gamma_2\xi}{r^2} - \left(\frac{f}{r} + \alpha\right) B_1\right]\Omega_R - \beta B_1 V_\Phi$$
$$+ \frac{n B_1}{r} V_Z + \frac{n\,(\gamma + 2\gamma_2)}{r^2}\Omega_\Phi - \frac{n\gamma}{r}\Omega_\Phi' - \beta\gamma\Omega_Z' = 0,$$
$$\gamma_2 \Omega_\Phi'' + \frac{\gamma_2}{r}\Omega_\Phi' - \left[\frac{\gamma n^2 + \gamma_2\xi}{r^2} - \left(f' + \alpha\right) B_2\right]\Omega_\Phi + B_2 V_Z' + \beta B_2 V_R$$
$$+ \frac{n\gamma}{r}\Omega_R' + \frac{n\,(\gamma + 2\gamma_2)}{r^2}\Omega_R - \frac{n\beta\gamma}{r}\Omega_Z = 0, \quad (31)$$

$$\gamma_2\Omega_Z'' - \left[\gamma\beta^2 + \frac{\xi-1}{r^2}\gamma_2 - B_3\left(f' + \frac{f}{r}\right)\right]\Omega_Z - B_3\left(\frac{n}{r}V_R + V_\Phi' + \frac{V_\Phi}{r}\right)$$
$$+ \frac{\gamma_2}{r}\Omega_Z' + \beta\gamma\left(\Omega_R' + \frac{\Omega_R}{r} - \frac{n}{r}\Omega_\Phi\right) = 0,$$
$$(\zeta_\pm + \mu_\pm)\left[\left(V_R^\pm\right)'' + \frac{\left(V_R^\pm\right)'}{r}\right] - \frac{1}{r^2}\left(\zeta_\pm + \xi\mu_\pm + \beta^2 r^2 B_2^\pm + n^2 B_3^\pm\right)V_R^\pm$$
$$+ \frac{n}{r}\left(\zeta_\pm - B_3^\pm\right)\left(V_\Phi^\pm\right)' - \frac{n}{r^2}\left(\zeta_\pm + 2\mu_\pm + B_3^\pm\right)V_\Phi^\pm + \beta\left(\zeta_\pm - B_2^\pm\right)\left(V_Z^\pm\right)' = 0,$$
$$\left(\mu_\pm + B_3^\pm\right)\left[\left(V_\Phi^\pm\right)'' + \frac{\left(V_\Phi^\pm\right)'}{r}\right] - \frac{1}{r^2}\left(n^2\zeta_\pm + \xi\mu_\pm + B_3^\pm + \beta^2 r^2 B_1^\pm\right)V_\Phi^\pm$$
$$- \frac{n}{r}\left(\zeta_\pm - B_3^\pm\right)\left(V_R^\pm\right)' - \frac{n}{r^2}\left(\zeta_\pm + 2\mu_\pm + B_3^\pm\right)V_R^\pm - \frac{n\beta}{r}\left(\zeta_\pm - B_1^\pm\right)V_Z^\pm = 0,$$
$$\left(\mu_\pm + B_2^\pm\right)\left(V_Z^\pm\right)'' + \left(\frac{\mu_\pm}{r} + \frac{1}{f_\pm' + \alpha}\left[B_1^\pm\frac{f_\pm + \alpha r}{r^2} - B_2^\pm f_\pm''\right]\right)\left(V_Z^\pm\right)'$$
$$- \beta\left(\zeta_\pm - B_2^\pm\right)\left(V_R^\pm\right)' - \beta\left(\frac{\zeta_\pm}{r} - \frac{1}{f_\pm' + \alpha}\left[B_1^\pm\frac{f_\pm + \alpha r}{r^2} - B_2^\pm f_\pm''\right]\right)V_R^\pm$$
$$- \frac{n\beta}{r}\left(\zeta_\pm - B_1^\pm\right)V_\Phi^\pm - \left(\left[\zeta_\pm + \mu_\pm\right]\beta^2 + \frac{n^2}{r^2}\left[\mu_\pm + B_1^\pm\right]\right)V_Z^\pm = 0.$$

Given the substitutions (23) and (24), the expressions for the linearized boundary conditions (18) take the form:

- on the inner and outer surfaces of the tube ($r = r_\pm$)

$$\left(\zeta_\pm + \mu_\pm\right)\left(V_R^\pm\right)' + \frac{\lambda_\pm + \alpha p_\pm}{r_\pm}\left(V_R^\pm + nV_\Phi^\pm\right) + \beta\left(\lambda_\pm + \frac{f_\pm}{r_\pm}p_\pm\right)V_Z^\pm = 0,$$
$$\frac{\alpha p_\pm + B_3^\pm - \mu_\pm}{r_\pm}\left(nV_R^\pm + V_\Phi^\pm\right) + \left(\mu_\pm + B_3^\pm\right)\left(V_\Phi^\pm\right)' = 0, \quad (32)$$
$$\beta\left(\frac{f_\pm}{r_\pm}p_\pm + B_2^\pm - \mu_\pm\right)V_R^\pm + \left(\mu_\pm + B_2^\pm\right)\left(V_Z^\pm\right)' = 0,$$

- on the interfaces between the coatings and the foam core ($r = c_\pm$)

$$\left(\zeta_\pm + \mu_\pm\right)\left(V_R^\pm\right)' + \frac{\lambda_\pm}{c_\pm}\left(V_R^\pm + nV_\Phi^\pm\right) + \beta\lambda_\pm V_Z^\pm - (\zeta + \tau)V_R'$$
$$- \frac{\lambda}{c_\pm}\left(V_R + nV_\Phi\right) - \beta\lambda V_Z = 0,$$
$$\frac{B_3^\pm - \mu_\pm}{c_\pm}\left(nV_R^\pm + V_\Phi^\pm\right) + \left(\mu_\pm + B_3^\pm\right)\left(V_\Phi^\pm\right)' - \tau V_\Phi'$$
$$+ \frac{\mu}{c_\pm}\left(nV_R + V_\Phi\right) - B_3\Omega_Z = 0, \quad (33)$$

$$\beta\left(B_2^{\pm}-\mu_{\pm}\right)V_R^{\pm}+\left(\mu_{\pm}+B_2^{\pm}\right)\left(V_Z^{\pm}\right)'+\mu\beta V_R-\tau V_Z'+B_2\Omega_\Phi=0,$$
$$(\gamma+\gamma_2)\,\Omega_R'+\frac{\gamma_1}{c_{\pm}}\left(\Omega_R-n\Omega_\Phi\right)-\gamma_1\beta\Omega_Z=0,$$
$$\frac{\gamma_3}{c_{\pm}}\left(n\Omega_R-\Omega_\Phi\right)+\gamma_2\Omega_\Phi'=0,\qquad \gamma_3\beta\Omega_R+\gamma_2\Omega_Z'=0,$$
$$v_R-v_R^{\pm}=0,\qquad v_\Phi-v_\Phi^{\pm}=0,\qquad v_Z-v_Z^{\pm}=0.$$

References

Altenbach J, Altenbach H, Eremeyev VA (2010) On generalized Cosserat-type theories of plates and shells: a short review and bibliography. Arch Appl Mech 80:73–92

Ashby MF, Evans AG, Fleck NA, Gibson LJ, Hutchinson JW, Wadley HNG (2000) Metal foams: a design guide. Butterworth-Heinemann, Boston

Banhart J (2000) Manufacturing routes for metallic foams. J Miner 52(12):22–27

Banhart J, Ashby MF, Fleck NA (eds) (1999) Metal foams and porous metal structures. Verlag MIT Publishing, Bremen

Cosserat E, Cosserat F (1909) Theorie des Corps Deformables. Hermann et Fils, Paris

Degischer HP, Kriszt B (eds) (2002) Handbook of cellular metals. Production, processing, applications. Wiley-VCH, Weinheim

Eremeyev VA, Pietraszkiewicz W (2012) Material symmetry group of the non-linear polar-elastic continuum. Int J Solids Struct 49(14):1993–2005

Eremeyev VA, Zubov LM (1994) On the stability of elastic bodies with couple-stresses. Mech Solids 29(3):172–181

Eringen AC (1999) Microcontinuum field theory I. Foundations and solids. Springer, New York

Gibson LJ, Ashby MF (1997) Cellular solids: structure and properties, 2nd edn. Cambridge solid state science series. Cambridge University Press, Cambridge

Green AE, Adkins JE (1960) Large elastic deformations and non-linear continuum mechanics. Clarendon Press, Oxford

Kafadar CB, Eringen AC (1971) Micropolar media - I. The classical theory. Int J Eng Sci 9:271–305

Lakes RS (1986) Experimental microelasticity of two porous solids. Int J Solids Struct 22:55–63

Lakes RS (1995) Experimental methods for study of Cosserat elastic solids and other generalized elastic continua. In: Muhlhaus H (ed) Continuum models for materials with micro-structure. Wiley, New York, pp 1–22

Lurie AI (1990) Non-linear theory of elasticity. North-Holland, Amsterdam

Maugin GA (1998) On the structure of the theory of polar elasticity. Philos Trans R Soc Lond A 356:1367–1395

Nikitin E, Zubov LM (1998) Conservation laws and conjugate solutions in the elasticity of simple materials and materials with couple stress. J Elast 51:1–22

Ogden RW (1997) Non-linear elastic deformations. Dover, Mineola

Pietraszkiewicz W, Eremeyev VA (2009a) On natural strain measures of the non-linear micropolar continuum. Int J Solids Struct 46:774–787

Pietraszkiewicz W, Eremeyev VA (2009b) On vectorially parameterized natural strain measures of the non-linear cosserat continuum. Int J Solids Struct 46(11–12):2477–2480

Sheydakov DN (2010) Stability of elastic cylindrical micropolar shells subject to combined loads. In: Pietraszkiewicz W, Kreja I (eds) Shell structures: theory and applications, vol 2. CRC Press, Boca Raton, pp 141–144

Sheydakov DN (2011) Buckling of elastic composite rod of micropolar material subject to combined loads. In: Altenbach H, Erofeev V, Maugin G (eds) Mechanics of generalized continua –

from micromechanical basics to engineering applications, vol 7. Advanced structured materials. Springer, Berlin, pp 255–271

Toupin RA (1964) Theories of elasticity with couple-stress. Arch Ration Mech Anal 17:85–112

Zubov LM (1997) Nonlinear theory of dislocations and disclinations in elastic bodies. Springer, Berlin

Zubov LM, Sheidakov DN (2008) Instability of a hollow elastic cylinder under tension, torsion, and inflation. Trans ASME J Appl Mech 75(1):011002

Frequency-Dependent Attenuation and Phase Velocity Dispersion of an Acoustic Wave Propagating in the Media with Damages

Anatoli Stulov and Vladimir I. Erofeev

Abstract In frame of the self-consistent mathematical model, which includes the dynamics of a material and the state of its defects, the particular qualities of acoustic wave propagation in the material with damage is considered. In this study a constitutive equation of the damaged medium is derived and the similarity between the models for damaged materials and the medium with memory is confirmed. The dispersion analysis of the model is carried out, and it is shown that the damage of the material gives rise to frequency-dependent attenuation and anomalous dispersion of phase velocity of acoustic wave propagating through that material. This makes it possible to estimate the damage of the material by means of a nondestructive acoustic method.

Keywords Medium with damages · Acoustic wave propagation · Dispersion analysis · Frequency-dependent attenuation · Anomalous dispersion

1 Introduction

Today, the mechanics of a damaged continuum is intensively developed by many authors. The first works in this field were the fundamental studies by L.M. Kachanov, which are summarized in his monograph (Kachanov 1986), and the detailed investigations and analysis by Yu.N. Rabotnov that are generalized in Rabotnov (1969). The significance of these pioneering works, which presently are recognized as classical, consists in the possibility of using a unified approach for description of the damage of elastic and elasto-plastic bodies.

A. Stulov (✉)
Institute of Cybernetics at Tallinn University of Technology,
Akadeemia 21, 12618 Tallinn, Estonia
e-mail: stulov@ioc.ee

V.I. Erofeev
Mechanical Engineering Research Institute of Russian Academy of Sciences,
85 Belinskogo st., 603024 Nizhny Novgorod, Russia
e-mail: erof.vi@yandex.ru

H. Altenbach and S. Forest (eds.), *Generalized Continua as Models for Classical and Advanced Materials*, Advanced Structured Materials 42,
DOI 10.1007/978-3-319-31721-2_19

The damage is usually understood as a reduction of an elastic response of the body due to decreasing of the effective area, through which the internal forces are transmitting from one part of the body to another. This phenomenon is caused by the appearance and spreading of the scattered field of microdefects (the microcracks in the case of elasticity, the dislocations in the case of plasticity, the micropores in the case of creep, and the surface microcracks in the case of fatigue, Maugin 1992).

The damage, i.e. the degradation of the mechanical properties of a solid material, cannot be measured directly in the same manner as, for example, velocity, force, or temperature. The damage can be detected indirectly only by analyzing the response of the elastic structure on the various external impacts. According to experimental knowledge, the presence of a damage field inside a solid material can be observed also by changing of physical features of the structure. For example, it may be the decreasing of velocity of ultrasonic signal propagation (Zuev et al. 1999; Hirao et al. 2000; Wang et al. 1998), a decrease in the Young's modulus ('the modulus defect') (Klepko et al. 2007), a decrease in material density ('loosening') (Volkov and Mironov 2005), a hardness change (Collins 1993), a decrease in the stress amplitude under the cyclic loading (Makhutov 1981; Romanov 1988), and an acceleration of the tertiary creep (Berezina and Mints 1976).

The purpose of the present study is the modeling of the process of acoustic wave propagation through the damaged material, and estimation of influence of damage on the phase velocity and attenuation of that wave.

2 Self-Consistent Model for Damage Description

In accordance with conventional assumptions, the measure of damage under deformation is taken to be a scalar damage parameter $\psi(x, t) > 0$ that characterizes the relative density of microdefects uniformly dispersed in a unit volume. This parameter is zero in the absence of damage and close to one at the instant of fracture. The process of the damage gain in the structure under study is calculated numerically step by step by solving the kinetic equation of damage at every stage of loading. This procedure is continued until the damage parameter $\psi(x, t)$ reaches an initially prescribed limiting value, which is close to one.

Generally, in mechanics of deformed solids, the dynamic problems and the problem of defects accumulation are considered separately. In the development of such approach, the usual practice is to postulate the relationship between the velocity of elastic wave and the value of damage by some kind of dependence in advance, and after that, it is assumed that the constant coefficients at this relation can be established on the basis of experimental data.

Usually Uglov et al. (2009), the phase velocity $V_{\text{ph}}(\omega)$ of propagating wave and its attenuation $\alpha(\omega)$ are chosen in the power polynomial form as functions of frequency ω, and as a linear functions of damage ψ as

$$V_{\text{ph}}(\omega) = C_0(1 - h_1\omega - h_2\psi\omega^2), \tag{1}$$

$$\alpha(\omega) = (h_3 + h_4\psi)\omega^4, \tag{2}$$

where $C_0 = \sqrt{E/\rho}$ is the velocity of the longitudinal elastic wave propagating in the material in the absence of defects, E is the Young's modulus, ρ is the density of the material, and h_{1-4} are the constant coefficients, which must be determined experimentally.

The evolution of damage is described by the kinetic equation derived in Volkov and Korotkikh (2008) in the form

$$\frac{\mathrm{d}\psi}{\mathrm{d}t} = f(\sigma, \psi), \tag{3}$$

where σ is the stress due to the external impact. In most cases, the function f is approximated by a linear function, or, in some cases, by a power polynomial dependence (Volkov and Korotkikh 2008). Although this approach has undoubted advantage such as simplicity, it has also some imperfections, which are typical for any approach that is not based on the physical models of the processes and systems.

Another novel method of materials with damage examination is presented in Erofeev and Nikitina (2010), Erofeev et al. (2010). In these papers the process of propagation of a longitudinal acoustic wave along a rod is considered. It is also assumed that the rod is subjected to the static or cyclic tests, and during the process of loading the damage may accumulate in the rod's material.

This work differs significantly from previous studies. In Erofeev and Nikitina (2010), Erofeev et al. (2010), the authors propose the idea that the problem under consideration is a self-consistent problem, and therefore, in addition to the damage evolution Eq. (3), which can be presented in the form

$$\frac{\partial\psi}{\partial t} + \frac{1}{\tau}\psi = \beta_2 E\frac{\partial u}{\partial x}, \tag{4}$$

an additional equation describing the dynamics of the rod

$$\frac{\partial^2 u}{\partial t^2} - C_0^2\frac{\partial^2 u}{\partial x^2} + \beta_1\frac{\partial\psi}{\partial x} = 0, \tag{5}$$

must be taken into account. Here we denote the particle displacement at the rod midline by $u(x, t)$, and the constants τ, β_1 and β_2 characterize the relations between the cyclic process of the rod loading and the speed of the damage accumulation.

Equation (4) may be rewritten in an equivalent form as

$$\psi(x, t) = \beta_2 E\int_0^t \frac{\partial u}{\partial x}(x, \xi)\mathrm{e}^{(\xi - t)/\tau}\mathrm{d}\xi = \beta_2 E R(t) * \frac{\partial u}{\partial x}(x, t), \tag{6}$$

where the sign $*$ denotes the convolution sign, and $R(t)$ is the relaxation function given by

$$R(t) = \mathrm{e}^{-t/\tau}. \tag{7}$$

Equation (6) describes the process of damage growth as a function of the strain ($\varepsilon = \partial u/\partial x$) history, and one can state that the constant $\tau > 0$ is the relaxation time. Here we assume that the history of the damage appearance starts at $t = 0$.

From Eq. (6), it follows that at the beginning of the process, if $t \ll \tau$, there are no defects ($\psi = 0$) in the rod material at all. In the opposite case, if $t \gg \tau$, from Eq. (6) one can obtain the dependence describing the process of damage growth for the case of slow changing of strain in the form

$$\psi = \tau\beta_2 E\frac{\partial u}{\partial x}. \tag{8}$$

Now, using Eqs. (5), (6) we can write

$$\rho\frac{\partial^2 u}{\partial t^2} = E\frac{\partial}{\partial x}\left(\frac{\partial u}{\partial x} - \rho\beta_1\beta_2 R * \frac{\partial u}{\partial x}\right). \tag{9}$$

Taking into account the classical equation of motion given by

$$\rho\frac{\partial^2 u}{\partial t^2} = \frac{\partial\sigma}{\partial x}, \tag{10}$$

we can derive the constitutive equation of the media with damage in the form

$$\sigma = E(1 - \rho\beta_1\beta_2 R*)\frac{\partial u}{\partial x} = E\left[\frac{\partial u}{\partial x} - \rho\beta_1\beta_2\int_0^t \frac{\partial u}{\partial x}(x,\xi)\mathrm{e}^{(\xi - t)/\tau}\mathrm{d}\xi\right]. \tag{11}$$

Materials described by this equation for which the exerted stress is determined by the history of the deformation are "materials with memory".

As indicated by Rabotnov (1969), a model of material with memory may be obtained by means of replacing constant elastic parameters of solids by time-dependent operators. So for the case of material with damage, the Young's modulus is now not a constant, but an operator

$$E_0(t) = E(1 - \rho\beta_1\beta_2 R*), \tag{12}$$

and thus the constitutive equation (11) of the media with damage one can rewrite in compact form as

$$\sigma(\varepsilon) = E_0(t)\varepsilon. \tag{13}$$

From Eq. (13), it follows that if $t \ll \tau$, then we obtain the constitutive equation for the fast loading in the form

$$\sigma = E\frac{\partial u}{\partial x} = E_{\mathrm{d}}\frac{\partial u}{\partial x}. \tag{14}$$

Here the constant $E_{\mathrm{d}} = E$ is the dynamic Young's modulus.

In the opposite case, if $t \gg \tau$, then we obtain the constitutive equation, which is valid for the slow loading

$$\sigma = E_{\mathrm{d}}(1 - \tau\rho\beta_1\beta_2)\frac{\partial u}{\partial x} = \delta E_{\mathrm{d}}\frac{\partial u}{\partial x} = E_{\mathrm{s}}\frac{\partial u}{\partial x}, \tag{15}$$

where the quantity $E_{\mathrm{s}} = \delta E_{\mathrm{d}}$ is the static Young's modulus of the material, and parameter $\delta = 1 - \tau\rho\beta_1\beta_2$ characterizes the material damage.

Due to the evident inequality $E_{\mathrm{d}} > E_{\mathrm{s}} > 0$, parameter $0 < \delta \leq 1$. The value of parameter $\delta = 1$ denotes the absence of damage, and the value of this parameter δ is close to zero at the instant of fracture.

We can notice that Eqs. (4) and (5) can be reduced to a single one by eliminating the damage parameter $\psi(x, t)$. In terms of displacement $u(x, t)$, it leads to an equation in the following form

$$\frac{\partial^2 u}{\partial t^2} - \delta C_0^2\frac{\partial^2 u}{\partial x^2} + \tau\frac{\partial^3 u}{\partial t^3} - \tau C_0^2\frac{\partial^3 u}{\partial x^2 \partial t} = 0. \tag{16}$$

Dimensionless form of the Eq. (16) is obtained by using the non-dimensional variables that are introduced by relations

$$U = u/\tau C_0, \quad X = \sqrt{\delta}x/\tau C_0, \quad T = \delta t/\tau. \tag{17}$$

Thus Eq. (16) in terms of non-dimensional displacement variable $U(X, T)$ takes the following form (Kartofelev and Stulov 2014)

$$\frac{\partial^2 U}{\partial T^2} - \frac{\partial^2 U}{\partial X^2} + \delta\frac{\partial^3 U}{\partial T^3} - \frac{\partial^3 U}{\partial X^2 \partial T} = 0, \tag{18}$$

and describes acoustic wave propagation in the medium with damage.

3 Dispersion Relations

The fundamental solution of Eq. (18) has the form of traveling waves

$$U(X, T) = U_0\,\mathrm{e}^{\mathrm{i}\kappa X - \mathrm{i}\omega T}, \tag{19}$$

where i is the imaginary unit, κ is the wavenumber, ω is the angular frequency, and U_0 is the amplitude. The dispersion law $\Phi(\kappa, \omega) = 0$ for Eq. (18) is defined by relation

$$\mathrm{i}\delta\omega^3 - \omega^2 - \mathrm{i}\kappa^2\omega + \kappa^2 = 0. \tag{20}$$

In the case of boundary value problems the general solution of Eq. (18) has the following form

$$U(X,T) = \frac{1}{2\pi}\int_{-\infty}^{\infty} \Theta(\omega)\mathrm{e}^{\mathrm{i}\kappa(\omega)X - \mathrm{i}\omega T}\mathrm{d}\omega, \tag{21}$$

where $\Theta(\omega)$ is the Fourier-transform of the boundary value of disturbance prescribed at $X = 0$

$$\Theta(\omega) = \int_{-\infty}^{\infty} U(0,T)\mathrm{e}^{\mathrm{i}\omega T}\mathrm{d}T. \tag{22}$$

In case of Cauchy problem the general solution of Eq. (18) has the following form

$$U(X,T) = \frac{1}{2\pi}\int_{-\infty}^{\infty} \chi(\kappa)\mathrm{e}^{\mathrm{i}\kappa X - \mathrm{i}\omega(\kappa)T}\mathrm{d}\kappa, \tag{23}$$

where $\chi(\kappa)$ is the Fourier-transform of initial disturbance prescribed at $T = 0$

$$\chi(\kappa) = \int_{-\infty}^{\infty} U(X,0)\mathrm{e}^{\mathrm{i}\kappa X}\mathrm{d}X. \tag{24}$$

In general case $\kappa = \kappa(\omega)$ and $\omega = \omega(\kappa)$ are the complex quantities and can be derived from dispersion relation (20). In order to provide the dispersion analysis in context with a boundary value problem we rewrite wavenumber $\kappa(\omega)$ in the form

$$\kappa(\omega) = k(\omega) + \mathrm{i}\lambda(\omega), \tag{25}$$

where $k = \Re(\kappa)$ and $\lambda = \Im(\kappa)$. Using this notation, expression (19) can be rewritten as follows

$$U(X,T) = U_0\mathrm{e}^{\mathrm{i}(k+\mathrm{i}\lambda)X - \mathrm{i}\omega T} = \mathrm{e}^{-\lambda X}U_0\mathrm{e}^{\mathrm{i}kX - \mathrm{i}\omega T}. \tag{26}$$

It is clear that for positive values of λ we can observe the exponentially decaying wave that propagates along the positive direction of the space axis. In other words the spectral components $k(\omega) = \Re(\kappa)$ decay exponentially as $x, t \to \infty$ for $\lambda(\omega) > 0$. On the other hand, if $\lambda(\omega) < 0$, then the amplitudes of the spectral components grow exponentially as they propagate further along the positive direction of the x-axis. In the latter case the solution of Eq. (18) becomes unstable for $T \gg 0$.

4 Dispersion Analysis

As discussed above, in order to study the wave propagation along the x-axis one needs to solve the dispersion relation (20) against wavenumber κ. This solution takes the form

$$\kappa(\omega) = \frac{\omega\sqrt{1-\mathrm{i}\delta\omega}}{\sqrt{1-\mathrm{i}\omega}}. \tag{27}$$

For real values of k and λ the dispersion relation (20) can be rewritten as follows

$$k^2 + 2\mathrm{i}k\lambda - \lambda^2 - \mathrm{i}k^2\omega + 2k\lambda\omega + \mathrm{i}\lambda^2\omega - \omega^2 + \mathrm{i}\delta\omega^3 = 0. \tag{28}$$

In order to study real and imaginary parts separately, the system of equations in the form

$$\begin{cases} k^2 - \lambda^2 + 2k\lambda\omega - \omega^2 = 0, \\ 2k\lambda - \omega(k^2 - \lambda^2) + \delta\omega^3 = 0 \end{cases} \tag{29}$$

is solved and analyzed. Solutions with respect to k and λ are

$$k(\omega) = LM\left(\sqrt{1+M^2} - 1\right)^{-1/2}, \tag{30}$$

$$\lambda(\omega) = L\left(\sqrt{1+M^2} - 1\right)^{1/2}, \tag{31}$$

where

$$L = \omega\sqrt{\frac{1+\delta\omega^2}{2(1+\omega^2)}}, \quad M = \frac{(1-\delta)\omega}{1+\delta\omega^2}. \tag{32}$$

The frequency dependencies $k(\omega) = \Re(\kappa)$ and $\lambda(\omega) = \Im(\kappa)$ of dispersion relation (20) are displayed in Fig. 1 for the various values of the material parameter δ. Parameter δ can have values on the interval $\delta = [0, 1]$.

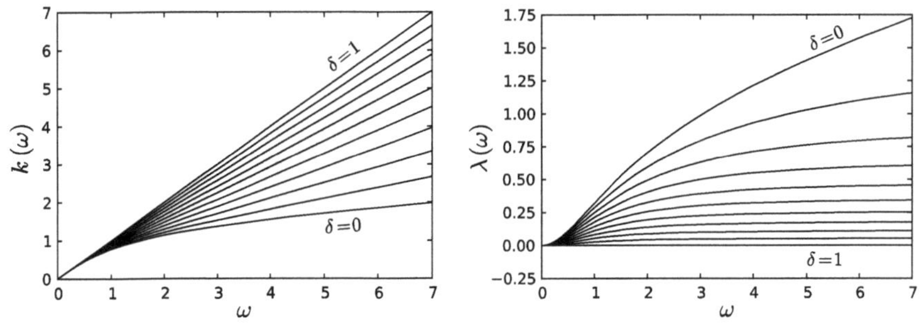

Fig. 1 Dispersion relations $k(\omega)$ and $\lambda(\omega)$ for various values of parameter δ in range [0.0, 1.0] with step 0.1

If $\delta = 1$, then from (30) and (31) one can find

$$k(\omega) = \omega, \quad \lambda(\omega) = 0. \tag{33}$$

These relations correspond to the ideal elastic material without damage, and in which the wave propagates without attenuation.

In case of $\omega \to \infty$ it is easy to see that $k(\omega) \to \omega\sqrt{\delta}$ and that

$$\lim_{\omega \to \infty} \lambda(\omega) = \frac{1-\delta}{2\sqrt{\delta}}. \tag{34}$$

For large frequencies, the exponential decay constant λ depends only on the parameter δ.

The phase velocity is defined as $v_{\text{ph}}(\omega) = \omega/k$, and it takes the following general form

$$v_{\text{ph}} = \frac{\sqrt{2(1+\omega^2)(N - \delta\omega^2 - 1)}}{(1-\delta)\omega}, \tag{35}$$

where

$$N = \sqrt{(1+\omega^2)(1+\delta^2\omega^2)}. \tag{36}$$

The frequency dependence $v_{\text{ph}}(\omega)$ for various values of parameter δ is shown in Fig. 2.

In case of $\delta = 1$, the phase velocity becomes $v_{\text{ph}}(\omega) = 1$ (cf. relationship (33)). For large frequencies, the phase velocity has a limit

$$\lim_{\omega \to \infty} v_{\text{ph}}(\omega) = \frac{1}{\sqrt{\delta}}. \tag{37}$$

The group velocity, which is defined as $v_{\text{gr}}(\omega) = \mathrm{d}\omega/\mathrm{d}k = (\mathrm{d}k/\mathrm{d}\omega)^{-1}$ takes in this case the following general form

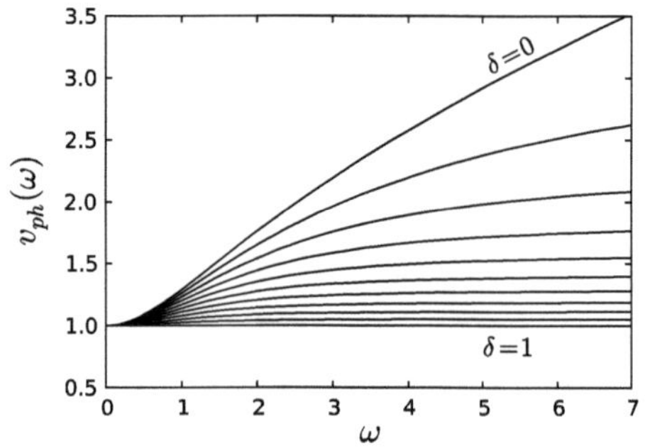

Fig. 2 Phase velocity as a function of frequency for various values of the parameter δ in range [0.0, 1.0] with step 0.1

$$v_{gr} = \frac{2(1+\omega^2)^2\sqrt{2(1+\delta^2\omega^2)}\left(N-\delta\omega^2-1\right)^{3/2}}{\omega(1-\delta)[(1+3\delta^2)\omega^4-(2N+2\delta N-3\delta^2-5)\omega^2-4(N-1)]}, \tag{38}$$

where N is defined by relation (36). The frequency dependence $v_{gr}(\omega)$ for various values of the parameter δ is presented in Fig. 3.

In case of $\delta = 1$, the group velocity $v_{gr}(\omega) = 1$ (cf. relationship (33)). For large frequencies the group velocity has the same limit as the phase velocity did

$$\lim_{\omega\to\infty} v_{gr}(\omega) = \frac{1}{\sqrt{\delta}}. \tag{39}$$

The essential difference between the behavior of phase and group velocities is that the phase velocity is a monotonic function of frequency, while the group velocity has a maximum. The maximum of different values of δ are located on the dashed line shown in Fig. 4.

Comparison of phase and group velocities for a single value of δ is presented in Fig. 4. In the material with damage the group velocity is always greater than the phase velocity for any frequency. This fact means that the material with damage is a medium with anomalous dispersion. This is true for any value of parameter $\delta < 1$. In case of $\delta = 1$, then $v_{gr} = v_{ph} = 1$, and we have the non-dispersive case.

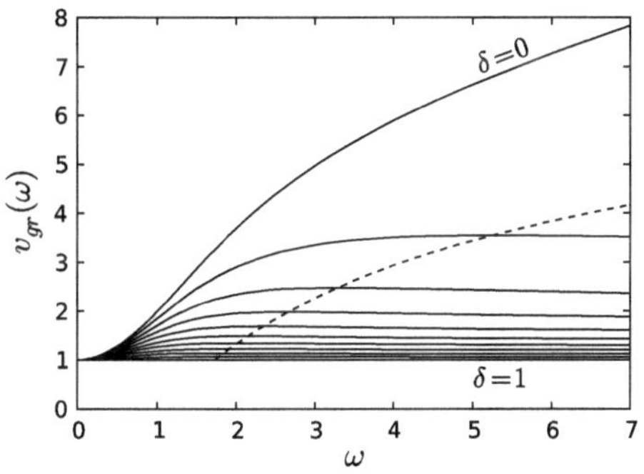

Fig. 3 Group velocity as a function of frequency for various values of the parameter δ in range [0.0, 1.0] with step 0.1. Maximum of v_{gr} for $\delta < 1$ is shown by *dashed line*

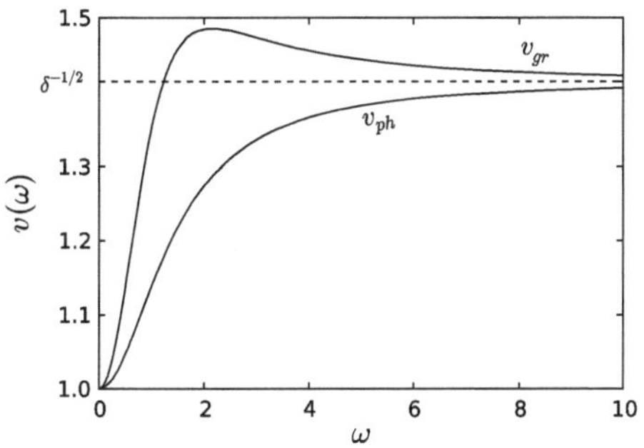

Fig. 4 Comparison of group and phase velocities for single value of the parameter $\delta = 0.5$. The *dashed line* shows the limit for the large frequencies

5 Conclusions

We have presented results for simulation of acoustic wave propagation in the medium with damage. Based on the self-consistent model for damage description, we have been able to demonstrate the similarity between the models for damaged materials and the medium with memory. We have derived the constitutive equation of the material with damage and examine the influence of the parameters of damage on the process of wave propagation in that medium.

The dispersion analysis of the model have been carried out, and the effect of the material damage on attenuation and phase velocity of propagating acoustic wave have also been estimated. It has been shown that the damage causes the anomalous dispersion and the frequency-dependent attenuation of the wave propagating through that material. The results obtained may be are useful for developing of nondestructive acoustic detection technique of damage in solids and structural elements.

Acknowledgments This research was supported by the European Regional Development Fund (Project TK124 (CENS)), by the Estonian Ministry of Education and Research (Project IUT33-24), and by the Russian Science Foundation (Grant N 14-19-01637).

References

Berezina TG, Mints II (1976) Heat-strength and heat-resistant of metallic materials. Nauka, Moscow (in Russ.)
Collins JA (1993) Failure of materials in mechanical design: analysis, prediction, prevention, 2nd edn. Wiley, New York
Erofeev VI, Nikitina EA (2010) The self-consistent dynamic problem of estimating the damage of a material by an acoustic method. Acoust Phys 56(4):584–587
Erofeev VI, Nikitina EA, Sharabanova AV (2010) Wave propagation in damaged materials using a new generalized continuum. In: Maugin GA, Metrikine AV (eds) Mechanics of generalized continua. One hundred years after the Cosserats. Advances in mechanics and mathematics, vol 21. Springer, Heidelberg, pp 143–148
Hirao M, Ogi H, Suzuki N, Ohtani T (2000) Ultrasonic attenuation peak during fatigue of polycrystalline copper. Acta Mater 48(2):517–524
Kachanov LM (1986) Introduction to continuum damage mechanics. Springer, New York
Kartofelev D, Stulov A (2014) Propagation of deformation waves in wool felt. Acta Mech 225(11):3103–3113
Klepko VV, Lebedev EV, Kolupaev BB, Kolupaev BS (2007) Energy dissipation and modulus defect in heterogeneous systems based on flexible-chain linear polymers. Polym Sci, Ser B 49(1–2):18–21
Makhutov NA (1981) Deformation criteria of fracture and calculation of construction elements for strength. Mashinostroenie, Moscow (in Russ.)
Maugin GA (1992) The thermomechanics of plasticity and fracture. Cambridge University Press, Cambridge
Rabotnov YN (1969) Creep problems in structural members. North-Holland series in applied mathematics and mechanics. North-Holland Publishing Company, Amsterdam
Romanov AN (1988) Fracture under small-cycle loading. Nauka, Moscow (in Russ.)

Uglov AL, Erofeev VI, Smirnov AN (2009) Acoustic control of equipment during its manufacture and operation. Nauka, Moscow (in Russ.)
Volkov IA, Korotkikh YG (2008) Equations of state of viscoelastoplastic media with damages. Fizmatlit, Moscow (in Russ.)
Volkov VM, Mironov AA (2005) United model of fatigue crack formation and growth in stress concentrations. Probl Strength Plast 67:20–25
Wang J, Fang Q, Zhu Z (1998) Sensitivity of ultrasonic attenuation and velocity change to cyclic deformation in pure aluminum. Phys Status Solidi (a) 169(1):43–48
Zuev LB, Murav'ev VV, Danilova YS (1999) Criterion for fatigue failure in steels. Tech Phys Lett 25(5):352–353

A Statistically-Based Homogenization Approach for Particle Random Composites as Micropolar Continua

Patrizia Trovalusci, Maria Laura De Bellis and Martin Ostoja-Starzewski

Abstract This article is focused on the identification of the size of the representative volume element (RVE) and the estimation of the relevant effective elastic moduli for particulate random composites modeled as micropolar continua. To this aim, a statistically-based scale-dependent multiscale procedure is adopted, resorting to a homogenization approach consistent with a generalized Hill's type macrohomogeneity condition. At the fine level the material has two phases (inclusions/matrix). Two different cases of inclusions, either stiffer or softer than the matrix, are considered. By increasing the scale factor, between the size of intermediate control volume elements (Statistical Volume Elements, SVEs) and the inclusions size, series of boundary value problems are numerically solved and hierarchies of macroscopic elastic moduli are derived. The constitutive relations obtained are grossly isotropic and are represented in terms of classical bulk, shear and micropolar bending moduli. The "finite size scaling" of these relevant elastic moduli for the two different material contrasts (ratio of inclusion to matrix moduli) is reported. It is shown that regardless the scaling behavior, which depends on the material phase contrast, the RVE size is statistically detected. The results of the performed numerical simulations also highlight the importance of taking into account the spatial randomness of inclusions which intersect the SVEs boundary.

Keywords Micropolar continua · Random composite · Scale-dependent homogenization · RVE

P. Trovalusci (✉)
Department of Structural Engineering and Geotechnics,
Sapienza University of Rome, via Gramsci 53, Rome, Italy
e-mail: patrizia.trovalusci@uniroma1.it

M.L. De Bellis
Department of Innovation Engineering, University of Salento,
Ecotekne Building - "Corpo O", Lecce, Italy
e-mail: marialaura.debellis@unisalento.it

M. Ostoja-Starzewski
Department of Mechanical Science and Engineering,
University of Illinois at Urbana-Champaign, Urbana 61801, IL, USA
e-mail: martinos@illinois.edu

H. Altenbach and S. Forest (eds.), *Generalized Continua as Models for Classical and Advanced Materials*, Advanced Structured Materials 42,
DOI 10.1007/978-3-319-31721-2_20

1 Introduction

The mechanical behavior of many materials of growing interest in materials science (such as composites, granular materials, alloys, liquid crystals and even rocks and masonry) is often strongly influenced by an existing or emergent microstructure (such as microcracks, voids, defects, dislocations, grains/phases in polycrystalline/multiphase materials). The discontinuous nature of these materials detectable at fine scales, smaller than the macroscopic scale, can be directly modeled (grain boundaries, dislocations, disclinations, joints, etc.) or modeled by deriving continuum field descriptions via multiscale approaches (Trovalusci et al. 2009; Sadowski and Trovalusci 2014; Trovalusci 2015), thus avoiding computationally cumbersome problems. In order to circumvent physical inadequacies and well-known theoretical/computational problems of microstructured material modelling it is widely acknowledged that continuum theories, suitable to retain memory of the microstructure taking into account the important role played by internal length scales, must have non-local character, i.e. must exhibit length scale parameters and spatial dispersion properties. Among these theories, from many years micropolar continua have been distinguished as suitable models for representing a wide class of materials with periodic structure, in particular for the possibility to take into account scale effects and non symmetrical shear behavior (Trovalusci and Masiani 1999, 2005; Bouyge et al. 2001; De Bellis and Addessi 2011; Eremeyev et al. 2012; Pau and Trovalusci 2012; Altenbach and Eremeyev 2013; Trovalusci and Pau 2014).

Attention is here focused on composite materials which display random morphologies, as particulate composites. These materials contain randomly distributed particles embedded into a matrix, typically made of metals, polymers or ceramics. Among numerous examples available, we cite tungsten carbide or titanium carbide embedded cobalt or nickel used to make cutting tools; aluminium alloy castings containing dispersed SiC particles widely used for automotive applications, including pistons and brake applications. Moreover, in the field of civil engineering, the concrete, made of cement, as binding medium, finely dispersed particulates of gravel, finer aggregates (sand) and water. A particular case of our interest is Roman concrete, made of hydraulic lime and pozzolan particles, widely diffused in constructions of Mediterranean regions. These composites are designed to produce unusual combinations of properties rather than to improve the strength of the plain matrix.

With the aim of investigating the gross mechanical response of this special class of random composites, we adopt a statistically-based multi-scale procedure which exploits the potentiality of a computational homogenization approach for evaluating the effective constitutive coefficients. The materials are modeled as linear elastic micropolar continua both at the fine heterogeneous level and at the gross homogenized level. The choice of such continuum models enriched by rigid local structure has advantages, with respect to the classical Cauchy continuum, related to the presence of length scale parameters and skew-symmetric strain and shear behaviour (Trovalusci and Masiani 1999; Pau and Trovalusci 2012; Trovalusci and Pau 2014).

A key issue in the case of materials with random microstructure is the circumstance that, differently from what happens in the case of periodic materials, it is not possible to 'a-priori' define a Representative Volume Element (RVE), this being an unknown of the problem. A possible way to solve this problem is to approach the RVE using finite-size scaling of intermediate control volume elements, named Statistical Volume Elements (SVEs), and solve the related Dirichlet and Neumann boundary value problems consistent with a generalized macrohomogeneity Hill's condition. In this way hierarchies of constitutive bounds for the material constitutive moduli can be defined, as proposed in the works (Ostoja-Starzewski 2006, 2008; Khisaeva and Ostoja-Starzewski 2006; Ostoja-Starzewski et al. 2007). This statistical approach developed for classical materials has been extended to micropolar materials (Trovalusci et al. 2015) also pointing out the unavailability of the standard periodic-type boundary conditions (Trovalusci et al. 2014).

At the fine level the material taken into account has two phases (inclusions/matrix). Two different cases of inclusions, either stiffer or softer than the matrix, are considered. By increasing the scale factor, between the size of SVEs and the inclusions size, series of boundary value problems are numerically solved. The constitutive relations obtained at the macroscopic level are isotropic and have been represented in terms of bulk, shear and micropolar bending moduli. The "finite size scaling" of these relevant elastic moduli has been represented for the two different material contrasts, paying particular attention to the differences in the solution obtained in the presence or not of inclusions that intersect the SVE's boundary.

The work is organized as follows. In Sect. 2 the main features of the proposed homogenization technique are addressed, with emphasis on the peculiar choice of micropolar continua at both levels. In Sect. 3.1 the statistical procedure is briefly described and attention is devoted to the numerical applications referred to two representative examples of particulate composites characterized by inclusions either stiffer or softer than the matrix (Sect. 3.2). In this context the results are interpreted by introducing properly conceived scaling measures, inspired to the work (Ranganathan and Ostoja-Starzewski 2008), and the convergence trend of these measures and of the relevant elastic moduli is investigated for both the material contrasts. It is shown that, regardless of the scaling behavior depending on the phase contrast in elastic moduli (ratio of inclusion to matrix moduli), the RVE size is statistically found. The importance of taking into account the spatial randomness accounting for non-homogeneous boundaries of the representative regions is also highlighted. In Sect. 4 some final remarks and ideas for future developments are discussed.

2 Micropolar Homogenization

The particulate composite is described at two levels: a 'fine' scale level, where the material is described as non homogeneous with particles of finite size (d) and given elastic parameters embedded in a matrix with different material properties, and a 'gross' scale level, in which the actual material is ideally replaced by an equivalent homogeneous material (with characteristic dimension L).

The micropolar continuum model is adopted for the description at both the fine and the gross level. At the gross level, namely the macroscopic or structural level, the use of a micropolar continuum is relevant when the dimension d is not negligible with respect to L. In this case, the principle of separation of scales does not hold and a standard local continuum cannot be adopted. Considering the fine level, here called mesolevel, we assume that each constituent is a microstructured material. In other words, we envisage the existence of a third underling finer scale (Forest et al. 1999, 2001; Onck 2002) where the microstructure is characterized by at least a further characteristic length, l_c, and can be effectively described either as a truss-like or a beam-like network. The mesoscale is then perceived as the result of an understood homogenization procedure from an underlying level, named microlevel, that in the former case leads to a classical continuum and in the latter case to a micropolar continuum. If l_c for instance characterizes the bending stiffness, this latter case is particularly suitable when microscopic bending deformation mechanism is predominant, as in the case of fiber-beam networks, polycrystals or metal matrix composites. In Fig. 1 a schematic of our basic assumptions is shown. In the following the attention will be focused on the transition from the mesoscopic to the macroscopic level, as we assume that the homogenization procedure linking the mesoscopic and

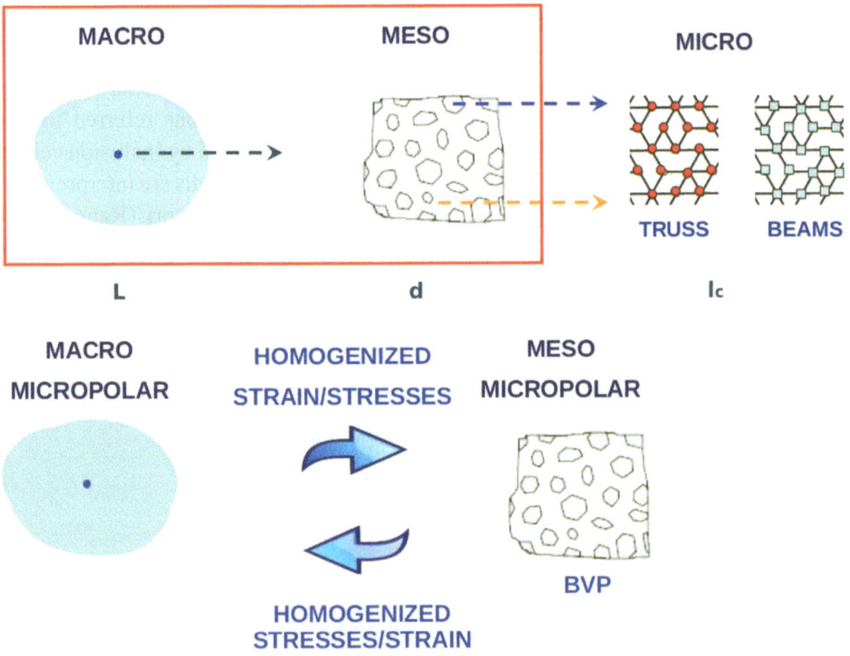

Fig. 1 Conceptual scheme of the proposed micropolar homogenization approach

the microscopic level is preparatory to this analysis and is beyond the scope of this work. In any case if at the microlevel l_c can be considered null for the constituent materials, and if the skew-symmetric shear stiffness is null, the classical continuum is recovered as model of the fine scale.

The idea is that a one to one correspondence between a macroscopic material point and a portion of the actual heterogeenous material holds. Whenever we want to estimate the homogenized constitutive response at the macroscopic level, we have to solve a properly defined boundary value problem at the mesoscopic level (problem that is driven by a macroscopic information).

The definition of the coupling conditions between levels depends on the so-called localization phase, i.e. the transfer of macroscopic information to the mesoscale, and the homogenization phase, when the local mesoscopic response is averaged and passed back to the macroscopic scale. It is worth noting that when the fine scale is represented by a continuum of the same kind of the gross scale continuum a generalized macrohomogenity condition of Hill's type can be adopted (Ostoja-Starzewski 2011; Li and Liu 2009), which ensures a one-to-one correspondence between the two scales without requiring the introduction of kind of internal constraints for the deformation mechanisms, as conversely occurs in the case of continua of different type (Forest and Sab 1998; De Bellis and Addessi 2011; Addessi et al. 2015). In the following these linking conditions, obtained via the solution of boundary value problems derived from macrohomogenity conditions generalized to the micropolar case, will be recalled in detail.

Let's start considering the mesoscopic level, where the heterogeneous actual material is described in detail assuming that each constituent is described as a linear elastic isotropic micropolar material. Within the framework of a linearized theory, the kinematics of this continuum is governed by the compatibility equations:

$$\gamma_{ij} = u_{i,j} + e_{kij}\varphi_k, \qquad \kappa_{ij} = \varphi_{i,j}, \tag{1}$$

where (u_i) and (φ_i) are the displacement and rotation vectors of each material point; (γ_{ij}) and (κ_{ij}) are the strain and curvature tensors, respectively, and (e_{ijk}) is the Levi–Civita tensor $(i, j, k = 1, 3)$.

The balance equations in the absence of body forces and couples are:

$$\tau_{ij,j} = 0, \qquad \mu_{kj,j} + e_{kji}\tau_{ij} = 0, \tag{2}$$

where (τ_{ij}) and (μ_{ij}) are respectively the stress and couple stress tensors. Denoting by (t_i) and (m_i) the tractions and surface couples on the boundary of a control volume of outward normal (n_i), we also have:

$$t_i = \tau_{ij}\, n_j, \qquad m_i = \mu_{ij}\, n_j. \tag{3}$$

In order to separately investigate the classical and micropolar components, we divide the strain and stress tensors into their symmetric and skew-symmetric part:

$$\gamma_{ij} = \varepsilon_{ij} + \alpha_{ij}, \qquad \tau_{ij} = \sigma_{ij} + \beta_{ij}, \tag{4}$$

where $\varepsilon_{ij} = \frac{1}{2}(u_{i,j} + u_{j,i})$, $\alpha_{ij} = \frac{1}{2}(u_{i,j} - u_{j,i}) - e_{kij}\varphi_k$ are the symmetric and skew-symmetric strain tensors, while $\sigma_{ij} = \frac{1}{2}(\tau_{ij} + \tau_{ji})$, $\beta_{ij} = \frac{1}{2}(\tau_{ij} - \tau_{ji})$ are the symmetric and skew-symmetric stress tensors.

From now on we restrict our calculations to the two-dimensional case (2D). In this framework the independent strain and stress components of the micropolar continuum can be ordered into the vectors:

$$\begin{array}{ll} \{\varepsilon\} = \{\varepsilon_{11} \quad \varepsilon_{22} \quad \varepsilon_{12}\}^T, & \{\sigma\} = \{\sigma_{11} \quad \sigma_{22} \quad \sigma_{12}\}^T, \\ \{\alpha\} = \{\alpha_{12}\}, & \{\beta\} = \{\beta_{12}\}, \\ \{\kappa\} = \{\kappa_{31} \quad \kappa_{32}\}^T, & \{\mu\} = \{\mu_{31} \quad \mu_{32}\}^T. \end{array} \tag{5}$$

The stress–strain relations for the 2D linear elastic isotropic micropolar material can then be written as:

$$\begin{bmatrix} \sigma_{11} \\ \sigma_{22} \\ \sigma_{12} \\ \beta_{12} \\ \mu_{31} \\ \mu_{32} \end{bmatrix} = \begin{bmatrix} \lambda + 2\mu & \lambda & 0 & 0 & 0 & 0 \\ \lambda & \lambda + 2\mu & 0 & 0 & 0 & 0 \\ 0 & 0 & 2\mu & 0 & 0 & 0 \\ 0 & 0 & 0 & -2\mu_c & 0 & 0 \\ 0 & 0 & 0 & 0 & 2\mu l_c^2 & 0 \\ 0 & 0 & 0 & 0 & 0 & 2\mu l_c^2 \end{bmatrix} \begin{bmatrix} \epsilon_{11} \\ \epsilon_{22} \\ \epsilon_{12} \\ \alpha_{12} \\ \kappa_{31} \\ \kappa_{32} \end{bmatrix}, \tag{6}$$

which involve four independent elastic constitutive parameters: the Lamé constants λ and μ, the Cosserat shear modulus μ_c, and the so-called characteristic length l_c, which is responsible for the bending stiffness.

At the macroscopic level no 'a-priori' hypotheses on the material symmetries are formulated and the homogenized elastic constitutive coefficients directly arise from the homogenization process, consistent with the so-called generalized macrohomogeneity condition which establishes an energetic equivalence between a portion of heterogenous material at the mesoscopic level and the material point at the macroscopic level. Let us now consider a portion of the heterogeneous material, i.e. a mesoscale window $\mathscr{B}_\delta$ of area A_δ (where $\delta = L/d$ is the scale factor, with d being the averaged inclusion size and L the size of the control area element). The generalized macrohomogeneity (Hill–Mandel's type) condition, accounting for the presence of classical and micropolar variables, here considered separately, can be expressed as:

$$\frac{1}{A_\delta} \int\limits_{\mathscr{B}_\delta} (\sigma_{ij}\varepsilon_{ij} + \beta_{ij}\alpha_{ij} + \mu_{ij}\kappa_{ij}) = \overline{\sigma}_{ij}\overline{\varepsilon}_{ij} + \overline{\beta}_{ij}\overline{\alpha}_{ij} + \overline{\mu}_{ij}\overline{\kappa}_{ij}, \tag{7}$$

where overbars denote macroscopic quantities obtained as surface averaged quantities. By exploiting the energy equivalence condition (Eq. 7) it is possible to derive the coupling conditions, both localization and homogenization conditions. At the localization step, Dirichlet-type and Neumann-type boundary conditions to enforce on the mesoscale are defined. The homogenization step is the evaluation of the homogenized constitutive moduli at the macroscopic level.

By ordering the independent strain and stress components of the 2D micropolar model into the following vectors:

$$\begin{array}{ll} \{\overline{\varepsilon}\} = \{\overline{\varepsilon_{11}}\ \ \overline{\varepsilon_{22}}\ \ \overline{\varepsilon_{12}}\}^T, & \{\overline{\sigma}\} = \{\overline{\sigma_{11}}\ \ \overline{\sigma_{22}}\ \ \overline{\sigma_{12}}\}^T, \\ \{\overline{\alpha}\} = \{\overline{\alpha_{12}}\}, & \{\overline{\beta}\} = \{\overline{\beta_{12}}\}, \\ \{\overline{\kappa}\} = \{\overline{\kappa_{31}}\ \ \overline{\kappa_{32}}\}^T, & \{\overline{\mu}\} = \{\overline{\mu_{31}}\ \ \overline{\mu_{32}}\}^T, \end{array} \tag{8}$$

the macroscopic stress–strain relations can be written as:

$$\begin{bmatrix} \overline{\sigma}_{11} \\ \overline{\sigma}_{22} \\ \overline{\sigma}_{12} \\ \overline{\beta}_{12} \\ \overline{\mu}_{31} \\ \overline{\mu}_{32} \end{bmatrix} = \begin{bmatrix} \overline{\mathbb{A}}_{1111} & \overline{\mathbb{A}}_{1122} & \overline{\mathbb{A}}_{1112} & \overline{\mathbb{D}}_{1112} & \overline{\mathbb{F}}_{1131} & \overline{\mathbb{F}}_{1132} \\ \overline{\mathbb{A}}_{2211} & \overline{\mathbb{A}}_{2222} & \overline{\mathbb{A}}_{2212} & \overline{\mathbb{D}}_{2212} & \overline{\mathbb{F}}_{2231} & \overline{\mathbb{F}}_{2232} \\ \overline{\mathbb{A}}_{1211} & \overline{\mathbb{A}}_{1222} & \overline{\mathbb{A}}_{1212} & \overline{\mathbb{D}}_{1212} & \overline{\mathbb{F}}_{1231} & \overline{\mathbb{F}}_{1232} \\ \overline{\mathbb{D}}_{1211} & \overline{\mathbb{D}}_{1222} & \overline{\mathbb{D}}_{1212} & \overline{\mathbb{B}}_{1212} & \overline{\mathbb{G}}_{1231} & \overline{\mathbb{G}}_{1232} \\ \overline{\mathbb{F}}_{3111} & \overline{\mathbb{F}}_{3122} & \overline{\mathbb{F}}_{3112} & \overline{\mathbb{G}}_{3112} & \overline{\mathbb{C}}_{3131} & \overline{\mathbb{C}}_{3132} \\ \overline{\mathbb{F}}_{3211} & \overline{\mathbb{F}}_{3222} & \overline{\mathbb{F}}_{3212} & \overline{\mathbb{G}}_{3212} & \overline{\mathbb{C}}_{3231} & \overline{\mathbb{C}}_{3232} \end{bmatrix} \begin{bmatrix} \overline{\epsilon}_{11} \\ \overline{\epsilon}_{22} \\ \overline{\epsilon}_{12} \\ \overline{\alpha}_{12} \\ \overline{\kappa}_{31} \\ \overline{\kappa}_{32} \end{bmatrix}. \tag{9}$$

The homogenized model is generally anisotropic and the components of the constitutive tensor of Eq. (9) are such that the major symmetries are preserved. If the material is centrosymmetric (i.e. non-chiral) the components $\overline{\mathbb{D}}_{ijhk}$, $\overline{\mathbb{F}}_{ijhk}$ and $\overline{\mathbb{G}}_{ijhk}$ are null. In this case, Eq. (9) can be rewritten as:

$$\begin{array}{l} \{\overline{\sigma}\} = [\overline{\mathbb{A}}]\{\overline{\varepsilon}\}, \\ \{\overline{\beta}\} = \overline{\mathbb{B}}_{1212}\{\overline{\alpha}\}, \\ \{\overline{\mu}\} = [\overline{\mathbb{C}}]\{\overline{\kappa}\}, \end{array} \tag{10}$$

the stress and strain components being collected as in Eq. (8).

The boundary conditions derived from the macro-homogeneity condition (7) are briefly recalled below.

- Dirichlet boundary conditions We consider a square-shaped mesoscale domain $\mathscr{B}_\delta$ whose center is fixed at the origin of the coordinate system. On account of the above condition in Eq. (7), we set up the Dirichlet boundary conditions:

$$u_i = \overline{\varepsilon_{ij}} x_j, \qquad \varphi_3 = \frac{1}{2} e_{ij3} \overline{\alpha}_{ij} + \overline{\kappa_{3i}} x_i \qquad \text{on } \partial\mathscr{B}_\delta,$$

$(i, j = 1, 2)$. The solution of the cell problem under various combinations of boundary conditions yields the homogenized stresses $(i, j = 1, 2)$:

$$\overline{\sigma}_{ij} = \frac{1}{A_\delta} \int_{\partial\mathscr{B}_\delta} (t_i x_j + t_j x_i), \qquad \overline{\beta}_{ij} = \frac{1}{2A_\delta} e_{ij3} \int_{\partial\mathscr{B}_\delta} m_3, \qquad \overline{\mu}_{3i} = \frac{1}{A_\delta} \int_{\mathscr{B}_\delta} \mu_{3i}.$$

- Neumann boundary conditions In the case of Neumann boundary conditions, on account of Eq. (7), we impose:

$$t_i = (\overline{\sigma}_{ij} + \overline{\beta}_{ij})n_j, \qquad m_3 = m_3^o + \overline{\mu}_{3i}n_i \qquad \text{on } \partial\mathscr{B}_\delta,$$

where

$$m_3^o = -\int_{\partial\mathscr{B}} e_{ij3}x_i\overline{\beta}_{jk}n_k$$

is the moment imposed to ensure the moment balance in the presence of skew–symmetric shear $(i, j, k = 1, 2)$. The resulting homogenized strains are $(i, j = 1, 2)$:

$$\overline{\varepsilon}_{ij} = \frac{1}{A_\delta}\int_{\mathscr{B}_\delta}\varepsilon_{ij} \qquad \overline{\alpha}_{ij} = \frac{1}{2A_\delta}\int_{\mathscr{B}_\delta}\alpha_{ij}, \qquad \overline{\kappa}_{3i} = \frac{1}{A_\delta}\int_{\mathscr{B}_\delta}\kappa_{3i}.$$

3 Statistical Homogenization Convergence

3.1 Computational Multiscale Procedure

We consider a simplified geometry of a two-dimensional particle composite material. We assume that all the inclusions are disks of equal size randomly distributed in the matrix phase. Some parameters are fixed: the inclusions density, defined as the ratio between the area of the inclusions with respect to the total area of the body $\rho = A_i/A_{tot}$; the diameter d of the inclusions and the material parameters of the two phases. The number and the position of the inclusions instead randomly vary.

The constitutive response of a non-periodic heterogeneous material requires the definition of the size of a Representative Volume Element (RVE), L_{RVE}, larger than the fine scale characteristic length, d, such that to nullify the influence of the boundary conditions on the RVE. This prescription ensures a homogenization limit in the sense of Hill which generically states that $L_{RVE} >> d$ (Ostoja-Starzewski 2006; Khisaeva and Ostoja-Starzewski 2006). According to the approach presented in the above mentioned papers, as well as in Ostoja-Starzewski et al. (2007), Ostoja-Starzewski (2008), the present statistical procedure requires the definition of a number of realizations of the microstructure, sampled in a Monte Carlo sense, allowing us to determine the statistics of scale-dependent upper and lower bounds for the overall elastic moduli and to specifically approach the RVE size for the micropolar continuum (Trovalusci et al. 2015).

In particular, fixed a value for the scale parameter $\delta = L/d$, we identify square test windows of side L (finite size Statistical Volume Elements—SVEs) as portions of the heterogenous material to homogenize. The realizations of the microstructure $\mathscr{B}_\delta(\omega)$, that is the number and position of inclusions within any test window (ω being an elementary event over a sample space), are generated by a hard-core Poisson's point field (i.e. not allowing for disks' overlaps), thus simulating a mesoscale window placed anywhere in the random medium. For any $\mathscr{B}_\delta(\omega)$, we perform the homogenization process by solving both classical and micropolar Dirichlet and Neumann BVPs. The estimation of an effective modulus $\overline{X}_\delta$ is obtained when its average value falls within the interval corresponding at a confidence level set at 95 % over a normal standard distribution. By increasing δ, the convergence is finally achieved when the number of realizations necessary to satisfy the above requirement is less than a proper fixed value (Trovalusci et al. 2015). The effective moduli are then estimated as the mean convergence values between the Dirichlet (upper) and Neumann (lower) bounds.

Note that the mesoscale window $\mathscr{B}_\delta(\omega)$ ideally corresponds to a portion of the actual random medium in which inclusions are not prevented from intersecting the window edges. Thus, the numerical simulations of Sect. 3.2 are performed by taking into account non-homogeneous boundaries (crossing inclusions). We also consider the less realistic case of homogeneous boundaries (non-crossing inclusions). The comparison between the homogenized responses obtained by performing numerical simulations for the two cases, either applying Dirichlet or Neumann boundary conditions, allows us to emphasize the influence of positions of the inclusions with respect to the window's boundary.

3.2 *Finite Size Scaling of Elastic Moduli. Numerical Simulations*

The multiscale statistically based procedure recalled in Sect. 3.1 has been implemented and the boundary problems have been numerically solved by using COMSOL Multiphysics® software. We consider two cases of particulate composites, referred as material (a) and material (b) (Fig. 2), characterized by different material contrasts defined by the relations: E_i/E_m (ratio between the Young modulus of inclusions and the Young modulus of the matrix) and l_c^i/l_c^m (ratio between the characteristic lengths of inclusions and matrix, respectively). In the material (a) stiffer inclusions are embedded into a softer matrix, while for the material (b) inclusions are softer than the matrix. In Table 1 we list the adopted material parameters in a dimensionless form, expressing the ratio between corresponding quantities of inclusions and matrix. Consistently with the above definitions, material (a) is characterized by an

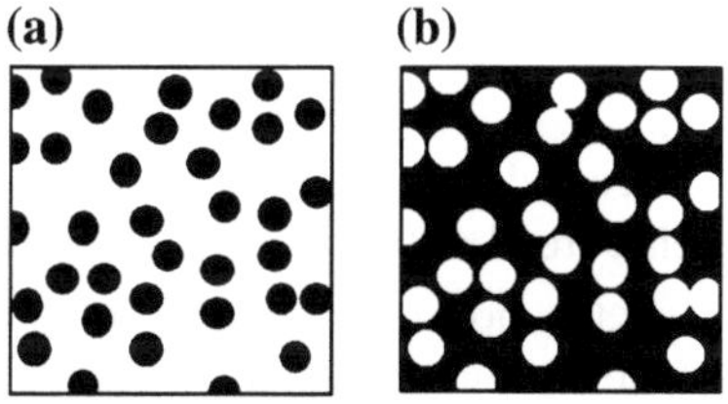

Fig. 2 Sketch of high contrast (**a**) and low contrast (**b**) material

Table 1 Ratios between material parameters of inclusions and matrix

Material	Parameters			
	λ_i/λ_m	μ_i/μ_m	μ_{c_i}/μ_{c_m}	l_{c_i}/l_{c_m}
(a)	46	4.93	4.93	10^1
(b)	0.021	0.202	0.202	10^{-1}

higher contrast value than material (b). Both the materials have circular inclusions and nominal nominal area fraction $\rho = 40\,\%$.

We numerically solve the two dimensional BVPs described in Sect. 2 considering increasing window sizes ranging from $\delta = 5$ to $\delta = 25$ ($\delta = L/d$, with the inclusions' diameter d fixed). In the Finite Element discretization, we adopt unstructured meshes of quadratic Lagrangian triangular elements.

We noticed that the homogenized material behaves with a good approximation like an isotropic linear elastic (micropolar) material, with the constitutive laws represented as in Eq. (6). Thus, by recognizing the spherical component of the stress, $\sigma^{sph} = (\sigma_{11} + \sigma_{22})/2$, and strain, $\varepsilon^{sph} = (\varepsilon_{11} + \varepsilon_{22})/2$, we can write:

$$\overline{\sigma}^{sph} = \frac{1}{2}(2\overline{\mathbb{A}}_{1122} + \overline{\mathbb{A}}_{1212})(\overline{\varepsilon}_{11} + \overline{\varepsilon}_{22}) = \overline{K}\cdot\varepsilon^{sph}, \tag{11}$$

where $\overline{K} = (2\overline{\mathbb{A}}_{1122} + \overline{\mathbb{A}}_{1212})$ is the bulk modulus in planar stress and strain state. The deviatoric part of the stress and strain are connected by the vectorial relation:

$$\overline{\sigma}^{dev} = \begin{bmatrix} \overline{\sigma}_{11} - \overline{\sigma}^{sph} \\ \overline{\sigma}_{22} - \overline{\sigma}^{sph} \\ \overline{\sigma}_{12} \end{bmatrix} = \frac{\overline{\mathbb{A}}_{1212}}{2} \begin{bmatrix} \overline{\varepsilon}_{11} - \overline{\varepsilon}^{sph} \\ \overline{\varepsilon}_{22} - \overline{\varepsilon}^{sph} \\ \overline{\varepsilon}_{12} \end{bmatrix} = \overline{G}\overline{\varepsilon}^{dev}, \tag{12}$$

where $\overline{G} = \overline{\mathbb{A}}_{1212}/2$ is the classical shear modulus. Thus we focus on the moduli $\overline{K}$ and $\overline{G}$ for investigating the convergence trend of the classical material response.

In the isotropic case the micropolar shear is a scalar term related to the relative rotation by the constitutive component $\mathbb{B}_{1212}$, while the couple stress is related to the curvature tensors by the modulus $\operatorname{tr}\mathbb{C}$:

$$\{\overline{\beta}\} = \overline{\mathbb{B}}_{1212}\{\overline{\alpha}\}, \tag{13}$$

$$\{\overline{\mu}\} = \frac{1}{2}\operatorname{tr}\mathbb{C}\{\overline{\kappa}\}. \tag{14}$$

We consider the bending modulus $\overline{l_C} = \sqrt{\operatorname{tr}\overline{\mathbb{C}}/\overline{\mathbb{B}}_{1212}}$, for representing the convergence trend of the micropolar material response.

Let us now consider a mesoscale window $\mathscr{B}_\delta$ and denote $\overline{K}_\delta^D$ and $\overline{K}_\delta^N$ the values of the average of the bulk moduli obtained by solving the boundary value problems on $\mathscr{B}_\delta$ by adopting Dirichlet and Neumann-type boundary conditions, respectively.

Analogously, we denote $\overline{G}^D_\delta$ and $\overline{G}^N_\delta$ the average of the shear moduli and $l^D_{c\,\delta}$ and $l^D_{c\,\delta}$ the average of the bending moduli, always obtained by respectively solving Dirichlet and Neumann BVPs.

For each window we evaluate the following constitutive scaling measures which, in the spirit of the work (Ranganathan and Ostoja-Starzewski 2008), give qualitative and quantitative information about the convergence trend of the classical solution:

$$f^K_\delta = \frac{\overline{K}^D_\delta}{\overline{K}^N_\delta} - \frac{\overline{K}^D_{hom}}{\overline{K}^N_{hom}}, \qquad f^G_\delta = \frac{\overline{G}^D_\delta}{\overline{G}^N_\delta} - \frac{\overline{G}^D_{hom}}{\overline{G}^N_{hom}}, \tag{15}$$

where $\overline{K}^D_{hom}$, $\overline{K}^N_{hom}$ ($\overline{G}^D_{hom}$, $\overline{G}^N_{hom}$) are the bulk moduli (shear moduli) obtained in the case of equivalent homogeneous material. The superscripts 'D' and 'N' always stand for Dirichlet and for Neumann-type boundary conditions, respectively. Similarly, we define an additional quantity suitable for investigating the convergence trend of the micropolar solution:

$$f^C_\delta = \frac{\overline{l}^D_{c\delta}}{\overline{l}^N_{c\delta}} - \frac{\overline{l}^D_{c\,hom}}{\overline{l}^N_{c\,hom}}. \tag{16}$$

where $\overline{l}^D_{c\,hom}$ and $\overline{l}^N_{c\,hom}$ are the bending moduli obtained in the case of equivalent homogeneous material. Note that, as for homogeneous materials it is:

$$\frac{\overline{K}^D_{hom}}{\overline{K}^N_{hom}} = 1, \qquad \frac{\overline{G}^D_{hom}}{\overline{G}^N_{hom}} = 1, \qquad \frac{\overline{l}^D_{c\,hom}}{\overline{l}^N_{c\,hom}} = 1, \tag{17}$$

the scaling measures f^K, f^G and f^C provide an estimate of the convergence error.

The results reported in Figs. 3, 4 and 5 show the convergence trend of the classical bulk, shear and micropolar bending moduli depending on the material contrast,

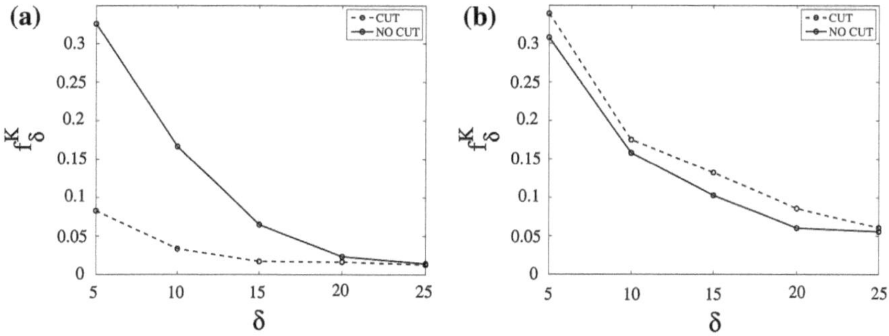

Fig. 3 Convergence trend of the classical bulk modulus. *Left side* material (**a**), *right side* material (**b**). Inclusions crossing (*dash lines*) and non-crossing (*solid lines*) the windows' edges

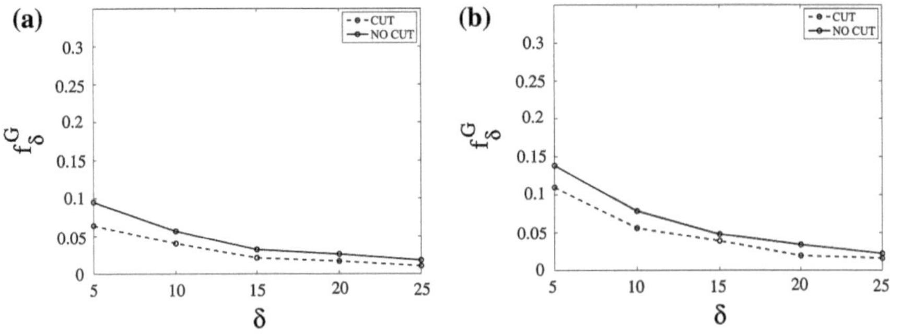

Fig. 4 Convergence trend of the classical shear modulus. *Left side* material (**a**), *right side* material (**b**). Inclusions crossing (*dash lines*) and non-crossing (*solid lines*) the windows' edges

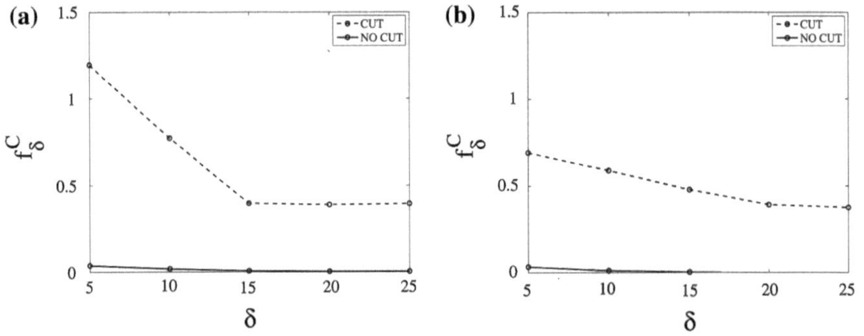

Fig. 5 Convergence trend of the micropolar bending modulus. *Left side* material (**a**), *right side* material (**b**). Inclusions crossing (*dash lines*) and non-crossing (*solid lines*) the windows' edges

highlighting the influence of the boundary conditions on the gross material response when accounting or not the intersections of the inclusions with the windows' edges. In particular, for the higher contrast material (a) f^K quite vanishes in both the cases of crossing and non crossing inclusions. In the lower contrast material (b) f^K_δ does not vanish, both in the case of crossing and more in the case of non crossing inclusions. For $\delta = 25$, in the material (a) the difference between the bulk moduli, evaluated via Dirichlet and Neumann BCs, is of 1.0 % while in the material (b) is of 6.0 % and 5.5 %, in the presence or not of inclusions crossing the test windows' boundary, respectively. Analogous considerations can be made for the values of f^G_δ. For $\delta = 25$, in the material (a) the difference between the shear moduli, evaluated via Dirichlet and Neumann BCs, is of 1.5 % (0.2 %), in the presence (or not) of crossing inclusions, while in the material (b) is of 2.0 % (2.5 %), in the presence (or not) of crossing inclusions.

The different trend between the case of crossing and non crossing inclusions is more evident for the micropolar value f^C_δ. In the material (a), for $\delta \geq 15$, the difference between the bending moduli evaluated via Dirichlet and Neumann BCs in the presence of crossing inclusions is of 38 %, while in the material (b), for $\delta \geq 20$, is of 42 %.f_δ^C vanishes for both materials when inclusions do not intersect the window's boundary.

Regarding the classical moduli, these results show that the contrast in the composite with particles softer than the matrix has the effect of slowing down the convergence to the RVE. More generally, it has been shown that simulations performed for composites with soft particles in a stiffer matrix seemed to indicate the need of very large scales for the RVE, thus compromising the possibility to homogenize the composite through upper (Dirichlet-BC) and lower (Neumann-BC) bounds for the constitutive moduli (Ostoja-Starzewski 2006; Ranganathan and Ostoja-Starzewski 2008). This discrepancy, related to the position of the boundary value problems based on affine displacements or uniform traction BC hypotheses, called for the need of different definitions of the boundary value problems, for instance by resorting to periodic boundary conditions (Terada et al. 2000).

We here accept the solution of upper/lower bound approach basing on the statistical criterion of convergence mentioned in Sect. 3, which does not require large scale representative windows for performing the homogenization process. In particular, we showed that the proposed statistical procedure converges when, by increasing the scale ratio δ, the results in terms of average elastic moduli do not change within a selected tolerance value. More precisely the number of realizations N, accounted for a given scale parameter δ, corresponding to randomly moving the window within the whole medium $\mathscr{B}_\delta(\omega)$ as mentioned above, is such that $1.96\,\sigma/(\overline{X}_\delta\sqrt{N}) \leq tol$, where: $\overline{X}_\delta$ is the average of an elastic modulus to estimate; σ its standard deviation and *tol* a given tolerance value depending on the dispersion of the data (Trovalusci et al. 2015). The convergence values of an effective modulus is obtained at a value δ for which N is less than a convenient small value. The effective moduli are then estimated as the mean convergence values between the Dirichlet (upper) and Neumann (lower) bounds. It is worth noting that in the case of inclusions that do not cross the windows' edges this value is comparable to the value achieved using periodic boundary conditions (Trovalusci et al. 2014).

In Fig. 6 the average of the bulk modulus $\overline{K}$ versus the scale parameter δ is shown for both materials (a) and (b). This value is normalized with respect to the corresponding modulus obtained for the RVE, $\overline{K}_{RVE}$, i.e. taking into account the average values of the coefficients evaluated at the convergence window in the case of crossing inclusions. The convergence trend to the RVE depends on whether inclusions cross or do not cross the windows' boundary. In particular, for the material (a) the RVE size, δ_{RVE}, is equal to 20 in the case of crossing inclusions, while it is equal to 25 in the case of non-crossing inclusions. The material (b) shows a slower convergence trend. Accordingly, the RVE is attained for $\delta_{RVE} = 25$ in the case of crossing inclusions, while in the case of non-crossing inclusions $\delta_{RVE} > 25$. It is worth noting that when crossing inclusions stiffer than the matrix (a) are accounted for the bulk modulus is higher with respect to the case of non crossing inclusions, while in the case of crossing inclusions softer than the matrix (b) the bulk modulus is lower than in the case of non crossing inclusions.

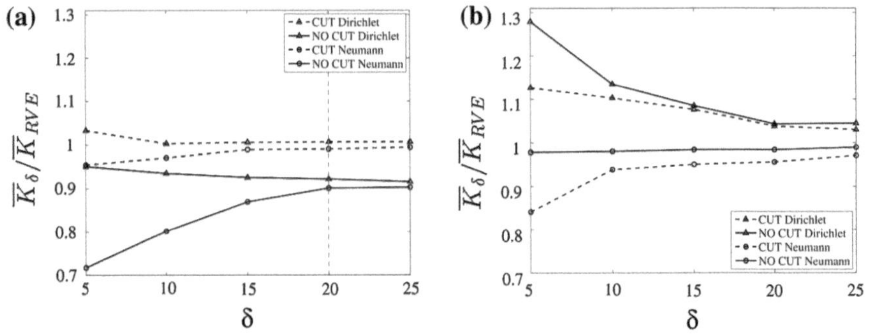

Fig. 6 Average of effective bulk modulus $\overline{K}$ (normalized to the RVE modulus $\overline{K}_{RVE}$) versus scale parameter δ, under Dirichlet (D-BC) and Neumann (N-BC) BVPs solutions. *Left side* higher contrast material (**a**). *Right side* lower contrast material (**b**). Inclusions crossing (*dash lines*) and non-crossing (*solid lines*) the windows' edges

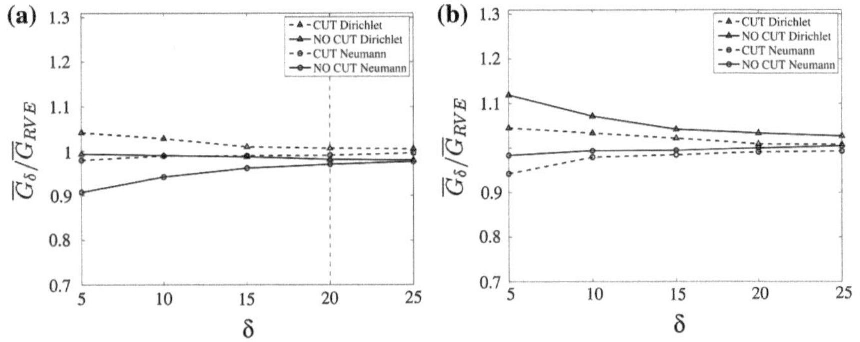

Fig. 7 Average of effective shear modulus $\overline{G}$ (normalized to the RVE modulus $\overline{G}_{RVE}$) versus scale parameter δ, under Dirichlet (D-BC) and Neumann (N-BC) BVPs solutions. *Left side* higher contrast material (**a**). *Right side* lower contrast material (**b**). Inclusions crossing (*dash lines*) and non-crossing (*solid lines*) the windows' edges

In Fig. 7 the average of the shear modulus $\overline{G}_\delta$ versus the scale parameter δ is shown for both materials (a) and (b). Also in this case the value is normalized with respect to $\overline{G}_{RVE}$, i.e. the average value of the modulus evaluated for the RVE in the case of crossing inclusions. The convergence trend and the RVE size is comparable to that obtained for the bulk modulus. Also in this case material (a) is stiffer when crossing inclusions are accounted for, while the material (b) is softer.

In Fig. 8 the average of the bending modulus $\overline{l}_{c\delta}$ versus the scale parameter δ, normalized with respect to the RVE value $\overline{l}_{cRVE}$ obtained in the case of crossing inclusions, is shown for both materials (a) and (b). The RVE for the material (a) is achieved for $\delta_{RVE} = 15$ in the case of crossing inclusions, while $\delta_{RVE} = 20$ in the case of non-crossing inclusions.

Overall the results show that the influence of crossing inclusions is more appreciable for the higher contrast material. The differences of the results obtained when

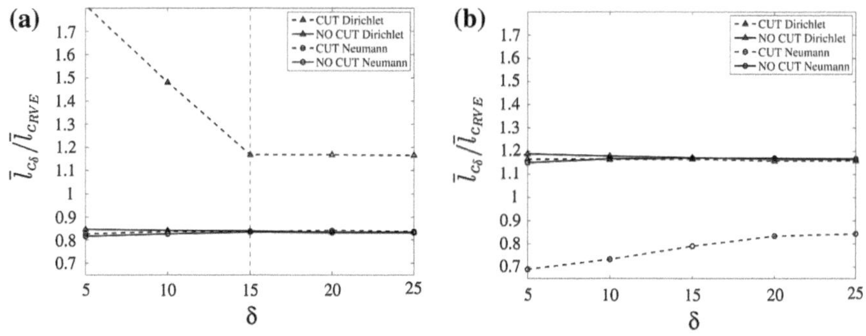

Fig. 8 Average of effective bending modulus $\bar{l}_c$ (normalized to the RVE modulus $\bar{l}_{cRVE}$) versus scale parameter δ, under Dirichlet (D-BC) and Neumann (N-BC) BVPs solutions. *Left side* higher contrast material (**a**). *Right side* lower contrast material (**b**). Inclusions crossing (*dash lines*) and non-crossing (*solid lines*) the windows' edges

crossing inclusions are accounted for, with respect to the case of non-crossing inclusions, are more significant in the micropolar case than in the classical case, exhibiting a stiffer behaviour for material (a) and a softer behaviour for material (b), respectively. In all the cases analyzed the RVE size has been detected and the classical and micropolar moduli have been estimated.

4 Final Remarks

A scale-dependent procedure, based on the derivation of hierarchies of upper and lower bounds, has been proposed to estimate the constitutive moduli of particulate random composite described as micropolar homogenized continua. The Representative Volume Element (RVE) size and the corresponding, classical and micropolar, effective moduli have been statistically detected for materials with different contrasts between elastic moduli (ratio of inclusion to matrix moduli). The convergence trend has been represented for all the significant moduli of an effective isotropic material in terms of scaling measures that quantify the error from the homogeneous solution. It has been shown that, regardless of the scaling behaviour depending on the phase contrast in elastic moduli, by virtue of the statistical approach, the RVE size and the homogenized solution can be achieved and the effective constitutive moduli identified. Both in the classical and, even more so, in the micropolar case, the estimation of the RVE size and the corresponding moduli strongly depends on the possibility of taking into account the presence of inclusions that cross the windows, an aspect inherent in spatial randomness of the material. We also found that the higher contrast medium (a) is slightly more sensitive than the medium (b) to the presence of inclusions which cross the windows' edges. In particular, the medium (a) when the stiff inclusions cross the boundary is stiffer than in the case in which inclusions are forced to not cross the boundary; conversely, the medium (b) with soft crossing inclusions is softer with respect to the case of non-crossing inclusions. Moreover, the differences in terms of average effective moduli and RVE sizes achieved in the presence of crossing or non crossing inclusions reduce but remain non-negligible when the window size increases. Overall, the statistical simulations show that it is not correct, both in the classical and the micropolar case, to neglect the presence of inclusions that intersect the windows' edges, even if the RVE size is large enough. To this regard, a subject of forthcoming research is to develop a scale-dependent statistical homogenization procedure resorting to the solution of scale-dependent periodic boundary conditions. This can be done specifically conceived periodized boundaries, by enforcing the periodicity of the material between corresponding edges of the windows, as for instance proposed by Gitman et al. (2007), Sab and Nedjar (2005), without forsaking the realistic hypothesis of non-homogeneous boundaries. It is expected that the advantages of specifically conceived periodic boundary conditions will significantly reduce the computational burden and accelerate the convergence to the RVE allowing us the estimation of the effective material constants.

References

Addessi D, De Bellis ML, Sacco E (2015) A micromechanical approach for the cosserat modeling of composites. Meccanica 51(3):569–592

Altenbach H, Eremeyev VA (2013) Cosserat media. In: Altenbach H, Eremeyev V (eds) Generalized continua from the theory to engineering application, CISM courses and lectures, vol 541. Springer, Berlin, pp 65–129

Bouyge F, Jasiuk I, Ostoja-Starzewski M (2001) A micromechanically based couple-stress model of an elastic two-phase composite. Int J Solids Struct 38:1721–1735

De Bellis ML, Addessi D (2011) A Cosserat based multi-scale model for masonry structures. Int J Multiscale Comput Eng 9(5):543–563

Eremeyev VA, Lebedev LP, Altenbach H (2012) Foundations of micropolar mechanics. Springer Science and Business Media, Heidelberg

Forest S, Sab K (1998) Cosserat overall modeling of heterogeneous materials. Mech Res Commun 25:449–454

Forest S, Dendievel R, Canova GR (1999) Estimating the overall properties of heterogeneous cosserat materials. Model Simul Materials Sci Eng 7:829–840

Forest S, Pradel F, Sab K (2001) Asymptotic analysis of heterogeneous Cosserat media. Int J Solids Struct 38:4585–4608

Gitman IM, Askes H, Sluys L (2007) Representative volume: existence and size determination. Eng Fract Mech 74:2518–2534

Khisaeva Z, Ostoja-Starzewski M (2006) On the size of RVE in finite elasticity of random composites. J Elast 85:153–173

Li X, Liu Q (2009) A version of Hill's lemma for Cosserat continuum. Acta Mech Sinica 25:499–506

Onck PR (2002) Cosserat modeling of cellular solids. Comptes Rendus Mec 330:717–722

Ostoja-Starzewski M (2006) Material spatial randomness: from statistical to representative volume element. Prob Eng Mech 21:112–132

Ostoja-Starzewski M (2008) Microstructural randomness and scaling in mechanics of materials., CRC series: modern mechanics and mathematicsTaylor & Francis, Boca Raton

Ostoja-Starzewski M (2011) Macrohomogeneity condition in dynamics of micropolar media. Arch Appl Mech 81:899–906

Ostoja-Starzewski M, Du X, Khisaeva Z, Li W (2007) Comparisons of the size of representative volume element in elastic, plastic, thermoelastic, and permeable random microstructures. Int J Multiscale Comput Eng 5:73–82

Pau A, Trovalusci P (2012) Block masonry as equivalent micropolar continua: the role of relative rotations. Acta Mech 223(7):1455–1471

Ranganathan S, Ostoja-Starzewski M (2008) Scale-dependent homogenization of inelastic random polycrystals. ASME J Appl Mech 75:1–9

Sab K, Nedjar B (2005) Periodization of random media and representative volume element size for linear composites. Comptes rendus de l'Academie des Sciences-Mecanique 333:187–195

Sadowski T, Trovalusci P (2014) Multiscale modeling of complex materials: phenomenological, theoretical and computational aspects. No. 556 in courses and lectures, CISM (International Centre for Mechanical Sciences), Springer, Vienna

Terada K, Hori T, Kyoya T, Kikuchi N (2000) Simulation of the multi-scale convergence in computational homogenization approach. Int J Solids Struct 37:2285–2311

Trovalusci P (ed) (2015) Multiscale and multifield modeling and simulation. Materials with internal structure. Springer Tracts in Mechanical Engineering, Springer Int. Publishing, Switzerland, Heidelberg

Trovalusci P, Masiani R (1999) Material symmetries of micropolar continua equivalent to lattices. Int J Solids Struct 36(14):2091–2108

Trovalusci P, Masiani R (2005) A multi-field model for blocky materials based on multiscale description. Int J Solids Struct 42:5778–5794

Trovalusci P, Pau A (2014) Derivation of microstructured continua from lattice systems via principle of virtual works. The case of masonry-like materials as micropolar, second gradient and classical continua. Acta Mech 225(1):157–177

Trovalusci P, Capecchi D, Ruta G (2009) Genesis of the multiscale approach for materials with microstructure. Arch Appl Mech 79:981–997

Trovalusci P, De Bellis ML, Ostoja-Starzewski M, Murrali A (2014) Particulate random composites homogenized as micropolar materials. Meccanica 49(9):2719–2727

Trovalusci P, Ostoja-Starzewski M, De Bellis ML, Murrali A (2015) Scale-dependent homogenization of random composites as micropolar continua. Eur J Mech A/Solids 49:396–407

Paradoxical Size Effects in Composite Laminates and Other Heterogeneous Materials

Marcus A. Wheel, Jamie C. Frame and Philip E. Riches

Abstract Size effects in which there is an apparent increase in stiffness with reducing size scale are forecast in those heterogeneous materials that have constitutive behaviour described by more generalized continuum theories such as couple stress, micropolar or micromorphic elasticity. This short paper considers possibly the simplest heterogeneous material exhibiting such size effects, a two phase composite laminate consisting of alternating layers of stiff and compliant material, and shows that when loaded in bending the nature of the size effect actually depends on the composition of the sample surfaces. The laminate material is apparently capable of exhibiting a diversity of size effects some of which are compatible with the predictions of generalized continuum theories while others are contradictory. Another heterogeneous material consisting of a periodic or regular array of voids within a classically elastic matrix is then considered. Detailed finite element analysis shows that the diversity of size effects encountered in the laminate material may also be observed in this more representative material thereby providing some insight into the contradictory size effects that have sometimes been reported elsewhere in the literature.

Keywords Generalized continua · Size effect · Laminate · Micropolar (Cosserat) elasticity

M.A. Wheel (✉)
Department of Mechanical and Aerospace Engineering, University of Strathclyde,
Glasgow G1 1XJ, UK
e-mail: marcus.wheel@strath.ac.uk

J.C. Frame · P.E. Riches
Department of Biomedical Engineering, University of Strathclyde,
Glasgow G4 0NW, UK

P.E. Riches
e-mail: philip.riches@strath.ac.uk

H. Altenbach and S. Forest (eds.), *Generalized Continua as Models for Classical and Advanced Materials*, Advanced Structured Materials 42,
DOI 10.1007/978-3-319-31721-2_21

1 Introduction

Loaded materials are often assumed to deform in a manner described by classical or Cauchy elasticity theory which presumes that the material stiffness, quantified by its modulus, will be independent of size scale. The fact that many engineering materials actually demonstrate such size independent behaviour across those size scales of interest has resulted in the almost unanimous acceptance of this theory. However, there are some materials that do exhibit size dependent behaviour when loaded. Such behaviour has been observed in fabricated materials like foams (Lakes 1983, 1986; Anderson and Lakes 1994) as well as biological tissues such as bone (Yang and Lakes 1982; Choi et al. 1990). The size dependency apparently results from the size scale of their microstructure which may be sufficient to influence their macroscopic behaviour.

More generalized continuum theories with the capacity to forecast size dependent behaviour do exist and while some of these incorporate higher derivatives of the deformation into the constitutive equations others contain additional degrees of freedom. Examples of the latter type include in ascending level of complexity: couple stress, Cosserat or micropolar and micromorphic elasticity theories (Eringen 1999). Common to all of these theories is the incorporation of additional constitutive parameters that must be identified through experimentation on real materials or simulation of virtual materials. Parameter identification invariably involves testing or simulating material samples of different sizes in loading modes that induce a non uniform state of stress such as torsion or bending to reveal any size dependency (Lakes 1995). The observed size effect can then be used to identify the additional constitutive parameters. The generalized continuum theories already mentioned predict size effects in which size scale reduction results in an apparent increase in stiffness. Behaviour like this has been observed in both polymeric foams (Lakes 1983, 1986; Anderson and Lakes 1994) and materials comprised of a two dimensional homogeneous matrix perforated by a regular array of circular voids (Beveridge et al. 2013; McGregor and Wheel 2014; Waseem et al. 2013). Micromorphic elasticity requires that the observed stiffness variation should remain finite as size diminishes while Cosserat elasticity permits singular behaviour. Although such behaviour is arguably physically less reasonable the experimental evidence is unable to discriminate since there is inevitably a practical limit to the minimum sample size that can be tested. Moreover, there are materials such as cortical bone, for which both increasing and decreasing stiffness with reducing size have been reported (Yang and Lakes 1982; Choi et al. 1990). This contradictory behaviour has been attributed to surface effects induced in sample preparation that increase compliance and result in a distorted size effect. Careful sample preparation has been highlighted as a requirement to avoid such distortion (Anderson and Lakes 1994). While sample preparation may give rise to unanticipated size effects, the influence of the material microstructure on surface behaviour might also be a possible cause. Indeed, size effects in transversely loaded beams were recently (Gao and Mahmoud 2014) shown to depend on the combined behaviour of bulk and surface where these were represented by a generalized continuum theory and a surface elasticity model respectively.

It has also been reported that composite or laminate materials demonstrate behaviour consistent with the predictions of micropolar or Cosserat elasticity including the dispersion of propagating elastic waves when loaded dynamically (Herrmann and Achenbach 1968) and also a size dependent stiffness when loaded statically (Forest and Sab 1998). This paper considers a simple laminate comprised of just two alternately layered constituents of differing moduli and demonstrates via a straightforward analysis that a rich variety of size effects might be anticipated. The paper then considers a material with regular or periodic heterogeneity, whose behaviour has previously been shown to be consistent with Cosserat elasticity theory, and demonstrates that some of the size effects exhibited by the laminate are also seen in this material. Finally, the size effects predicted for the laminate material are briefly compared to those reported for both virtual and real materials in other literature.

2 Size Effects in a Two Phase Laminated Beam

Figure 1 shows the cross sections of slender rectangular beam samples comprised of a simple laminated material consisting of alternating layers or plies of two different materials of Young's moduli E_1 and E_2 respectively. For simplicity all plies of both materials are assumed to be of the same thickness, t, and thus all internal layers of

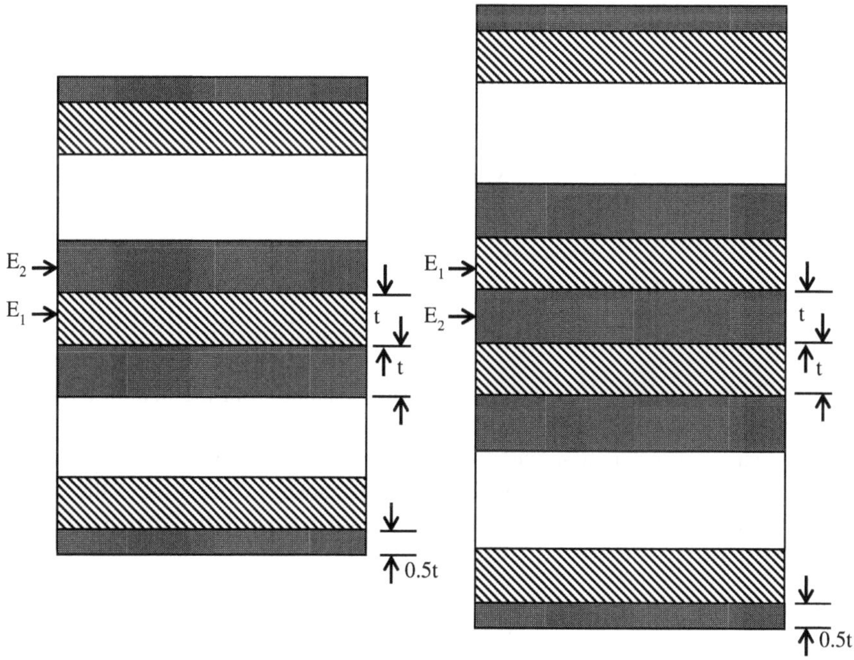

Fig. 1 Laminated beam model of heterogeneous material

the laminate are of the same thickness as illustrated. The central or core layer may be comprised of either of the two materials hence the two variants of the cross section shown in Fig. 1. However, the upper and lower surface layers of the sample always consist of the second material with modulus E_2 and, in addition, the thickness of these surface layers is always half that of the internal plies. The sample cross section is thus both symmetric, with the neutral axis of bending midway through the section, and comprised of equal proportions of each constituent material. Now if there are n plies of the first material then there will be $n - 1$ internal plies together with the two surface plies of the second material so that the cross section is comprised of $2n + 1$ layers in total. The case where n is odd corresponds to the central layer of the beam consisting of the first material while when n is even the central ply will be comprised of the second material. Both cases are illustrated in Fig. 1. Bernoulli Euler beam theory allows the flexural rigidity, D, of the laminated beam section to be obtained by firstly evaluating the products of the moduli and second moments of area about the section neutral axis of each of the individual plies and then summing these products. When n is odd this summation is represented thus:

$$D = 2\int_0^{t/2} E_1 b y^2 \mathrm{d}y + \sum_{i=1}^{(n-1)/2} 2 \int_{(i-1/2)t+it}^{(i+1/2)t+it} E_1 b y^2 \mathrm{d}y + \sum_{i=1}^{(n-1)/2} 2 \int_{(i-1/2)t+(i-1)t}^{(i-1/2)t+it} E_2 b y^2 \mathrm{d}y + 2 \int_{(n/2)t+[(n-1)/2]t]}^{nt} E_2 b y^2 \mathrm{d}y \quad (1)$$

and when n is even the summation becomes:

$$D = 2\int_0^{t/2} E_2 b y^2 \mathrm{d}y + \sum_{i=1}^{n/2} 2 \int_{(i-1/2)t+it}^{(i+1/2)t+it} E_1 b y^2 \mathrm{d}y + \sum_{i=1}^{n/2-1} 2 \int_{it+(i-1/2)t}^{it+(i+1/2)t} E_2 b y^2 \mathrm{d}y + 2 \int_{(n/2)t+[(n-1)/2]t]}^{nt} E_2 b y^2 \mathrm{d}y \quad (2)$$

where b is the breadth of the beam and y the distance from the neutral axis. The core layer is accounted for by the first term in each of these expressions while the summation terms are associated with all remaining internal layers and the final integrals account for the surface layers which are always comprised of the second material. Interestingly, when these summations are evaluated the flexural rigidity is given by following single expression:

$$D = \frac{E_1 b n t^3}{12}[4n^2 - 3] + \frac{E_2 b n t^3}{12}[4n^2 + 3] \quad (3)$$

in both cases.

The depth, d, of the beam is:

$$d = 2nt \tag{4}$$

and if the length to depth aspect ratio of the beam is a it follows that the length, L, of the beam is given by

$$L = 2ant \tag{5}$$

When loaded in three point bending the stiffness, K, of the beam will therefore be:

$$K = \frac{E_1 bn}{2a^3 n^3}[4n^2 - 3] + \frac{E_2 bn}{2a^3 n^3}[4n^2 + 3] \tag{6}$$

which can be rearranged thus:

$$K = \frac{4(E_1 + E_2)b}{2a^3} + \frac{3(E_1 - E_2)b}{2a^3} \tag{7}$$

The motivation for considering three point bending in particular is its practical simplicity which frequently renders it the loading mode of choice; the stiffness of the beam when loaded in other bending configurations could of course be derived similarly.

Now, given that the ply thicknesses of both constituent materials are the same then if the samples were loaded in uniaxial tension parallel to the plies then the material modulus, E^*, would simply be:

$$E^* = \frac{(E_1 + E_2)}{2} \tag{8}$$

and thus the first term in (7) could be simplified so that the expression for the stiffness of the beam becomes:

$$K = \frac{4E^* b}{2a^3} + \frac{3(E_1 - E_2)b}{2a^3} \tag{9}$$

In (9) the first term can be regarded as the stiffness of a slender homogeneous beam of modulus E^* loaded in three point bending. The second term however gives rise to a size effect resulting from the heterogeneous nature of the laminated beam. Equation (9) can be compared to the expression for the stiffness of a slender micropolar beam:

$$K = 4E^* b \left(\frac{d}{L}\right)^3 \left[1 + \left(\frac{l_c}{d}\right)^2\right] \tag{10}$$

which was derived by assuming that on every cross section of the beam a linear variation in bending stress and a uniform state of couple stress act (Beveridge et al. 2013) and that across the breadth of the beam any transverse deformations can be ignored. Equation (10) represents a simplification of the more general solution for the deformation quoted previously (Lakes 1995). The constitutive parameter, l_c that is usually termed the characteristic length, quantifies the length scale associated with the couple stresses. It should be noted that this definition of the characteristic length is a factor of $\sqrt{24}$ greater than the usual definition denoted l_b (Lakes 1995) and termed the characteristic length in bending. Equation (10) implies that both the modulus and the the characteristic length can be determined from the size effect that may be observed when the stiffness of material samples of differing sizes but with the same aspect ratio, L/d, and breadth, b, are loaded in three point bending. Experimental testing and detailed finite element analysis of slender beam samples of a heterogeneous material comprised of periodically distributed circular voids within a homogeneous matrix confirmed the validity of Eq. (10) (Beveridge et al. 2013). Beam stiffness was found to increase linearly with the reciprocal of beam depth squared, $1/d^2$, in accordance with Eq. (10). Characteristic length values were determined from the gradient of the stiffness variation while modulus values were identified from the interception of this variation with the stiffness axis. Subsequently, comparable behaviour was observed in slender ring samples with similar heterogeneity that were loaded diametrically (Waseem et al. 2013) and in ring samples in which the topology of the void array constituting the heterogeneity was varied (McGregor and Wheel 2014).

Equation (10) predicts that as sample size is reduced there will be a corresponding increase in stiffness; a size effect that agrees with the forecasts of more generalized continuum theories such as micropolar and micromorphic elasticity and with observed behaviour as already noted. However, in the case of the laminate Eq. (9) implies that size effects may be more elaborate since the second term depends on the relative magnitudes of the ply moduli, E_1 and E_2 and thus it may either amplify or modulate the stiffness at any particular sample size.

Figures 2 and 3 show how the predicted stiffness varies with sample size for different combinations of constituent material moduli. The stiffness variations shown in these figures assume an aspect ratio, a, of 10 and a common breadth, b, of unity.

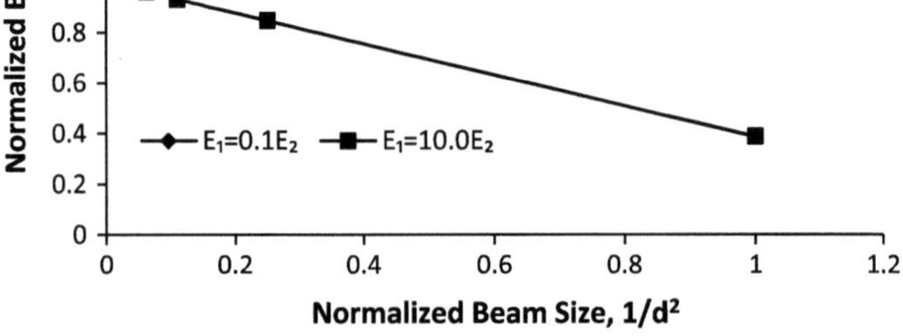

Fig. 2 Variation in stiffness with beam size for cases where $E_1 = 0.1E_2$ and $E_1 = 10.0E_2$

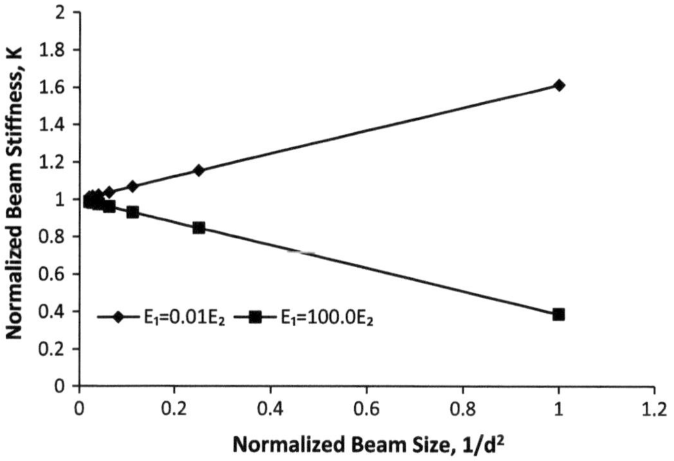

Fig. 3 Variation in stiffness with beam size for cases where $E_1 = 0.01E_2$ and $E_1 = 100.0E_2$

In both figures the sample size is quantified by the reciprocal of depth squared, $1/d^2$, to enable the predicted size effects to be compared directly with the forecasts of Eq. (10). Furthermore, this size measure, $1/d^2$, has been normalized with respect to that of the thinnest possible beam, this being $1/(2t)^2$ while the stiffness has been normalized with respect to that of a beam of almost infinite depth, that is, a beam for which n is very large implying that the second term in Eq. (9) diminishes to zero and any size effect becomes insignificant.

Figure 2 shows the size effects predicted by Eq. (9) when one of the materials is 10 times stiffer than the other. Evidently, when the material constituting the surface layers is the stiffer of the two, that is $E_1 = 0.1E_2$, a positive size effect in which stiffness increases as size reduces is predicted. Furthermore, this positive size effect is apparently linear as anticipated by Eq. (10). While extrapolation of this size effect to smaller size scales implies unbounded stiffness increase with diminishing size, any such extrapolation is doubtful since the size scales involved are then less than the minimum laminate thickness of $2t$. Thus it is uncertain whether the behaviour of the laminate is best described by Cosserat or micromorphic elasticity. Figure 2 also shows that when $E_1 = 10E_2$ and the surface is therefore formed from the more compliant material a contrasting, negative size effect is seen. This negative effect appears to vary linearly in accordance with its positive counterpart. Interestingly, although these size effects are entirely different in nature the magnitude of the rate at which the stiffness varies appears to be similar in both cases.

Figure 3 shows the corresponding stiffness variations when the ratio of the material moduli is increased by a further factor of 10. When $E_1 = 0.01E_2$ implying the surfaces are comprised of the stiffer material, a positive size effect is again seen while when $E_1 = 100E_2$ and the surfaces are more compliant a negative size effect results once more. It is also interesting to note that when presented in this non dimensional manner both of the size effects seen at this ratio of material moduli reflect those shown in Fig. 2. By utilizing Eq. (4) to compare Eqs. (9) and (10) it can be shown that the characteristic length parameter, l_c, present in the latter is related to the thickness of the layers, t, by:

$$l_c^2 = \frac{3[(E_2/E_1) - 1]}{[1 + (E_2/E_1)]} t^2 \quad (11)$$

implying l_c varies linearly with t when the moduli ratio, (E_2/E_1), is greater than unity. For the positive size effects depicted in Figs. 2 and 3 in which the moduli ratios are 10.0 and 100.0 respectively, the corresponding values of l_c are $1.57t$ and $1.72t$. As the moduli ratio is increased further, signifying greater material heterogeneity, the value of l_c asymptotically approaches $\sqrt{3}t$ according to Eq. (11) indicating an upper bound in the value of this parameter.

3 Size Effects in a Two Dimensional Material with Periodic Heterogeneity

Figure 4 illustrates a material with regular heterogeneity created by introducing a periodic array of circular voids into an otherwise classically elastic matrix material. The constitutive behaviour of materials of this type has been investigated previously (Beveridge et al. 2013) in the context of generalized continua and the behaviour shown to be consistent with the predictions of micropolar elasticity theory. Prescribing the void radius, V_R, together with the separation of the void centres, S_x and S_y, in the indicated x and y directions respectively is sufficient to fully define the geometry of the heterogeneity. The void centres thus lie on a triangular grid. When $S_y = \sqrt{3}S_x/2$ a detailed finite element analysis of a representative piece of material incorporating a sufficiently large number of voids revealed that the material exhibited approximate planar isotropy. Thus the material is transversely isotropic. The analysis assumes that the matrix material exhibits classically elastic behaviour as quantified by its Young's modulus and Poisson's ratio.

This material was shown to exhibit a size dependent stiffening consistent with Eq. (10) through finite element analysis of slender beam samples of different depths but the same aspect ratio (Beveridge et al. 2013). In producing the mesh required to represent each beam sample the geometric details of the heterogeneity were explic-

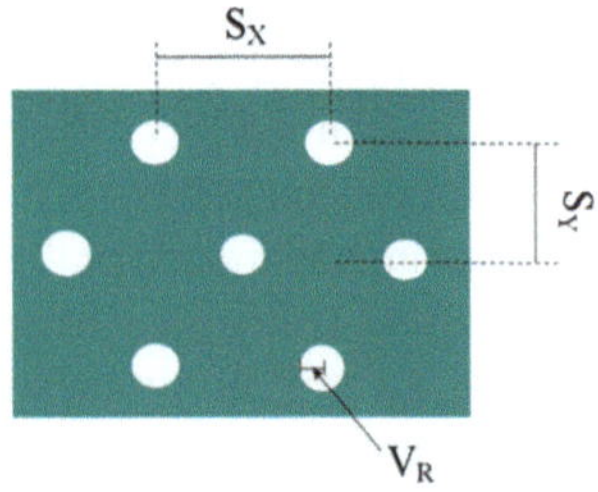

Fig. 4 Two dimensional material with regular, periodic heterogeneity investigated previously within the context of micropolar elasticity theory

itly incorporated by firstly paving the rectangular region around a particular void with the structured array of quadratic quadrilateral elements illustrated in Fig. 5 and then repeatedly regenerating this array of elements at suitable spatial increments to represent an entire sample. One consequence of generating the entire sample representation in this way was that the upper and lower sample surfaces were each located midway between adjacent rows of voids as illustrated in Fig. 6 and thus these surfaces intersected none of the voids. However, this is not the only manner in which a finite sized beam sample could be identified from within an infinite sheet of the perforated material. A sample could just as easily be identified in alternative ways and in some of these the surfaces may intersect the voids. Figure 6 shows one such alternative in which the structured mesh shown in Fig. 7 was employed to represent the region of matrix material located between a specific void and its neighbours. Consequently, when this mesh is repeatedly regenerated to represent an entire sample the sample surfaces periodically bisect all voids in a given row as shown in Fig. 6. Both means of mesh generation were therefore used to analyse beams of increasing depth, this being

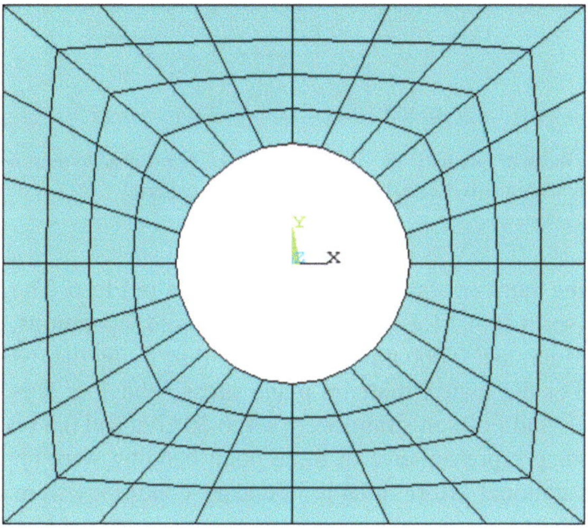

Fig. 5 Structured mesh of quadratic quadrilateral finite elements used to represent rectangular region around a particular void within two dimensional heterogeneous material

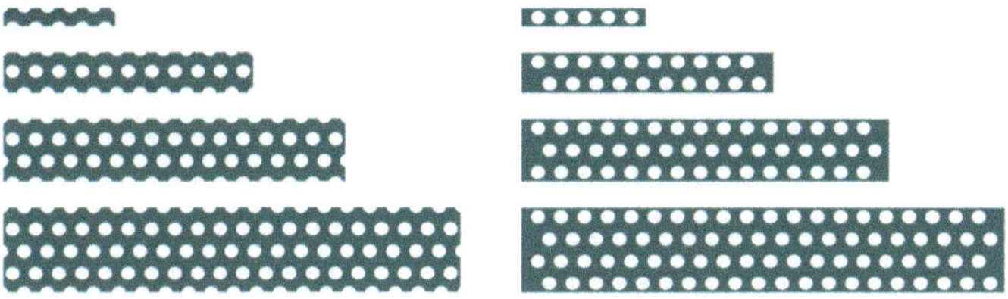

Fig. 6 Representation of beam samples of increasing size generated by finite element meshes shown in Figs. 5 (*right*) and 7 (*left*)

Fig. 7 Alternative mesh of quadratic quadrilateral finite elements used to represent region between neighbouring voids within two dimensional heterogeneous material

determined by the number of rows of voids as shown in Fig. 6. The length to depth aspect ratio was set at 10.4:1 for all beams while the void separations, S_x and S_y, were prescribed at 1.0 mm and 0.866 mm respectively thus fixing the overall dimensions of each beam. For the matrix material plane stress behaviour was assumed while Young's modulus and Poisson's ratio were set to 20 GPa and 0.3 respectively. Constraints and loading representative of three point bending were applied. However, to reduce computational effort suitable boundary conditions were imposed at the central loading plane to exploit the symmetries in geometry and loading and thereby facilitate analysis of only one half of each beam.

Figure 8 shows the predicted variations in beam stiffness with size for different void volume fractions, V_f, when the sample surfaces do not intersect the array of voids. The stiffness variation appears to be approximately linear at any given void size. The characteristic length of each material can be obtained from the gradient of the corresponding variation while the modulus of can be derived from the intercept according to Eq. (10). Table 1 lists values of each of these constitutive parameters as a function of void radius and volume fraction. This table also lists data derived from the stiffness variations determined for more slender samples with an increased aspect ratio of 20.8:1. These data imply that as void radius increases the material modulus decreases, as might be expected since there is less material to support the

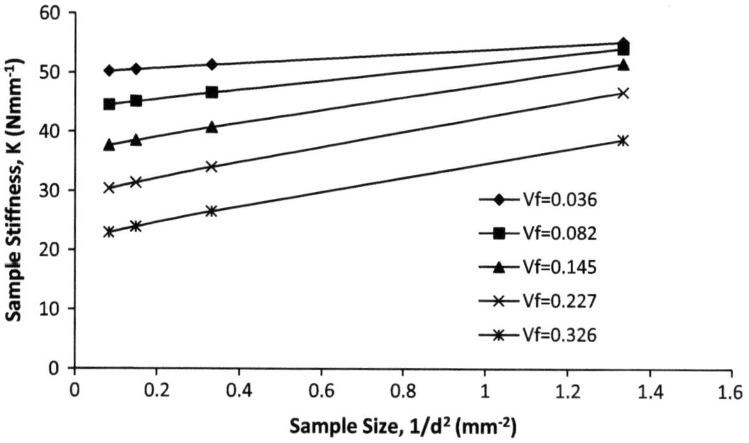

Fig. 8 Stiffness against the reciprocal of depth squared for beams with smooth surfaces at a 10.4:1 length to depth aspect ratio for various void volume fractions, V_f

Table 1 Comparison of the characteristic lengths for different void radii at 10.4:1 and 20.8:1 length to depth aspect ratios

Void diameter V_d (m)	Void fraction V_f	Normalised void radius V_R/S_y	Young's Modulus (GPa)		Characteristic length (mm)	
			10.4:1 aspect ratio	20.8:1 aspect ratio	10.4:1 aspect ratio	20.8:1 aspect ratio
0.2	0.036	0.12	17.47	17.87	0.28	0.28
0.3	0.082	0.17	15.37	15.71	0.42	0.43
0.4	0.145	0.23	12.90	13.16	0.55	0.57
0.5	0.227	0.29	10.31	10.50	0.66	0.70
0.6	0.326	0.35	7.74	7.83	0.75	0.82

applied loading, while the characteristic length increases. Furthermore, the values of this latter parameter obtained at aspect ratios of 10.4:1 and 20.8:1 only vary slightly implying that although slender beam behaviour is assumed in Eq. (10) the lower, 10.4:1, aspect ratio beams are sufficiently slender enough to provide very reasonable estimates of the characteristic length since they appear to satisfy this underlying assumption. Figure 9 shows the relationship between void radius and characteristic length for the higher, 20.8:1, aspect ratio beams. This relationship is evidently linear as anticipated by both a previous theoretical prediction (Bigoni and Drugan 2007) and the laminate model through Eq. (11).

Figure 10 shows variations in beam stiffness with size for the same void radii when the aspect ratio is set at 10.4:1 but the surfaces now bisect the voids. These variations each show a decrease in stiffness with reducing size and therefore no longer concur with Eq. (10). However, each variation is nonetheless linear which does accord with

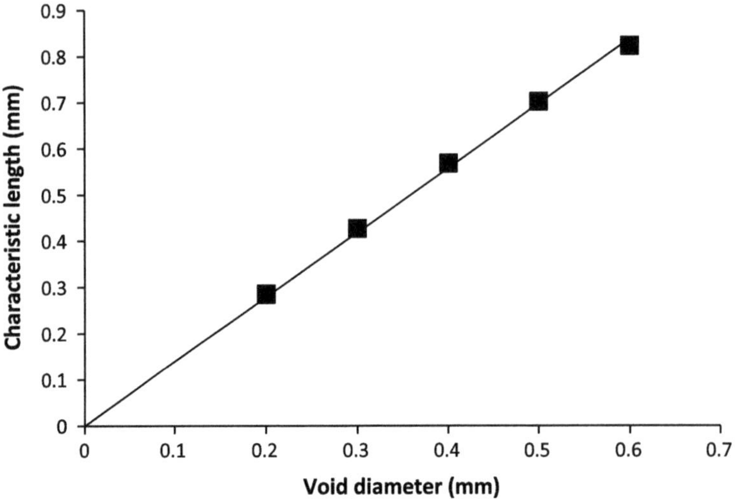

Fig. 9 Variation in micropolar characteristic length with void diameter for 20.8:1 aspect ratio beams

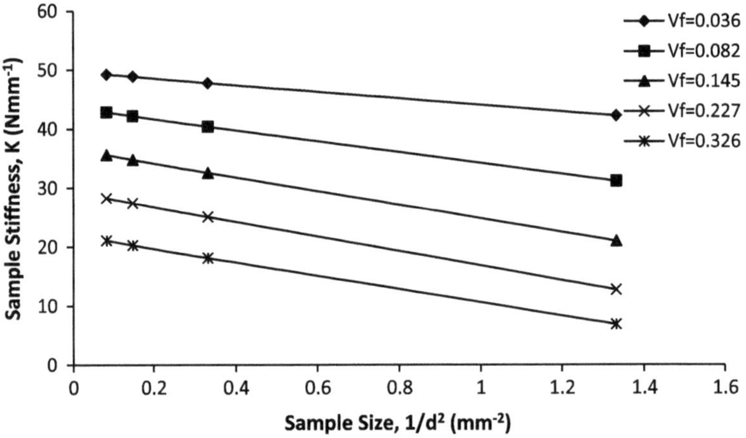

Fig. 10 Stiffness against the reciprocal of depth squared for beams with intersected surfaces at a 10.4:1 length to depth aspect ratio for various void volume fractions, V_f

Eq. (9) for the case where the sample surfaces are comprised of the more compliant material. Additionally, the intercept of a given negative size effect seen in Fig. 10 corresponds to that obtained from the positive effect shown in Fig. 8 for the equivalent void radius. This correspondence thus reflects the convergence of both size effects seen in the laminate material at large beam depths and illustrated in Fig. 2. Figure 11 shows how the magnitudes of both the positive size effects seen in Fig. 8 and their negative counterparts shown in Fig. 10 vary with void volume fraction. This figure suggests that the magnitude of the negative effect broadly reflects that of the positive

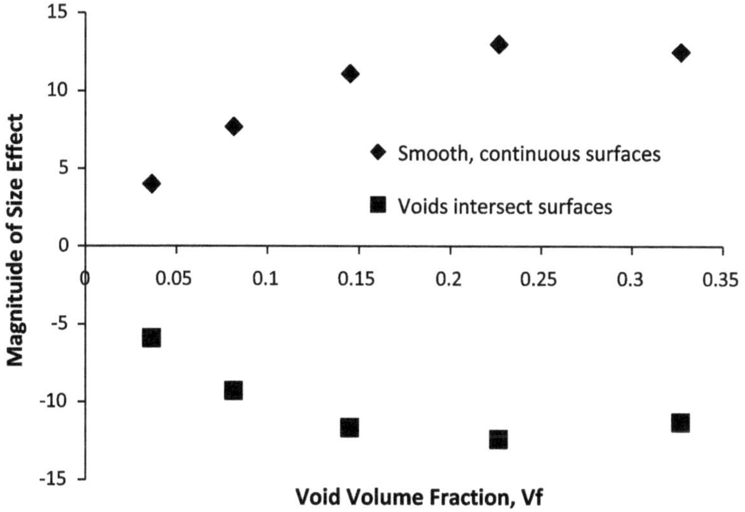

Fig. 11 The magnitude of the size effects as a function of void volume fraction V_f for beams with both smooth and intersected surfaces and a 20.8:1 length to depth aspect ratio

effect. Moreover, when the void volume fraction is low then both size effects are small since the beam samples are predominantly comprised of matrix material. The magnitudes of each effect then increase as void volume fraction increases. However, both magnitudes reach a maximum beyond which they then diminish as the samples become predominantly comprised of voids.

4 Discussion and Conclusions

Size effects forecasts were made previously (Tekoglu and Onck 2008) for virtual two dimensional foams. The stochastic cellular microstructure was represented as random Voronoi tessellations with Timoshenko beam finite elements being used to represent individual cell wall sections. Closed polygons were used to represent all internal cells within an elongated rectangular region. However, for those cells intersecting the boundary no elements were located coincident to the boundary to facilitate their closure and they remained open. The finite element representations were loaded in pure bending of the major axis and multiple analyses were conducted using a different randomly generated finite element mesh on each occasion in order to capture the behaviour of the intrinsically stochastic microstructure. Bending stiffness was found to vary with beam depth which was altered by changing the lesser dimension of the representative rectangular region. At small depths forecast stiffness was less than that anticipated by classical elasticity theory. However, as depth increased stiffness was predicted to rise and asymptotically approach the classical result. This behaviour appears to arise from the rupture of those cells adjacent to the boundaries which results in a local increase in compliance thereby compromising bending stiffness and ultimately resulting in a negative size effect rather than a positive one that might be expected for a generalized continuum. Thus the negative size effect forecast for the stochastic foam reflects that seen in both the present laminate and voided materials when stiffness of the material adjacent to the surface is deliberately compromised.

Contradictory size effects have been reported previously in human cortical bone (Yang and Lakes 1982; Choi et al. 1990) where samples were loaded in bending experiments in each case. While the earlier work reported a positive size effect the later reported an opposite, negative effect. In the earlier work beam specimens with depths down to 1.4 mm were tested but in the later samples with substantially smaller depths were also investigated. While a mild positive size effect was reported earlier the most significant effect was observed in the small samples investigated in the later work and, moreover, this effect was unequivocally negative. The investigators suggested a qualitative explanation of this size effect based on the exposure of the major microstructural feature, the vascular channel or Haversian canal system, at the sample surfaces giving rise to an increase in compliance of the material located in the vicinity. However, the effect was interpreted quantitatively as a reduction in the apparent modulus of the material rather than the behaviour of a more generalized continuum. Nevertheless, the negative effect reported for cortical bone appears to reflect that seen in the laminate and voided materials considered here.

These negative size effects seen in both stochastic foams and cortical bone have not been interpreted in the context of generalized continuum theories. Presumably since these effects appear to contradict the predictions of such theories identification of relevant constitutive properties was not attempted. However, constitutive property identification is paramount since it provides a rational basis for comparing the practical performance of materials when loaded. Apparently this cannot be realized in the case of a heterogeneous material that displays a negative size effect. Nevertheless, the correspondence between the positive and negative size effects forecast for both the simple laminated and the more involved perforated materials considered suggest a pragmatic, albeit empirical, resolution to this dilemma since the correspondence suggests that it may be possible to infer constitutive property data from observed effects in cases where these are negative.

References

Anderson W, Lakes R (1994) Size effects due to Cosserat elasticity and surface damage in closed-cell polymethacrylimide foam. J Materials Sci 29(24):6413–6419

Beveridge A, Wheel M, Nash D (2013) The micropolar elastic behaviour of model macroscopically heterogeneous materials. Int J Solids Struct 50:246–255

Bigoni D, Drugan W (2007) Analytical derivation of Cosserat moduli via homogenization of heterogeneous elastic materials. Trans ASME J Appl Mech 74:741–753

Choi K, Kuhn J, Ciarelli M, Goldstein S (1990) The elastic moduli of human subchondral, trabecular, and cortical bone tissue and the size-dependency of cortical bone modulus. J Biomech 23(11):1103–1113

Eringen AC (1999) Foundations and solids. Microcontinuum field theories. Springer, New York

Forest S, Sab K (1998) Cosserat overall modeling of heterogeneous materials. Mech Res Commun 25(4):449–454

Gao X, Mahmoud F (2014) A new Bernoulli–Euler beam model incorporating microstructure and surface energy effects. Zeitschrift für angewandte Mathematik und Physik (ZAMP) 65:393–404

Herrmann G, Achenbach J (1968) Applications of theories of generalized Cosserat continua to the dynamics of composite materials. In: Kröner E (ed) Proceedings of IUTAM symposium mechanics of generalized continua, Springer, Berlin, pp 69–79

Lakes R (1983) Size effects and micromechanics of a porous solid. J Materials Sci 18:2572–2580

Lakes R (1986) Experimental microelasticity of two porous solids. Int J Solids Struct 22:55–63

Lakes R (1995) Experimental methods for study of Cosserat elastic solids and other generalized elastic continua. In: Mühlhaus H (ed) Continuum models for materials with micro-structure. Wiley, New York, pp 1–25

McGregor M, Wheel M (2014) On the coupling number and characteristic length of micropolar media of differing topology. Proceedings of the royal society A 470

Tekoglu C, Onck P (2008) Size effects in two-dimensional Voronoi foams: a comparison between generalized continua and discrete models. J Mech Phys Solids 56:3541–3564

Waseem A, Beveridge AJ, Wheel M, Nash D (2013) The influence of void size on the micropolar constitutive properties of model heterogeneous materials. Eur J Mech A: Solids 40:148–157

Yang J, Lakes R (1982) Experimental study of micropolar and couple stress elasticity in bone in bending. J Biomech 15:91–98

MIX
Papier aus verantwortungsvollen Quellen
Paper from responsible sources
FSC® C105338
www.fsc.org

If you have any concerns about our products,
you can contact us on
ProductSafety@springernature.com

In case Publisher is established outside the EU,
the EU authorized representative is:
Springer Nature Customer Service Center GmbH
Europaplatz 3, 69115 Heidelberg, Germany

Printed by Libri Plureos GmbH
in Hamburg, Germany